International Conference on Fluid Mechanics and Industrial Applications (FMIA 2017)

Journal of Physics: Conference Series Volume 916

Taiyuan, China
21 – 22 October 2017

ISBN: 978-1-5108-5069-9
ISSN: 1742-6588

Printed from e-media with permission by:

Curran Associates, Inc.
57 Morehouse Lane
Red Hook, NY 12571

CURRAN ASSOCIATES INC. proceedings.com

Some format issues inherent in the e-media version may also appear in this print version.

Printed by Curran Associates, Inc. (2017)

For permission requests, please contact the Institute of Physics
at the address below.

Institute of Physics
Dirac House, Temple Back
Bristol BS1 6BE UK

Phone: 44 1 17 929 7481
Fax: 44 1 17 920 0979

techtracking@iop.org

Additional copies of this publication are available from:

Curran Associates, Inc.
57 Morehouse Lane
Red Hook, NY 12571 USA
Phone: 845-758-0400
Fax: 845-758-2633
Email: curran@proceedings.com
Web: www.proceedings.com

Table of contents

Volume 916

International Conference on Fluid Mechanics and Industrial Applications (FMIA 2017)

21–22 October 2017, Taiyuan, China

Accepted papers received: 19 October 2017
Published online: 29 October 2017

Preface

Papers

Aerodynamics

Combustion and Emissions

Fluid Properties, Performance and Applications

Modeling and Simulation

Thermodynamics and Heat Transfer

IOP Conf. Series: Journal of Physics: Conf. Series **916** (2017) 011001 doi:10.1088/1742-6596/916/1/011001

Preface

2017 International Conference on Fluid Mechanics and Industrial Applications (FMIA 2017: http://www.icfmia.org) was held on October 21-22, 2017 in Taiyuan, China. This conference was sponsored by Asian Union of Information Technology, and was attended by academicians, researchers, entrepreneurs, government agencies and policy-makers, postgraduates from China and all over the world.

FMIA 2017 is the annual international conference devoted to the discussion of recent developments and applications of parallel computing in the field of fluid mechanics, CFD, fluid power, fluid flow and related disciplines. The field of fluid mechanics is vast and covers numerous and diverse applications. The conference covered a wide range of topics, including basic formulations and their computer modelling as well as the relationship between experimental and analytical results. The meeting provided a forum for discussing new work on fluid mechanics and in particular for promoting the interchange of novel ideas and the presentations of the latest developments in this field. This conference will also provide an ideal environment to develop new collaborations and meet experts on the fundamentals, applications, and products of the mentioned fields.

FMIA 2017 received 291 manuscripts, and 54 submissions have been accepted by our reviewers. By submitting a paper to FMIA 2017, the authors agreed to the review process and understood that papers would undergo a peer-review process. Manuscripts were reviewed by appropriately qualified experts in the field selected by the conference committee, who took detailed comments and-if the submission was accepted-the authors would submit a revised version that took into account this feedback. All papers were reviewed using a double-blind review process: authors declared their names and affiliations in the manuscript for the reviewers to see, but reviewers did not know each other's identities, nor did the authors receive information about who had reviewed their manuscript. The committees of FMIA 2017 invested great efforts in reviewing the papers submitted to the conference and organizing the sessions to enable the participants to gain maximum benefit.

With our warmest regards,

Jun Xiao

Wuhan, China

IOP Conf. Series: Journal of Physics: Conf. Series **916** (2017) 011002 doi:10.1088/1742-6596/916/1/011002

Committees

Chairmen
Prof. O. Somei, Hokkaido University, Japan
Prof. T.W. Fu, Nanjing University, China
Dr. J. Xiao, Wuhan University of Technology, China

Members
Prof. H. Zhou, Wuhan University, China
Prof. P.L. Liu, National Chung-Hsing University, Taiwan
Prof. J.P. Wang, Beijing Institute of Technology, China
Prof. S.D. Tan, Beijing Institute of Technology, China
Prof. J.J. Xu, Northeast Petroleum University, China
Prof. F. Liu, Huazhong University of Science and Technology, China
Prof. C.J. Jia, Huazhong University of Science and Technology, China
Prof. F. Yang, Huazhong University of Science and Technology, China
Prof. X. Long, Zhejiang University, China
Prof. C.H. Li, Zhejiang University, China
Prof. L.L. Deng, Zhejiang University, China
Prof. Y.D. Shi, Zhejiang University, China
Dr. X.L. Li, Peking University, China
Dr. J.W. Yan, Wuhan University of Technology, China
Prof. H. Chen, Shanghai University of Engineering Science, China
Associate Prof. S. Zhu, Wuhan University, China
Prof. G.S. Liu, Hunan University, China
Prof. P. Wang, Guangxi college of Education, China
Dr. H.Y. Su, Central South University, China
Dr. T.Q. Liu, Central South University, China
Dr. J.P. Fang, Chongqing University, China
Dr. G.P. Pan, Chongqing University, China
Dr. X.B. Wang, Chongqing University, China
Dr. D.M. Li, Shandong University of Science and Technology, China
Mr. W. Jiang, Xiangchen Cultural Development Co., Ltd, China
Mr. C.P. Wang, Xiangchen Cultural Development Co., Ltd, China
Mr. G.F. Tan, Xiangchen Cultural Development Co., Ltd, China
Ms. Z.L. Liu, Xiangchen Cultural Development Co., Ltd, China

IOP Conf. Series: Journal of Physics: Conf. Series **916** (2017) 011003 doi:10.1088/1742-6596/916/1/011003

Peer review statement

All papers published in this volume of *Journal of Physics: Conference Series* have been peer reviewed through processes administered by the proceedings Editors. Reviews were conducted by expert referees to the professional and scientific standards expected of a proceedings journal published by IOP Publishing.

Aerodynamic Shape Optimization Design of Wing-Body Configuration Using a Hybrid FFD-RBF Parameterization Approach

Yuefeng Liu [1,*], Zhuoyi Duan [1], Song Chen [2]

[1]The First Aircraft Institute of AVIC, Xi'an 710089, China;
[2]Northwestern Polytechnical University, Xi'an 710072, China

*Corresponding author e-mail: lyfcrazy@163.com

Abstract. Aerodynamic shape optimization design aiming at improving the efficiency of an aircraft has always been a challenging task, especially when the configuration is complex. In this paper, a hybrid FFD-RBF surface parameterization approach has been proposed for designing a civil transport wing-body configuration. This approach is simple and efficient, with the FFD technique used for parameterizing the wing shape and the RBF interpolation approach used for handling the wing body junction part updating. Furthermore, combined with Cuckoo Search algorithm and Kriging surrogate model with expected improvement adaptive sampling criterion, an aerodynamic shape optimization design system has been established. Finally, the aerodynamic shape optimization design on DLR F4 wing-body configuration has been carried out as a study case, and the result has shown that the approach proposed in this paper is of good effectiveness.

1. Introduction

With the rapid development in the computer science and technology, numerical optimization has been widely used as a powerful tool in research and development of aircraft industries. Aerodynamic shape optimization design is of the applications aiming at improving the efficiency of an aircraft by redesigning its aerodynamic shape. There are several key components that significantly influence the utilization and result of the aerodynamic shape optimization design, including computational fluid dynamics (CFD) flow solver for aerodynamic performance evaluation, surface parameterization technique for shape definition and manipulation, and the optimization algorithms for obtaining an optimum solution. By now Reynolds Averaged Navier-Stokes equations (RANS equations) CFD solvers have been a relatively mature technique in aircraft engineering research and development. Thus the other two components, surface parameterization technique and optimization algorithm have been the bottlenecks in efficient aerodynamic shape optimization design.

The surface parameterization not only determines whether the design space of an aerodynamic optimization design contains an optimum solution, but also determines that whether this solution can be obtained with an acceptable computational cost. Due to the limit of computers, researchers in the period from 1970s to 1990s tend to use 2D parameterization method to modify wing sections instead of a three-dimensional wing for designing the wing shape. Parameterization methods handling wing sections are, for example, PARSEC [1], Hicks-Henne bump function [2]. Then some interpolation or spline based methods would be used for combining the several wing sections into a three-dimensional wing shape. Kulfan proposed the CST (class function/ shape function transformation) technique [3] for aircraft component parameterization, and titled it as a universal parametric geometry representation method. However, CST cannot be used for modification of a specified initial aerodynamic shape, for it cannot represent a specified shape absolutely exactly. With the development in geometry technique and computer performance, after late 1990s, researchers have begun to use direct three-dimensional surface parameterization methods for aerodynamic shape optimization design. Hicken and Zingg used a patched B-spline surfaces approach for shape modification in aerodynamic shape design [4]. Yamazaki et.al. used physics-based direct-manipulation FFD method for geometry parameterization of both airfoils and wings [5]. Samareh used a NURBS based approach to parameterize the shape perturbation of the wing geometry while using FFD to modify the planform [6]. Though there are quite a number of surface parameterization methods that can be applied in the three-dimensional aerodynamic shape design, the discussion about handling the junction part between the wing and the body in aerodynamic optimization design of wing-body configuration is relatively rare. Nakahashi et.al. have proposed a surface parameterization method which can handle wing-body junction by neighbor-to-neighbor searching and geodesic distance calculating[7]. However within Nakahashi's method, when the wing geometry is modified, the fuselage geometry would also be deformed obviously, which is often unacceptable in practical aerodynamic shape design.

In this paper, we have proposed an efficient surface parameterization method using an hybrid FFD-RBF parameterization approach, which can handle the junction part between the wing and body by interpolation based updating. In the implementation of this method, FFD technique [8] has been used for paramterizing the wing shape, including wing shape deformation and geometric twist. After the wing shape is modified, including wing rood section deformed, the junction part between the wing and the body would be updated by a RBF based interpolation method to ensure preservation of original geometry continuity.

Apart from the surface parameterization method, optimization algorithm is another key component in aerodynamic shape optimization problem. With the development in computers, nature-inspired heuristic optimization algorithms such as genetic algorithm (GA) and particle swarm optimization (PSO) have already been widely used in aerodynamic shape optimization, related work can be found in literatures[9,10]. Nature-inspired algorithms have many advantages. One key advantage is their ability to handle non-smooth functions since most nature-inspired algorithms are gradient-free methods. Though there may not be guarantee for global optimality due to the stochastic nature of these algorithms, they usually can find global optimal solutions in practice. In this paper we use a novel optimization algorithm – Cuckoo Search (CS) as the optimum searching engine. Recent studies have shown that Cuckoo Search algorithm can be more efficient than Genetic Algorithm and Particle Swarm Optimization, with better optimized results and a relatively small number of function evaluations [11].

Another important issue in aerodynamic shape optimization design on complex shapes is the application of surrogate model. In aerodynamic shape optimization design, especially when the problem is high-dimensional, the optimization process would require enormous number of function evaluation. To remedy this, Kriging surrogate model has been used for approximation of the aerodynamic parameters with expected improvement adaptive sampling criterion. With integration of the hybrid FFD-RBF geometry paramterization with CS algorithm and Kriging model, an aerodynamic shape optimization design system has been established and tested on a case of DLR F4 wing-body model.

2. Hybrid FFD/RBF Surface Parameterization Approach

The success of an aerodynamic shape optimization design largely depends on the choice of design variables, which would be represented in terms of efficient geometrical parameterization. In this paper, the FFD technique has been chosen for surface parameterization in aerodynamic shape design. When handling wing-body design, especially when the wing root section is deformed during optimization, the junction part between wing and body has to be updated or reconstructed in order to preserve the original connectivity. An RBF based interpolation method has been adopted in this paper for updated the surface of the wing root junction area with preservation of surface connectivity when the wing root section has been changed in optimization process. The DLR F4 wing-body configuration would be redesigned in this paper using the hybrid FFD-RBF surface parameterization approach proposed in this paper as a study case.

2.1. *Free Form Deformation Surface Parameterization*

FFD is suitable for the parameterization of arbitrary three-dimensional spatial geometrical property, regardless of the representation of the objective. Without the need to fit or interpolate the original shape, the objective geometry could be parametrically deformed by using FFD with the conservation of its original smoothness and continuity.

The procedures of using the FFD technique for parametrically deforming an objective geometry (the wing of DLR F4 wing-body configuration in this paper) are as follows. At first, a parallelepiped box frame, also called FFD control frame, is established around the wing as shown in Figure 1. Then the Cartesian coordinates of the surface points of the wing shape are transformed into local coordinates that are defined by the vertices of the FFD control frame using the following formula:

$$\vec{x}(s,t,u) = \sum_{i=0}^{l}\sum_{j=0}^{m}\sum_{k=0}^{n}\left[B_l^i(s)B_m^j(t)B_n^k(u)\right]\cdot \vec{P}_{i,j,k} \tag{1}$$

where $\vec{x}$ is the position vector of a surface point of the wing shape, and (s,t,u) are the corresponding local coordinates of the point, which is defined inside the FFD control frame by formula (1). $\vec{P}_{i,j,k}$ is the position vector in the Cartesian system of every vertices of the FFD control frame, with l, m, n being the order of the FFD control frame along each of the three directions. Furthermore $B_l^i(s)$, $B_m^j(t)$ and $B_n^k(u)$ are the Bernstein basis functions, defined as follows:

$$\begin{aligned} B_l^i(s) &= \frac{(l)!}{(i)!(l-i)!}s^i(1-s)^{l-i} \\ B_m^j(t) &= \frac{(m)!}{(j)!(m-j)!}t^j(1-t)^{m-j} \\ B_n^k(u) &= \frac{(n)!}{(k)!(n-k)!}u^k(1-u)^{n-k} \end{aligned} \tag{2}$$

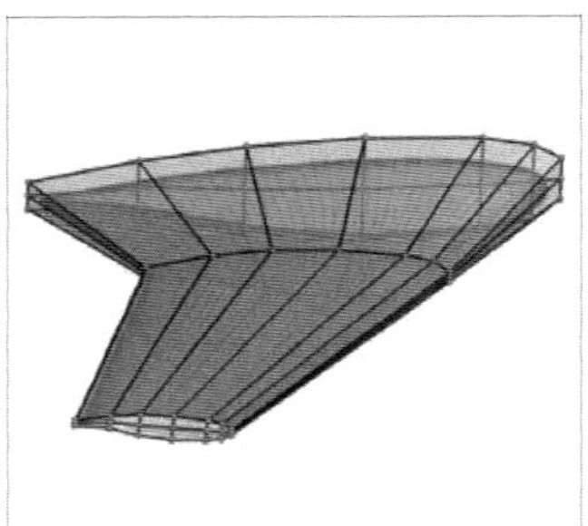

Fig. 1 The FFD Control Frame around the Wing

2.2. RBF based Interpolation Approach Handling the Wing Root Junction Part

In this paper, the FFD technique is used for parameterizing the wing surface only, as in fig 2, FFD frame would only determine the deformation of the wing, the mesh color of which is red in fig 2. The wing body junction area, the mesh color of which is green, would not be influenced by the FFD frame.

After the wing has been deformed, the wing root section has been changed, thus the intersection between the wing and the body, together with the wing root junction area, have to be updated or reconstructed to adjust to the deformed wing shape. Here we adopt an RBF interpolation based method to calculate the displacements of every mesh point of the wing root junction part after the wing deformation.

Here we use the root section of the deformed wing and the outer boundary of the wing root junction part (green in fig 3) as the set of interpolation centers to realize the total interpolation process so as to update the mesh points on the wing body junction part after wing deformation. Fig 3 illustrates the different parts of the wing body surface, which are, wing mesh parameterized by FFD, body mesh which is fixed during the design process, and the wing body junction part area which is to be updated by RBF interpolation.

The procedures for updating the wing body junction part are illustrated as below. First, after the wing has been deformed as the design variables changes, the displacements of the interpolation centers, $\Delta\mathbf{X}_c$, $\mathbf{X}_c \subseteq (B_{wr} \cup B_j)$ are calculated, where $\mathbf{X}_c$ are the positions of the interpolation centers, B_{wr} and B_j are the point set on wing root section and the junction part outer boundary respectively. Since that the updated junction part mesh have to match the fixed junction part boundary and the deformed wing root section, so that the RBF interpolation function is subject to:

$$\Delta\mathbf{X}_{cj} = \sum_{1}^{n} \gamma_i \varphi(\|\mathbf{X}_{cj} - \mathbf{X}_{ci}\|) \tag{3}$$

Where $\mathbf{X}_{cj}$ and $\mathbf{X}_{ci}$ are the positions of mesh points in the set of interpolation centers, the number of which is n . $\varphi(\bullet)$ is the radial basis function, which is chosen as follows in this paper:

$$\varphi(\xi) = \begin{cases} (1-\xi)^2 & \xi \le 1 \\ 0 & \xi > 1 \end{cases}, \quad \xi = \frac{\|\mathbf{X} - \mathbf{X}_c\|}{r} \tag{4}$$

Where r is the support radius which is specified in the interpolation process, and $\|\bullet\|$ is the Euclidian distance in three dimensional space. γ_i in formula (3) is the interpolation parameters, which are to be computed by solving the linear system dominated by formula (3). After the interpolation parameters have been obtained, then the RBF interpolation function has been defined, as follows:

$$\Delta\mathbf{X} = \sum_{1}^{n} \gamma_i \varphi(\|\mathbf{X} - \mathbf{X}_{ci}\|) \tag{5}$$

The displacements of the mesh points of wing root junction area can be computed by the formula above. Then the new position of a mesh point on the wing body junction part area after the wing deformation could be computed by:

$$\mathbf{X}_{new} = \mathbf{X}_{old} + \sum_{1}^{n} \gamma_i \varphi(\|\mathbf{X}_{old} - \mathbf{X}_{ci}\|) \tag{6}$$

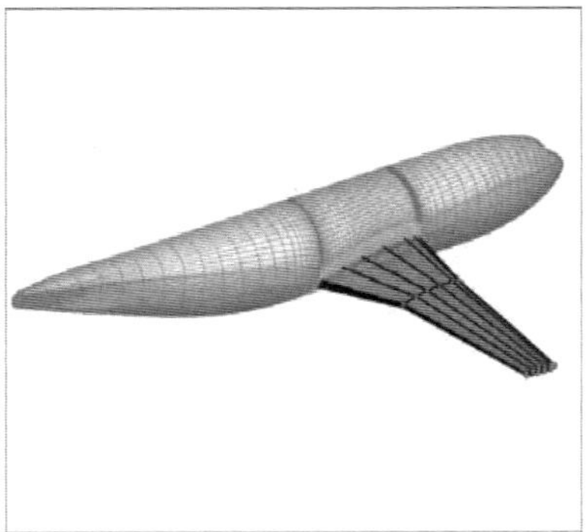

Fig. 2 FFD Control Frame and the Wing Body Shape

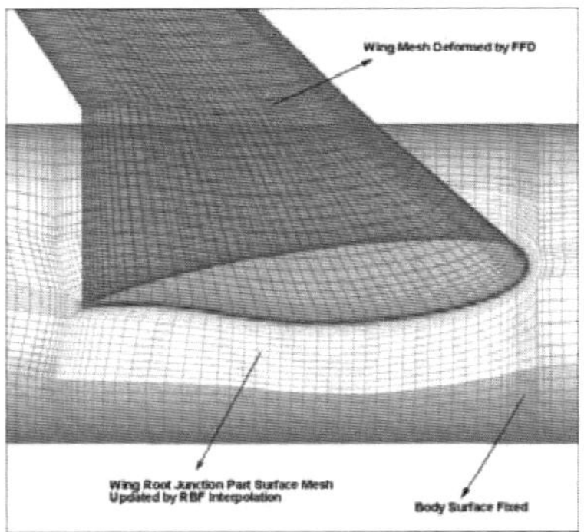

Fig. 3 The Different Parts of the Wing-Body within the Deformation

3. Cuckoo Search Algorithm

Cuckoo Search algorithm (CS) is one of the novel nature-inspired based algorithms developed by Yang and Deb[12] in 2009. Its basic methodology is to mimic the brood parasitism behavior of some cuckoo species. This algorithm is further enhanced by Lévy flight [13] based on Lévy distribution, which can generate solutions both locally and at far field, and thus has been proven to be a more efficient searching strategy than those using Gaussian isotropic random walks. Recent studies have shown that CS algorithm has better efficiency compared with the GA and PSO [11].

In the implementation of Cuckoo Search algorithm, a balanced combination of a local random walk and the global explorative random walk can ensure that the entire design space can be explored by both local and global search capabilities. The switching between local and global searching is essentially controlled by a switching parameter p_a. The local random walk can be denoted as:

$$\vec{x}_i^{t+1} = \vec{x}_i^t + \alpha s \otimes H(p_a - \varepsilon) \otimes (x_j^t - x_k^t) \tag{7}$$

where x_j^t and x_k^t are two different solutions selected by random permutation, $H(u)$ is a Heaviside function and $\otimes$ represents the entry-wise multiplication operator. Here ε is a random number drawn from a uniform distribution, and s is the step size. Technically speaking, this step can also become a global search step if the step size is large enough by controlling the scaling factor α.

The variations of the step length of random walks based on a Gaussian distribution are typically small. In contrast, the steps of the Lévy flight usually differ significantly from each other . In Cuckoo Search algorithm, the global random walk is implemented by using Lévy flights strategy:

$$\vec{x}_i^{t+1} = \vec{x}_i^t + \alpha s \otimes Le'vy(\beta) \qquad 0 < \beta \le 2 \tag{8}$$

Where $\alpha > 0$ is a scaling factor of the step size, and is related to the scales of the particular characteristics of the problem. $Le'vy(\beta)$ denotes the Lévy distribution, given in terms of a simple power law formula[14].

4. Aerodynamic Performance Computation by RANS Solver

The aerodynamic performance of the wing-body cnfiguration is evaluated by solving Reynolds Averaged Navier-Stokes (RANS) equations solver using multi-block structured grid with cell center scheme. The spatial discretization scheme chosen is 2nd order Roe scheme and the turbulence model is Spalart-Allmaras turbulence model. A approximation factor decomposition approach has been adopted to accelerate the convergence.

For validating the accuracy of the solver, simulations on the DLR F4 wing-body model have been carried out and the results have been compared with the experimental data. The test conditions are: Mach=0.75, Reynolds Number=3.0E6, and Left Coefficient=0.5. The computational mesh size is 1.2E6 in mesh point's number. The computational results of the pressure coefficient distribution of three sections are compared with the experimental data, which are shown in Figure 8. Considering the mesh size is relatively small in optimization process, which requires a large number of CFD evaluations, the computed pressure coefficient distribution obtained by the RANS solver matches well to the experimental data along the wing span, which shows that the solver adopted in this paper has good accuracy for evaluating the aerodynamic performance.

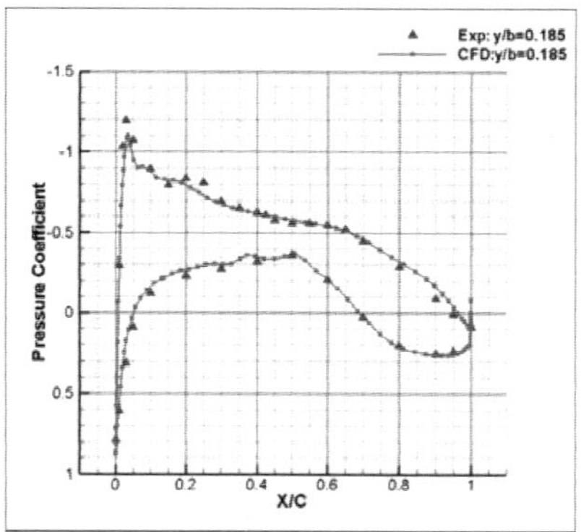

Fig. 4 Cp Distribution at the 44% Semi Span

5. Kriging Model with Expected Improvement Adaptive Sampling Criterion

The Kriging metamodel, originating form geostatistics, has widely been used in engineering optimization for its good accuracy and efficiency, the detailed description of using Kriging model in optimization design could be found in literatures[15]. The most powerful feature of the Kriging model is its capability for estimating the variance or error at a position in design space for approximation. Thus, in optimization using Kriging as approximation model, the uncertainty of probability of the optimum function value could be taken into consideration. With this methodology, the solution at which the global optimum most likely to occur would be determined, in other words, we want to obtain a solution with the largest probability of being the global optimum. The probability of being the global optimum is evaluated using the expected improvement (EI) criterion [16]. The expected improvement of Kriging model is defined as below:

$$E\left[I(x)\right] = (f_{\min} - \hat{y})\Phi\left(\frac{f_{\min} - \hat{y}}{s}\right) + s\phi\left(\frac{f_{\min} - \hat{y}}{s}\right) \tag{9}$$

where the $I(x)$ is the improvement of $f_{\min}$, which is the minimum true objective function value obtained so far. And $\Phi(\bullet)$ is the normal cumulative distribution function, and $\phi(\bullet)$ is the normal probability density function. s is the root mean squared error (RMSE) predicted by the Kriging model.

$I(x)$ is defined as:

$$I(x) = \max\left[f_{\min} - \hat{y}(x), 0\right] \tag{10}$$

Expected improvement criterion serves as an adaptive sampling infilling method, for additional sampling at the location where the global minimum most likely to occur.

With integration of the hybrid FFD-RBF parameterization approach in section 2, cuckoo search algorithm in section 3 and the Kriging model with expected improvement adaptive sampling criterion above, an aerodynamic shape optimization design system aiming at drag reduction of wing-body configuration has been established. The flowchart of the optimization design procedures are as follows:

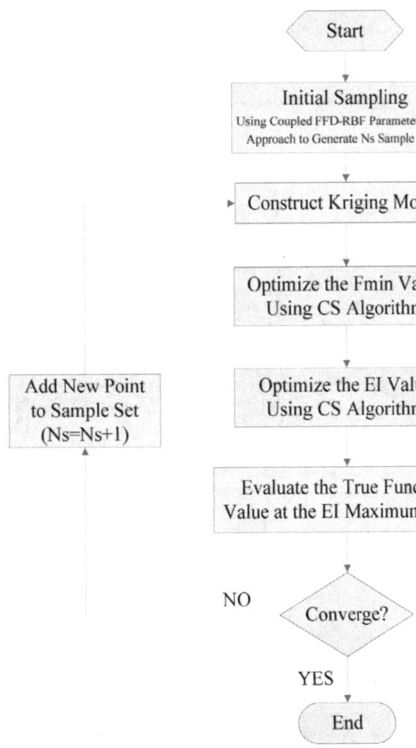

Fig. 5 Flow Chart of Aerodynamic Optimization Design Process using Hybrid FFD-RBF Surface Parameterization, Krigin Metamodel and CS

6. Study Case Verification

A study case using the aerodynamic optimization design system established in this paper has been carried out on the DLR F4 wing-body model, the wing-body shape and the corresponding FFD control frame are as the figure 2 shows, the total number of the design variables is 38 (36 ffd control points' vertical displacements together with 2 sections' rotation). The objective function is drag reduction at fixed lift coefficient of 0.5, with the constraint that the wing volume should be no less than the initial one. The optimization model could be formulated as:

$$\begin{aligned} \min \quad & Cd \\ s.t. \quad & Cl = 0.5 \\ & Volume_{optimized} \geq Volume_{original} \end{aligned} \tag{11}$$

The Cuckoo Search Algorithm with a population size of 300 has been used with 100 generations to search the objective function minimum value solution and EI value maximum solution in every optimization cycle. The parameters in CS are $P_a = 0.5$, $\alpha = 0.01$ and $\beta = 1.5$. The comparison of the final optimized wing-body with the original DLR F4 is as the Table 1 shows. The comparisons of the wing upper surface pressure coefficient distribution and several wing sections pressure coefficient distribution are as the figure 6 to 7 show.

Table 1 Comparison of the optimized wing-body and the original DLR F4 (at fixed Cl=0.50).

Wing	Cd	Cl/Cd	ΔCd	ΔCl/Cd
Original	300.2E-4	16.656		
Optimized	286.3E-4	17.464	4.63%	4.85%

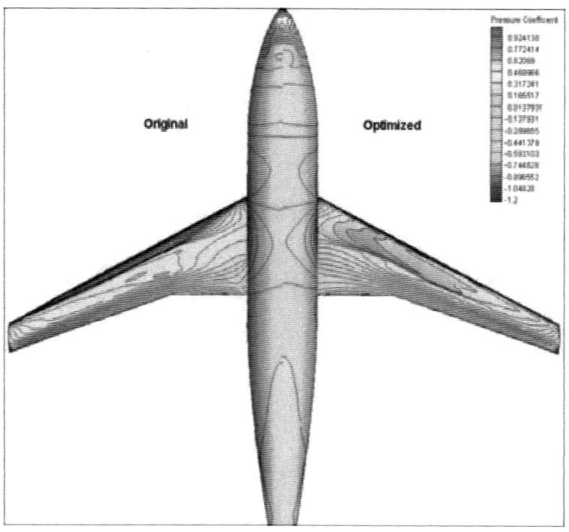

Fig. 6 Comparison of the Wing Upper Surface Pressure Coefficient Distribution of the Optimized and the Original.

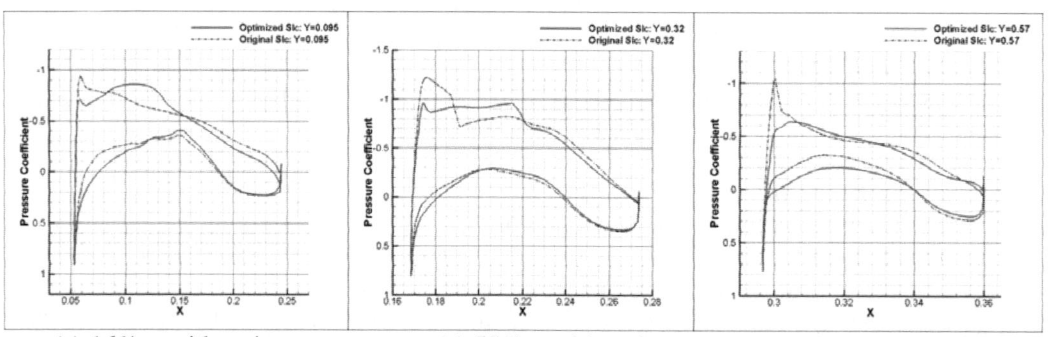

(a) 16% semi location (c) 55% semi location (c) 98% semi location

Fig. 7 Comparison of the Wing Section Pressure Coefficient Distributions of the Optimized and the Original

The comparisons above have shown that, after optimization design, the drag coefficient has been reduced 4.63% and the lift-drag-ratio has been improved 4.85% compared to the original one, while the lift coefficient is at fixed Cl=0.50. Figure 6 has shown that the strength of the shock wave has been alleviated apparently, though the shock wave has not been eliminated after optimization. Figure 7 (a) to (c) have shown that the pressure coefficient at the suction peaks have been reduced after optimization, which are all nearly below the line of Cp=-1.0.

7. Conclusion

A methodology that combines the FFD technique with RBF interpolation approach has been proposed as a hybrid FFD-RBF surface parameterization approach for wing-body configuration. The FFD technique is used for parametrically definition of the wing surface and the RBF approach is used for handling the updating the wing-body junction part while the wing has been deformed by interpolation. Together with a novel heuristic optimization algorithm – Cuckoo Search, and Kriging model with expected improvement adaptive sampling criterion, an aerodynamic design optimization study case has been carried out on a standard test model – DLR F4 wing-body. The result of the study case has shown that the hybrid approach proposed in this paper is of good feasibility and effectiveness of drag reduction, and the shock wave over the upper surface of the wing has been alleviated dramatically.

Further research will focus on shape optimization problems on much more complicated configurations with more engineering practical geometrical constraints, like a wing-body-nacelle-pylon configuration.

References

[1] Sobieczky.H. Parametric Airfoils and Wings. Notes on Numerical Fluid Mechanics, Vieweg, Braunschweig, 1998, pp:71–88.
[2] Hicks.R.M, Henne.P.A. Wing Design by Numerical Optimization. Journal of Aircraft, Vol.15, No.7, 1978, pp:407–412.
[3] Kulfan.M.B. Universal Parametric Geometry Representation Method. Journal of Aircraft, Vol.45, No.1, 2008, pp:142–158.
[4] Hicken.J, Zingg,D. Aerodynamic Optimization Algorithm with Integrated Geometry Parameterization and Mesh Movement. AIAA JOURNAL Vol. 48, No. 2, 2011 pp:400-413
[5] Yamazaki.W, Mouton.S, Carrier.G. Geometry Parameterization and Computational Mesh Deformation by Physics-Based Direct Manipulation Approaches. AIAA JOURNAL Vol. 48, No.8, 2010, pp:1817-1832
[6] Samareh. J. Novel Multidisciplinary Shape Parameterization Approach. JOURNAL OF AIRCRAFT Vol.38, No.6, 2001, pp:1015-1024
[7] Kim.H.J, Nakahashi.K. Surface Mesh Movement for Geometric Modification of Body-Installation Junctions. Proceedings of the Third International Conference on Computational Fluid Dynamics (ICCFD3), July 2004.Toronto, Canada.
[8] Sederberg T, Parry S. Free-form deformation of solid geometric models. Computer Graphics, 20(4):151–160, 1986.
[9] Obayashi.S, Oyama.A. Three-Dimensional Aerodynamic Optimization with Genetic Algorithms. Proceedings of the Third ECCOMAS Computational Fluid Dynamics Conference, John Wiley & Sons, Ltd, Chichester, U.K., 1996, pp:420-424.
[10] Duvigneau.R. A multi-level particle swarm optimization strategy for aerodynamic shape optimization. Proceedings of EUROGEN 2007, Evolutionary Methods for Design, Optimization and Control, Jyvaskyla, Finland, June 2007.
[11] Yang X S, Deb S. Engineering optimization by Cuckoo Search. International Journal of Mathematical Modeling and Numerical Optimization, 2010, 1(4):330-343.
[12] Yang X-S, Deb.S. Cuckoo search via Lévy Flights. Proceedings of World Congress on Nature & Biologically Inspired Computing (NaBIC2009, India), IEEE Publications, USA 2009, pp: 210–214.

[13] Pavlyukevich.I. Lévy Flights, Non-local Search and Simulated Annealing. Computational Physics, 226, 1830-1844 (2007)
[14] Mantegna.R. Fast, Accurate Algorithm for Numerical Simulation of Lévy Stable Stochastic Process Physics. Review. Vol.49, No.5, pp:4677–4683, 1994.
[15] Simpson.T, Korte.J, Mauery.T, Mistree.F. Comparison of Response Surface and Kriging Models for Multidisciplinary Design Optimization. Proceedings of the 7th AIAA/USAF/NASA/ISSMO Symposium on Multidisciplinary Analysis and Optimization, St. Louis, MO, AIAA, Vol. 2, Sep. 2-4, 1998, pp:381~391. (AIAA-98-4755).
[16] Fang K T, Lin D K J. Uniform experimental designs and their applications in industry. Handbook of Statistics, vol. 22. North-Holland, Amsterdam, pp:131–170, 2003

Aerodynamic Simulation Analysis of Unmanned Airborne Electronic Bomb

Jiaoying Yang*, Yachao Guo

School of Mechatronical Engineering, Beijing Institute of Technology University, Beijing 100081, China

*Corresponding author e-mail: yangjiaoying@bit.edu.cn

Abstract. For microelectronic bombs for UAVs, on the basis of the use of rotors to lift the insurance on the basis of ammunition, increased tail to increase stability. The aerodynamic simulation of the outer structure of the ammunition was carried out by FLUENT software. The resistance coefficient, the lift coefficient and the pitch moment coefficient under different angle of attack and Mach number were obtained, and the aerodynamic characteristics of the electronic bomb were studied. The pressure line diagram and the velocity line diagram of the flow around the bomb are further analyzed, and the rationality of the external structure is verified, which provides a reference for the subsequent design of the electronic bomb.

1. Simulation model

This paper designs a miniature electronic bomb that can be used for UAV bombs, which, on the basis of the original ammunition, adds tail fins for low altitude bombs. With the FLUENT software analysis of its trajectory characteristics, to understand the force in the ammunition flight, the calculation of aerodynamic parameters. Its structure and shape as shown below:

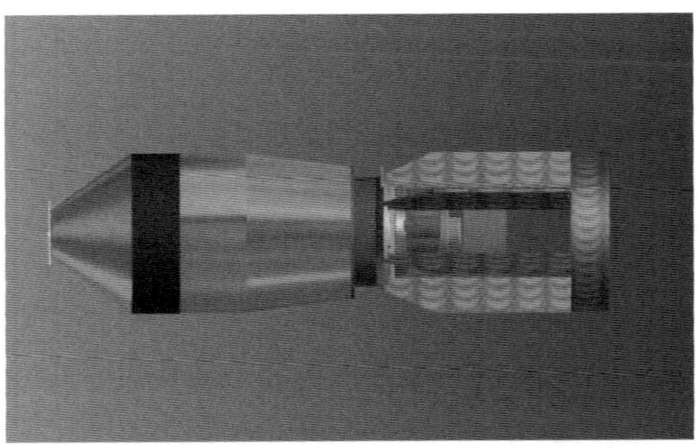

Figure 1. Unmanned Airborne Miniature Electronic Bomb Pneumatic Structures

Electronic bombs through the UAV projectile, so need to rely on the tail to increase stability. Grid division using FLUENT fluid analysis software Gambit preprocessor. As the ammunition structure is more complex, so in Gambit selected non-structural grid to divide the flow field, the projectile head, tail and other special parts to take a grid encryption [1, 4]. Since the submunition is similar to the cylinder, the simulated flow field is established as a cylinder with a radius of 30 times the radius of the field, about 12 times from the head of the ammunition, about 17 times from the bottom of the ammunition. The mesh near the missile is shown in the following figure:

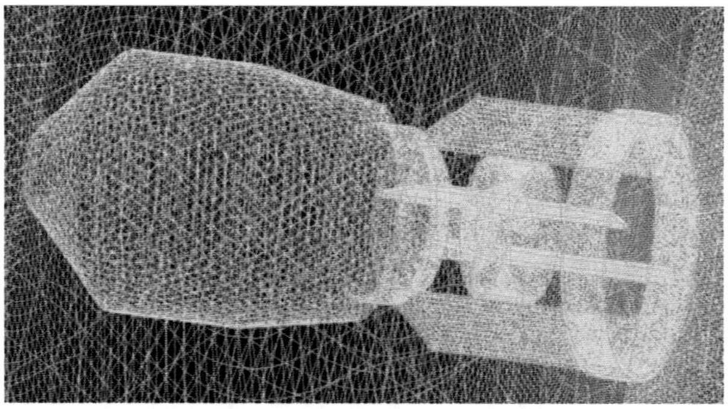

Figure 2. Electronic bomb grid division

The total number of meshes is 1119277, and the mesh evaluation is based on the EquiAngle Skew quality evaluation algorithm. The grid quality data is shown in the following figure. The minimum and maximum values of the mesh are 1.59656×10^{-6}, 0.940966.

From value	To value	Count in range	% of total count (1119277)
0	0.1	110539	9.88
0.1	0.2	86944	7.77
0.2	0.3	211959	18.94
0.3	0.4	282090	25.20
0.4	0.5	193103	17.25
0.5	0.6	124138	11.09
0.6	0.7	85791	7.66
0.7	0.8	24685	2.21
0.8	0.9	10	0.00
0.9	1	18	0.00
0	1	1119277	100.00

Measured minimum value: 1.59656e-006
Measured maximum value: 0.940966

Figure 3. Grid quality data

2. Parameter setting

In this paper, the submunition flying speed is less than a Mach number, so the pressure-based solver is used. The viscous model adopts the S-A model [2]. Boundary conditions are set to slip-free conditions, roughness 0.5, and other scalars are set to impermeable wall conditions. Select the SIMPLE algorithm in the pressure-velocity coupling mode to solve, that is, for a given pressure field, solve the discrete form of the momentum equation, and then calculate the velocity field. Gradient select Green-Gaus Cell Based, the Pressure is selected using the presto! Format and select the Second Order Upwind scheme for the momentum equation. Pressure relaxation factor 0.3, density relaxation factor 1, mass force relaxation factor 0.7, turbulent viscosity relaxation factor 0.7.

3. Simulation results and analysis

The velocity of the projectile is subsonic, and the resistance is frictional resistance and eddy current resistance. As the unmanned aerial vehicle bomb height is low, the bomb flight speed is not high, so the simulation conditions for the Mach number 0.2 ~ 1.0Ma, angle of attack 0 ~ 8 degrees. The relationship between the force and the moment coefficient and the angle of attack and the Mach number is obtained by the above simulation， the data is shown below.

Table 1. Three Scheme comparing.

Mach number	coefficient	Angle of attack				
		0	2	4	6	8
0.2	c_x	0.36003	0.36315	0.37306	0.38997	0.41317
	c_y	0	0.09668	0.19379	0.29135	0.38962
	m_x	0	0.01885	0.03758	0.05606	0.07428
0.4	c_x	0.35839	0.36152	0.37129	0.38825	0.41138
	c_y	0	0.09636	0.19325	0.29078	0.38890
	m_x	0	0.01889	0.03761	0.05611	0.07437
0.6	c_x	0.35096	0.36357	0.36917	0.38723	0.41059
	c_y	0	0.09564	0.19286	0.29066	0.38924
	m_x	0	0.01887	0.03773	0.05633	0.07463

The data is fitted with the following relationship:

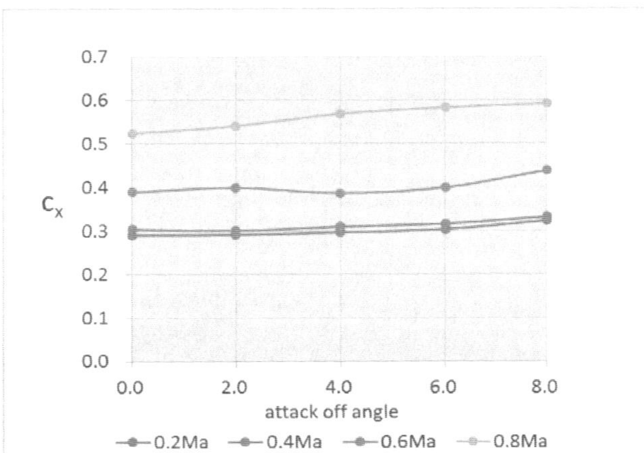

Figure 4. The relationship between the resistance coefficient and the angle of attack under the different Mach number

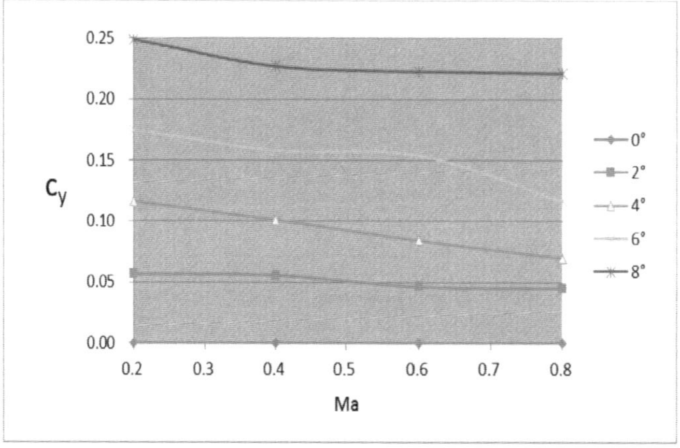

Figure 5. The lift coefficient varies with the Mach number under the different angle angle of attack

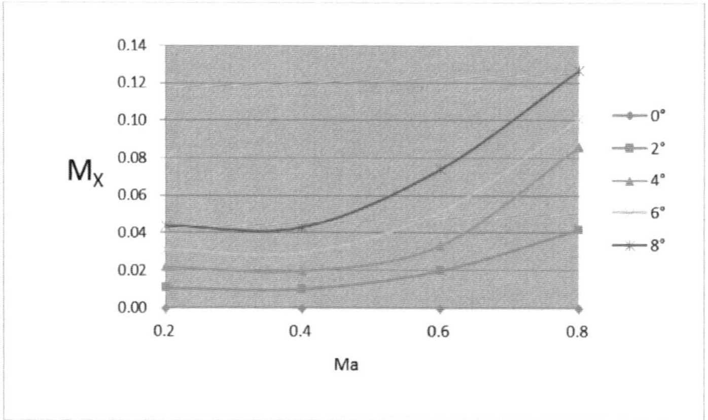

Figure 6. The Change of Moment Moment Coefficient with Mach number under different angle of attack

It can be seen from Fig. 4, Fig. 5 and Fig. 6 that under the same Mach number, the lift coefficient increases linearly with the increase of the angle of attack. Under different angles of attack, the lift coefficient varies with the Mach number. At the angle of attack, the lift coefficient decreases with the Mach number, and the lift coefficient of other angles of attack does not change significantly.
Next, selected the pressure, speed concours of some special points to analysis.

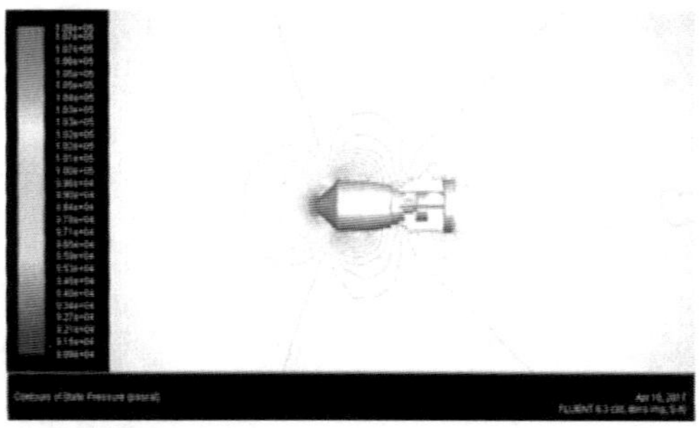

Figure 7. 0°, 0.3ma pressure contours

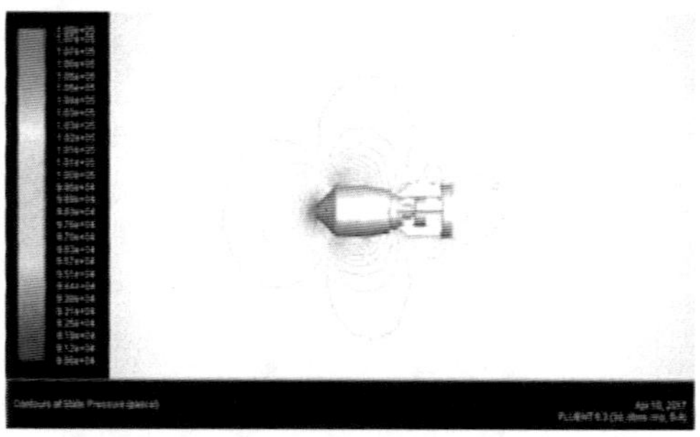

Figure 8. 2°, 0.3ma pressure contours

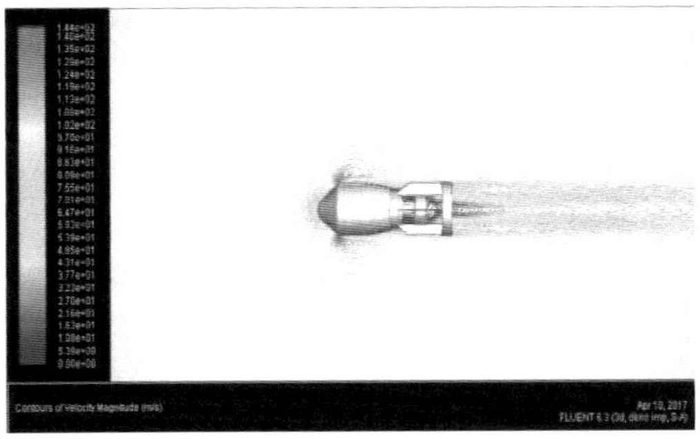

Figure 9. 0°, 0.3ma speed contours

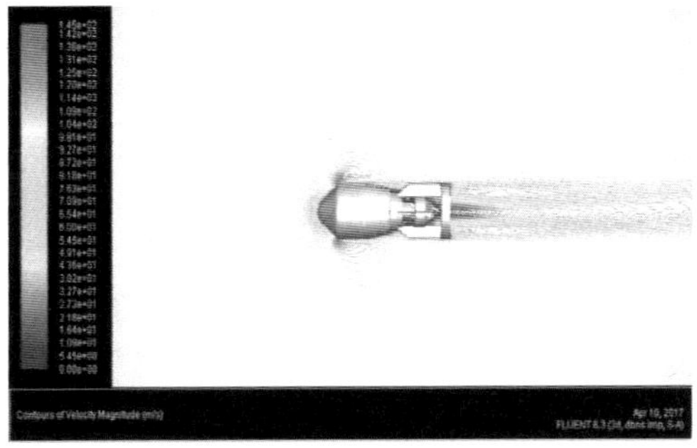

Figure 10. 2°, 0.3ma speed

It can be seen from the pressure contours: 1) The pressure concentration in the head of the projectile;

2) The projectile will be from the air friction resistance and the formation of the pressure difference between the projectile. It can be seen from the velocity contours diagram that the air near the projectile will flow along the contours of the projectile, and the airflow is separated from the projectile in the tail.

4. Conclusion

In this paper, FLUENT software is used to calculate and analyze the aerodynamic parameters of the miniature electronic bombs for UAVs, and the data are fitted with the post-processing function of the software, and the different angles of attack and Mach number are obtained. The influence of ammunition lift coefficient, drag coefficient and pitch moment coefficient, and further analyzes the situation of air flow in the process of sub-ammunition flight, and validates the rationality of submodule appearance, which provides a reference for subsequent structural design. As a result of the division of the grid, the corners of the corners and other parts of the simplified, there may be errors, still need to follow-up experiments to verify.

References

[1] Jiang Fan, Huang Peng FLUENT Advanced Application and Case Analysis [M]. Beijing: Tsinghua University Press, 2008.

[2] Shen Zhongshu, Liu Yafei. Projectile aerodynamics [M]. Beijing: national defense Industrial Press, 1984.

[3] Numerical study of aerodynamic characteristics of subsonic airfoil based on FLUENT [J]. Guo Jiangtao, Xing Cunzhen, Tan Xianzhong, Li Peng, Wang Ke. Journal of Zhongyuan Institute of Technology. 2012 (01)

[4] ANSYS FLUENT User''s Guide for Release 12.1. ANSYS,Inc. . 2009

Research on Aerodynamic Characteristics of Composite powered Unmanned Airship

Yu Chen*, Yun Wang, Lu Wang, Chengyu Ma and Jun Xia

School of Aircraft Engineering, Nanchang Hangkong University, Nanchang, Jiangxi 330063, China

*Corresponding author e-mail: 745317205@qq.com

Abstract. The main structure of the composite powered unmanned airship is consists of airbags and four-rotor system, which airbag increases the available lift, and has more advantages in terms of load and flight when compared with the traditional four-rotor .In order to compare the aerodynamic performance of the composite powered unmanned airship and the traditional four-rotor, the SIMPLE algorithm and the RNG k-epsilon model method are be used. The energy consumption of the composite powered unmanned airship is lesser than the traditional four-rotor under the same load and range was found.

1. Introduction

Today, four-rotor craft has become more common in our field, which covering the use of aerial photography, aerial surveys, plant protection and flight experience. Because of its superior handling performance and stability, especially in the field of aerial photography has made the industry's favorite. But due to the battery capacity and its own structure, the traditional four rotor is not good enough in the flight and load than other types of aircraft [1]. To this end, the industry developed a variety of improved programs, such as adding airbag to provide an additional upward forcedly[2]using the airbag in the four-rotor structure ,which greatly increasing the payload of the aircraft. Due to the whole of four rotor structure hanging under the airbag that the overall center of gravity of the aircraft move up, Resulting in the aircraft has some difficulties to control In the flight. In order to solve this problem, the connection of traditional four-rotor and airbag has been improved [3].

In order to analyze the performance advantages of the improved aircraft, choose the way that Compare the energy consumption of the two kinds of aircraft under the same load and range to see which one is better.

2. Model design

2.1. Three - dimensional model

The main structure of the compound powered unmanned airship includes airbags and four rotor systems. The traditional four rotor is made up of two carbon fiber rod as the main structure which contact under each other. Each carbon rod at both ends of the installation of a rotor motor used to provide lift, and the concrete model is shown in Fig1.The compound powered unmanned airship has been added a spherical airbag on the basis of the traditional four-rotor system, and the geometric center of the whole four-rotor structure coincides with the geometric center of the airbag, The concrete model is shown in Fig2.

The main geometric parameters of the study object are as follows: Four rotor shaft arm is round bar, which length is 5m, and outer diameter is 50mm. The vertical distance between the upper and lower shafts is 52mm, the airbag is a balloon which diameter is 3.6m. The coordinates of the center of the rotor in the geometric center of the overall structure are the coordinates of the origin:(2460 mm, 0, 14 mm), (- 2460 mm, 0, 14 mm), (0,2460 mm, 66 mm), (0, -2460 mm, 66 mm).

The two leaves "2880" model rotor is used to calculate, which diameter of the rotor is 711.2 mm, and pitch is 20.32mm. The rotor is used more in the market, that experimental data more comprehensive, and easy to carry out analytical research. For easier to calculate the following assumptions:

(1) In all states, the rotational speed of the four rotors is the same;

(2) The flow is constant and stable [4];

(3) Does not take into account the inclination of the torque generated by the problem;

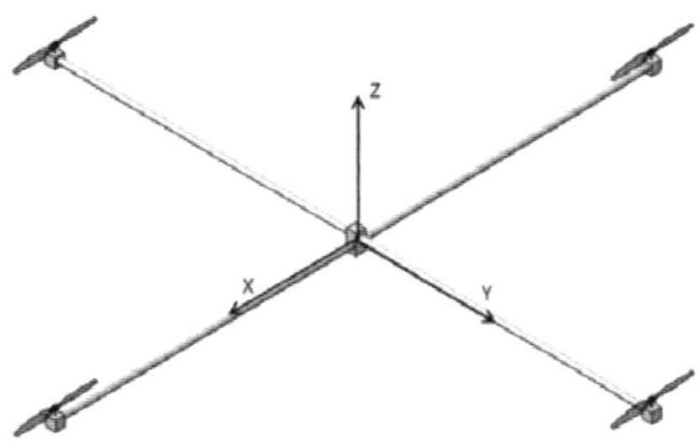

Figure 1. The model of the traditional four-rotor

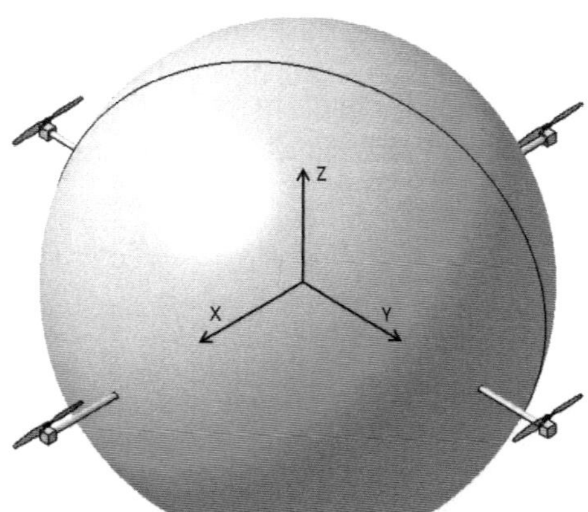

Figure 2. The model of Composite powered unmanned airship

2.2. *Formula introduction*

Due to the special principle of the rotor structure flight, using the differences tilt angles of rotor to produce different forward pull force. So different tilt angles represent different flight speeds. As a result of the flight speed increases, the air resistance increases. In each flight inclination corresponds to a flat speed V, at the speed of the aircraft forward direction of the resultant force is zero, do uniform motion. The aircraft is flying in an "X" flight and the angle between the XY plane of the body coordinate system in the flight state and the XY plane in the spatial coordinate system is the angle of inclination, denoted by the angle θ.

IOP Conf. Series: Journal of Physics: Conf. Series **916** (2017) 012003 doi:10.1088/1742-6596/916/1/012003

In order to ensure that the load of the two aircraft is the same during the flight, the air lift of the composite powered unmanned airship is calculated [5].The airbag is filled with helium that provides buoyancy. The formula is as follows:

$$F_{He} = K_{He}V \tag{1}$$

$$M = \rho S \tag{2}$$

$$F_T = F_{He} - Mg \tag{3}$$

F_{He}: The total buoyancy provided for helium;

K_{He}: Is the unit volume lift coefficient of helium, here take $10.35N/m^3$;

V: The helium volume;

M: The quality of airbag skin;

ρ: The density of the skin material;

S; The surface area of the skin (here selected imported Kevlar matrix polyurethane coating material [6], the surface density is $240g/m^2$);

F_T: The net lift of the airbag;

g: Gravity acceleration, here take. $9.8N/kg$.

The airbag provides a net lift of 157 N by calculated, so the combined powered unmanned airship rotor under the same load provides a lift of 157 N smaller than that of the traditional four-rotor. The simple thrust method is used to determine the flight speed of the aircraft. Increase the flow velocity, and obtain the rotor speed and flow velocity that satisfy the flat fly condition under the inclination angle, by simulating the iterative method. The flow velocity is the flat speed. The formula for the flight conditions is as follows:

$$\left.\begin{array}{l} T = D \\ L = W \end{array}\right\} \tag{4}$$

T: The aircraft forward thrust;

D: The flight state resistance;

L: The aircraft to lift force;

W: The overall gravity of the aircraft.

2.3. Calculation method

The RNG k-epsilon model is an improved version of the k-epsilon turbulence model. The RNG k-epsilon considers the low Reynolds number flow viscosity and improves the high Reynolds number properties of the standard k-epsilon model, taking into account the Turbulence swirl, and therefore applicable to the low value of Reynolds number and a strong vortex of the numerical simulation.

The SIMPLE algorithm is used to separate the velocity component and the pressure equation. The second order upwind scheme is used for the interpolation of the convective terms. The node-based Green-Gaussian method is used to interpolate the diffusion terms. The multi - reference coordinate system model (MRF) is used to numerically solve the three - dimensional N - S equation [7].

3. Results and Analysis

3.1. Verification of the calculation method

By comparing the CFD calculation data and experimental data to verify the reliability of the method. The net tension of the rotor at different speeds in two cases is compared. As can be seen from Figure 3, the simulation model calculation data and specific experimental data look similar. Some of the minor differences in the latter data are due to the fact that the density or viscosity coefficient is somewhat different under experimental conditions. It can be concluded that the calculation method is correct and reliable and can be used to study the aerodynamic characteristics of the physical model of the aircraft.

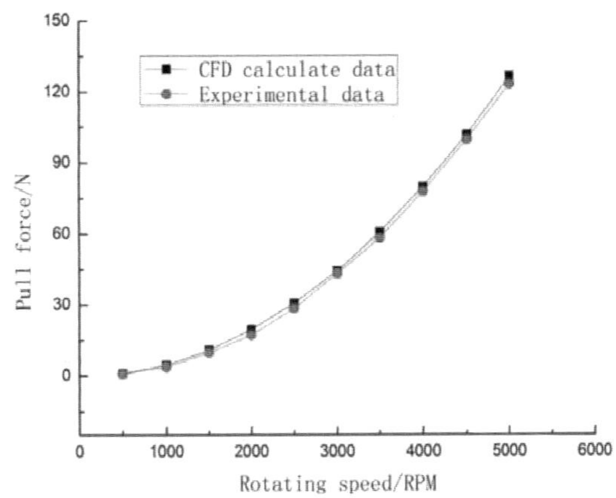

Figure 3. CFD calculation data and experimental data comparison chart

3.2. Analysis and Research on Parameters of Parameter

The experimental method is carried out by studying the difference between the flight speed and the rotor speed of the two aircraft at the same flight angle, the same load and range. Assuming that the total load is 400 N, the lift required for conventional aircraft is 400 N, and the lift for the combined powered unmanned airship is 243 N. Using the above calculation method, the two kinds of aircraft were simulated and analyzed to determine the energy consumed by the two aircraft in the horizontal flight state when the angle of inclination is $20°$.

By calculating the current consumption at a motor rotation speed, the total energy dissipation under the load W and the range S is obtained. From formula (5), it is concluded that the total energy consumption is proportional to $\frac{IU}{WV}$ in a certain voyage, where it is defined as "the energy consumed per unit load"

$$\left.\begin{array}{l} E = UIt \\ S = Vt \end{array}\right\} \Rightarrow E = \frac{I}{V} US \tag{5}$$

E : The motor power consumption;
U : The motor operating voltage;
I : The motor operating current;
S : The distance;
V : The flat speed;

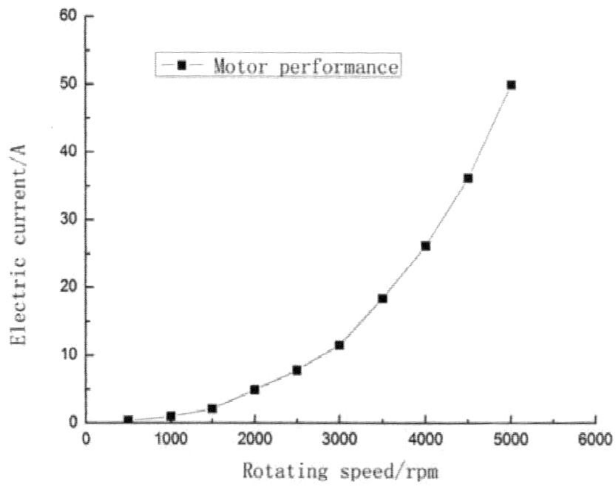

Figure 4. The speed and current of motor

In order to know the operating current of a motor at each speed, the motor is tested at constant voltage to obtain the current curve when the voltage is 44.8V, as shown in Figure 4.When the motor operating voltage is constant, the motor current increases with the rotor speed, and each motor speed corresponds to a working current. Combined with the formula (5), it can be drawn that the greater of the flying speed, the less of the total energy consumption.

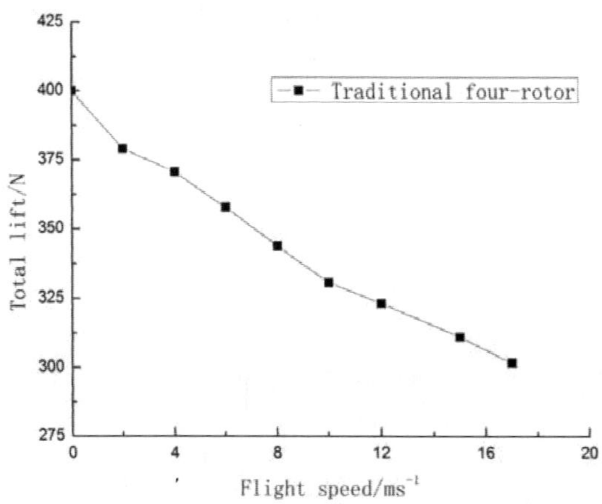

Figure 5. Performance characteristics of traditional four-rotor

Test the size of the lift of the traditional four-rotor at different flight speeds, at the rotor speed of 4500rpm.Observe the changes in lift. It can be seen from Fig. 5 that the total lift of the traditional four rotor decreases with the increase of the flight speed when the inclination angle is constant, which is in accordance with the law of the maximum static tension of the rotor [8].

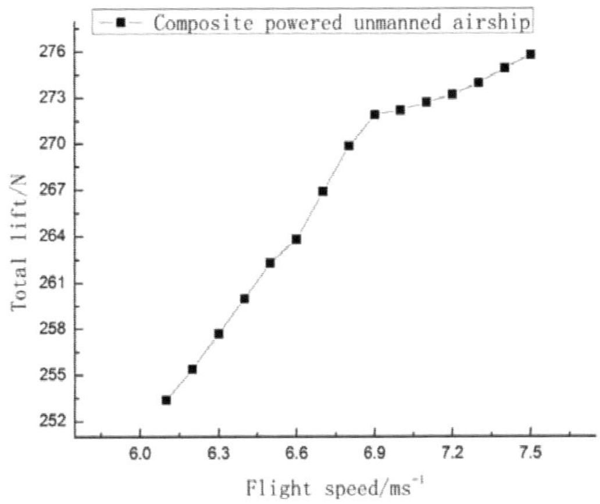

Figure 6. Performance characteristics of composite powered unmanned airship

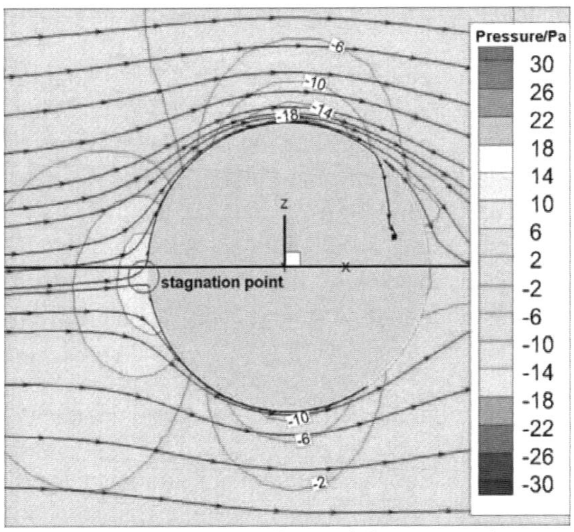

Figure 7. The pressure nephogram and streamlines of the composite powered unmanned airship

As can be seen from Figure 6, when the composite powered unmanned airship at a certain angle, as the flight speed increases, the total lift increases, this is different from the traditional four-rotor. Figure 7 shows the pressure reprogram and streamlines of the composite powered unmanned airship at the Y=0 section, it can be seen that the flow is affected by the rotor, so the position of the fore stagnation point of the airship is lowered when the airship flies forward, this made the airflow in the upper half of the composite powered unmanned airship is faster when the airflow through the composite powered unmanned airship, then the pressure drop causes the airbag produce lift, which allows the composite powered unmanned airship to carry more loads[9].

Based on the analysis of the flight characteristics of the two aircraft and the adjustment of the flight speed and the rotor speed, the flight speed and the corresponding speed of the minimum energy consumption under the load requirement are obtained. As the maximum speed of the selected motor is 5280 rpm, the rotor speed is selected between 0 and 5000 rpm for easy energy consumption calculation. As the traditional four-rotor with the increase in flight speed lift reduced, so under a certain load there must be a maximum flight speed to ensure the minimum energy consumption. So the minimum energy consumption of the selection point should be found in the test speed to meet the lift when the maximum flight speed of 400N.

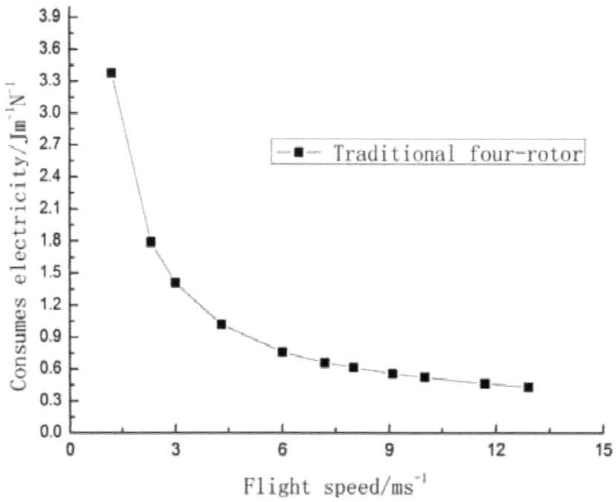

Figure 8. Traditional four rotor flight speed and consumption of electricity relationship line chart

According to the data analysis in Fig. 8, when the load is 400N, the traditional four-rotor reaches a maximum flight speed of 12.9m / s at a speed of 5000rpm, and the power consumption per unit voyage is 0.434J / MN. Due to the flying characteristics of the compound powered unmanned airship, at the inclination of 20 degrees, it is the point where the resultant force satisfies the load and satisfies the horizontal direction, that is, the optimal flat speed. By the iterative calculation, the optimal flying speed of the composite unmanned airship is 6.6m/s at an inclination of 20° , and the corresponding speed is 3250rpm.The electricity consumed under the unit voyage is 0.246J / MN <0.434J / MN. So in this flight state, the composite power unmanned airship energy efficiency is higher.

4. Conclusion

(1) The composite powered unmanned airship is higher in energy efficiency than the traditional four-rotor in the same load.

(2) By the rotor air flow, the composite power unmanned airship can get additional lift.

Acknowledgments

This work was financially supported by the National Natural Science Foundation of China (51266012), And Jiangxi Province Graduate Innovation Fund Project (YC2016-S352).

References

[1] Y. Wang, S. L. Le, Z.W. Fan. A New Type of Rotor + Airbag Hybrid Unmanned AerialVehicle J. International Journal of Energy Science, 2013, pp.183-187.
[2] S.L.Liu . Rotor + airbag composite power unmanned aerial system design analysis [D]. Nanchang , 2014.
[3] Y. Chen and Y. Wang , CHN Patent CN205469759U, (2016)
[4] C.T.DIAO H.J.WANG . CFD calculation and analysis of aerodynamic characteristics of dish-shaped wing aircraft C .China Aviation Society, 2012,pp.55-61.
[5] Z.M.Gu . Stratospheric airship skin material research J. Space return and remote sensing, 2007, pp.62-66.
[6] Z.F.Liu , J.Y.Xiao , Y.Wang , J.Liu , J.C.Zeng . Preparation of Aramid Carpenton Airbag J. New Technology Fiber and Application, 2006, pp.26-28
[7] Wilcox D. C. Tubulence modeling for CFD [M]. California, 1994,pp. 77-83.
[8] Alexander Lov. Air propeller M. Y.C.Wang , P.S.Wang , Y.W.Peng translation. Beijing:

Higher Education Press, 1954: 173.
[9] Y.Nie . Stratospheric airship propeller design and experimental research D. Chinese Academy of Sciences Graduate School (Space Science and Applied Research Center), 2008.

IOP Conf. Series: Journal of Physics: Conf. Series **916** (2017) 012004 doi:10.1088/1742-6596/916/1/012004

Experimental Study on the Propulsion Performance of the M-shape flapping wing's bending angle

Jingxian Chen[a], Xiaofang Nie[b,*], Ximing Zhou[c]

Lilienthal laboratory, Beijing Implant Aircraft Co., Ltd., 6 Chaoqian Road, Changping, Beijing, 102200, China

[a]79834096@qq.com,
[b] *Corresponding author e-mail: 2377811041@qq.com
[c]930326939@qq.com

Abstract. To study the the effect of flapping wing with different bending angles α on the thrust, in this paper, 9 M-shape flapping wing models with different bending angles, ranging for 0° to 22°, were designed. The rotating arm experiment was adopted to conduct the thrust test on the flapping wing models with different bending angels under the wind speed of 15m/s. The result shows that the span-wise flapping wing's curvature could rectify the airflow, the proper curvature could prevent the span-wise airflow at the surface the flapping wing and leads the airflow towards backward, the amount of air pushed backwards by the flapping wing is larger, therefore the value of thrust is increased; As well as the rectification of M-shape flapping wing increases the thrust value, the flapping wing's form drag also increased due to the bending angle. According to the results of the experiment, when the bending angle is less than 12°, the increment of the thrust is larger than the decrease of the form drag, so the thrust value increases gradually. However, when the bending angle is larger than 12°, the increment of the thrust is less than the decrease of the form drag, so the thrust value decreases. The thrust value is the largest when the bending angle is 12°.

1. Introduction

Unlike fixed wing aircraft, the flapping-wing flight could generate both lift and thrust force at the same time. Therefore, domestic and foreign scholars have carried out plenty of theoretical and experimental studies on the mechanism of flapping wing propulsion. By observation of micro structures of feathers from birds with different weight and flight altitude, researchers have found that the three are open, semi-open or closed and gradual contracting nozzles in birds first primaries (Fig.1).During down-stroke, air enters the tubular structure from the plume and is turned to the trailing edge and also accelerated; therefore, a strong thrust is generated [1]. Besides, a wind tunnel test has shown that flapping wings with wing blades could prevent the span-wise airflow beneath the wings (Fig.2), which results in the improvement of the flapping wing's aerodynamic performance [1].

Figure 1. Nozzle system for bird feathers.

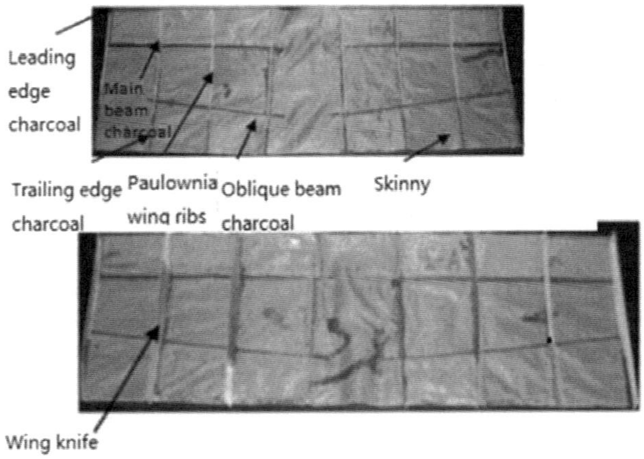

Figure 2. The flapping wing model with wing blades

A flapping wing model with span-wise curvature (Fig. 3) is made by Lilienthal laboratory of Beijing Implant Aircraft Co., Ltd. An air blowing test is carried out with the flapping wing model fluttering. As shown in the result, compared with the planer flapping wing, the airflow of flapping wing with a span-wise curvature is rectified, thus, the velocity vector of the airflow is turned backwards.

Figure 3. Flapping wing model with span-wise curvature

In this paper, a large size flapping wing model with span-wise curvature as M-shape is designed (Fig. 4). The cross section of the flapping wings perpendicular to the axis is isosceles triangle (Fig. 5). The sharp angel formed by the hypotenuse and the base of the triangle is defined as the bending angle α of the flapping wing. Nine M-shape flapping wing models with different bending angles are designed, ranging for 0° to 22°. The aerodynamics test of the flapping wing model was carried out by means of the rotating arm experiment. Through the test, the effect of flapping wing with different bending angles on the average thrust at the speed of 15m/s was studied.

Figure 4. M-shape flapping wing model (left) and planar flapping wing model(right)

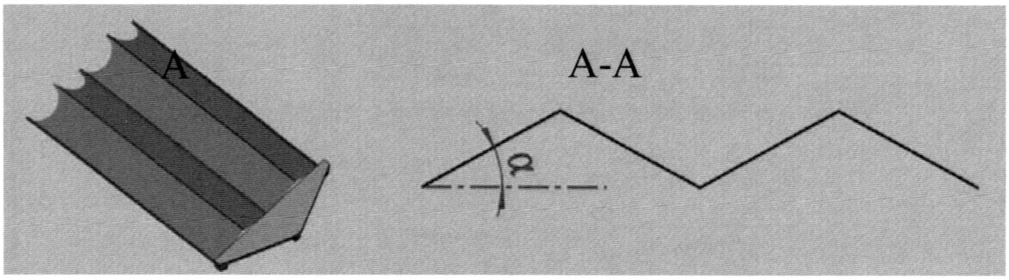

Figure 5. Sketch of bending angle α

2. Experimental equipment and methodology

In this paper, the aerodynamics test of the flapping wing model was carried out by means of the rotating arm experiment. The equipment is a rotating platform. The rotating platform consists of fixed frame, rotating shaft, rotating power supply circuit and model installation plate (Fig.6). The rotation platform is equipped with a transmission mechanism (Fig.7) which could realize the movement of flapping wings. The transmission mechanism is equipped with a crank slider structure and driven by a 1kw motor. The crank slider could drive the flapping rod reciprocating up and down to achieve the flapping of the flapping wings. A cantilever is fixed on the installation plane of the rotation platform, and a 3kw motor-driven propeller is installed on the cantilever.

Figure 6. Rotation Platform

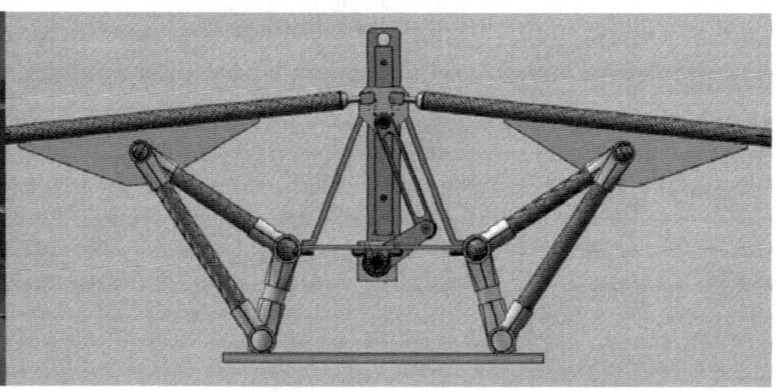

Figure 7. Transmission Structure

The installation plane of the propeller is in the rotating direction perpendicular to the rotating direction of the platform. The 3kw motor is powered by a DC power supply through a rotating circuit on the rotation platform. After the propeller is started, the thrust generated by the propeller would drive the platform to rotate. The speed of the propeller could be regulated by the remote control. By changing the thrust of the propeller, the speed of the rotation platform could be controlled. The experimental flapping wings are installed on the flapping structure of the flapping wings, rotating along with the platform. And the wings achieve forward movement while flapping in order to simulate the straight-line flight of flapping wings.

One flapping wing model was respectively placed on the flapping wing rods of both sides. The placement position of the model was 1.35m away from the center of the shaft. The two flapping wing models were the same, with consistent installation direction with the platform rotation direction. The thrust was positively correlated to the platform rotation rate. The greater the rotation rate was, the greater the thrust would be. The line speed of the sector mounting point was defined as the inlet velocity. Through calculation, the line speed of the sector mounting point would reach 106r/min, when the platform speed was 15m/s. During the experiment, the flapping wing was made to flap at the frequency of 2.5Hz first, to drive the platform rotate. And then, the remote control was used to make the propeller driving the platform to accelerate to $\omega 0 = 106$r/min. The propeller speed was kept unchanged, and the flapping of the flapping wings was stopped. In this case, the platform speed became $\omega 1$, with the speed change $\Delta\omega(\Delta\omega = \omega_0 - \omega_1)$ namely the action quantity on the platform rotating speed when the flapping wing was flapping at the speed of 15m/s.

$$F_T = F_w + F_H = F_D \quad (1)$$

$$F_W = F_D - F_H \quad (2)$$

In the formula, F_T is the thrust, F_D is the aerodynamic drag, F_H is the force onto the platform applied by the propeller, and F_w is the force on the platform applied by the flapping wing. The greater $\Delta\omega$ is, the larger F_W will be. That means, the thrust generated by the flapping of the flapping wing will reach the maximum when F_w is 15m/s.

The experimental air speed was set at 15m/s, namely, the platform speed was 106r/min. The arm length of the flapping wing was 0.8 meters, the installation angle was 0°, the flapping frequency was 2.5Hz, and the flapping angle was 45°.

3. Experimental results and analysis

The effect of the flutter of the M-shape flapping wings with different bending angle on the platform rotating speed at the wind speed of 15m/s was obtained through the rotation platform experiment. The experimental data are as shown in Tab. 1.

Table 1. Experiment Data

Bending angle/°	Flapping Wing + Propeller Platform Rotation rate/(r·min^{-1})	Platform rotation rate after Flapping Stops/(r·min-1)	Change of rotation rate/(r·min-1)
0	105.805	101.971	3.834
7	105.986	101.830	4.156
10	106.204	101.580	4.624
12	106.188	99.776	6.412
15	105.593	100.577	5.016
17	106.000	101.546	4.455
18	106.311	101.926	4.385
20	106.652	102.093	4.559
22	105.962	101.468	4.494

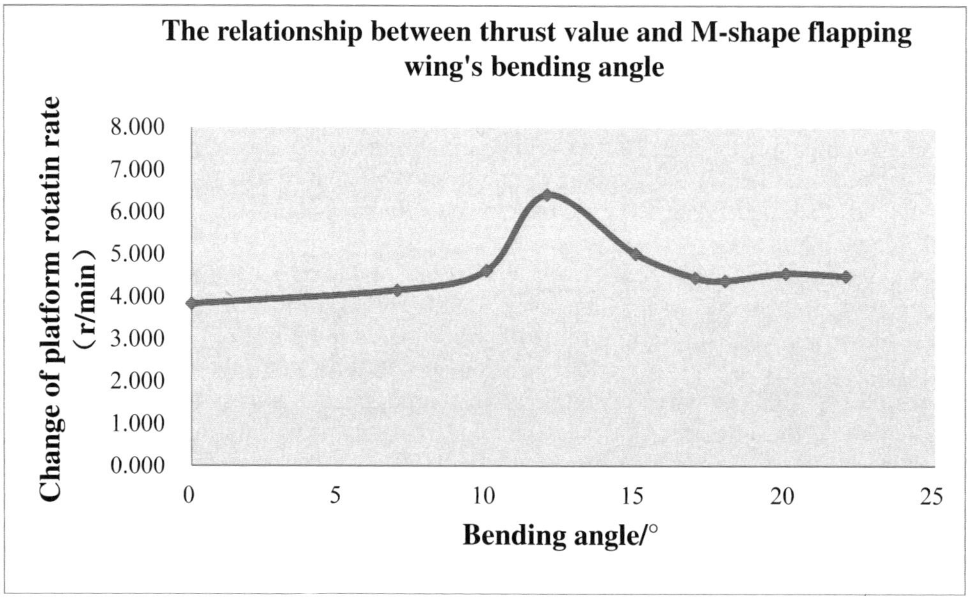

Figure 8. The relationship between thrust value and M shape flapping wing's bending angle

Fig. 8 shows that the with the augment of M-shape flapping wing's bending angle, the value of thrust force increases first, and then decreases with the augment of the bending angle. On the one hand, the span-wise flapping wing's curvature could rectify the airflow; a proper curvature could prevent the span-wise airflow at the surface the flapping wing and leads the airflow towards backward, the amount of air pushed backwards by the flapping wing is then larger, therefore the value of thrust is increased. On the other hand, flexible deformation exists during flapping wing's flutter. The flexible deformation produces thrust as well as resistance. With the increase of the bending angle, the area of flapping wing increases, which leads to the increase of airfoil shape resistance. According to the experiment's results, when the bending angle is less than 12°, the increment of thrust caused by the curvature's rectification is larger than the increment of resistant; thus the thrust of the flapping wing increase gradually. However, when the bending angle is larger than 12°,the increment of thrust caused by the curvature's rectification is smaller than the increment of resistant; therefore, the thrust of the flapping wing decreases gradually.

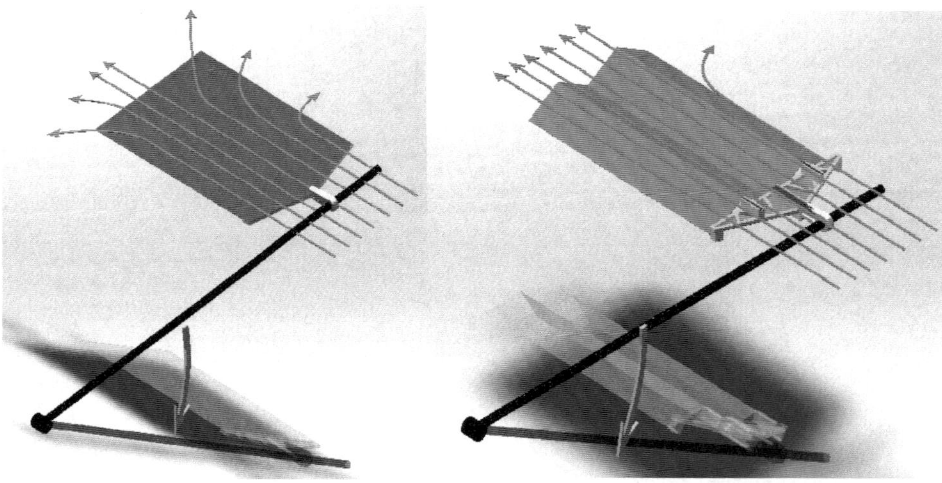

Figure 9. Path line of planar flapping wing (left) and M-shape flapping wing (right)

4. Conclusion

In this paper, 9 M-shape flapping wing models with different bending angles, ranging for 0° to 22°, were designed. The rotating arm experiment was adopted to conduct the thrust test on the flapping wing models with different bending angels, ranging from 0° to 22°, and the following conclusion and inference have been drawn:

(1)The span-wise flapping wing's curvature could rectify the airflow; the proper curvature could prevent the span-wise airflow at the surface the flapping wing and leads the airflow towards backward, the amount of air pushed backwards by the flapping wing is larger, therefore the value of thrust is increased.

(2)As well as the rectification of M-shape flapping wing increases the thrust value, the flapping wing's form drag also increased due to the bending angle. According to the results of the experiment, when the bending angle is less than 12°, the increment of the thrust is larger than the decrease of the form drag, so the thrust value increases gradually. However, when the bending angle is larger than 12°, the increment of the thrust is less than the decrease of the form drag, so the thrust value decreases. The thrust value is the largest when the bending angle is 12°.

References

[1] Li Bing; A Preliminary Study on The Microstructure of Bird Primaries and Their Significance For Flight[D]; Northeast Forest University; 2008

[2] ZHANG Yafeng, SONG Wenping; Experimental Investigation on Wing Aerodynamic Characteristics of Flapping-Wing Aerocraft with Wing-Blade[J]; Journal of Experimental Mechanics,2011,26 (2) :146-150

Numerical simulation of aerodynamic characteristics of multi-element wing with variable flap

Hongyan Lv, Xinpeng Zhang [a], Jianghong Kuang

College Of Air Transportation, Shanghai University of Engineering Science, Shanghai 201600, china.

[a]zxp_auto@163.com

Abstract: Based on the Reynolds averaged Navier-Stokes equation, the mesh generation technique and the geometric modeling method, the influence of the Spalart-Allmaras turbulence model on the aerodynamic characteristics is investigated. In order to study the typical configuration of aircraft, a similar DLR-F11 wing is selected. Firstly, the 3D model of wing is established, and the 3D model of plane flight, take-off and landing is established. The mesh structure of the flow field is constructed and the mesh is generated by mesh generation software. Secondly, by comparing the numerical simulation with the experimental data, the prediction of the aerodynamic characteristics of the multi section airfoil in takeoff and landing stage is validated. Finally, the two flap deflection angles of take-off and landing are calculated, which provide useful guidance for the aerodynamic characteristics of the wing and the flap angle design of the wing.

Key words: multi-element wing; mesh generation; numerical simulation.

1. Introduction

In the course of flying in the air, the lift of a plane is mainly generated by the wing, and the resistance is also produced on the wing. The magnitude and direction of the lift and drag values of the wing depend to a great extent on the shape, characteristics and dimensions of the wing, i.e. airfoil geometry (wing profile), wing geometry, etc..

The modern aircraft aerodynamic design tools selection design and optimization design can not only accurately predict the aerodynamic characteristics of aircraft layout absolute quantity, relative quantity can also accurately predict caused by small changes in design parameters on the aerodynamic characteristics of change. Wang Yuntao, with "sub trans CFD software platform" (TRIP2.0) numerical simulation of complex flow of 30P-30N multi-element airfoil, the main purpose is to affect the assessment of turbulence models and transition position on the pressure distribution and airfoil multi segment speed type [1] typical stations. Liu Peiqing, on a plane wing airfoil mounted control different height Gurney simulation results show that the installation of Gurney flap can increase the lift coefficient and drag coefficient of multi-element airfoil, but will strengthen the wake flow instability [2]. Mao Jun and so on, using computational fluid dynamics method of civil aircraft two-dimensional lifting device, that is, the design of multi section airfoil [3]. The analysis and calculation of aerodynamic characteristics of multi section airfoils has become one of the focuses of computational fluid dynamics research.

2. The establishment of 3D wing model based on FLUENT

2.1. The three-dimensional structure of the wing

In this paper, the DLR F11 wing of the German Aerospace Center is taken as the object of study. The reference geometry parameters of the DLR F11 airfoil (wing shape as shown in Figure. 1.1.1) are calculated:

Reference area of the semi-span model = Sref/2 = 419130 mm2

Mean aerodynamic chord (MAC) = 347 mm

Wing semi-span = 1400 mm

Moment reference center (MRC): x=1430 mm, y=0.0 mm, z=-42 mm

The 3D model of DLR F11 wing is established by using CATIA 3D modeling software. Three dimensional models of flat flight, take-off and landing are established, such as 1.1.2, flat flight model / take-off model / landing model.

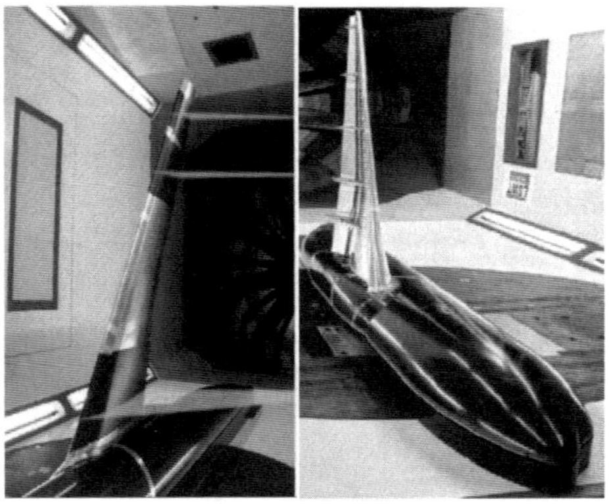

Figure 1. Wing profile

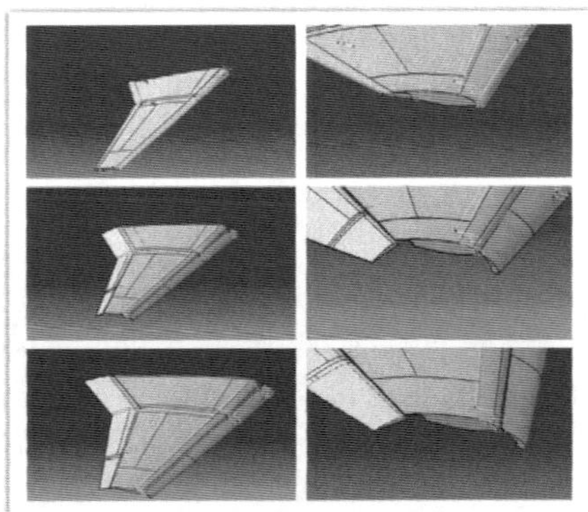

Figure 2. DLR F11 wing plane takeoff / takeoff / landing 3D model diagram

2.2. Mesh of wings

The flow around an airfoil is an external flow problem, and the boundary of the basin is far from the wing. But this is not possible for the simulation. Generally, the tail boundary of the basin should be more than 10 times the length of the wing rotation. In the simulation, the structured grid without the wings, and structured grids in the aspect of the inheritance of the calculation accuracy and efficiency of grid and grid generation standards than other types have obvious advantages, need to use ANSYS software in the ICEM series CFD software to build the flow field structure and geometry of mesh generation, and to facilitate further calculation. Structured mesh is adopted, and the mesh number is about 10 million. The grid is divided into plane flight, take-off and landing respectively, such as DLR, F11, wing mesh. Figure 1.2.1 shows a schematic diagram of the whole grid and the grid section; figure 1.2.2 shows a schematic diagram of the front and back grid grid, fly level before and after the flap is closed, take off and landing different before and after the open angle of flap.

Figure 3. DLR F11 wing overall frame mesh diagram

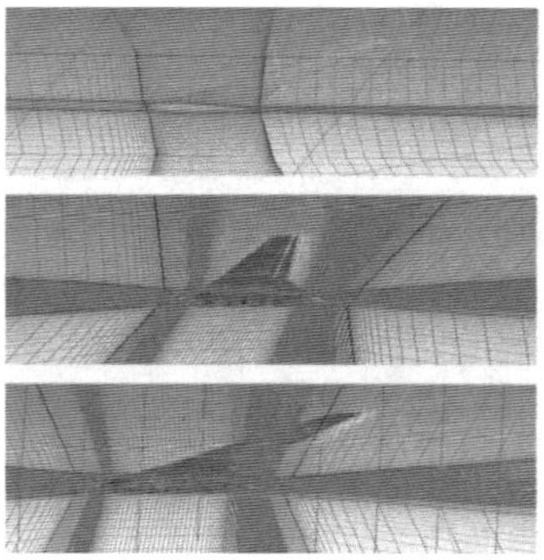

Figure 4. DLR F11 wing flat takeoff / landing / landing grid diagram

3. Simulation test calculation

In order to study the typical configuration of aircraft, a similar DLR-F11 wing is adopted to validate the prediction capability of the method. The flap angle is 32 degrees, and the angle of the flap is 26.5 degrees. The experimental data are compared with the structural mesh and unstructured mesh simulation. Using the S-A turbulence model, the Maher number is 0.175, and the corresponding Reynolds number is 15.1 * 106. Angle of attack ranged from -3 degrees to -24 degrees. The contrast test data is ETW (European, Transonic, WindTunnel) RUN238 data. As shown in Figure 2.1, the DLR-F11 simulation pressure nephogram, shown in figures 2.2.1 and 2.2.2, shows the comparison of lift and drag coefficients between the structural mesh, unstructured mesh and the experimental simulation.

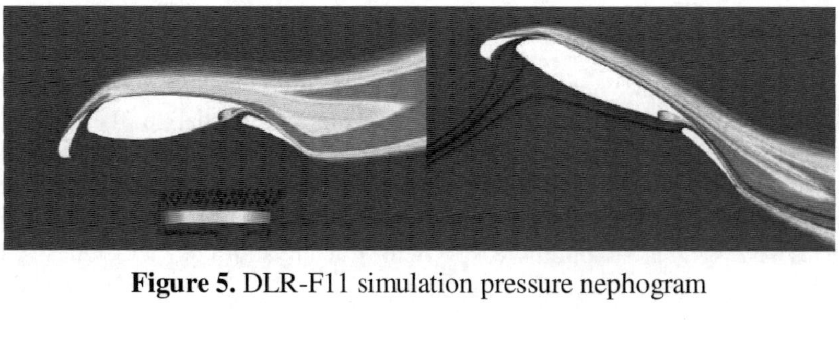

Figure 5. DLR-F11 simulation pressure nephogram

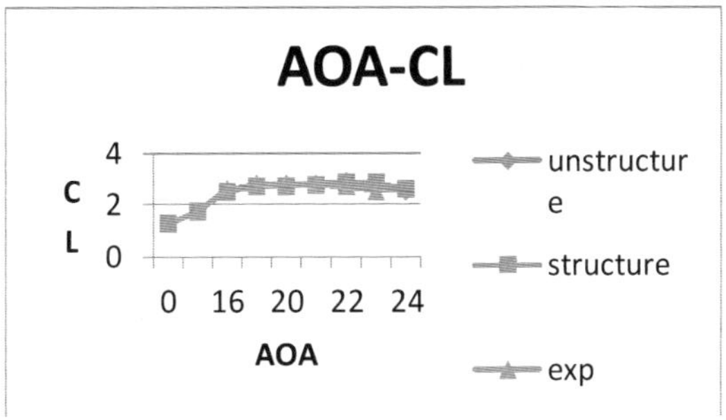

Figure 6. Lift coefficients between structured mesh, unstructured mesh and experimental simulation

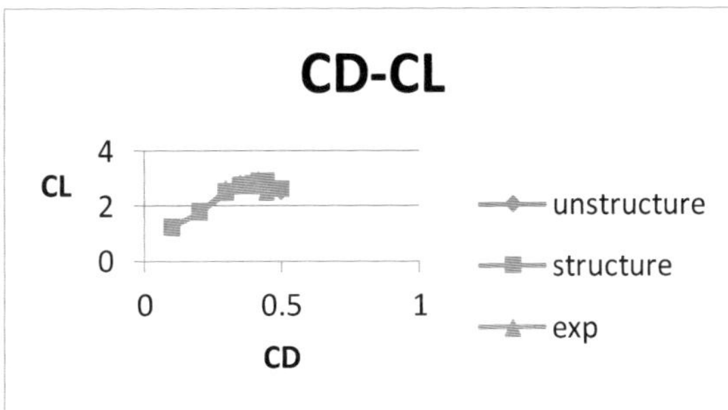

Figure 7. Lift to drag ratio between structured mesh, unstructured mesh and experimental simulation

The results show that the slope of the lift coefficient curve between the numerical simulation results and the experimental results is slightly different, which may be due to the differences between the two models of numerical simulation and experiment. The calculated results are in good agreement with the experimental results at low and mid attack angles, but in the nonlinear region, the lift coefficient is small and the stall delay is compared with the experiment. When the stall coefficient is 20 degrees, the maximum lift coefficient is 2.873. The maximum lift coefficients of the structured and unstructured grids are 2.827 and 2.893, respectively, and the stall angles are 24 and 22 degrees respectively. The structure network is closer to the experimental value, which shows that the structure network is more effective. (Green: experiment: Red: structured mesh; blue: unstructured grid)

The computed lift and torque curves are compared with the results of the literature [4] and the wind tunnel test results, as shown in Figure 2.3.1 and figure 2.3.2. From figure 2.3.1 and figure 2.3.2 shows that the calculated results are in good agreement with the literature [4] in the stall angle before the lift coefficient of two forecast is slightly less than the test value, which results in [5-6] and consistent, which may be due to not considering the influence of transition, for the turbulence, and the model in the wind tunnel test there will be deformation, but not considering the factors of stall angle, between the results between experimental values and the results in the literature; the same stall attack angle of lift coefficient calculation values and test values of maximum deviation is 2%; the torque curve calculation compared with literature results and experimental data show that the prediction is more consistent. The model is consistent with the wind tunnel test load distribution.

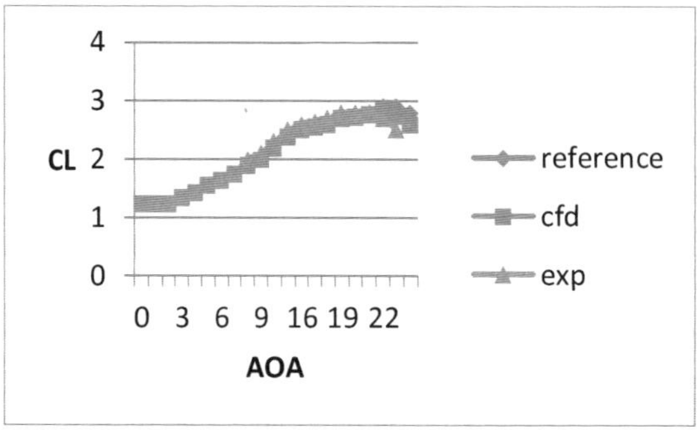

Figure 8. Curves of lift coefficient versus angle of attack

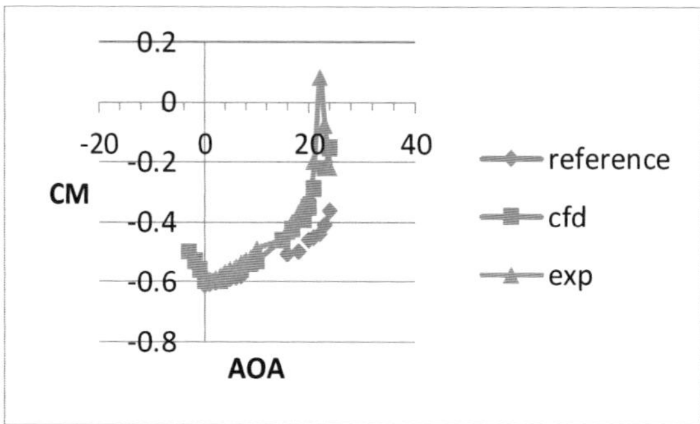

Figure 9. Variation curve of pitching moment with angle of attack

In summary, the numerical calculation method and grid strategy on the aerodynamic performance prediction precision has reached advanced level at the present stage and in good agreement with the experimental results, the prediction for the wing aerodynamic characteristics is feasible.

4. Calculation results and discussion

Calculation of configuration is installed in the body of the simplified large chord, medium aspect ratio, slat / wing / trailing edge flap three configuration. The calculation model shows that the leading edge wing angle is 30. Angle of takeoff and landing, trailing edge flaps were selected two groups of angle comparison, leading edge slats and trailing edge flaps from the wing tip extends to the root and into the body. Using the S-A turbulence model, the Maher number is 0.175, and the corresponding Reynolds number is 3 * 10^6. Angle of attack range from 0 degrees to 28 degrees. Using ANSYS FLUENT software import mesh model, numerical simulation.

4.1. Aerodynamic characteristics change with attack angle

The angle of attack for fixed wing aircraft is the angle in which the wing moves in advance (the direction of the airflow) and the chord (different from the axis of the fuselage), which is the basis for determining the attitude of the wing in the airflow. Under the condition of keeping the Maher number unchanged, the numerical simulation method is used to change the angle of attack to observe the changes of aerodynamic characteristics, and compare with the wind tunnel experimental data. The wind tunnel test data are from two two stage high lift prediction workshops (hiliftpw) sponsored by the American aerodynamic Committee ([7-8]). As shown in figures 3.1.1, 3.1.2 and 3.1.3, the lift coefficient, drag coefficient and lift drag ratio change with the angle of attack.

The following changes of stall angle, lift coefficient and drag coefficient was basically constant variation down torque coefficient with increasing angle of attack and gradually reduce, calculated the aerodynamic characteristics of the variation of the test result and the value and the same trend.

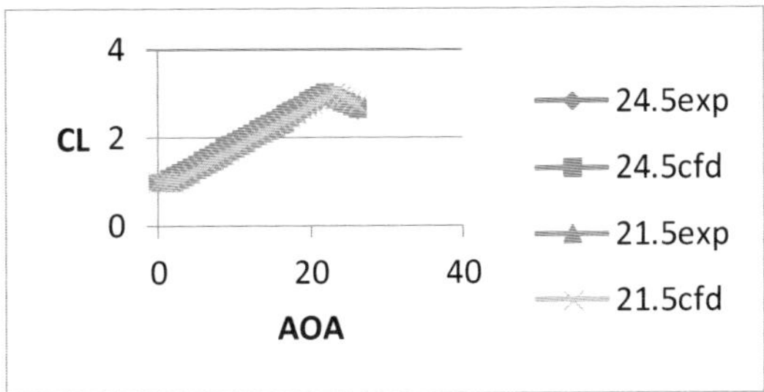

Figure 10. Sketch of change of lift coefficient with angle of attack

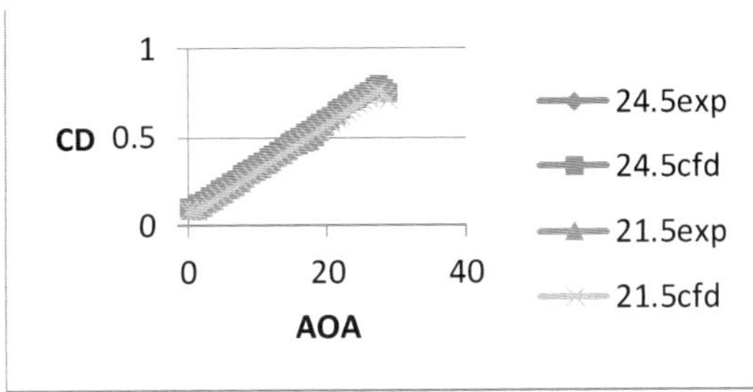

Figure 11. Sketch of variation of drag coefficient with angle of attack

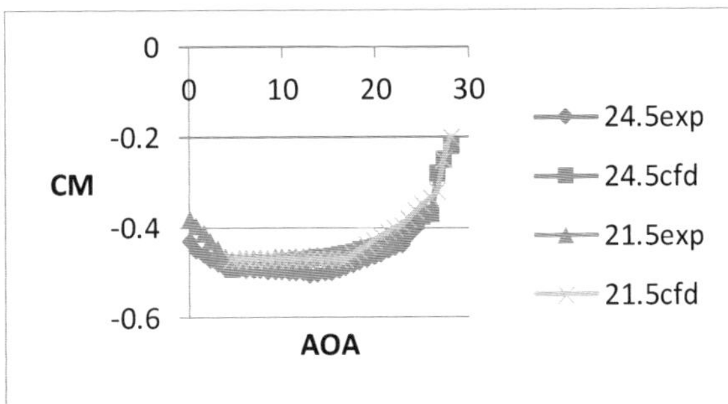

Figure 12. Sketch of change of lift drag ratio with attack angle

4.2. The influence of deflection angle

Gaps in multi section airfoils can produce many effects such as ring volume, suction, pressure recovery, and boundary layer. Therefore, it is necessary to make a simulation calculation of the flap deflection angle of the multi wing. In this section, the effects of different deflection angles on aerodynamic forces during take-off and landing are investigated. The takeoff angle is chosen as 21.5 degrees and 24.5 degrees as the flap deflection angle, and 31.5 and 36.5 degrees are chosen as the calculation deflection angle. Figure 3.2.1 and Fig. 3.2.2 show the variation of the lift drag coefficient versus angle of attack at different deflection angles.

It can be seen that the trailing edge flaps are deflected 24.55 degrees, the lift coefficient the same maximum angle of attack; angle less than 24.5 degrees, with the increase of deflection angle, the same attack angle the lift increases gradually; the deflection angle is greater than 24.5 degrees, the state of negative angle of attack and zero angle of attack angle the greater the corresponding lift flap the smaller the positive attack angle state deflection angle of trailing edge flap is corresponding to lift more. As for the drag coefficient, the deflection angle is less than 24.5 degrees, the drag coefficient increases with the deflection angle; angle of 31.5 degree, the drag coefficient increases smaller, but the maximum negative attack angle, angle of 4 degrees but less than 24.5 degrees and 36.5 degrees angle. When the deflection angle is 36.5 degrees, the resistance is the maximum. Comprehensive analysis, for the study of multi wing, the recommended take-off stage flaps the best deflection angle of 24.5 degrees, the landing phase recommended the best deflection angle of 36.5 degrees.

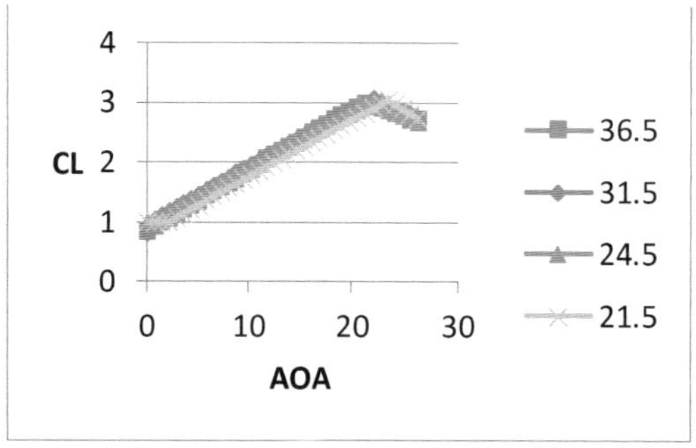

Figure 13. Sketch of variation of lift coefficient with angle of attack at different flap angles

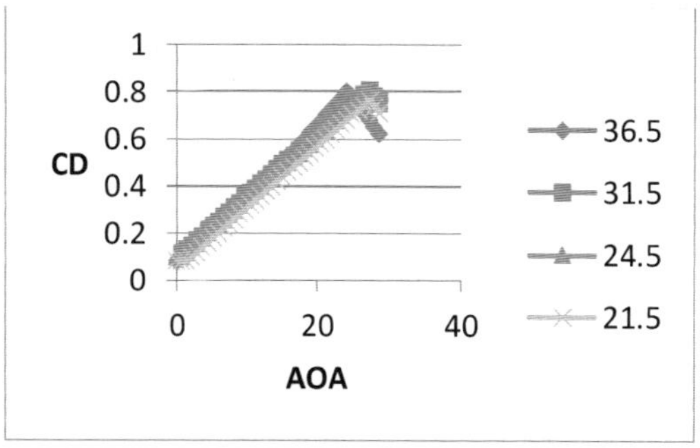

Figure 14. Sketch of variation of resistance coefficient of flap with angle of attack

5. Conclusion

The Reynolds averaged Nervier-Stokes equations, structural grid technique and geometric modeling method based on using Spalart-Allmaras turbulence model, the influence of DLR-F11 wing, wing flap angle on aerodynamic characteristics.

(1)The aerodynamic performance prediction accuracy has reached the advanced level at the present stage by using numerical calculation method and grid method, and it is in good agreement with the experimental data. It is feasible to predict the aerodynamic characteristics of the multi section wing.

(2)The aerodynamic characteristics of multi section wings vary with the angle of attack. The variation of aerodynamic characteristics calculated by the deflection angle is basically the same as the values and trends of the experimental results.

(3) For the study of the multi section wing, the optimum deflection angle of the flap at the takeoff stage is 24.5 degrees, and the optimum deflection angle is 36.5 degrees at the landing stage. The results provide a useful reference for the study of flap deflection.

References

[1] Wang Yuntao, Wang Guangxue, Zhang Yulun. 30P-30N. Numerical simulation of complicated flow field of multi section airfoil [J]. Journal of aerodynamics, 2010,01:99-103.
[2] Mao Jun, Chen Yingchun, Li Yalin. Design of two-dimensional lifting device for civil aircraft [J]. civil aircraft design and research, 2009,03:4-7.
[3] Liu Peiqing, Deng Xueying. Numerical study of flow structures around airfoils [J]. Journal of the society of Aeronautics and electronics, 1997,18 (4)

[4] gopalakrishna n, balakrishnan n high lift flow computations using the code hifun[R]. alaa - 2014 – 2569,2014
[5] kedar c, michel r, jeffrey m, et al. finite element flow simulations of the eurolift dlr - fll high lift configuratin[J]. Eprint arxiv,2014(3):749-752
[6] jeremy h, prashanth s deryl s numerical simulation of dlr – f11 high lift configuration from hiliftpw - 2 using STAR-CCM+[C].AIAA2014-0914,2014
[7] Slotnick j p hannon j a, chaffin m. overview of the first alaa cfd high lift prediction workshop. aiaa 2011-0862[R]. Reston:AIAA,2011
[8] Rumsey c i, long m, stuever r a summary of the first aiaa cfd high lift prediction workshop, aiaa-2011 0939 r. reston aiaa, 2011

IOP Conf. Series: Journal of Physics: Conf. Series **916** (2017) 012006 doi:10.1088/1742-6596/916/1/012006

Numerical study on influence of single control surface on aero elastic behavior of forward-swept wing

Ning Wang [a], Xinbing Su [b], Binlin Ma [c] and Xiaofei Zhang *

Aeronautics and Astronautics Engineering College of Air Force Engineering University, Xi'an, China

[a]1289603317@qq.com,
[b]sxinbing@sohu.com,
[c]mbinlin001@163.com,
*Corresponding author e-mail:343305484@qq.com

Abstract. In order to study the influence of elastic forward-swept wing (FSW) with single control surface, the computational fluid dynamics/ computational structural dynamics (CFD/CSD) loose coupling static aero elastic numerical calculation method was adopted for numerical simulation. The effects of the elastic FSW with leading- or trailing-edge control surface on aero elastic characteristics were calculated and analysed under the condition of high subsonic speed. The result shows that, the deflection of every single control surface could change the aero elastic characteristics of elastic FSW greatly. Compared with the baseline model, when leading-edge control surface deflected up, under the condition of small angles of attack, the aerodynamic characteristics was poor, but the bending and torsional deformation decreased. Under the condition of moderate angles of attack, the aerodynamic characteristics was improved, but bending and torsional deformation increased; When leading-edge control surface deflected down, the aerodynamic characteristics was improved, the bending and torsional deformation decreased/increased under the condition of small/moderate angles of attack. Compared with the baseline model, when trailing-edge control surface deflected down, the aerodynamic characteristics was improved. The bending and torsional deformation increased under the condition of small angles of attack. The bending deformation increased under the condition of small angles of attack, but torsional deformation decreases under the condition of moderate angles of attack. So, for the elastic FSW, the deflection of trailing-edge control surface play a more important role on the improvement of aerodynamic and elastic deformation characteristics.

1. Introduction

Compared with the traditional aircraft aerodynamic layout, Forward-Swept Wing (FSW) has significant advantages in the configuration of the aircraft aerodynamic performance and operating performance [1, 2], but the aircraft static aero elastic divergence problem of FSW seriously limits its practical application and development. The flexibility of structure in nowadays aircrafts increases

IOP Conf. Series: Journal of Physics: Conf. Series **916** (2017) 012006 doi:10.1088/1742-6596/916/1/012006

gradually, which also exacerbates the aircraft static aero elastic divergence problem, so it is of great significance to study on static aero elastic divergence problem of FSW.

Before the advent of numerical calculation method, the aero elastic simulation is mainly through experiments. This method is of high cost, high risk and difficult to operate. With the computational fluid dynamics (CFD) and computational structural dynamics (CSD) development and the rapid improvement of computer hardware, the CFD/CSD coupling method for aircraft static aero elasticity has developed rapidly[3]. On this basis, the scholars at home and abroad have done a lot of researches for the static aero elasticity of forward-swept wing aircraft. Polli [4] has analyzed aerodynamic thermal elastic response and stability of the composite wing with coupling numerical method, Zhang [5] with his partners has analyzed the aerodynamic characteristics of the flexible wing with elastic deformation.

Active Aero elastic Wing (AAW) technology is a new generation of technology for aircraft design in the United States raised since 1980s[6]. A large number of AAW technology studies have been carried out in recent years at home and abroad [7-9], the results show that the vehicle performance is greatly improved with multi-control surfaces, and it also takes the design of aerodynamic aspects, active control and structure design into consideration. As a result, the unfavorable factors of the flexible structure are transformed into favorable factors. In this paper, the calculation and analysis of static aero elastic characteristics of forward-swept wing with leading- and trailing-edge control surfaces are performed based on the CFD / CSD loosely coupled numerical method. Due to the application of AAW technology, the deflection of multi-control surfaces will become the main future advanced fighter control. As a result, the study of multi-control surface deflection is also important for aircraft static aero elastic response effect.

2. The calculation method and verification

The three-dimensional Reynolds-averaged Navier-Stokes (RANS) equation was used to calculate aerodynamics. In this paper, the integral form of the conservative three-dimensional N-S equations could be written as follows:

$$\frac{\partial}{\partial t}\iiint_{\Omega}\boldsymbol{Q}dV+\iint_{S}(\boldsymbol{F}^{\mathrm{I}}-\boldsymbol{F}^{\mathrm{V}})\cdot\boldsymbol{n}dS=0 \tag{1}$$

Where: Q, FI, FV, Ω and S were the solution vector, in viscous flux vectors, viscous flux vectors, control volume and surface of control volume, respectively; and n respected the surface normal.

Shear stress transport (SST) K-Omega model was used as turbulence model. Flow control equations was presented by nonlinear partial differential equations. The exact analytical solution generally could not be given except doing discretization. Therefore, the finite volume method and finite element method were used to disperse the flow field and the structures into finite small units, respectively. The polyhedral mesh was more efficient for calculation and had high convergence precision[10], so it was chosen for mesh generation of the flow field.

The transmission of aerodynamic force and displacement data between CFD mesh and CSD mesh node was carried out by constant volume transformation (CVT) method. Besides the radial basis function interpolation method was used to deform and update the CFD mesh to improve the efficiency of calculation[11].

The coupling methods were used to resolve the coupling problem between CFD and CSD calculation in time domain. At present, the coupling methods mainly included tight coupling and loose coupling methods. In order to improve the computational efficiency and reduce the computational complexity, CFD/CSD loose coupling static aero elastic calculation method was adopted for numerical simulation. The calculation process was shown in Figure1.

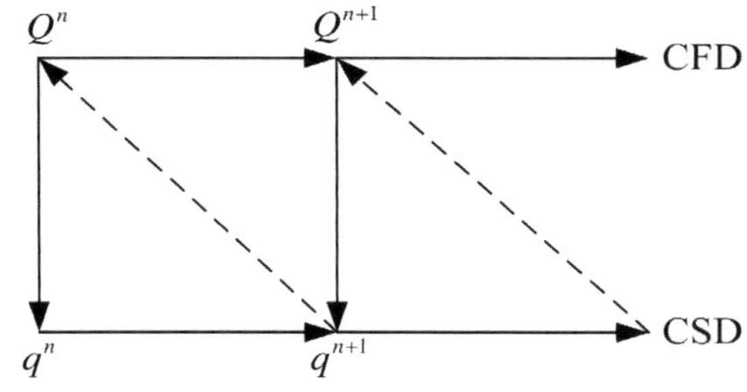

Figure 1. The calculation process of coupling Method for static aero elasticity

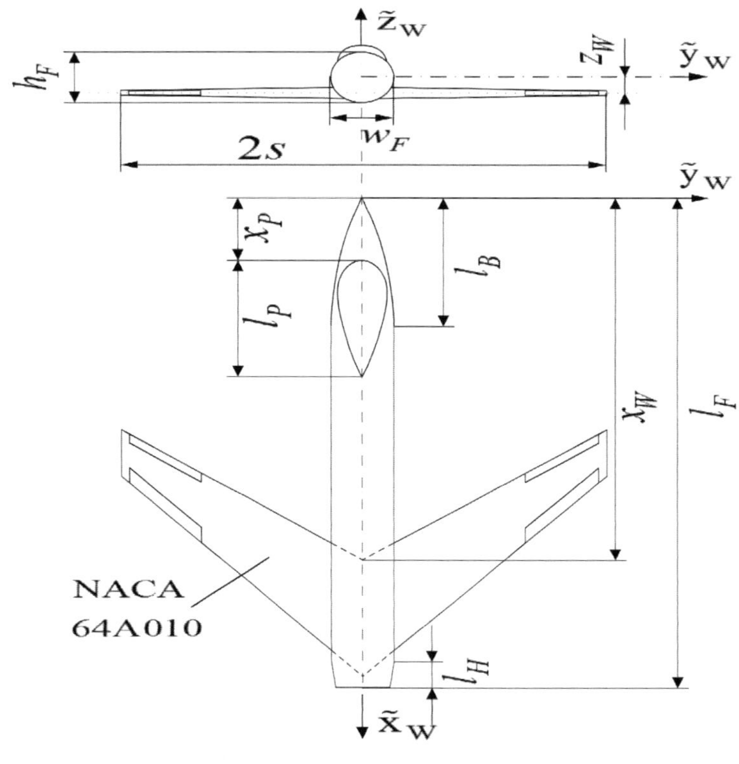

Figure 2. Geometric sketch of model wing

In order to verify the accuracy of the calculation method, HIRNASD aero elastic example was used to verify the method[12]. Firstly, the flow field and the model aircraft HIRENASD wing structure were generated by polyhedral mesh and tetrahedron mesh, respectively. The initial conditions for calculation were $\alpha = 2°$, Ma=0.8, Re=2.35×107 and q/E=4.8×10-7. Where q was the dynamic pressure in the calculation and E was Young's elastic modulus of structure model.

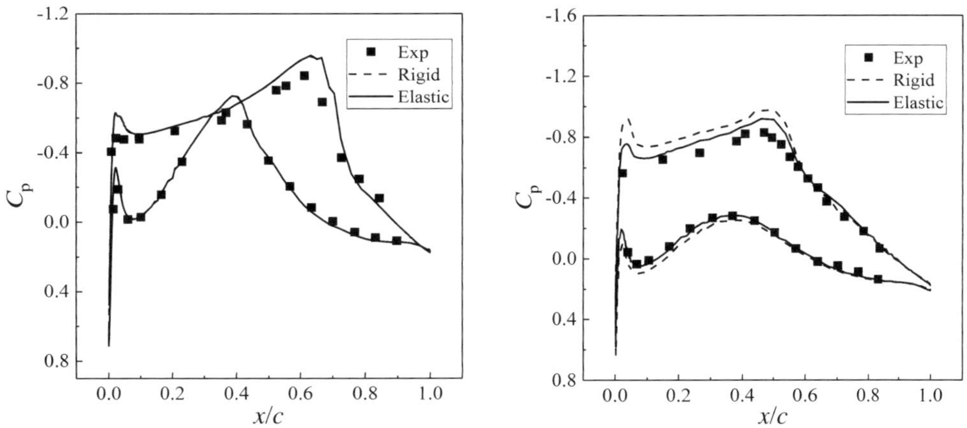

(a) y/b=0.14 (b) y/b=0.14

Figure 3. Pressure coefficient of different positions

As shown from Figure 3, the numerical method could fit the experimental value better and meet the calculation requirements of this paper.

3. Calculation model

The model of FSW layout for computation was close to reference [13]. In order to study the influence of static elastic response of the control surfaces, the aircraft wing tip was additionally provided with a leading-edge control surface (LC) and trailing-edge control surface (TC). The geometric outline was shown in Figure 2.

The specific control surface model size was shown in table 1. Where c_{TW} and S were span wise length of wingtip chord length and semi-span. δ_L And δ_T were the definitions of the leading- and the trailing-edge control surface skewness, and δ_L and δ_T were positive when the LC deflected up and TC down.

Table 1. Geometry size of control surfaces

Item	Chord length	Span wise length	Span wise position
LC	$0.1c_{TW}$	$0.2s$	$0.97s$
TC	$0.2c_{TW}$	$0.2s$	$0.97s$

The AAW technology was usually used in two ways[8]: the same deflection(LC deflected up and TC down) and different deflection (LC and TC deflected down). Therefore, it was studied the effects of the deflection of single control surface as the basis of the study of FSW static elastic response. The Brief introduction of computational models used in this paper were shown in Table 2.

Table 2. Brief introduction of computational models

Abbreviation	δ_L	δ_T
Baseline	0°	0°
LC deflecting alone	-30°~30°	0°
TC deflecting alone	0°	10°~30°

The computation conditions of flow field boundary were Ma=0.6, Re=2.32×106, ρ=1.23 kg/m3, T∞=288.15 K and P∞=1.01×105 Pa. The wing was fixedly connected in the body. Different materials were used for the main body[14], wing and control surfaces. The structural material properties for the wing and main body were E1=0.89 GPa, E2=1.54 GPa, ν=0.31, G=2.6 GPa, ρ1=381.98 kg/m3; and the structural material properties for control surfaces were E=72 GPa, ν=0.34, ρ2=2700 kg/m3. Where E1 and E2 were elastic modulus of the wing along the chord and stretch. G, ν and ρ were the shear modulus, Poisson's ratio and density, respectively.

The control with the longitudinal aerodynamic characteristics of forward swept wing and elastic deformation characteristics were only studied, so half model was used to simplify the following analysis. The number of polyhedral mesh of the half model in fluid domain was 4 million, and the number of tetrahedral mesh in structural domain was 100,000.

4. Aerodynamic characteristic

The elastic FSW model was only studied. Therefore, elastic and rigid aerodynamic characteristics of FSW models must be compared. Taking the baseline model, which the control surfaces didn't deflect, as an example, when Ma=0.6, Re=2.32×106, the calculated elastic and rigid aerodynamic characteristics of forward swept wing model were shown in Figure 4.

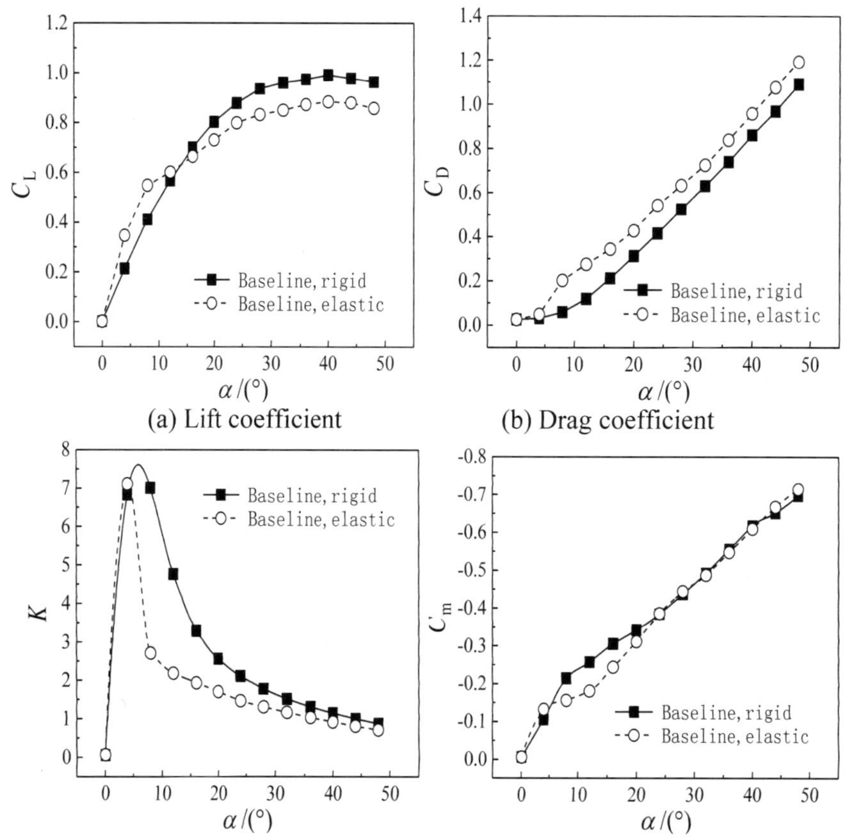

(a) Lift coefficient (b) Drag coefficient

(c) Lift-drag ratio (d) Pitching moment coefficient

Figure 4. Aerodynamic characteristics of elastic and rigid Baseline models

As shown in Figure 4 (a), when $\alpha < 14°$, the lift coefficients of elastic wings were greater than those of rigid wings, but the result was opposite when $\alpha > 14°$; When $\alpha < 8°$, the C_L^α of elastic wings were greater than those of rigid wings, which was smaller when $\alpha > 8°$; In addition, the rigid wing and elastic wings had little difference in the stall angle of attack, both at about 40°; Compared with the rigid wing, the maximum lift coefficient elastic wing decreased about 10.78%. Therefore, the bending and torsional deformation of the elastic wing led to the increase of the effective angle of attack, the lift characteristics at the smaller angle of attack were better than those of the rigid wing; But at higher angle of attack, the separation of air flow on the wing surface was more serious, and the lift characteristics became worse; In addition, as the elastic wing produced a forward torsion, the frontal area increased, which caused that drag characteristic was worse than that of the rigid wing; As shown in Figure 4 (b), the drag coefficients of the elastic wing were always greater than that of the rigid wing at the same angle of attack.

According to figure 4 (c) , the aerodynamic efficiency of elastic forward-swept wing was slightly higher than that of the rigid wing at small angle of attack, but it was far lower than that of the rigid wing at high angle of attack. The maximum lift-drag ratio of the elastic wing was about 7% lower than that of the rigid wing, and the angle of attack at the maximum lift-drag ratio Kmax was achieved about 2° ahead of that of the rigid wing. From Figure 4 (d), when $\alpha < 4°$, the pitching moments of elastic wing were greater than rigid wing, which were opposite when $4° < \alpha < 24°$, the pitching moments of elastic wing and rigid wing were almost equal when $\alpha > 24°$.

From the analysis above, the aerodynamic characteristics of elastic and rigid forward-swept wing models were obviously different.

When LC deflected alone, the aerodynamic characteristics of the elastic LC model were shown in Figure 5 under the conditions of Ma=0.6 and Re=2.32×106.

According to figure 5 (a), when LC deflected up, compared with the baseline model, it caused the lift coefficient decrease under the condition of $\alpha < 12°$, and the greater the skewness was, the greater the magnitude of lift coefficient decreased; It caused the lift coefficient increase when $\alpha > 12°$, and the increasing magnitude decreased with the increase of the skewness. The increase of effective angles of attack at outside partial wing was caused by the upward deflection of LC, the stall angle of attack of wing outer segment was ahead at a low angle of attack, so the lift decreased. The stronger leading edge vortex formed under the condition of high angles of attack, which caused the increase of lift.

Compared with the baseline model, the upward defection of LC led to the increase of lift coefficient. As the skewness increased, the increasing magnitude of lift coefficient would increase first and then decrease. The downward defection of LC made the suction peak of the wing leading edge decrease, adverse pressure gradient decrease and the flow separation delay. As a result, the lift increased [15].

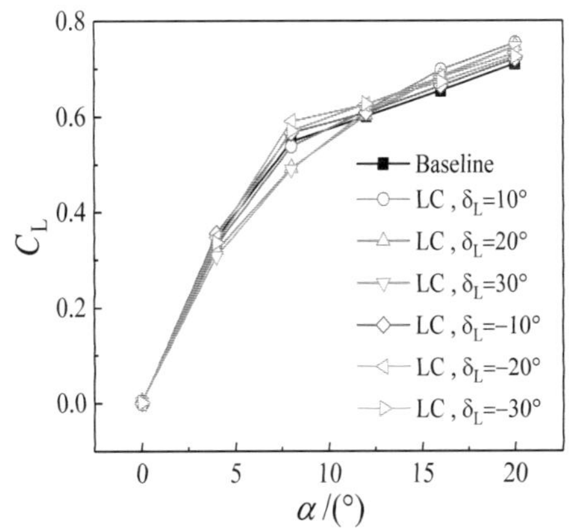

(a) Lift coefficient

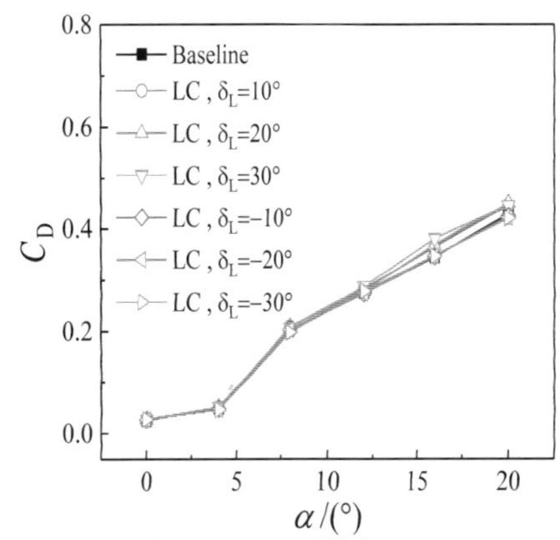

(b) Drag coefficient

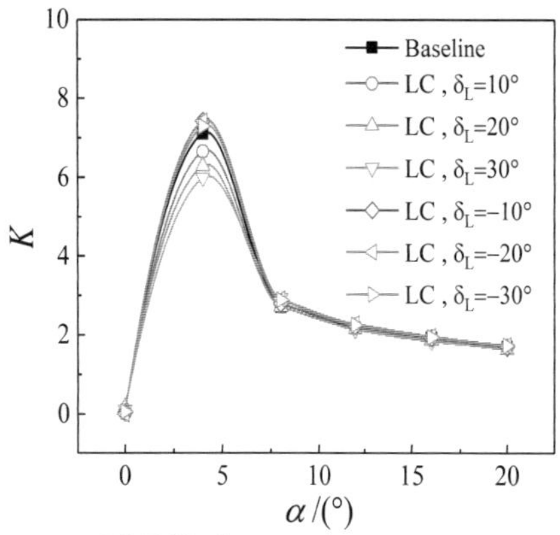

(c) Lift-drag ratio

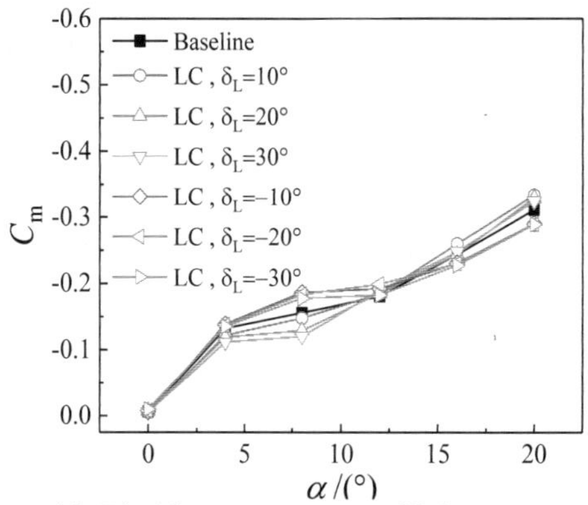

(d) Pitching moment coefficient

Figure 5. Aerodynamic characteristics of LC model

According to figure 5 (b), compared with the baseline model, the frontal area increased when LC deflected up. As a result, drag coefficient increased slightly. The magnitude increased slightly with the increase of skewness; On the contrary, the downward deflection of LC had little impact on drag coefficient.

According to Figure 5 (c), compared with the baseline model, the upward deflection of LC made the lift-drag ratio decrease, and the greater the skewness was, the higher decrease magnitude of lift-drag ratio would be. The situation would be opposite when LC deflected down.

According to figure 5 (d), compared with the baseline model, when LC deflected up/down, the pitching moments would decrease/increase under the condition of $\alpha < 12°$. The result would be opposite when $\alpha > 12°$.

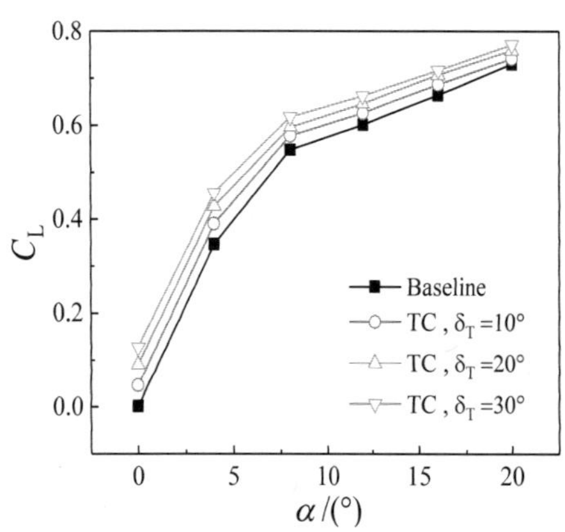

(a) Lift coefficient

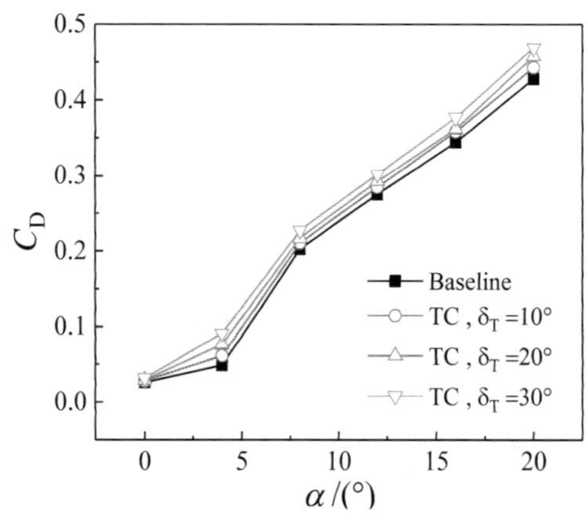

(b) Drag coefficient

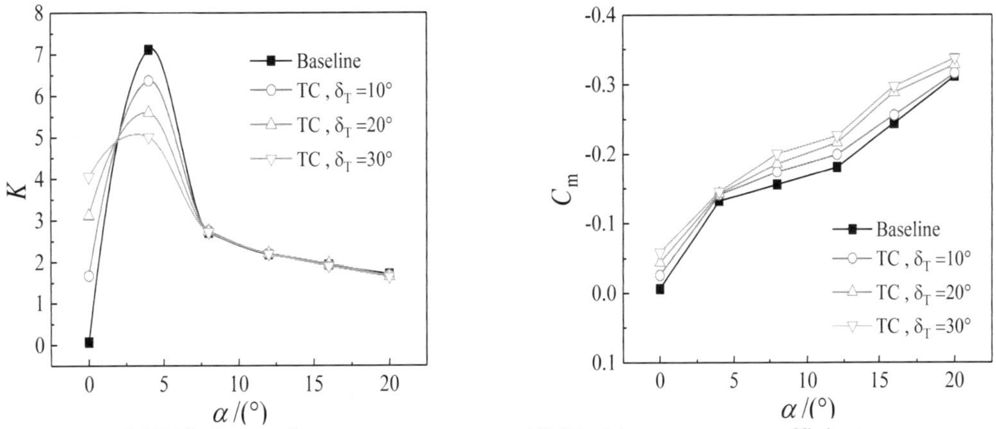

(c) Lift-drag ratio (d) Pitching moment coefficient

Figure 6. Aerodynamic characteristics of TC model

When the TC deflected alone, the aerodynamic characteristics of the elastic TC model were calculated under the conditions of Ma=0.6 and Re=22.32×106, which is shown in Figure 6.

According to figure 6 (a) , compared with the baseline model, the downward deflection of TC led to the increase of lift coefficient, and the magnitude would increase with the increase of skewness, so did the camber of wing's outer section and the lift.

According to Figure 6 (b), the downward deflection of TC made the frontal area of the wing's outer section increase. As a result, the drag coefficient increased. The magnitude increasd with the increase of skewness.

According to figure 6 (c), compared with the baseline model, when TC deflected down, the maximum lift-drag ratio decreased, the aerodynamic efficiency was reduced, and the decreasing magnitude of Kmax increased with the increase of skewness. According to figure 6 (d), compared with the baseline model, when TC deflected down, the pitching moment increased, and the increasing magnitude of Cm increased with the increase of skewness.

5. Elastic deformation characteristics

When $\alpha = 4^\circ$, the deflections of the wing in Z direction caused by the deflections of control surfaces were shown in Figure 7.

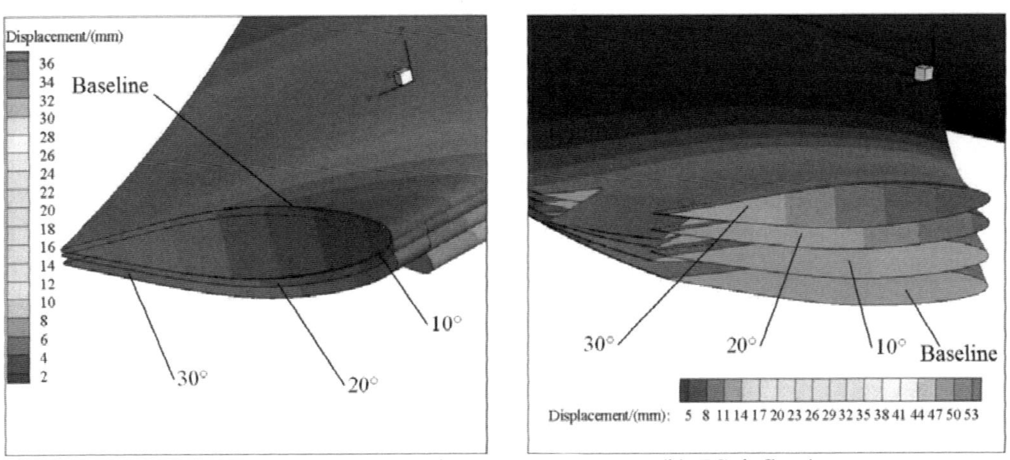

(a) LC deflecting (b) TC deflecting

Figure 7. Elastic deformation contour of models when $\alpha = 4^\circ$

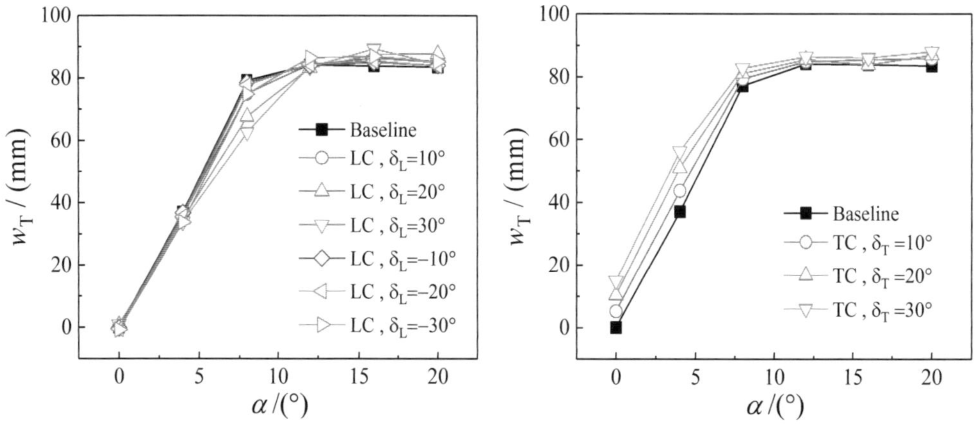

(a) LC (b) TC

Figure 8. Variations of displacement of wingtip leading-edge of LC and TC models with angle of attack

When the control surface deflected alone, the calculated leading wing tip deflection w_T changed with the increase of the angle of attack and the result was shown in Figure 8. When $\alpha < 8°$, the leading wing tip deflections of LC and TC models increased linearly with the increase of the angle of attack sharply; when $\alpha > 8°$, the change of w_T with the increase of angle of attack was very small.

According to Figure 8 (a), when LC deflected up or down, compared with the baseline model, w_T decreased under the condition of $\alpha < 12°$, and the decreasing magnitude increased with the increase of the skewness; w_T increased slightly when $\alpha > 12°$. According to Figure 8 (b), compared with the baseline model, with the downward deflection of TC, w_T would increase. In addition, the increasing magnitude of w_T increased with the increase of skewness.

When the control surface deflected alone, the wing tip torsional angle θ_T changed with the angle of attack. The result was shown in Figure 9. When the control surface deflected alone, under the condition of $\alpha < 8°$, θ_T increased linearly with the angle of attack sharply, which was similar to the changing rule of w_T. Besides, θ_T was basically unchanged when $\alpha > 8°$.

According to figure 9 (a), compared with the baseline model, when $\alpha < 12°$, the deflection of LC caused the decrease of wing tip torsional angle. While, the magnitude increased with the increase of skewness; the situation was opposite when $\alpha > 12°$. According to Figure 9 (b), compared with the baseline model, under the condition of $\alpha < 12°$, when TC deflected down, w_T increased. The magnitude increased with the increase of the skewness. When TC deflected down, under the condition of $\alpha < 12°$, θ_T decreased and the magnitude increased slightly with the increase of skewness.

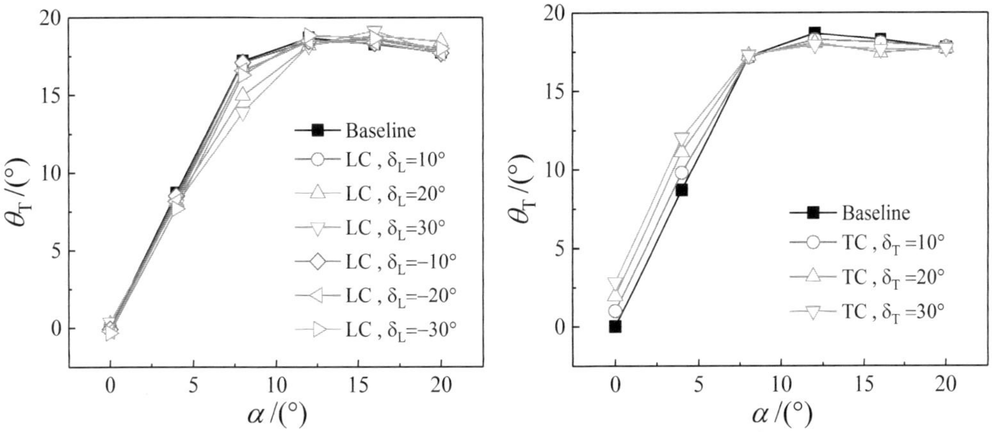

(a) LC (b) TC

Figure 9. Variations of torsional angle of wingtip leading-edge of LC and TC models with angle of attack

Therefore, compared with the baseline model, when $\alpha < 12°$, the deflection of control surface slowed down the bending and torsional deformation, but it would exacerbated that when $\alpha > 12°$; The downward deflection of TC exacerbated the bending and torsional deformation when $\alpha < 8°$,and it exacerbated the bending deformation, but slowed down the torsional deformation when $\alpha > 8°$.

6. Conclusion

1) The aerodynamic characteristics of elastic and rigid forward-swept wing had obvious difference, the influence of structural flexibility on the aerodynamic performance must be fully considered of in design of forward-swept wings.

2) Compared with the baseline model, when LC deflected up, under the condition of small angles of attack, the aerodynamic characteristics was poor, but the bending and torsional deformation decreased. Under the condition of moderate angles of attack, the aerodynamic characteristics was improved, but bending and torsional deformation increased; When LC deflects down, the aerodynamic characteristics was improved, the bending and torsional deformation decreased/increased under the condition of small/moderate angles of attack.

3) Compared with the baseline model, when TC deflected down, the aerodynamic characteristics was improved. The bending and torsional deformation increased under the condition of small angles of attack. The bending deformation increased under the condition of small angles of attack, but torsional deformation decreased under the condition of moderate angles of attack.

References

[1] J S E, W H J. In-flight-lift drag characteristics for a forward swept wing aircraft, NASA TP-3414 [R]. Washington, D.C.: NASA, 1994:15-16.
[2] Zhang B Q, LASCHKA B. On forward-swept wing's aerodynamic characteristics[J]. Journal of Northwestern Poly technical University, 1989, 7(3):321-328.
[3] Liu Y, Bai J Q, Hua J, et al. A high-fidelity static aeroelastic analysis method for complex configuration and its application[J]. Journal of Northwestern Polytechnical University, 2015, 33(1): 14-20.
[4] Polli G M, Mastroddi F. Aero thermo elastic stability of composite aircraft wings subjected to head inputs. AIAA Paper 2006-1731,2006.
[5] Zhang H Ma Di Ma T. Analysis of aerodynamics characteristic of flexible wing caused by deflection[J]. Journal of Beijing University of Aeronautics and Astronautics, 2008, 34 (5):487-490.

IOP Conf. Series: Journal of Physics: Conf. Series **916** (2017) 012006 doi:10.1088/1742-6596/916/1/012006

[6] Andersen G, Forster E, Kolonay R. Multiple control surface utilization in active aeroelastic wing technology[J]. Journal of Aircraft, 1997, 34(4): 552-557.
[7] Dowell E H, Bliss D B, Clark R L. Aeroelastic wing with leading-and trailing-edge control surfaces[J]. Journal of Aircraft, 2003, 40(3): 559-565.
[8] Gaspari A D, Ricci S, Riccobene L, et al. Active aeroelastic control over a multisurface wing: Modeling and wind-tunnel testing[J]. AIAA journal, 2009, 47(9): 1995-2010.
[9] Chen L, Wu Z G, Yang C, et al. Active control and wind tunnel test verification of multi-control surfaces wing for gust alleviation. Acta Aeronautica et Astronautica Sinica, 2009, 30(12): 2250-2256.
[10] Feng H Y, SU X B, MA B L, et al. Static Aero-elastic Computation with Polyhedron Grid[J]. Flight Dynamics, 2016, 34(4):24-28.
[11] Gumerov N A, DURAISWAMI R. Fast radial basis function interpolation via preconditioned Krylov iteration[J]. SIAM Journal on Scientific Computing, 2007, 29(5): 1876-1899.
[12] Ballmann J, DSafnis A, Korsch H, et al. Experimental Analysis of High Reynolds Number Aero-Structural Dynamics in ETW[C]// The 46th AIAA Aerospace Sciences Meeting. Reno, Nevada: AIAA, 2008, 841:7-10.
[13] Breitsamter C, Laschk B. Vortical Flowfield Structure at Forward Swept-Wing Configurations [J]. Journal of Aircraft, 2001, 38 (2): 193-207.
[14] Feng H Y. Numerical study on static aeroelastic characteristics of forward-swept wing based on polyhedral mesh[D]. Xi'an: Air Force Engineering University, 2016:69-71.
[15] Fang B R. Aircraft aerodynamic configuration design[M].Beijing:Aviation Industry Press, 1997:934-935.

IOP Conf. Series: Journal of Physics: Conf. Series **916** (2017) 012007 doi:10.1088/1742-6596/916/1/012007

Numerical study on the aerodynamic characteristics of both static and flapping wing with attachments

Lingwang Xie, Xingwei Zhang*, Pan Luo, Panpan Huang

College of Engineering, Shantou University, Shantou 515063, China

*Corresponding author e-mail: zhangxw@stu.edu.cn

Abstract. The purpose of this paper is to investigate the aerodynamic mechanism of airfoils under different icing situations which are different icing type, different icing time, and different icing position. Numerical simulation is carried out by using the finite volume method for both static and flapping airfoils, when Reynolds number is kept at 135000. The difference of aerodynamic performance between the airfoil with attachments and without attachments are be investigated by comparing the force coefficients, lift-to-drag ratios and flow field contour. The present simulations reveal that some influences of attachment are similar in the static airfoil and the flapping airfoil. Specifically, the airfoil with the attachment derived from glaze ice type causes the worse aerodynamic performance than that derived from rime ice type. The longer the icing time, the greater influence of aerodynamic performance the attachment causes. The attachments on the leading-edge have the greater influence of aerodynamic performance than other positions. Moreover, there are little differences between the static airfoil and the flapping airfoil. Compared with the static airfoil, the flapping airfoil which attachment located on the trailing edge causes a worse aerodynamic performance. Both attachments derived from rime ice type and glaze ice type all will deteriorate the aerodynamic performance of the asymmetrical airfoils. Present work provides the systematic and comprehensive study about icing blade which is conducive to the development of the wind power generation technology.

1. Introduction

An increasing demand for renewable energy in countries all over the word has stimulated a growing number of studies on wind energy. The use of wind energy has had a remarkable progression in the past decades. However, to make full use of the wind energy, some challenges should be further investigated. For example, the blades of wind turbines often freeze because of the cold weather when it is used in the high latitude area (Zhang and Ren, 2007). The ice forms attachments on the surface of blades and has a great influence about the performance and safety of wind turbines. This mechanism still is not fully understood. Many researchers have been devoted to the aerodynamic analysis of blade icing problem, both experimentally and theoretically. Bose (1992) carried out the experiment of icing blade in the natural situation. He found that the ice mainly located in the leading edge and the trailing edge, and was little in the suction surface. Jasinski et al. (1998) made an experimental measurement on the special airfoil NREL-S809 in an icing wind tunnel. Their results revealed that the ice on the airfoil surface made the airfoil's aerodynamic characteristics become worse and significantly changed the

IOP Conf. Series: Journal of Physics: Conf. Series **916** (2017) 012007 doi:10.1088/1742-6596/916/1/012007

performance of the wind turbines. Clement Hochart (2007) did an experiment to investigate the ice shape of airfoil. He got some ice shapes by changing the liquid water content and temperature. After the aerodynamic analysis of these icing airfoils, he found that the lift decreased and drag increased when glaze or rime accreted on the blade profile. Deng et al (2010) simulated the ice accretion process on the airfoil by solution the Navier-Stokes equations. They found that the attachment formed by icing can advance the stalling angle of attack. Zhu (2011) used the Lagrange solution to calculate the ice accretion. Their results show that the glaze ice could enlarge the drag coefficient and aggravated the separation of flow. However, most above study focus on just one situation. The systematic and comprehensive study on both static and flapping airfoils with many kinds of attachment formed by icing process is infrequent. Accordingly, we aim to increase our understanding on the aerodynamic mechanism of airfoils under different icing situations. The attachments on the surface of airfoils are formed by three conditions which are different icing type, different icing time, and different icing position. A finite volume method is used to simulate the flow characteristics around both static airfoil and flapping airfoils. The difference of the aerodynamic forces among the series of flow field is examined and how difference attachments affect the fluid field is discussed in detail.

2. Description of the Problem

Current works propose to investigate the attachments on the aerodynamic mechanisms of airfoil in a general-purpose numerical method. Both NACA4412 and NACA0012 airfoils with the same thickness and infinite span are used as the smooth models. The attachments on the surface were formed in three situations which are different icing type, different icing time and different icing position.

The geometric shapes of surface attachments on the leading-edge of airfoil are derived from two icing types. One is the glaze ice, the other one is the rime ice, as shown in Figure 1. It is seen that the attachments formed by the two icing condition is very different. In the glaze ice condition, the ice grows mainly along longitudinal direction and the chord length of airfoil is almost no increased. In contrast, rime ice grows along transverse direction and its geometric shapes are more smooth and regular in the rime ice condition. Thereout, some modeled airfoils on base of NACA4412 and NACA0012 when icing time is 10min are respectively shown in Figure 2(a) and (b). Moreover, the change of the attachment shapes at different icing times is depicted in Figure 3.

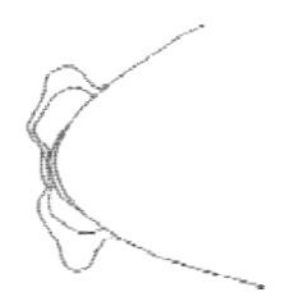

(a) The glaze ice's geometric shapes in 2min, 6min, 10min

(b) The rime ice's geometric shapes in 2min, 6min, 10min

Figure 1. The glaze ice and the rime ice's geometric shapes in different icing time [7]

(a) The glaze ice airfoil of ice 10min

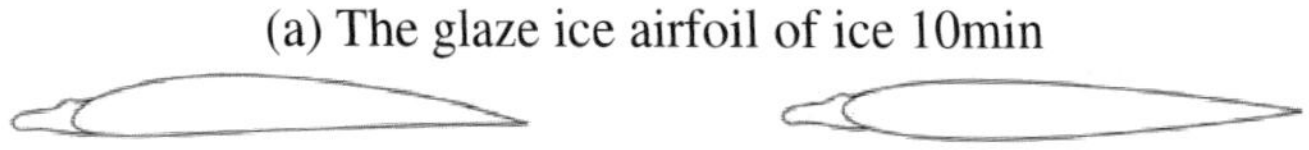

(b) The rime ice airfoil of ice 10min

Figure 2. NACA4412, NACA0012 airfoil with surface attachments derived from different icing types

(a) The rime ice airfoil and the glaze ice airfoil of icing 2min

(b) The rime ice airfoil of icing 6min

(c) The glaze ice airfoil of icing 6min

Figure 3. The modeled airfoils in different icing time

Figure 4 show the experimental measurement of an airfoil in a real icing situation. It can be seen that icing attachments mainly locates at the leading edge, the trailing edge, upper and lower surfaces where are near the leading edge. Accordingly, to further investigate the aerodynamic characteristics of these icing attachments at different positions, the modeled airfoils are built, as shown in Figure 5. Four cases which the icing attachments distribute on leading edge, trailing edge, upper surface and lower surface's, are respectively analyzed in this paper.

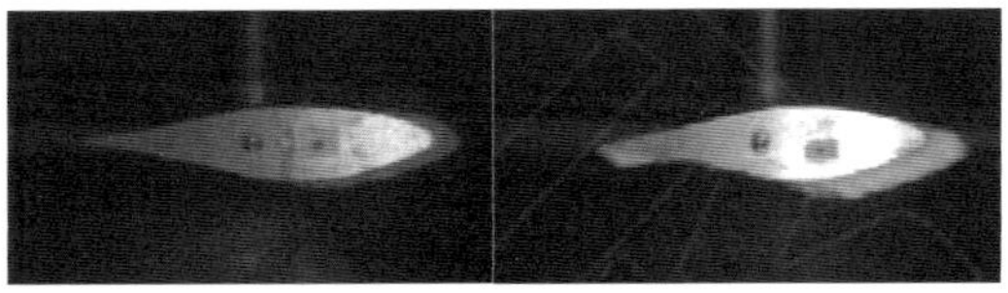

Figure 4. The airfoil's complete icing graph [8]

(a) The airfoil's ice on the upper surface

(b) The airfoil's ice on the lower surface

(c) The airfoil's ice on the trailing edge

Figure 5. The modeled airfoils in different icing position

3. Method of the Solution

The two-dimensional, unsteady, incompressible Reynolds-averaged Navier-Stokes equations are expressed as follows:

$$\frac{\partial u_i}{\partial x_i}=0,\ \frac{\partial u_i}{\partial t}+u_j\frac{\partial u_i}{\partial x_j}=-\frac{1}{\rho}\frac{\partial p}{\partial x_i}+\frac{\partial}{\partial x_j}\left[v\frac{\partial u_i}{\partial x_j}-\overline{u_i^{'}u_j^{'}}\right],\ \overline{u_i^{'}u_j^{'}}=\frac{2}{3}k\delta_{ij}-v_t\left(\frac{\partial u_i}{\partial x_j}+\frac{\partial u_j}{\partial x_i}\right) \quad (1)$$

Where u is the velocity vector, the pressure P is normalized by ρU^2, $\rho\overline{u_i^{'}u_j^{'}}$ is the Reynolds stress, δ_{ij} is the Kronecker Delta number. v_t Is eddy viscosity determined by the turbulence model, k is the turbulent kinetic energy. $k-\omega$ Standard turbulence model is used in the present work. The Reynolds number is 135000. The equations are solved in the finite volume method in the inertial frame. The first order time integrations are used. The SIMPLEC algorithm is employed to deal with the coupling between the pressure and velocity. The convective flux and diffusive flux terms are evaluated using the second-order accurate central difference scheme, respectively. The computational domain which is 19c×16c in two directions is shown in Fig 6(a). After a grid independent test, the total number of triangular meshes in the computational domain is about 60000, as shown in Figure 6(b).

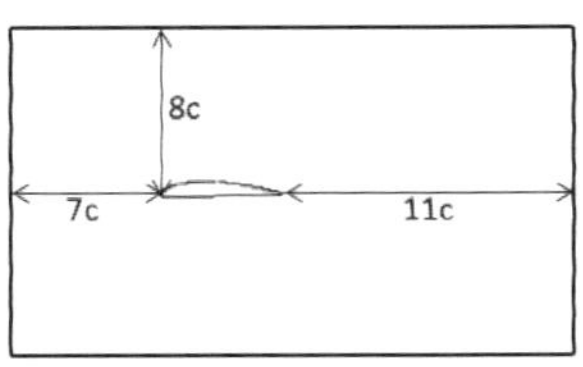

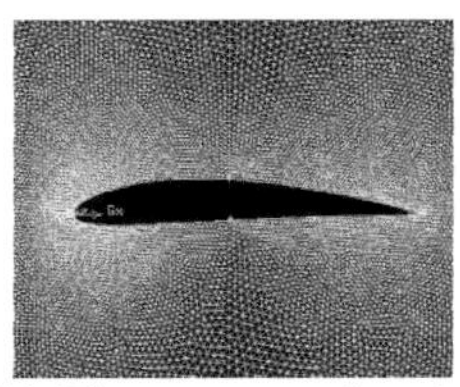

(a) The compute domain

(b) The computational grid

Figure 6. The compute domain and the grid partition

The inlet boundary located at 7c upstream of the leading-edge point of airfoils, an uniform flow is prescribed

$$u1=10,\ u2=0,\ u3=0 \tag{2}$$

The outlet boundary located at 11c downstream of the rear of airfoils, the outflow boundary conditions is given, i.e.

$$\frac{\partial u_1}{\partial x}=\frac{\partial x_2}{\partial x}=0 \tag{3}$$

The upper and down boundaries is 8c far away from the surface of the airfoils. A symmetry boundary conditions is given as follows

$$u_2=0, \frac{du_1}{dy}=0, \frac{dp}{dy}=0 \tag{4}$$

The no-slip boundary conditions are prescribed at the wall. The influence of these conditions have been investigated and found to be sufficiently adaptive. The simulation starts assuming the fluids is initially at rest.

The lift and drag are scaled here by velocity, so that the lift and drag coefficients are computed from:

$$C_L=\frac{2F_L}{\rho U_\infty^2 c},\ C_D=\frac{2F_D}{\rho U_\infty^2 c} \tag{5}$$

Where FL and FD are lift and drag forces, respectively.

For the dynamic cases, a harmonic motion $h = A\sin 2\pi ft$ is employed by the airfoils in this study. The value of A is 0.1c and f is 1.59 rad/s, respectively. Then, the period-averaged lift and drag coefficients of the flapping airfoils are calculated as follows:

$$\overline{C}_L=\frac{1}{T}\int_0^T C_L dt,\ \overline{C}_D=\frac{1}{T}\int_0^T C_D dt \tag{6}$$

This study uses the airfoil NACA0012 to validate the numerical algorithm when Reynolds number is kept at 135000. The comparison between our calculated values and the experimental values (Lee and Greontakos, 2004) of a static airfoil is shown in Figure 7. It can be seen that the present calculation is in a good agreement before stalling. Although the stalling angle of attack advances 2° and the peak value is smaller, which the general difference can be consider very small. The validation of the numerical algorithm for the flapping cases has been done in our previous work (Zhang and Zhou, et al, 2013). That numerical code is reliable for qualitatively analyzing flow characteristics around static and flapping airfoils.

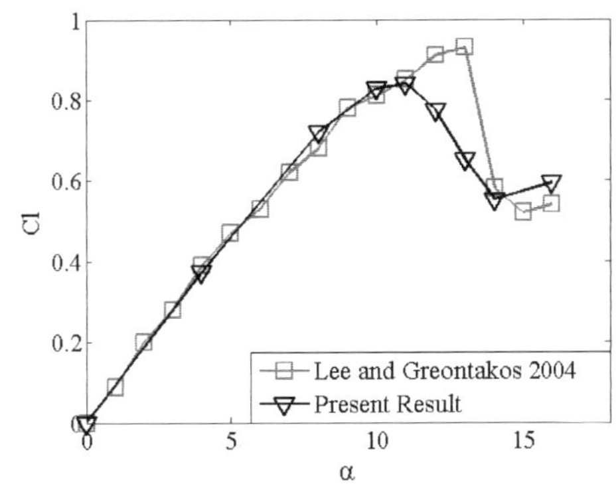

Figure 7. The comparison between the analog values and the experimental values

4. Result and Discussion

4.1. The effect of attachments on the flow characteristics of static airfoils

Different attachment models formed in different icing conditions are implemented in current study. For the static cases, simulations are carried out for the attack angle from 0° to 18°. The force coefficients and flow field structures as observed in this study are presented here systematically.

Figure 8 gives the force coefficients and lift-to-drag coefficients of airfoil with surface attachments deriving from different icing types. It is seen that the stall attack angle of smooth NACA4412 airfoil is about 10° and the maximum lift coefficient is about 1.06. The lift to drag ratio reach a peak and then reduce as the attack angle increases. The attachments derived from both glaze ice and rime ice conditions, enhance the drag, and the lift to drag ratios are all smaller than those of no attachment airfoils. Further, current results show that the attachments derived from glaze ice condition have a more dramatic influence on the aerodynamic characteristics of the airfoil. The forces and lift-to-drag coefficients of airfoil with attachments deriving from glaze ice condition change more significant than those driving form rime ice condition. Take the results of attack angle at 4° as example, the lift to drag ratio of airfoil with attachments deriving from glaze ice condition is reduced about 53.8% than that of rime ice result. Attachments deriving from glaze ice condition make aerodynamic characteristics of airfoil become much worse and the stall angle of attack become smaller. This is likely due to the attachments effectively changes the vortex shedding mechanism at the separation point of airfoil. Figure 9 gives the flow field contour of airfoil with attachments deriving from different icing types. It can be seen that a significant difference between the three conditions appears at higher attack angles. The attachments induce the flow field separation more intense than those of no attachments. Moreover, current results also show that the flow around the airfoil the attachments deriving from glaze ice condition, are more complex at the trailing-edge than that of rime ice condition.

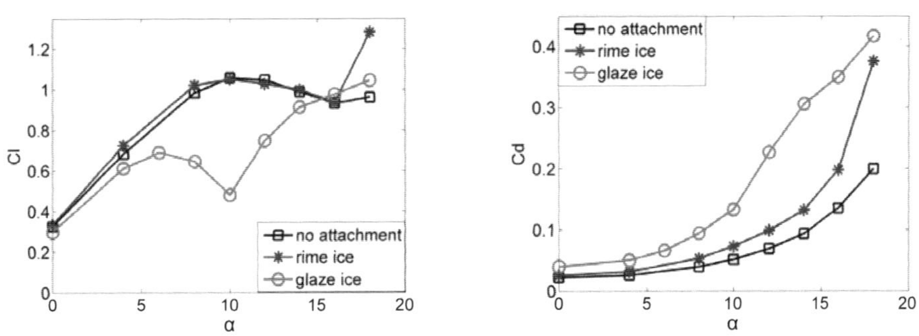

(a) The lift coefficient (b) The drag coefficient

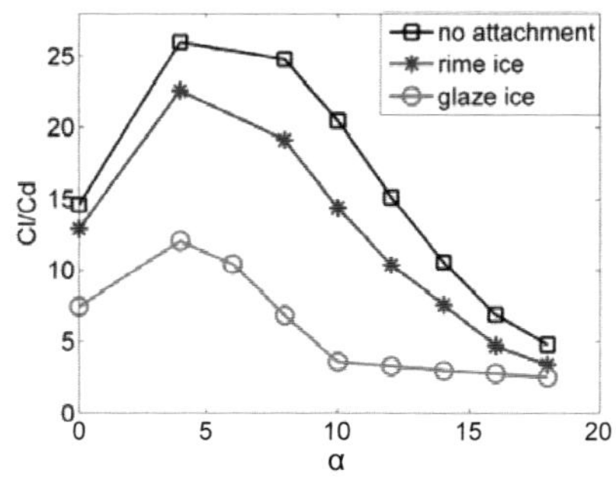

(c) The lift to drag ratio

Figure 8. The force coefficients and lift-to-drag ratios of static NACA4412 airfoil with attachments derived from different icing type.

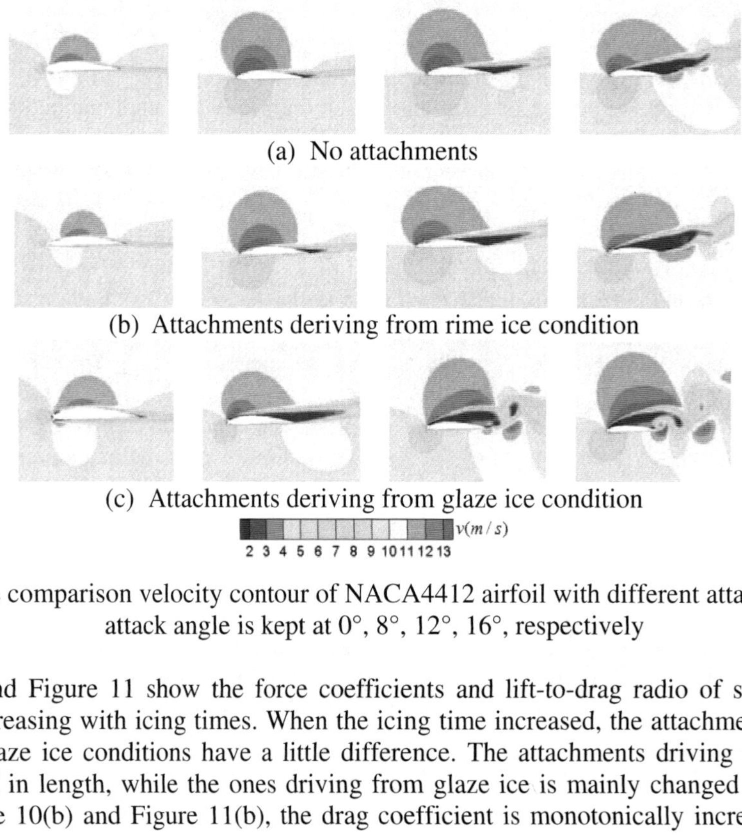

(a) No attachments

(b) Attachments deriving from rime ice condition

(c) Attachments deriving from glaze ice condition

Figure 9. The comparison velocity contour of NACA4412 airfoil with different attachments when attack angle is kept at 0°, 8°, 12°, 16°, respectively

Figure 10 and Figure 11 show the force coefficients and lift-to-drag radio of static airfoil with attachments increasing with icing times. When the icing time increased, the attachments driving from rime ice and glaze ice conditions have a little difference. The attachments driving from rime ice is mainly changed in length, while the ones driving from glaze ice is mainly changed in thickness. As shown in Figure 10(b) and Figure 11(b), the drag coefficient is monotonically increasing with icing time. The lift coefficients and lift to drag ratios reach a peak and then reduce as the icing time. The longer the icing time, the greater the attachments affect the aerodynamic performance of airfoils. It is also seen that the aerodynamic performance of airfoil with attachments driving from glaze ice condition become much worse that the others. Especially, the lift to drag ratios reduce to the smaller values as icing time increase as shown in Figure 10(c) and Figure 11(c).

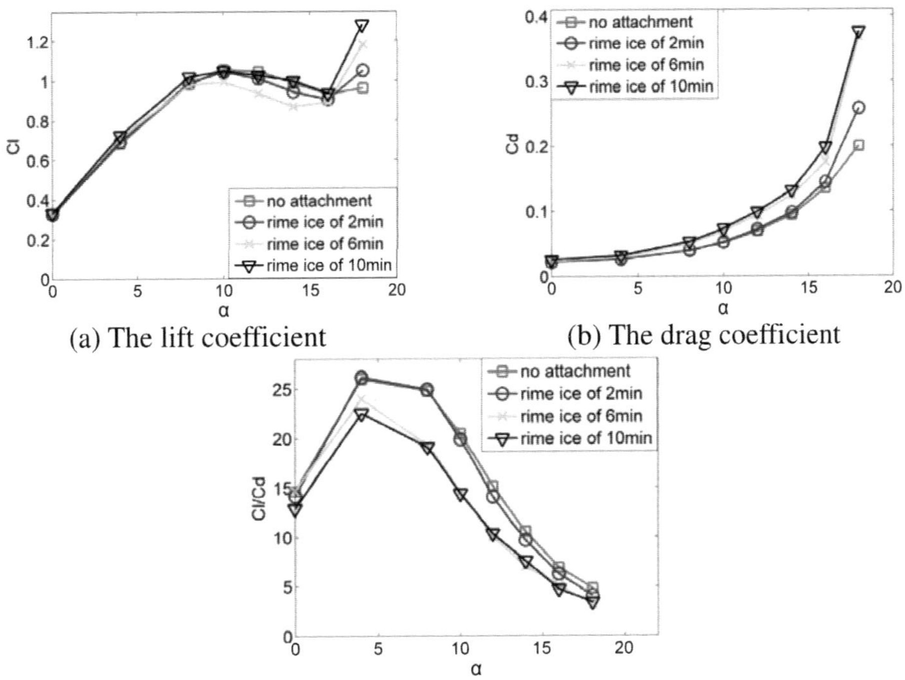

(a) The lift coefficient

(b) The drag coefficient

(c) The lift to drag ratio

Figure 10. The force coefficients and lift-to-drag ratios of airfoil with attachments derived from rime ice at different icing times

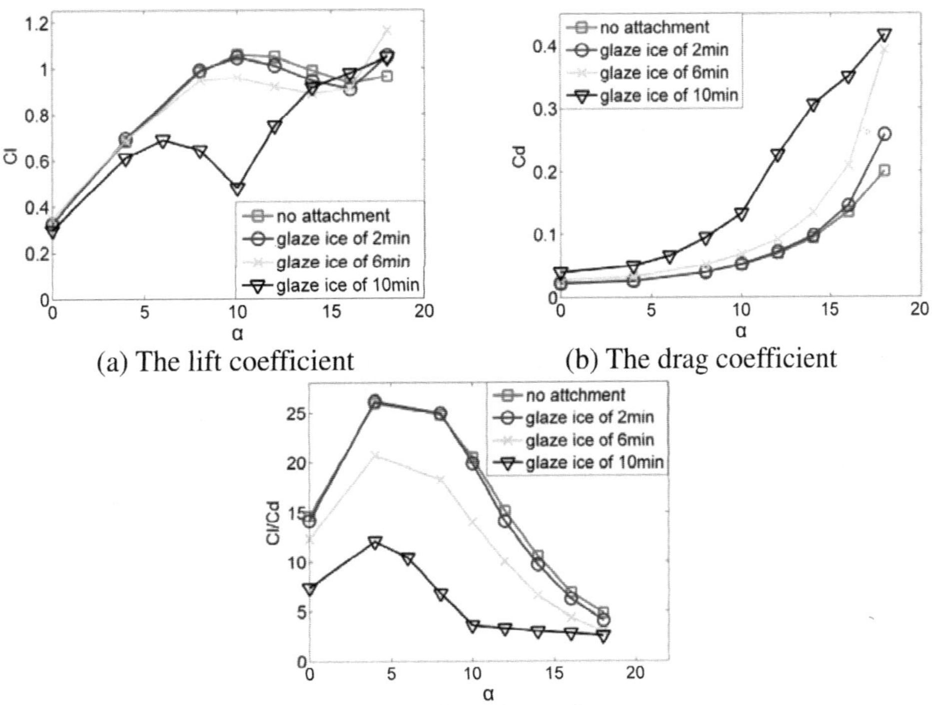

(a) The lift coefficient

(b) The drag coefficient

(c) The lift to drag ratio

Figure 11. The force coefficients and lift-to-drag ratios of airfoil with attachments derived from glaze ice at different icing times

Figure 12 respectively shows the force coefficients and lift-to-drag ratios of the static NACA4412 airfoil with attachments at different icing positions. The icing position on the surface of airfoil include leading-edge, trailing-edge, upper surface and lower surface. Among, the attachment located at the leading-edge is used the glaze ice type. Current results show that the positions of attachments indeed significantly change the aerodynamic performance of the airfoil. Wherever attachments positions locate, the lift to drag ratios is reduced as shown in figure 12(c). It is also found that the attachment on the leading-edge has the greater effect than others. When the attachments deriving from glaze ice condition locate at the leading-edge of airfoil, the lift to drag radios become the most small at every attack angle, and the stalling appear at a little smaller attack angle than other conditions. Furthermore, when attachment located on the upper surface of airfoil, the stalling attack angle appeared in advance. In contrast, when attachment located on the lower surface, the stalling attack angle is delayed. The phenomenon is more obviously at high attack angles. Figure 13 shows the pressure contour when attachments located at trailing-edge and lower surface of airfoil at 8°attack angle. It can be seen that attachments on lower surface of airfoil can create a bigger high pressure zone near the lower surface and significantly affect the distribution of the pressure gradient, as shown in figure 13(a) and (b). As a result, the lift is enhanced and the stalling is delayed. When the attachments is on the trailing-edge of the airfoil, the high pressure zones near the lower surface is smaller than that of no attachment case, as shown in figure 13(a) and (c). The lift and lift-to-drag ratios are all reduced, and the aerodynamic performance becomes worse.

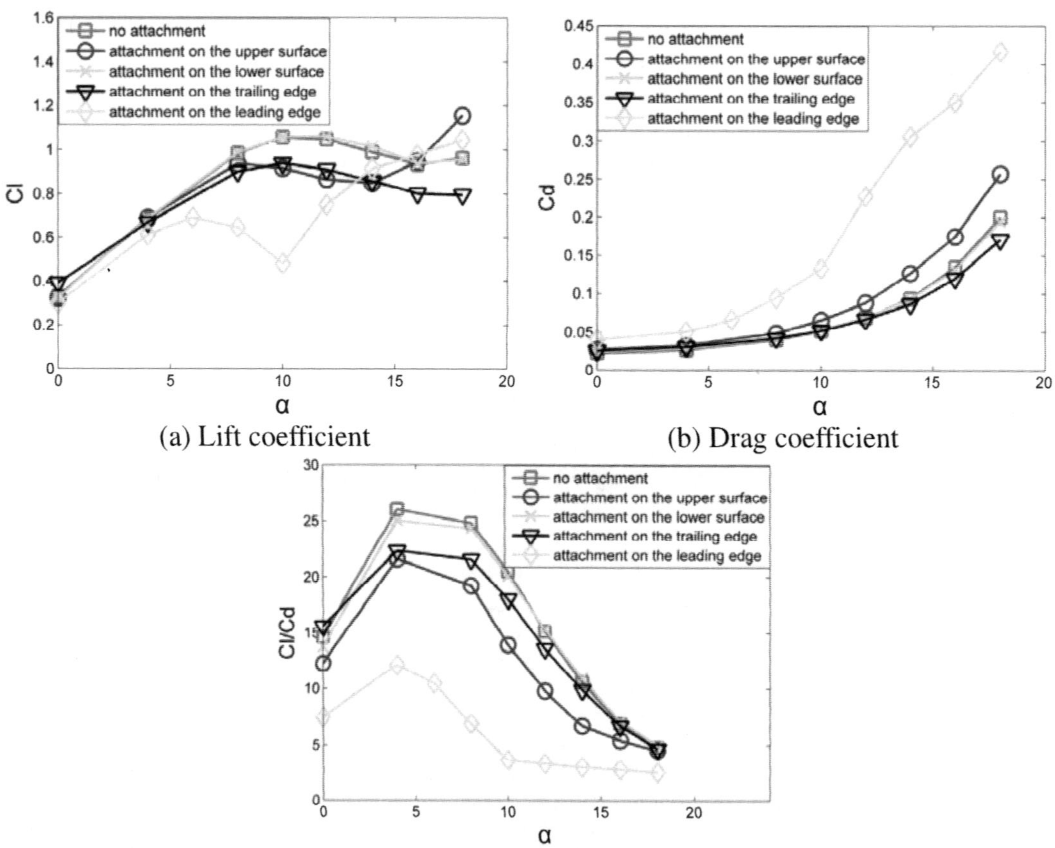

(a) Lift coefficient (b) Drag coefficient

(c) Lift to drag ratio

Figure 12. The force coefficients and lift-to-drag ratios of static NACA4412 airfoil with attachments on different icing positions.

(a) no attachment (b) attachment on the lower surface (c) attachment on the trailing-edge

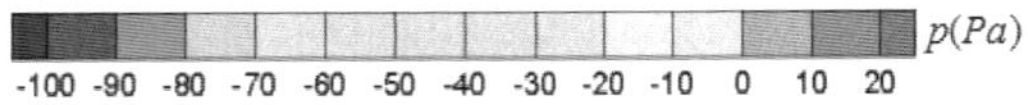

Figure 13. The pressure contour when attach angle is 8°

To investigate the effect of the airfoil's basic shape, the numerical calculation on the aerodynamic forces of two types of airfoils with the same attachments, are carried out. A symmetrical NACA0012 airfoil and an asymmetrical NACA4412 airfoil are respectively selected as airfoil's basic shape. The results are compared in detail and systematically analyzed. Figure 14 gives the lift to drag ratios of NACA0012 and NACA4412 airfoils with the same attachments. The lift to drag ratios of both NACA4412 and NACA0012 airfoils are all reduced by the attachments. Nonetheless, the extent of influence on two airfoils is not the same. For example, the largest reductions of lift to drag ratio between NACA4412 and NACA0012 airfoils with the same attachment derived from rime ice condition, is respectively 24.5% and 7.8%. Similarly, the largest reductions of lift to drag ratio between the two airfoils with the same attachment derived from glaze ice condition, is respectively 72.4% and 90.2%. These results reveal that the airfoil's basic shape is one of key impact factors on the aerodynamic characteristics of airfoil with icing attachments. More specifically, it is seen that the symmetrical airfoil is more easily influenced by the attachment of glaze ice type and the asymmetrical airfoil is more easily influenced by the attachment of rime ice type.

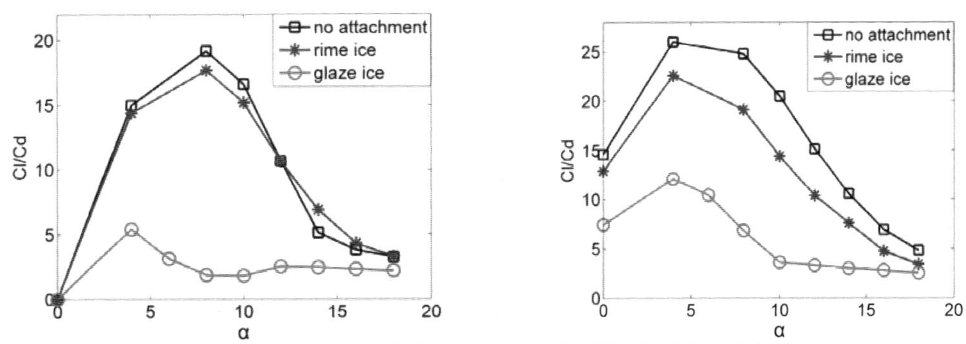

(a) NACA0012 airfoil (b) NACA4412 airfoil

Figure 14. Lift to drag ratios of NACA0012 and NACA4412 airfoils with the same attachments

4.2. The effect of attachments on the flow characteristics of flapping airfoils

For the flapping cases, simulations are carried out for the attack angle from 0° to 20°. The period-averaged force coefficients and flow field structures are systematically presented. Figure 15 shows the period-averaged lift coefficient, drag coefficient and lift to drag ratio of a flapping NACA4412 airfoil with surface attachments deriving from different icing types. All the attachments deriving from either rime ice or glaze ice condition raise the period-averaged drag coefficients and reduced period-averaged lift to drag ratios of the airfoil as shown in figure 15(b) and (c). The vorticity snapshot around the flapping airfoil with attachments deriving from different icing types are illustrated in figure 16. It is seen that the vortex separation mechanism of flapping airfoils is significantly changed by the surface attachments. Moreover, similar to results of static airfoil, the attachment deriving from glaze ice condition provides the lowest period-averaged lift to drag ratio and causes the worst aerodynamic performance than others. This is mainly due to the surface attachments derived from glaze ice condition, give rise to the larger vortexes at both leading-edge and trailing-edge of the airfoil, and cause the worse aerodynamic performance.

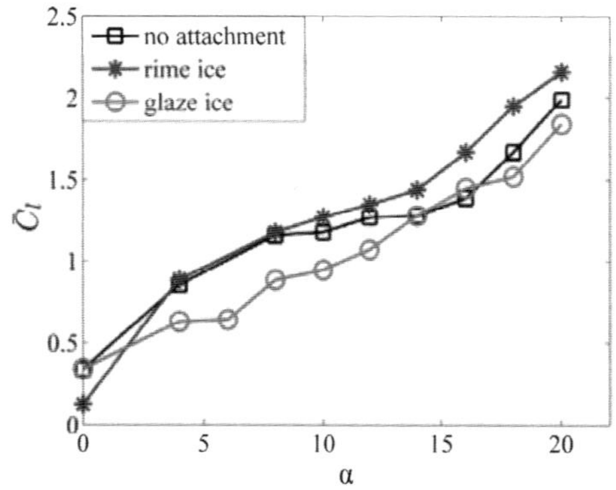

(a) Period-averaged lift coefficient

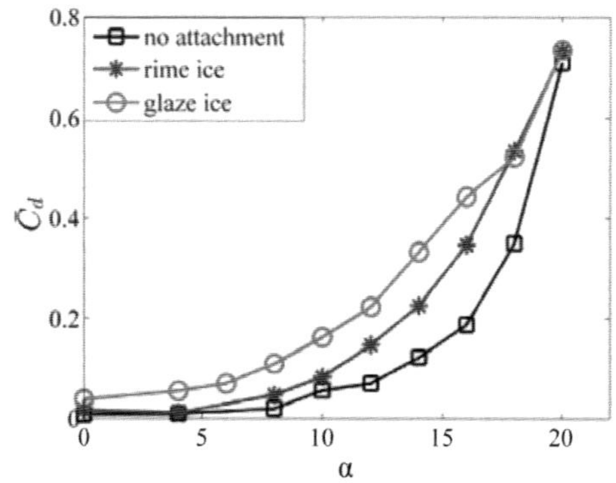

(b) Period-averaged drag coefficient

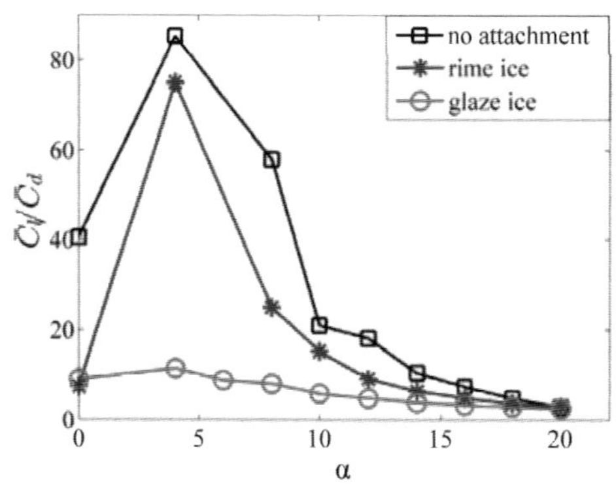

(c) Period-averaged lift todrag ratio

Figure 15. The period-averaged lift coefficient, drag coefficient and lift to drag ratio of a flapping NACA4412 airfoil with attachments deriving from different icing types.

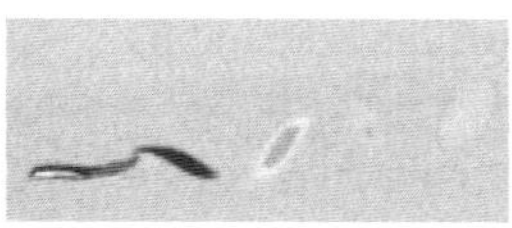

(a) The airfoil with attachment

(b) The glaze ice airfoil

(c) The rime ice airfoil

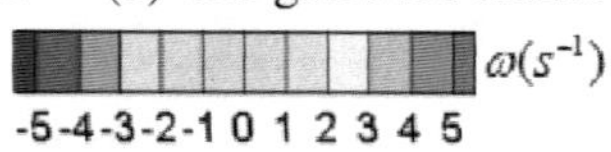

Figure 16. The vorticity contour of a NACA4412 airfoil with attachements deriving from different icing types, when attack angle is kept at 8°.

Figure 17 and Figure 18 give the period-averaged force coefficients and lift-to-drag radio of the flapping airfoil with attachments increasing with icing times. The period-averaged drag coefficient is monotonically increasing with icing time. The lift to drag ratios reaches a peak and then reduces as the icing time. The longer the icing time, the greater the attachments affect the aerodynamic characteristics of the flapping airfoil. This observation is similar to the static airfoil.

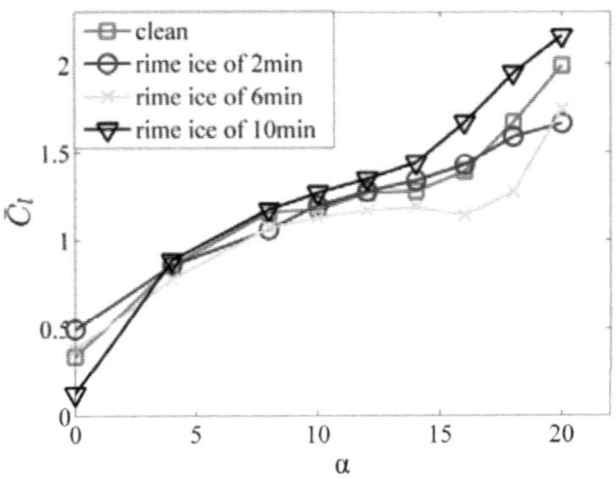

(a) Period-averaged lift coefficient

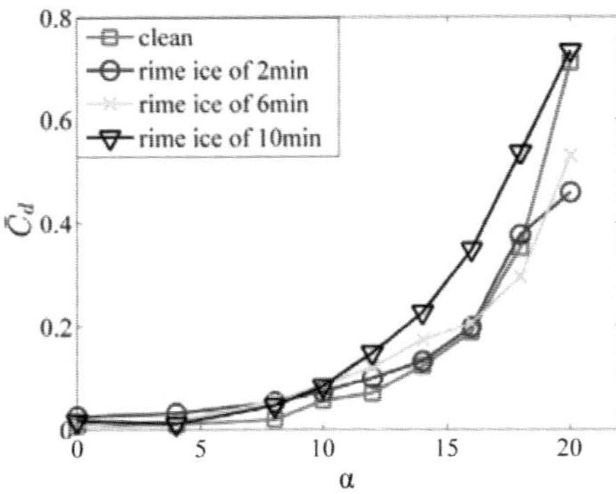

(b) Period-averaged drag coefficient

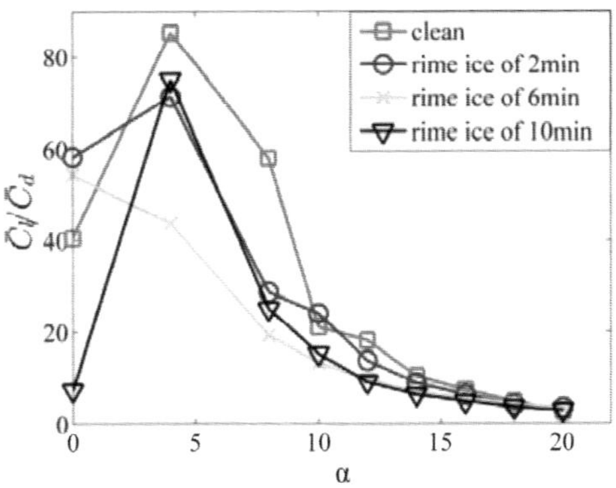

(c) Period-averaged lift todrag ratio

Figure 17. Period-averaged force coefficients and lift-to-drag radio of the flapping airfoil with attachments derived from rime ice at different icing times

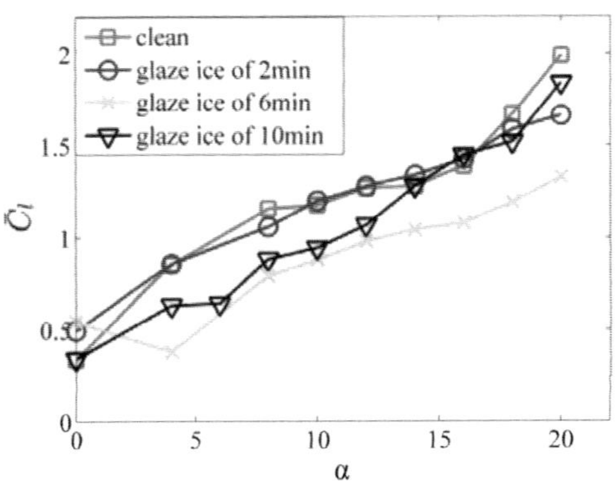

(a) Period-averaged lift coefficient

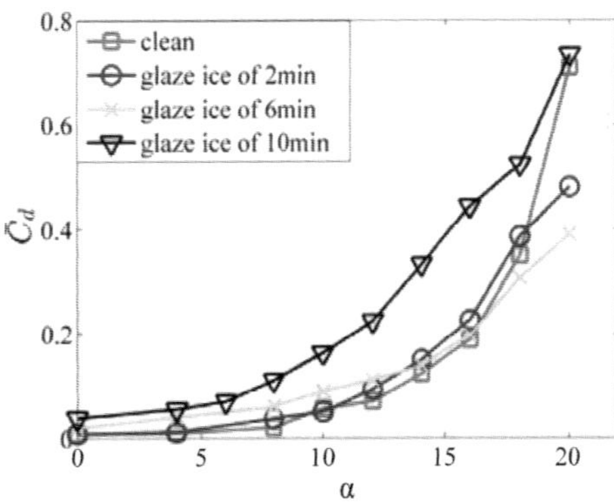

(b) Period-averaged drag coefficient

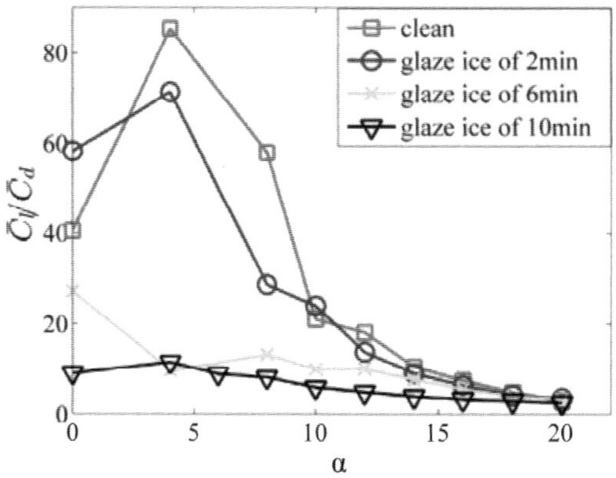

(c) Period-averaged lift todrag ratio

Figure 18. Period-averaged force coefficients and lift-to-drag radio of the flapping airfoil with attachments derived from glaze ice at different icing times

Figure 19 shows the period-averaged force coefficients and lift-to-drag ratios of the flapping NACA4412 airfoil with attachments at different icing positions. When attachment located on the trailing edge of airfoil, the period-averaged force coefficients and lift-to-drag ratios are both small. Compared with the static airfoil, it is seen that attachment located on the trailing edge cause the worse aerodynamic performance in the flapping airfoil. In addition, it is similar to the results of static airfoil that the attachments on the leading-edge have the greater effect of aerodynamic performance than others. Figure 20 shows the vorticity snapshot when attachments located on different positions at 10°attack angle. It can be seen that the direction of vortex shedding is more horizontal when attachment is located on the trailing edge of airfoil, as shown in figure 20(b). It is the possible reason that why the period-averaged force coefficients and lift-to-drag ratios are both small. The airfoil with attachment at the upper surface has the larger leading-edge's vortexes than the airfoil with attachment at the lower surface, as shown in figure 20(c) and (d). So, the aerodynamic performance of airfoil which attachment located on the upper surface is worse.

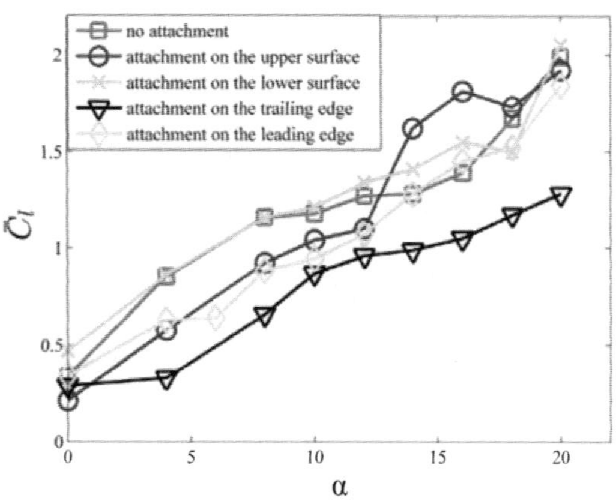

(a) Period-averaged lift coefficient

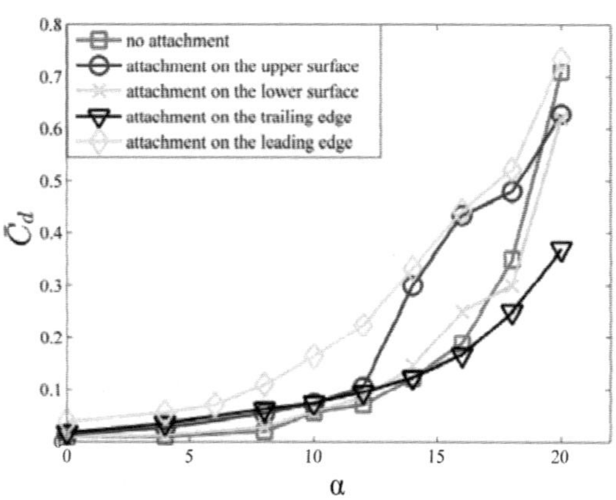

(b) Period-averaged drag coefficient

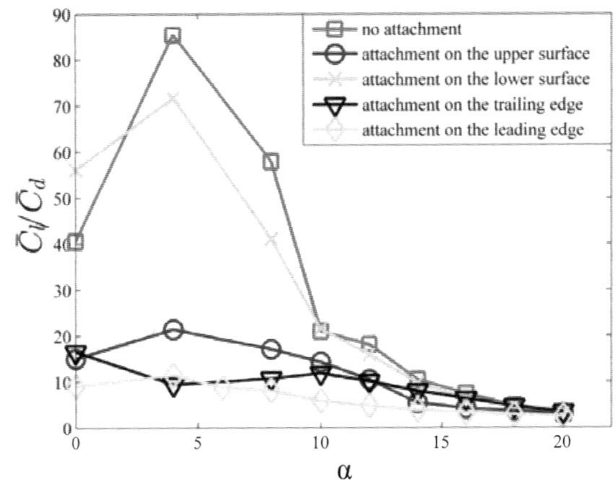

(c) Period-averaged lift to drag ratio

Figure 19. The period-averaged lift coefficient, drag coefficient and lift to drag ratio of a flapping NACA4412 airfoil with attachments on different positions

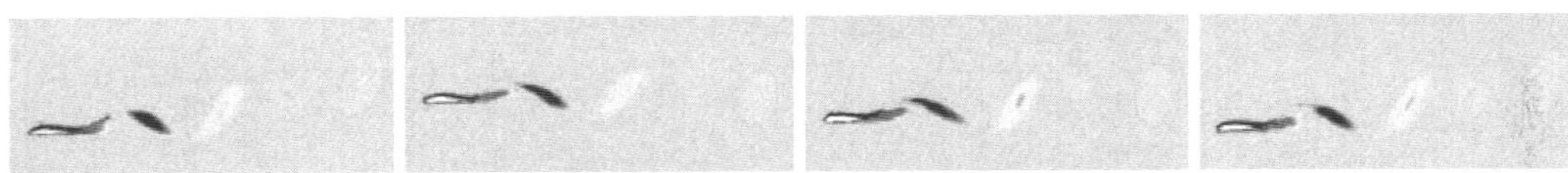

(a) No attachment (b) Attachment on the trailing edge (c) Attachment on the upper surface (d) Attachment on the lower surface

$\omega(s^{-1})$

-6 -5 -4 -3 -2 -1 0 1 2 3 4 5 6

Figure 20. The vorticity contrast map of different icing position in the 10°angle of attack

Figure 21 shows the period-averaged lift-to-drag ratios of flapping NACA4412 airfoil and flapping NACA0012 airfoil with the same attachments. Similar to the results of static airfoils, the basic shapes of airfoil have a significant influence. The largest reductions of period-averaged lift to drag ratio between NACA4412 and NACA0012 airfoils with the same attachment derived from rime ice condition, is respectively 56.82% and 35.88%. Similarly, the largest reductions of lift to drag ratio between the two airfoils with the same attachment derived from glaze ice condition, is respectively 86.6% and 79.4%. It seems that the asymmetrical airfoil has the greater effect of aerodynamic performance in both icing types of attachment. Figure 22 shows the vorticity snapshot of different airfoil with the same attachments. In general, flapping NACA4412 airfoil gives rise to the larger vortexes at both leading-edge and trailing-edge of the airfoil. That is probably reason that why flapping NACA4412 airfoil has the worse aerodynamic performance than that of NACA0012 airfoil.

IOP Conf. Series: Journal of Physics: Conf. Series **916** (2017) 012007 doi:10.1088/1742-6596/916/1/012007

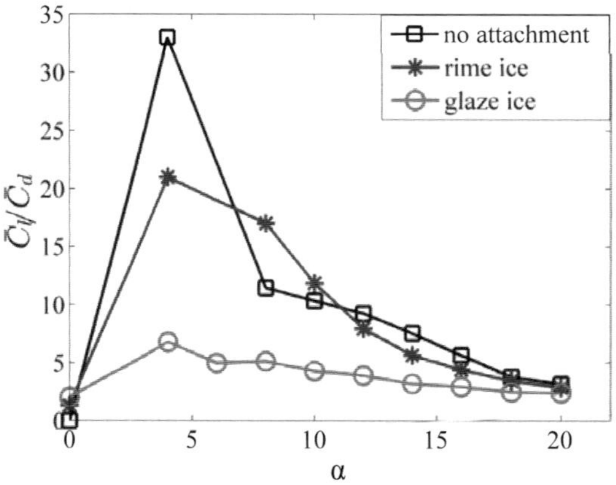

(a) Period-averaged lift to drag ratio of NACA0012

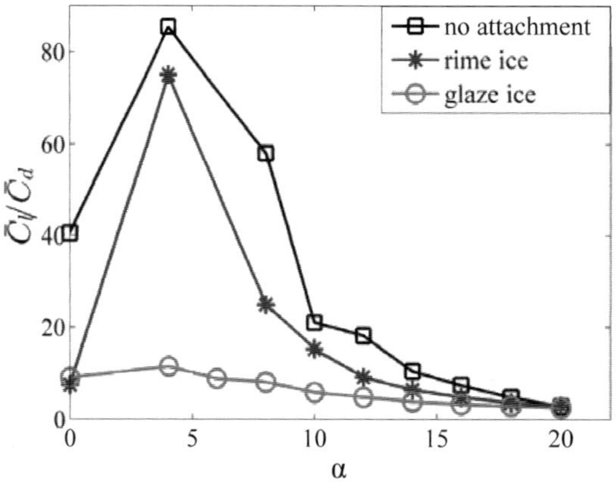

(b) Period-averaged lift to drag ratio of NACA4412

Figure 21. Period-averaged lift to drag ratios of NACA0012 and NACA4412 airfoils with the same attachments

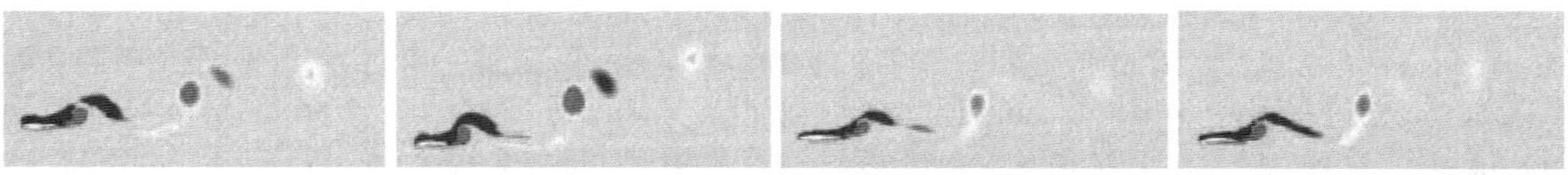

(a) The glaze ice airfoil of NACA0012 (b) The glaze ice airfoil of NACA4412 (c) The rime ice airfoil of NACA0012 (d) The rime ice airfoil of NACA4412

Figure 22. The vortices contrast map of different icing airfoil in the 12°angle of attack

5. Conclusion

In this paper, how aerodynamic characteristics affected by the attachment on the surface of airfoil, is systematically investigated. In the situation of static airfoil, according to the different icing situations, there are some conclusions can be draw in follows. First, the airfoil with the attachment of glaze ice type causes the worse aerodynamic performance than the airfoil with the attachment of rime ice type. Second, the longer the icing time, the greater influence of aerodynamic performance the attachment causes. Third, the attachments on the leading-edge have the greater effect of aerodynamic performance than other positions. Fourth, the symmetrical airfoil is more easily influenced by the attachment of glaze ice type and the asymmetrical airfoil is more easily influenced by the attachment of rime ice type.

Among, the first three conclusions are also in accordance with the flapping airfoil. However, there are some differences between the flapping and static airfoils. Specifically, the attachments located on the trailing edge of flapping airfoil cause aerodynamic performance become worse. The asymmetrical airfoil with the attachments derived from either rime ice type or glaze ice type all has a worse aerodynamic performance. In general, the attachments formed in various icing conditions deteriorate the aerodynamic performance of airfoil in most of situation. Hence, it is very essential to further develop the effective deicing technology to improve the operation life and performance of the wind machinery in the next work.

Acknowledgments
This study was supported by the National Youth Natural Science Foundation of China (Grant No. 51405279). The financial support is gratefully acknowledged.

References

[1] L. d. Zhang and L.C. Ren, Analysis of the Influence of Poor Weather Condition on Wind Power Generating Unit, J. Hydroelectric Engineering. 33 (2007) 67-69.
[2] N. Bose, Icing on a small horizontal-axis wind turbine-part 1: glaze ice profiles, J. Wind Engineering and Industrial Aerodynamics. 45 (1992) 87-96.
[3] W.J. Jasinski, S.C. Noe and M.S. Selig, et al., Wind turbine performance under icing conditions, J. Solar Energy Engineer. 120 (1998) 60-65.
[4] H. Clement and F. Guy, Icing Simulation of Wind-Turbine Blades, Paper presented at The AIAA 45th Aerospace Science Meeting, Reno, Nevada. (2007)
[5] X. H. Deng, X. X. LU and L. P. LI, et al., Numerical Simulation of Airfoil Ice Accretion Process on Horizontal-axis Wind Turbine Blade, J. Energy Technology. 31 (2010) 266-271.
[6] C. X. Zhu, L. Wang, Z. G. Sun, et al., Numerical Study of Wind Turbine Blade Ice Accretion. J. Acta Aerodynamica Sinica. 29 (2011) 522-528.
[7] J. J. Yan, Numerical Simulation and Aerodynamic Performance of Ice Accretion on the Wind Turbine Airfoil. North China Electric Power University, BeiJing. (2014)
[8] Y. Li, X. L. Wang, F. Feng, et al., An icing wind tunnel experiment on the icing distribution of a blade airfoil rotating around a shaft, J. Harbin Engineer University. 38 (2017) 545-553.
[9] G. Tie and W. J. Qi, Numerical Simulation on the Aerodynamic Performance of Aerofoil of Horizontal Wind Turbine, J. Fluid Machinery. 41 (2013) 29-33.
[10] T. Lee and P. Greontakos, Investigation of Flow over an Oscillating Airfoil, J. Fluid Mechanics, 512 (2004) 313-341.
[11] X. W. Zhang and C. Y. Zhou, et al., Numerical study on effect of leading-edge tubercles, J. Aircraft Engineering and Aerospace Technology. 85 (2013) 247 – 257.

Optimization design and dynamic analysis on the drive mechanisms of flapping-wing air vehicles based on flapping trajectories

Lingwang Xie, Xingwei Zhang*, Pan Luo, Panpan Huang

Shantou University, College of Engineering, Shantou 515063, China

*Corresponding author e-mail: zhangxw@stu.edu.cn

Abstract. The optimization designs and dynamic analysis on the driving mechanism of flapping-wing air vehicles on base of flapping trajectory patterns is carried out in this study. Three different driving mechanisms which are spatial double crank-rocker, plane five-bar and gear-double slider, are systematically optimized and analysed by using the Mat lab and Adams software. After a series debugging on the parameter, the comparatively ideal flapping trajectories are obtained by the simulation of Adams. Present results indicate that different drive mechanisms output different flapping trajectories and have their unique characteristic. The spatial double crank-rocker mechanism can only output the arc flapping trajectory and it has the advantages of small volume, high flexibility and efficient space utilization. Both planar five-bar mechanism and gear-double slider mechanism can output the oval, figure of eight and double eight flapping trajectories. Nevertheless, the gear-double slider mechanism has the advantage of convenient parameter setting and better performance in output double eight flapping trajectory. This study can provide theoretical basis and helpful reference for the design of the drive mechanisms of flapping-wing air vehicles with different output flapping trajectories.

1. Introduction

Because of the high concealment and maneuverability, the practical flapping wing air vehicles can be broadly used in the military industry and livelihood industry in the near future (Fang and Wu, 2013). In the past decades, the research on the flapping wing air vehicles has a remarkable progress. However, some challenges still exist and needed to be further studied. One of them is the design of drive mechanisms of the flapping wing air vehicles. Many scholars have done lots of related studies and get some useful results. Michelson (2000) used the crank-slider mechanism to achieve the simple flapping-wing movement. Kinkade (2002) used the gear- bar mechanism to achieve the flapping trajectory of arc. Madangopal (2004) used the plane crank-slider mechanism to achieve a two-dimensional flapping-wing movement. They used the spring to provide more energy and improve the lift. Wang (2006) used the piezoelectric actuator and the four-bar mechanism to make a biomimetic and flexible flapping wing air vehicle. Their drive mechanism has some special advantages such as little friction and high sensitivity. Zhu (2007) designed the seven-bar mechanisms to achieve the flapping trajectory of "eight" and class oval. These researches and achievements provide the huge contribution to the development of flapping wing air vehicles. However, most above study focus on the movement performance and the feasibility of drive mechanisms. The investigation of flapping trajectory is infrequent, especially lacked of the flapping trajectory of double "eight". In addition, their drive mechanisms' flapping trajectory is single. Accordingly, we aim to increase our understanding of flapping trajectory and find a drive mechanism which can achieve varied flapping trajectories expediently. Three mechanisms which are spatial double crank-rocker, plane five-bar and gear-double slider, are employed by the drive mechanisms. The type and characteristics of flapping trajectories of these drive mechanisms will be analyzed in detail.

2. Analytical method

The pattern of flapping trajectory of flying creatures found by previous studies mainly contain oval, 'eight' and double 'eight' (Daltion S., 2004). Especially, the waving patterns of most insect wings are 'eight' and double 'eight'. As an example, the flapping trajectories of dragonfly are "eight" and double "eight", and the flapping trajectory of cicada is class oval (Sudo S. et al., 2005), as shown in Figure 1 and 2. This is likely related to the excellent flying skills of insects and can provide inspired ideas for the artificial flying vehicles. Therefore, it is worthy of further study about this trajectory mechanism. In this study, the optimized design on three drive mechanisms can output three trajectories are firstly carried out. The detail components of mechanisms are optimized by using MATLAB based on four output flapping trajectories which are arc, oval, and 'eight' and double 'eight'. Then, ADMAS is used to simply simulate and validate the feasibility of mechanism.

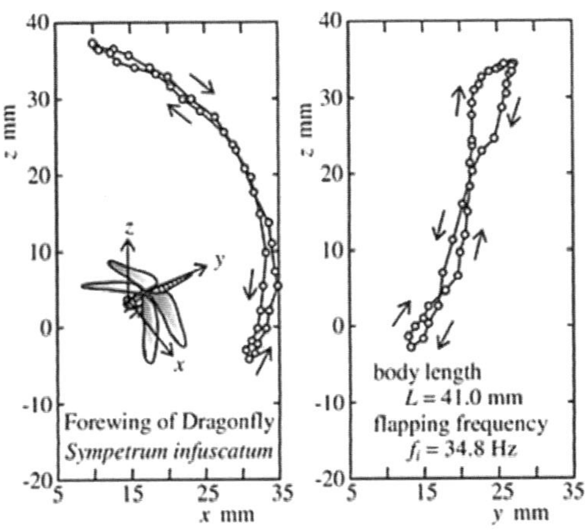

Figure 1. The eight trajectory [7]

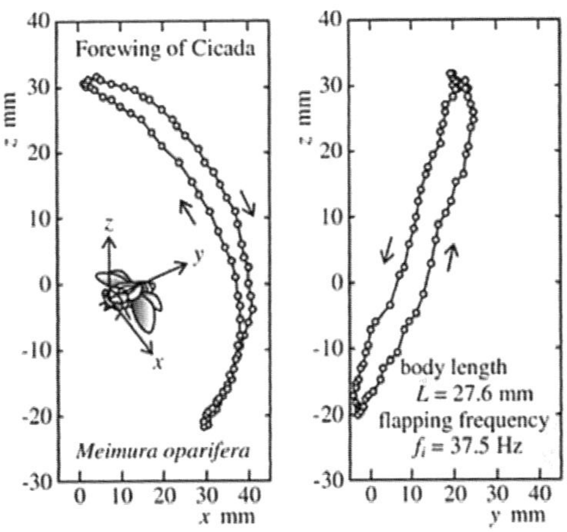

Figure 2. The double eight trajectory [7]

3. Mechanism modeling

3.1. The spatial double crank-rocker mechanism for outputting an arc flapping trajectory

First, a spatial double crank-rocker structure is used by the drive mechanism to produce the arc flapping trajectory. The schematic illustration of the drive mechanism is shown in figure 3. A fixed Cartesian reference coordinate frame is attached at A. The bar AE, AB, BC, CD and ED are represented by l1, l2, l3, l4 and r. The angle between the r and the plane XAY is α, and the angle between the l2 and l3 is β. l1 and l2 are fixed, and r is the drive bar. The coordinates of the points B, C, D and E are (0, 0, z2), (0, yc, zc), (xd, y1, zd) and (0, y1, 0). Among, yc=l3sinβ, zc=z2-l3cosβ, xd=-rcosα, zd=rsinα.

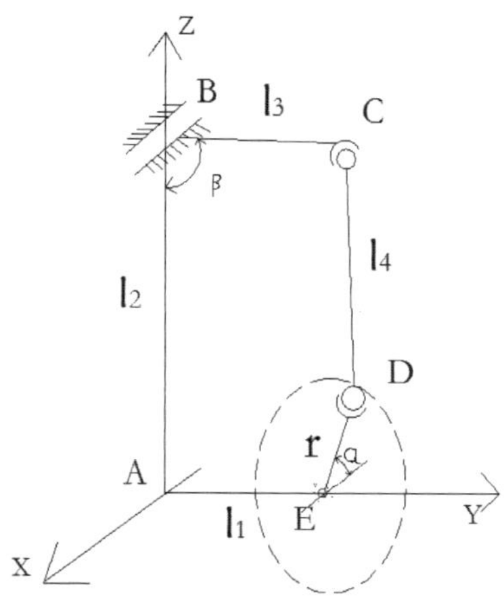

Figure 3. Schematic illustration of spatial double crank-rocker mechanism

According to the length of l4, the solve equation can be built as:

$$(0-x_d)^2+(y_c-y_1)^2+(z_c-z_d)^2=l_4^{\,2} \tag{1}$$

Further, equation (1) becomes:

$$(r\cos\alpha)^2+(l_3\sin\beta-y_1)^2+(z_2-l_3\cos\beta-r\sin\alpha)^2=l_4^{\,2} \tag{2}$$

Equation (2) becomes:

$$l_3^{\,2}+(r\cos\alpha)^2+y_1^{\,2}+(z_2-r\sin\alpha)^2-l_4^{\,2}=2l_3y_1\sin\beta+2(z_2-r\sin\alpha)l_3\cos\beta \tag{3}$$

The equation of auxiliary angle is given by

$$a\sin x+b\cos x=\sqrt{a^2+b^2}\sin\left(x+\arctan\frac{b}{a}\right) \tag{4}$$

Then, the solution of β can be got as follows.

$$\beta=\arcsin\left(\frac{l_3^{\,2}+(r\cos\alpha)^2+y_1^{\,2}+(z_2-r\sin\alpha)^2-l_4^{\,2}}{2l_3\sqrt{y_1^{\,2}+(z_2-r\sin\alpha)^2}}\right)-\arctan\frac{(z_2-r\sin\alpha)}{y_1} \tag{5}$$

When the α is definitive, the β can be solved, so the coordinates of the points C can be got. Finally, the output trajectory of drive mechanism (i.e. movement of point C) can be solved.

There are two constraint conditions in this drive mechanism. One is the restrict on the flapping angle of l3. So the equality constraints can be set as follows. The maximal flapping range is 70°. When the maximize downward flapping range of the mechanism is 40°, β is kept at 50°. The other constraint condition is that l4 should not contact with the runner. Then the inequality constraint can be set as yc>y1.

The max angle and min angle of l3 in the coordinate system is described by βs and βx. When the α is 90°, the l3 has the maximal upward flapping angle, Then the equation (5) becomes:

$$\beta_s = \arcsin\left(\frac{l_3^{\,2} + y_1^{\,2} + (z_2 - r)^2 - l_4^{\,2}}{2l_3\sqrt{y_1^{\,2} + (z_2 - r)^2}}\right) - \arctan\frac{(z_2 - r)}{y_1} \tag{6}$$

When the is 270°, the l3 has the maximal downward flapping angle, Then the equation (5) becomes:

$$\beta_x = \arcsin\left(\frac{l_3^{\,2} + y_1^{\,2} + (z_2 + r)^2 - l_4^{\,2}}{2l_3\sqrt{y_1^{\,2} + (z_2 + r)^2}}\right) - \arctan\frac{(z_2 + r)}{y_1} \tag{7}$$

Because $\beta_x = 50° = 5\pi/18$ and $\beta_s = 120° = 2\pi/3$.
So, the equality constraints are as follows

$$g_1(x) = \arcsin\left(\frac{l_3^{\,2} + y_1^{\,2} + (z_2 + r)^2 - l_4^{\,2}}{2l_3\sqrt{y_1^{\,2} + (z_2 + r)^2}}\right) - \arctan\frac{(z_2 + r)}{y_1} - \frac{5}{18}\pi = 0 \tag{8}$$

$$g_2(x) = \arcsin\left(\frac{l_3^{\,2} + y_1^{\,2} + (z_2 - r)^2 - l_4^{\,2}}{2l_3\sqrt{y_1^{\,2} + (z_2 - r)^2}}\right) - \arctan\frac{(z_2 - r)}{y_1} - \frac{2}{3}\pi = 0 \tag{9}$$

Another, yc>y1 and equation (5), then

$$y_c = l_3 \sin\beta \ge y_1 \tag{10}$$

$$l_3 \sin\left(\arcsin\left(\frac{l_3^{\,2} + (r\cos\alpha)^2 + y_1^{\,2} + (z_2 - r\sin\alpha)^2 - l_4^{\,2}}{2l_3\sqrt{y_1^{\,2} + (z_2 - r\sin\alpha)^2}}\right) - \arctan\frac{(z_2 - r\sin\alpha)}{y_1} \right) \ge y_1 \tag{11}$$

Finally, the equation of inequality constraints are:

$$g_3(x) = y_1 - l_3 \sin\left(\arcsin\left(\frac{l_3^{\,2} + (r\cos\alpha)^2 + y_1^{\,2} + (z_2 - r\sin\alpha)^2 - l_4^{\,2}}{2l_3\sqrt{y_1^{\,2} + (z_2 - r\sin\alpha)^2}}\right) - \arctan\frac{(z_2 - r\sin\alpha)}{y_1} \right) \le 0 \tag{12}$$

For the linkage mechanisms, the bigger the transmission angle, the better performance of the drive mechanisms can produce (Yan and Jin, 2008). The transmission angle should bigger than 40° (Sun, et al, 2007). The pressure angle is complementary with transmission angle. So a smaller pressure angle is better. For this reason, the optimized object of this mechanism is to make the maximum pressure angle as small as possible. In current study, the pressure angle locating between the straight DC and the tangential line of straight BC's velocity direction can be calculated by

$$\cos\langle\vec{a},\vec{b}\rangle = \frac{\vec{a}\cdot\vec{b}}{\left|\vec{a}\right|\cdot\left|\vec{b}\right|} = \frac{a_1b_1 + a_2b_2 + a_3b_3}{\sqrt{a_1^{\,2} + a_2^{\,2} + a_3^{\,2}}\cdot\sqrt{b_1^{\,2} + b_2^{\,2} + b_3^{\,2}}} \tag{13}$$

Where the $\vec{a}$ describes the vector of straight DC, and the $\vec{b}$ describes the vector of tangential line of straight BC's velocity direction. Further,
$\vec{a} = (0, -l_3 \sin\beta, kl_3 \sin\beta - 2l_3 \cos\beta + 2z_2)$; $\vec{b} = (r\cos\alpha, l_3 \sin\beta - y_1, z_2 - l_3 \cos\beta - r\sin\alpha)$.

And then, the mechanism may have the biggest pressure angle when α is 90° or 270°, the optimized object is the sum of pressure angles in the two situation. Furthermore, the objective function is given by

$$f(x) = \min\left(\left(\frac{\vec{a}\cdot\vec{b}}{\left|\vec{a}\right|\cdot\left|\vec{b}\right|} \right)_{\alpha=90^\circ} + \left(\frac{\vec{a}\cdot\vec{b}}{\left|\vec{a}\right|\cdot\left|\vec{b}\right|} \right)_{\alpha=270^\circ} \right) \tag{14}$$

Finally, the optimized three-dimensional model of spatial double cranks-rockers drive mechanism is given in figure 4. After numerical optimization, the detail parameters of each component of the drive mechanism described in section 3.1 are r=7.5992, l1=5.0021, l2=15.0021, l3=12.8606, l4=10.089. The sum of all components is 50.553. These values are small enough and good for the miniaturization of flapping-wing air vehicles. When α=270°, the mechanism has the biggest pressure angle which is 43.9°, so the smallest transmission angle is 90°-43.9°=46.1°>40°. Hence, the mechanism meets the requirements.

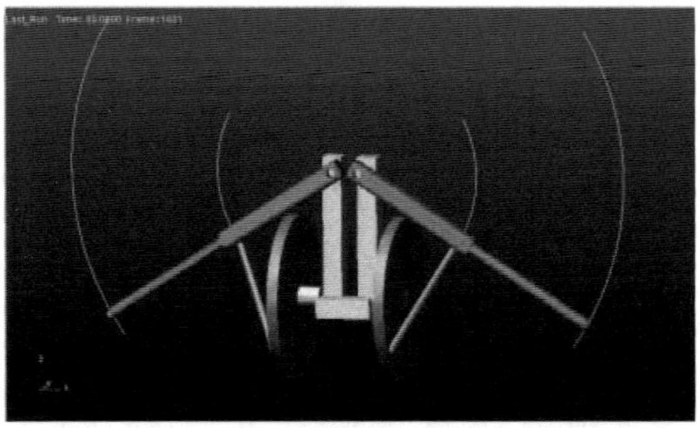

Figure 4. The 3D model of spatial double cranks-rockers drive mechanism

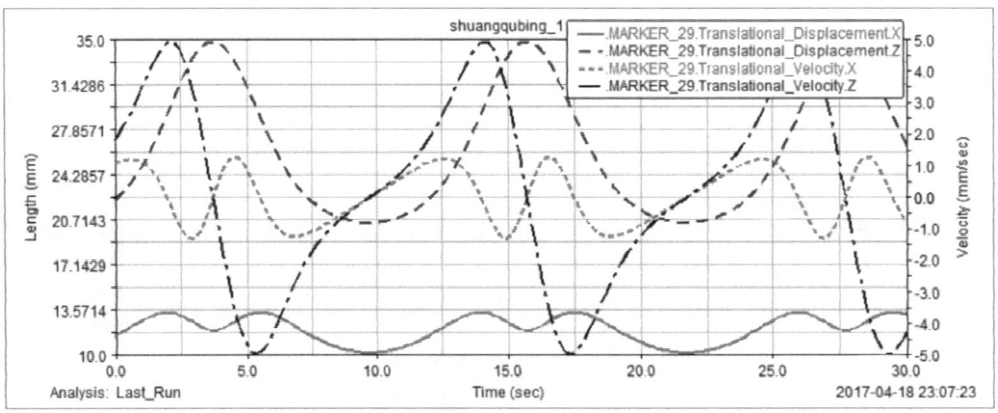

Figure 5. The velocity and displacement graph of the output terminal

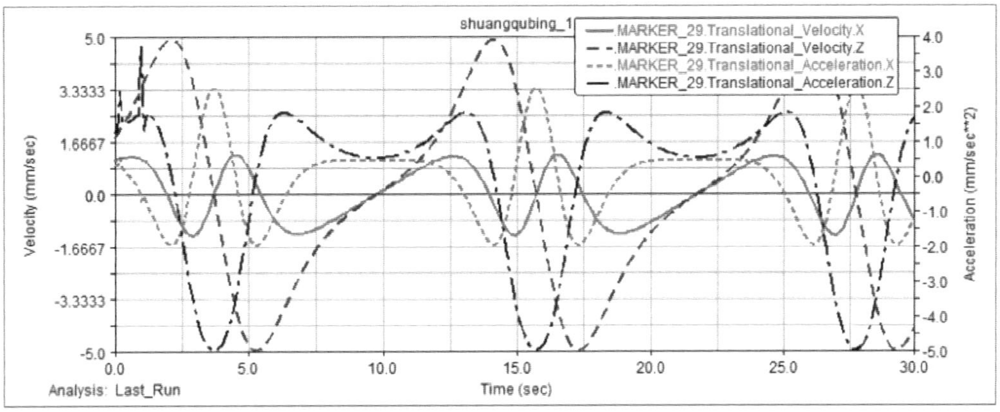

Figure 6. The velocity and acceleration graph of the output terminal

Figure 4 shows the flapping trajectory of the spatial double cranks-rockers mechanism. Its flapping trajectory is arc which is consistent to the expectation. Figure 5 and 6 show the dynamical parameters of this drive mechanism. It can be seen that the curves are smooth and don't have sharp variation, so it confirm the mechanism is feasible and meets the requirements.

3.2. The planar five-bar mechanism for outputting oval, figure of eight and double eight flapping trajectories

To output figure of both oval and eight trajectory, of "eight", a planar five-bar mechanism is used by the drive mechanism. Figure 7 gives the schematic illustration of the drive mechanism. In this fixed Cartesian reference coordinate frame, the bar AB, BC, CD, DE and AE are respectively described by l1, l2, l3, l4and l5. α is the angle between l1 and l5. β is the angle between l4 and the positive half of the x-axis. The degree of freedom of plane five bars mechanism is two. In order to reduce the number of motor, gear mesh is used, and ω1 and ω2 are the angular velocity of gears.

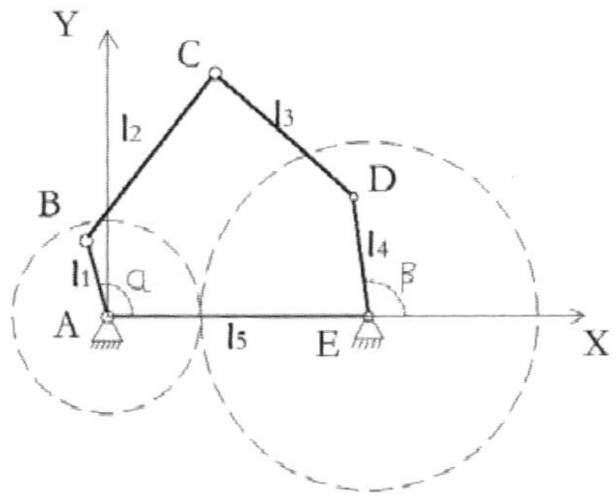

Figure 7. Schematic illustration of plane five bars mechanism

The coordinates of the points B, C, D and E are (xb, yb), (xc, yc), (xd, yd) and (l5, 0). The solve equation can be built as follows.

$$\begin{cases}(x_c - x_b)^2 + (y_c - y_b)^2 = l_2^{\ 2} \\ (x_c - x_d)^2 + (y_c - y_d)^2 = l_3^{\ 2}\end{cases} \tag{15}$$

Another, among, xb=l1cosα, yb=l1sinα; xd=l5+l4cosβ, yd=l4sinβ. Then,

$$\begin{cases} (x_c - l_1\cos\alpha)^2 + (y_c - l_1\sin\alpha)^2 = l_2^{\,2} \\ (x_c - l_5 - l_4\cos\beta)^2 + (y_c - l_4\sin\beta)^2 = l_3^{\,2} \end{cases} \tag{16}$$

So the coordinates of point C is given by

$$x_c = x_b + l_2\cos\left(\arccos\left(\frac{l_2^{\,2} + (x_b - x_d)^2 + (y_b - y_d)^2 - l_3^{\,2}}{2l_2\sqrt{(x_b - x_d)^2 + (y_b - y_d)^2}}\right) + \arctan\left(\frac{y_d - y_b}{x_d - x_b}\right)\right) \tag{17}$$

$$y_c = y_b + l_2\cos\left(\arccos\left(\frac{l_2^{\,2} + (x_b - x_d)^2 + (y_b - y_d)^2 - l_3^{\,2}}{2l_2\sqrt{(x_b - x_d)^2 + (y_b - y_d)^2}}\right) + \arctan\left(\frac{y_d - y_b}{x_d - x_b}\right)\right) \tag{18}$$

This section investigates the flapping trajectory from two situation which are ω1=ω2 and ω1=2ω2. When the modulus is same, the pitch diameter decides the angular velocity of gear. Hence, when ω1=ω2, there are two inequality constraints.

$$g_1(x) = l_1 - \frac{1}{2}l_5 < 0,\ g_2(x) = l_4 - \frac{1}{2}l_5 < 0 \tag{19}$$

Similarly, there are two inequality constraints when ω1=2ω2.

$$g_3(x) = l_1 - \frac{1}{3}l_5 < 0,\ g_4(x) = l_4 - \frac{2}{3}l_5 < 0 \tag{20}$$

In addition, the gears should have a complete circle. There are two extreme situations in the following figure 8 and 9. According to the existence theorem of triangle, there are two inequality constraints:

$$g_5(x) = l_1 + l_5 + l_4 - l_2 - l_3 < 0 \tag{21}$$

$$g_6(x) = |l_2 - l_3| - (l_5 - l_1 - l_4) < 0 \tag{22}$$

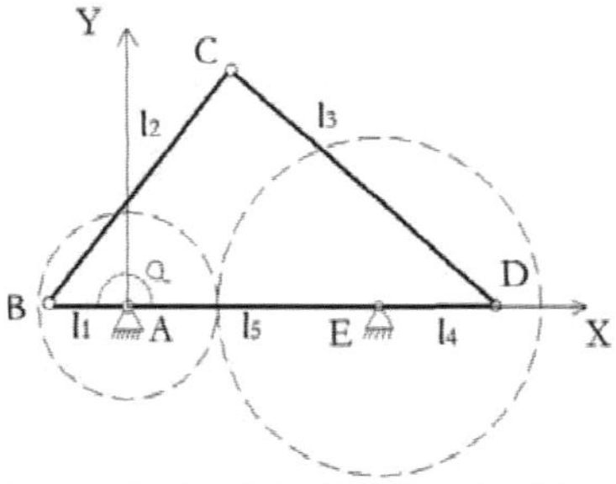

Figure 8. Straight BD has the Maximum

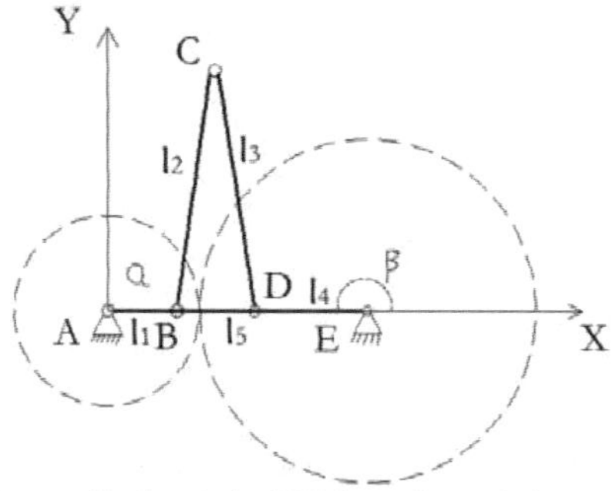

Figure 9. Straight BD has the minimum

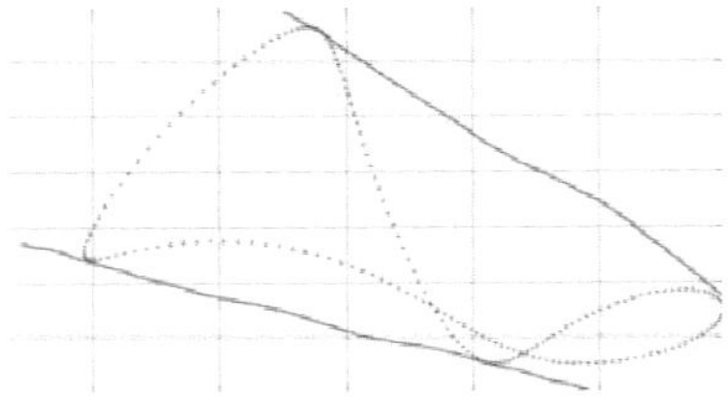

Figure 10. Sample of large width trajectory

Then, it is also found that the aforementioned constraint conditions are not enough because the width of trajectory is large, as shown in figure 10.

So, the additional constraint conditions are needed and defined as follows.

$$\begin{cases} g_7(x) = l_2 + l_4 - l_5 < 0 \\ g_8(x) = l_3 + l_4 - l_5 < 0 \\ g_9(x) = l_1 + l_2 - l_5 < 0 \\ g_{10}(x) = l_1 + l_3 - l_5 < 0 \end{cases} \tag{23}$$

The MAV (micro air vehicle) has strict requirements of size, so it's better to reduce the mechanism' size. Hence, the objective function is defined by

$$f(x) = \min(l_1 + l_2 + l_3 + l_4 + l_5) \tag{24}$$

After numerical optimization, the parameters of each component are: l1=7.5, l2=27.5, l3=37.5, l4=15, l5=52.5. The sum of all components is 150. Although, its sum is bigger than spatial double cranks-rockers mechanism's, it also can achieve miniaturization of flapping-wing air vehicles. The 3d model which ω1=ω2 is shown in figure 11.

Figure 11. The 3D model of planar five-bar mechanism

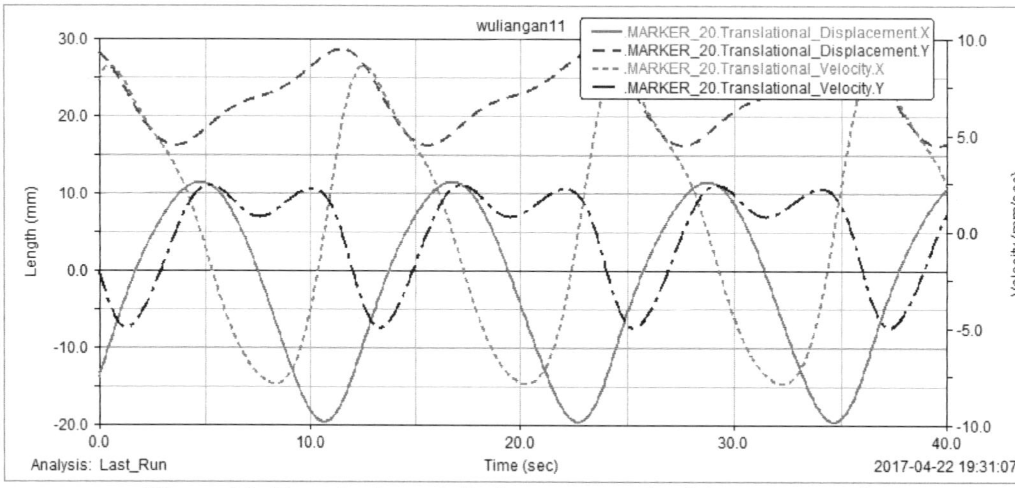

Figure 12. The velocity and displacement graph of the output terminal

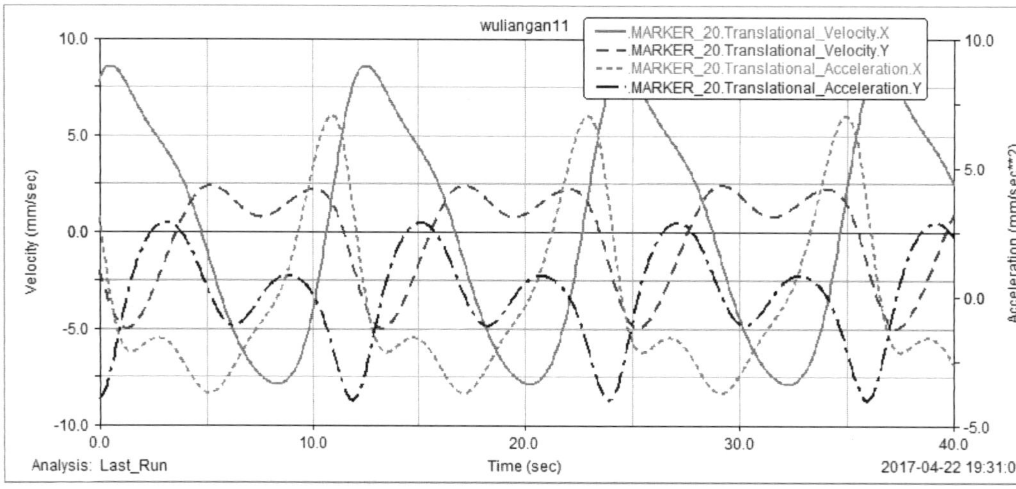

Figure 13. The velocity and acceleration graph of the output terminal

Through many debugging, there is an ideal "eight" trajectory can be found when the initial conditions are $\omega 1=\omega 2$, $\alpha=180°$and $\beta=180°$, just like the Figure 11 shows. Its turnings are smooth that help the mechanism work stably and efficiently. Also, the width of flapping trajectory is narrow which is similar with the insect. Figure 12 and 13 show the dynamical parameters of this drive mechanism. It can be seen that the curves are smooth and don't have sharp variation, so it confirms the mechanism is feasible and meets the requirements.

Furthermore, this mechanism also can achieve other flapping trajectories when change the initial condition. When $\omega 1=\omega 2$, $\alpha=180°$and $\beta=15°$, the flapping trajectory of class oval can be got as the following figure 14.

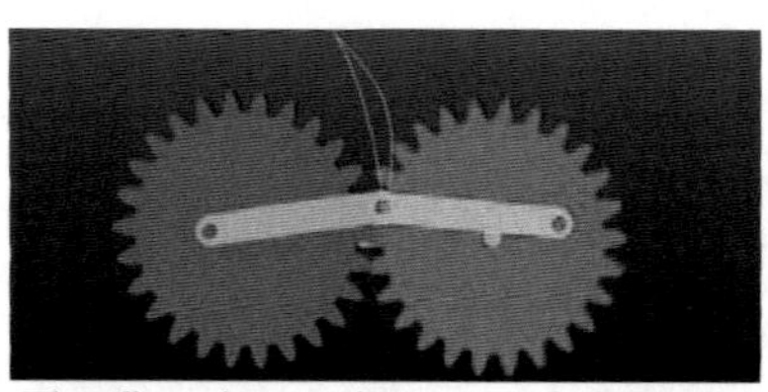

Figure 14. The flapping trajectory of class oval in Adams

Figure 15. The flapping trajectory of double "eight" in Adams

In addition, when the ω1=2ω2, α=120°and β=100°, the flapping trajectory of double "eight" can also be got as shown in figure 15. However, after dynamic analysis by Adams, this drive mechanism's performance is not good when it output the flapping trajectory of double eight.

3.3. The gear-double slider mechanism for outputting oval, figure of eight and double eight flapping trajectories

This paper uses the gear-double slider mechanism to output double 'eight' flapping trajectory. Figure 16 gives a schematic illustration of drive mechanism. A shown in figure 16, the bar O1C, AC, BD, O2D and O1O2 are respectively described by l1, l2, l3, l4, l5. α is the angle between l4 and the positive half of the x-axis. β is angle between l1 and l5. Point O1 and O2 are fixed. The mechanism also uses gear mesh to reduce the number of motor. The ω1 and ω2 are the angular velocity of gears. The coordinates of the points O1, O2, A, B, C and D are (0, 0), (l5, 0), (xa, ya), (xb, yb), (xc, yc), and (xd, yd). Further,

$$x_c = l_1 \cos\beta,\ y_c = l_1 \sin\beta;\ x_d = l_5 + l_4 \cos\alpha,\ y_d = l_4 \sin\alpha,\ x_a = x_b = l_5 + l_4 \cos\alpha - \sqrt{l_3^2 - (l_4 \sin\alpha)^2}\ .$$

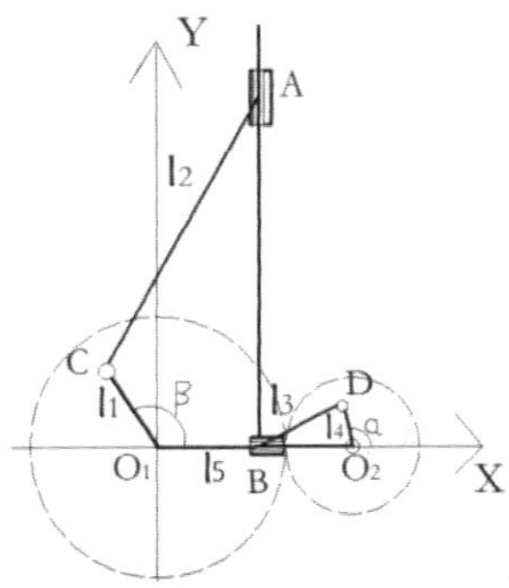

Figure 16. Schematic illustration of gear-double sliders mechanism

The length between point C and straight AB is calculated by

$$\left|CB\right|_x = x_b - x_c = l_5 + l_4 \cos\alpha - \sqrt{l_3^{\ 2} - (l_4 \sin\alpha)^2} - l_1 \cos\beta \tag{25}$$

The ordinate of a point is given by

$$y_a = y_c + \sqrt{l_2^{\ 2} - \left|CB\right|_x^{\ 2}} \tag{26}$$

For this mechanism, the output trajectories can be controlled by changing the rotating speed ratio. This section investigates three conditions. When ω1=ω2, the output trajectory is oval. When ω1=2ω2,

IOP Conf. Series: Journal of Physics: Conf. Series **916** (2017) 012008 doi:10.1088/1742-6596/916/1/012008

the trajectory is figure of eight. When ω1=3ω2, the trajectory is double eight. From figure 16, there are a constraint condition that point B should be on the right of point O_1, so:

$$g_1(x) = 0 - l_5 + l_4 \cos\alpha - \sqrt{l_3^{\ 2} - (l_4 \sin\alpha)^2} < 0 \tag{27}$$

Similar with last section, there are two inequality constraints in each rotating speed ratio.
For ω1=ω2

$$g_2(x) = l_1 - \frac{1}{2} l_5 < 0,\ g_3(x) = l_4 - \frac{1}{2} l_5 < 0 \tag{28}$$

For ω1=2ω2

$$g_4(x) = l_1 - \frac{2}{3} l_5 < 0,\ g_5(x) = l_4 - \frac{1}{3} l_5 < 0 \tag{29}$$

For ω1=3ω2

$$g_6(x) = l_1 - \frac{3}{4} l_5 < 0,\ g_7(x) = l_4 - \frac{1}{4} l_5 < 0 \tag{30}$$

The optimized object of this mechanism is that make the maximum pressure angle as small as possible. In this mechanism, the pressure angle is between the straight AC and straight AB. From figure 16, the mechanism has the biggest pressure angle when α=0 and β=π. Due to the sine function is monotonic in the interval of 0 to 0.5π, so it can be replaced the pressure angle as the optimization target. The sine function of the pressure angle is defined by

$$\sin\theta = \frac{|CB|_x}{l_2} \tag{31}$$

Hence, the objective function is defined by

$$f(x) = \min\left(\frac{|CB|_x}{l_2}\right)_{\alpha=0,\beta=\pi} = \min\left(\frac{\left|l_5 + l_4\cos\alpha - \sqrt{l_3^{\ 2} - (l_4\sin\alpha)^2} - l_1\cos\beta\right|}{l_2}\right)_{\alpha=0,\beta=\pi} \tag{32}$$

After numerical optimization, the parameters of each component are: l1=12.0007, l2=29.9997, l3=23.3582, l4=3.0001, l5=26.3587. The sum of all components is 94.7. Its value is small and good to the miniaturization of flapping-wing air vehicles. In addition, the mechanism has the biggest pressure angle which is 29°<50°, so the performance of the gear-double sliders mechanism is good.

 doi:10.1088/1742-6596/916/1/012008

Figure 17. The flapping trajectory of double "eight" in Adams

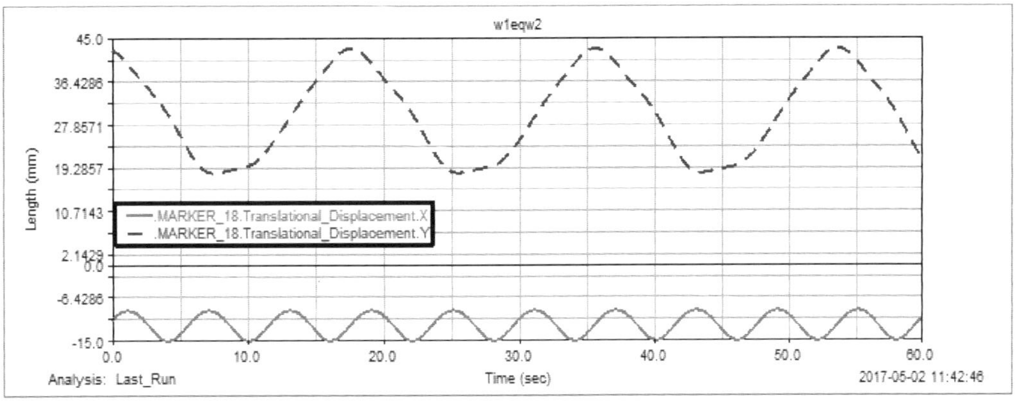

Figure 18. The displacement graph of the output terminal

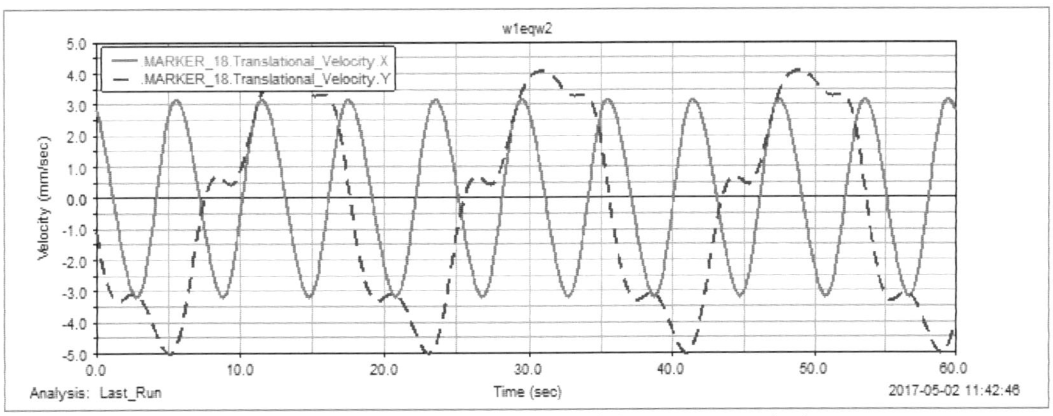

Figure 19. The velocity graph of the output terminal

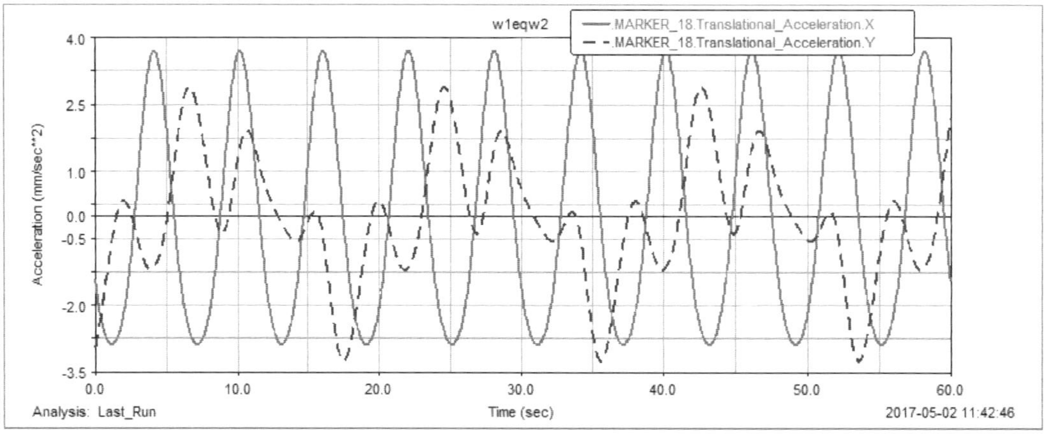

Figure 20. The acceleration graph of the output terminal

Through many debugging, there is an ideal double "eight" trajectory can be found when the initial conditions are ω1=3ω2, α=70°and β=90°, just like the figure 17 shows. Compared with the double "eight" trajectory of plane five bars mechanism, its trajectory is more regular and smooth. Hence, it has the better performance. Figure 18 to 20 show the dynamical parameters of this drive mechanism. It can be seen that the curves are smooth and don't have sharp variation, so it confirm the mechanism is feasible and meets the requirements.

Furthermore, this mechanism also can achieve other flapping trajectories when the rotating speed ratio is changed. When ω1=ω2, the flapping trajectory of class oval can be got as the following figure 21. When ω1=2ω2, the flapping trajectory of "eight" can be got as the following figure 22. After dynamic analysis by Adams, this drive mechanism's performance meets the requirements when the flapping trajectory is class oval or "eight".

Fig.21 The flapping trajectory of class oval in Adams

Fig.22 The flapping trajectory of "eight" in Adams

4. Conclusion

In this paper, the optimization and analysis about three flapping-wing air vehicles' drive mechanisms which are based on the flapping trajectory has been carried out. Different drive mechanisms with their unique characteristics can be used to output different flapping trajectories. The spatial double crank-rocker mechanism can output the flapping trajectory of arc. It has the advantages of small volume, high flexibility and efficient space utilization. Hence, it is suitable for the flapping-wing air vehicles which need the high flapping frequency. The plane five-bar mechanism can output the flapping trajectory of "eight", double "eight" and class oval. It has the advantages of small friction. However, the complicated parameter setting and unsmooth flapping trajectory of double "eight" are its limits. As for the gear-double slider mechanism, it can output the flapping trajectory of "eight", double "eight" and class oval, too. Compared with the plane five-bar mechanism, its flapping trajectory of double "eight" is smoother and parameter setting is more convenient. Hence, this helpful drive mechanism can be used to provide the varied flapping trajectories for the MAV.

Acknowledgments
This study was supported by the National Youth Natural Science Foundation of China (Grant No. 51405279). The financial support is gratefully acknowledged.

References

[1] R. J. Fand and F. J. Wu, The development of the MVA, J. Mechanical & Electrical Technology. 5 (2013) 158-160.
[2] R. C. Michelson, Entomopter and Method for Using Same, U. S. Patent 6,082,671. (2000)
[3] A. S. Kinkade, Ornithopter, U. S. Patent 5,899,408. (2002)
[4] R. Madangopal, Z. A. Khan and S. K. Agrawal, Energerics Based Design of Small Flapping Wing Air Vehicles. Proceeding of the 2004 IEEE, International Conference on Robotics and Automation, New Orieans, 2004, pp. 2367-2372.
[5] S. X. Wang and G. P. Chen, Study on insect-based flapping-wing system driven by piezoelectric bimorph, J. Optics and Precision Engineering, 14 (2006) 617-622.
[6] B. L. Zhu, H. S. Ang and L. Guo, Design and Analysis of New 3D Insect-Like Flapping-Wing Mechanism, J. Nanjing University of Aeronautics & Astronautics, 39 (2007) 457-460.
[7] S. Dalton, The miracle of flight, McGraw –Hill Book Company, New York, 2004
[8] S. Sudo, K. Tsuyuki, and K. Kanno, Wing characteristics and flapping behavior of flying insects. J, Experimental Mechanics. 45 (2005) 550-555.
[9] S. M. Yan and G. G. Jin, A study of average transmission angle of planar linkage and its application, J. Machinery. 35 (2008) 19-22.
[10] Q. H. Sun, E. G. Dong, L. Zhang and S. R. Gao, Parameters analysis and optimal design of four-bar steering mechanism, J. Liaoning Technical University, 26 (2007) 278-280.

IOP Conf. Series: Journal of Physics: Conf. Series **916** (2017) 012009 doi:10.1088/1742-6596/916/1/012009

A study on the aerodynamic characteristics of airfoil in the flapping adjustment stage during forward flight

Pan Luo, Xingwei Zhang *, Panpan Huang, Lingwang Xie

Shantou University, College of Engineering, Shantou 515063, China

*Corresponding author e-mail: zhangxw@stu.edu.cn

Abstract. The aim of this study is to investigate the aerodynamic characteristics of a flapping airfoil in the adjustment stage between two specific flight patterns during the forward flight. Four flapping movement models in adjustment stage are firstly established by using the multi-objective optimization algorithm. Then, a numerical experiment is carried out by using finite volume method to solve the two-dimensional time-dependent incompressible Navier-Stokes equations. The attack angles are selected from -5°to 7.5° with an increase of 2.5°. The results are systematically analyzed and special attention is paid to the corresponding changes of aerodynamic forces, vortex shedding mechanism in the wake structure and thrust efficiency. Present results show that output aerodynamic performance of flapping airfoil can be improved by the increasement of amplitude and frequency in the flapping adjustment stage, which further validates and complements previous studies. Moreover, it is also show that the manner using multi-objective optimization algorithm to generate a movement model in adjustment stage, to connect other two specific plunging motions, is a feasible and effective method. Current study is dedicated to providing some helpful references for the design and control of artificial flapping wing air vehicles.

1. Introduction

Birds and insects fly by flapping their wings in the air to get enough aerodynamic force to support and propel them. Even though the external aerodynamic environment has a sudden changed, they can still keep the flight state steady by varying the flapping motion of wing. It has been unanimously recognized that disclosing these flight mechanisms of flying creatures is be very helpful for the development of artificial Micro Air Vehicle (MAV). Numerous experimental studies on flapping wings of birds or insects in a forward flight have been conducted [1-4], pertaining to aerodynamic forces, control-abilities and propulsive characteristics of flapping wings. Schouveiler et al (2005) experimentally investigated the influences of Strouhal number and the maximum attack angle on propulsive performance of a harmonically plunging airfoil [5]. Many numerical investigations on oscillating airfoil have been also carried out. Tuncer and Platzer (1996) investigated the mechanisms of thrust generation for a single flapping airfoil by employing a multiblock Navier-Strokes solver, where the unsteady flow filed around the flapping airfoil was analyzed [6]. Lewin and Haj-Hariri (2003) simulated the flow characteristics and power coefficients of an elliptical airfoil heaving sinusoidally over a range of frequencies and amplitudes [7]. However, most previous numerical studies on flapping flight are concerned about the aerodynamic performance of a certain specific stable flapping motion during forward flight. While the investigation on the transition stage between two stable flapping motion states are not so much. In actual flapping flight, the moment of wings needs to be adjusted in terms of the output needs and external flight environment. The aerodynamic characteristics of the wings during the adjust stage have a significant effect of on stability of the flapping flight. Accordingly, this study focus on revealing the aerodynamic performance of plunging airfoil in adjustment stage and the changing of aerodynamic characteristics while the flapping motion changed. Multi-objective optimization algorithm used to optimize the flapping modes of plunging airfoil in adjustment stage. The power coefficients changing at different attack angles with different flapping mode in adjustment stage are also studied in this work. A NACA0014 airfoil is employed and the two-dimensional governing equations of fluid flow is solved in the finite volume method. Though the effect on spanwise flow is not considered in this work, the present results still can be useful to reveal the fundamental aerodynamic characteristics of flapping wing in the adjustment stage during forward flight.

2. Solution Method

2.1. *Problem description*

A two-dimensional NACA0014 airfoil with chord c is employed in this study. The plunging airfoil with attack angles range from -5°to 7.5°, is located in a uniform flow field. Fig.1 gives a schematic configuration of the flapping airfoil in the forward flight. h0 is dimensionless amplitude measured from the mean position. U_∞ is the velocity of the uniform flow.

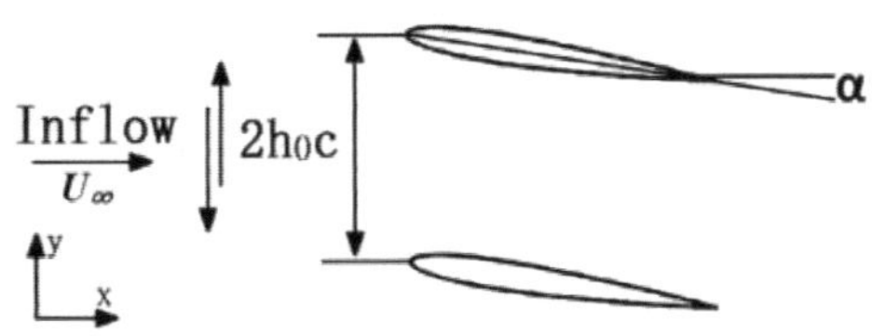

Fig.1 Schematic configuration of a rigid airfoil undergoing a plunging motion in a uniform flow.

2.2. *Method of generating motion in adjustment stage*

Before and after the adjustment stage, the airfoil plunge in the specific stable motions, and are all described by the sinusoidal equation $h(t)=A_i\sin(2pf_it)$, i=1, 2. A_1, A_2 are the amplitude of motion before and after adjustment stage, respectively. f_1, f_2 are the frequency of motion before and after adjustment stage, respectively. $T_1=1/f_1$, $T_2=1/f_2$ are the frequency of motion before and after adjustment stage, respectively. In current study, the transition time in the adjustment stage is set as $0.5T_2$. Figure 2 give a Kinematics specification of three stages.

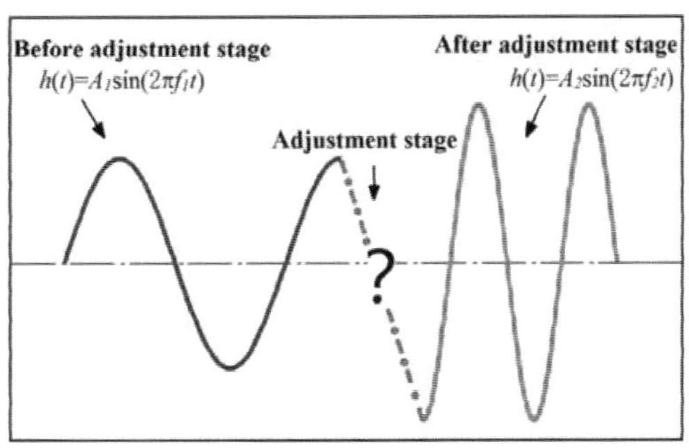

Fig.2 Kinematics specification of three stages.

In this study, to ensure the three stages can be effective continuous and the sudden step change do not exist, the Fourier equation is used to describe the movement mode of airfoil in the adjustment stage. The Fourier equation is as follows.

$$h(t) = a_0 + \sum_{i=1}^{n} (a_i \cos(i\omega t) + b_i \sin(i\omega t)) \tag{1}$$

Where a_0, a_i and b_i are coefficients of Fourier equation.

In order to establish the Fourier plunging motion of airfoil in the adjustment stage, NAGA-II which is one of solid effective multi-objective optimization algorithms [8-9] is selected to optimize the trajectory.

Two objective functions have been established in this work. One is use to keep both displacement and velocity of the airfoil continue. In adjustment stage, the motion of airfoil should be unidirectional and no stationary. The continuous of airfoil's displacement can be described as a continuous curve between start point and ending point of adjustment stage in the Cartesian coordinate system, and the motion's unidirectional can be measured by the track length of the curve. So function $F_1(a_0,a_i,b_i)$ is used to measure the track length of the airfoil, and defined as:

$$F_1(a_0,a_i,b_i) = \int_t^{t+T/2} \sqrt{h'(t)^2+1}dt \qquad (i= 1,2,3,...,n) \tag{2}$$

In order to simplify calculation, using function min $f_1(a_0,a_i,b_i)$ to take place of function $F_1(a_0,a_i,b_i)$ as a objective function and min $f_1(a_0,a_i,b_i)$ can be written as :

$$\min f_1(a_0,a_i,b_i) = \int_t^{t+T/2} (h'(t)^2+1)dt \qquad (i= 1,2,3,...,n) \tag{3}$$

The other objective function derives from the control of the amplitude of the acceleration in the adjustment stage. For the artificial flapping wing vehicles, the flap of wings is driven motor. The acceleration of the wings is related to the output of the motor. A huge sudden change on the output of motor is detrimental, and sometimes can't be achieved. So a constraint function is used to limit the magnitude of the acceleration in the adjust stage.

The acceleration equation is derived from above displacement equation (1), and is as follows:

$$a(t) = \sum_{i=1}^{n} (-a_i i^2\omega^2 \sin(i\omega t) - b_i i^2\omega^2 \cos(i\omega t)) \tag{4}$$

Another,

$$a\sin(x) + b\cos(x) = \sqrt{a^2+b^2}\sin(x+\varphi) \tag{5}$$

And

$$\sin\varphi = \frac{b}{\sqrt{a^2+b^2}}, \cos\varphi = \frac{a}{\sqrt{a^2+b^2}} \tag{6}$$

Then,

$$a(t) = -\omega^2 \sum_{i=1}^{n} (i^2\sqrt{a_i^2+b_i^2}\sin(i\omega t+\varphi_i)) \tag{7}$$

Where

$$\sin\varphi_i = \frac{b_i}{\sqrt{a_i^2 + b_i^2}},\quad \cos\varphi_i = \frac{a_i}{\sqrt{a_i^2 + b_i^2}} \tag{8}$$

The constraint condition is

$$-\omega^2 \sum_{i=1}^{n} (i^2 \sqrt{a_i^2 + b_i^2} \sin(i\omega t + \varphi_i)) \le a(t) \le \omega^2 \sum_{i=1}^{n} (i^2 \sqrt{a_i^2 + b_i^2} \sin(i\omega t + \varphi_i)) \tag{9}$$

Present work established another function $F_2(a_i, b_i)$ to measure the maximum acceleration as the another objective function to keep the acceleration amplitude within reasonable bounds in adjustment stage. And $F_2(a_i, b_i)$ defined as:

$$F_2(a_i, b_i) = \omega^2 \sum_{i=1}^{n} (i^2 \sqrt{a_i^2 + b_i^2}) \tag{10}$$

So present work established function min $f_2(a_i, b_i)$ as the other objective function to control the amplitude of the acceleration in the adjustment stage, and objective function min $f_2(a_i, b_i)$ defined as:

$$\min f_2(a_i, b_i) = \omega^2 \sum_{i=1}^{n} (i^2 \sqrt{a_i^2 + b_i^2}) \tag{11}$$

2.3. Numerical equations and solver

The governing equations for unsteady, incompressible and viscous fluid flow can be written in the following form

$$\nabla \cdot U = 0 \tag{12}$$

$$\frac{\partial U}{\partial t} + \nabla \cdot F = 0 \tag{13}$$

With the two-dimensional Cartesian components, F_{Kj} is given by

$$F_{kj} = (U_j - \hat{U}_j)U_k + \delta_{jk}P - \frac{1}{\mathrm{Re}}\frac{\partial U_k}{\partial x_j} + \overline{u_j u_k} \tag{14}$$

Where $U_j U_k$ is the Cartesian velocity component, $\hat{U}_j$ is the mesh velocity, P is the pressure and d_{jk} is the Kronecker Delta number. The Reynolds stress tensor $\overline{u_j u_k}$ is given by:

$$\overline{u_j u_k} = \frac{2}{3} k \delta_{jk} - v_t \left(\frac{\partial U_j}{\partial x_k} + \frac{\partial U_k}{\partial x_j}\right) \tag{15}$$

IOP Conf. Series: Journal of Physics: Conf. Series **916** (2017) 012009 doi:10.1088/1742-6596/916/1/012009

Where v_t is the eddy viscosity determined by the turbulence model and k is the turbulent kinetic energy. The above time-dependent Naver-stokes equation are solved by using the finite volume method [10,11] and the Shear Stress Transport (SST) k-omega. Mode in the present work. One can refer to Menter (1993) for a detailed description of the SST model [12]. The Pressure Implicit with Splitting of Operators (PISO) algorithm [13, 14] is employed to deal with the coupling between the pressure and velocity. The third-order accurate QUICK scheme [15] is used to discrete the convective terms. Fig.3 shows a schematic diagram of the computational domain round the airfoil. The whole computational domain is 18c (L_x) long and 10c (L_y) wide. A conformal-hybrid grids system that consists of quadrangular meshes in an internal domain around the airfoil and triangular meshes in the outer field is shown in Fig.4. The complete inner mesh moves according to the wing kinematic and a dynamic deformable remesh method in accordance with Geometric conservation law method [16] is used in the outer field. A serial tests for grid independent and time step independent were conducted before actual calculations were carried out. Based on these tests, element number about 4.2×10^4 and time step 0.0025T are selected for all the cases.

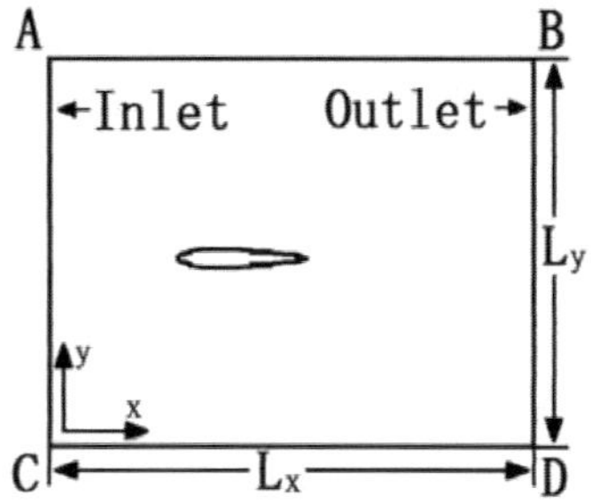

Fig.3 The boundaries around the computational domain

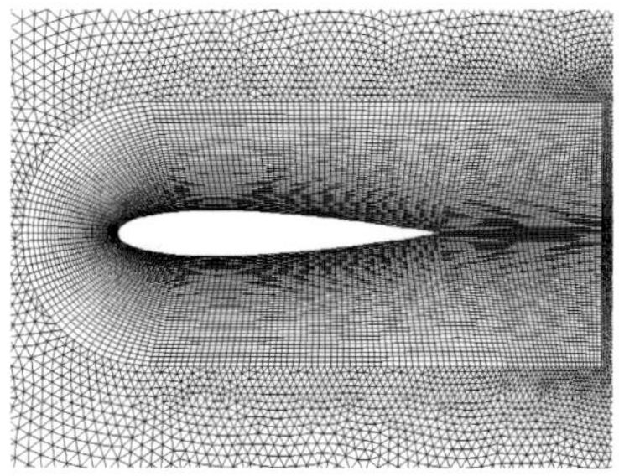

Fig.4 Computational grid around the plunging airfoil.

The incoming flow runs from the left to right. At the inlet of the computational domain, a uniform profile is prescribed:

$$U=U\infty \tag{16}$$

The outlet of the computational domain is far away from the airfoil, and a boundary condition of fully developed flow is given by：

$$\frac{\partial U}{\partial x}=0 \tag{17}$$

The no-slip boundary condition is specified at the airfoil surface and a symmetry boundary condition [17] is applied at the lateral faces of the computational domain. The influence of this symmetry condition has been investigated and found to be sufficiently adaptive. The simulation starts assuming the fluids is initially at rest.

The Reynolds number, which characterizes the effects of viscous, is defined as:

$$Re= \quad U\infty c/ \tag{18}$$

Where, is the density of fluid and is the dynamic viscosity.

The Strouhal number, which characterizes the periodicity in a flow field, is defined as:

$$St=2h0cf/U=kh0/ \tag{19}$$

Where f is the frequency of plunging motion. $k=$ c/U_∞ is the reduced frequency. Current results use the quantity kh_0 to categorize.

The lift and drag coefficients are computed from:

$$C_l = \frac{2F_y}{\rho c s U_\infty^2},\ C_d = \frac{2F_x}{\rho c s U_\infty^2} \tag{20}$$

Where s is the surface area per unit span-wise length of the airfoil and F_y and F_x are instantaneous lift and drag forces, respectively. These forces are calculated from the integration of the pressure and the viscous force on the surface of the airfoil.

The validation of the numerical algorithm for the flapping cases has been conducted in our previous work [18,19] and the results have been compared with related experimental and numerical results published in the literature. That numerical code is reliable for qualitatively analyzing flow characteristics around static and flapping airfoils.

2.4. Other important parameters

The averaged thrust and lift can be evaluated as:

$$\bar{F}_{Th} = -\frac{1}{T}\int_0^{T/2} F_x(t)dt \tag{21}$$

$$\bar{F}_L = \frac{1}{T}\int_0^{T/2} F_y(t)dt \tag{22}$$

Where F_x and F_y and represent the instantaneous force components in x and y directions respectively. Then, the mean vertical force coefficient C_v is defined as:

$$C_v = \frac{2\bar{F}_L}{\rho c s U_\infty^2} \tag{23}$$

The mean thrust coefficient C_{Thm} is defined as:

$$C_{Thm} = \frac{2\bar{F}_{Th}}{(\rho U_\infty^2 c s)} = \frac{1}{T}\int_0^{T/2} (-C_d)dt \tag{24}$$

The averaged consumption power $\bar{p}_i$ and power coefficient δ are given by

$$\bar{P}_i = \frac{1}{T}\int_0^{T/2} F_y(t)V_0(t)d(t),\ \delta = \frac{2\bar{P}_i}{\rho U_\infty^3 c} \tag{25}$$

Where $F_y(t)$ represents the instantaneous lift in the y direction and $V_0(t)$ is the instantaneous heaving velocity of the wing is: $V_0(t) = \partial h/\partial t$.

The averaged out power and power coefficient from thrust force are written by

$$\bar{P}_0 = -\frac{1}{T}\int_0^{T/2} F_x(t) U_\infty d(t),\quad \delta_T = 2\bar{P}_0 \big/ \rho U_\infty^3 c \tag{26}$$

Where $F_x(t)$ is the instantaneous lift in the x direction.
The thrust efficiency is defined by

$$\eta = \bar{P}_0 \big/ \bar{P}_i = \delta_T / \delta \tag{27}$$

3. Results And Disscussion

3.1. Optimization results

In this study, the adjustment stage is classified as four cases, which are decrease flapping amplitude, increase flapping amplitude, decrease flapping frequency and increase flapping frequency. In each case, the plunging motion of the airfoil is first established by using the NAGA-II optimization. After series of multi-objective optimization, the results four cases are respectively depicted in Fig.5 (a) to (d). It should be noted that the every point in Figure 5 represent an array of kinematic equation of the plunging airfoil in the adjustment stage. Then, we selected reasonable kinematics form the optimization results as shown in Fig. 6, and used them in next aerodynamic investigation.

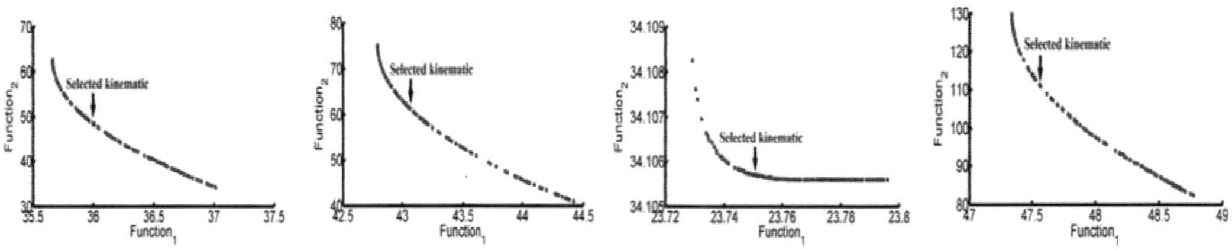

(a) Decrease Flapping Amplitude (b) Increase Flapping Amplitude (c) Decrease Flapping frequency (d) Increase Flapping frequency

Fig.5 The optimized Result of the plunging motion in four cases of adjustment stage.

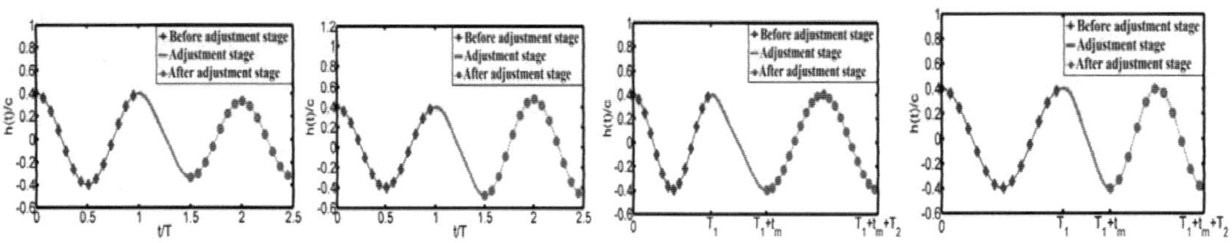

(a) Decrease Flapping Amplitude (b) Increase Flapping Amplitude (c) Decrease Flapping frequency (d) Increase Flapping frequency

Fig.6 The kinematic displacement equations of three stage.

3.2. Comparison on aerodynamics characteristics

In this study, the Reynolds number is kept at 10^4 and kh_0 in the adjustment stage ranges from 0.67 to 1.04, which results in the Strouhal number range from 0.213 to 0.331. To investigate the aerodynamic performance of plunging airfoil in adjustment stage, the drag coefficient and lift coefficient of four cases are illustrated in Fig. 7 to Fig. 10. The negative drag coefficients indicate that the output thrust is positive. First, it can be seen that the instantaneous time signal of force coefficients in the adjustment stage are remarkably different from the other stages. The variation on amplitude and frequency of plunging airfoil will affect the maximum effective output load in the forward flight. Either flapping amplitude decreases or flapping frequency reduction, can both reduce the peak values of both lift and drag coefficients. When the flapping frequency increased, the maximum lift coefficient is also increased. Moreover, the first peak of time signal of lift coefficient is remarkably enhanced. It should be also noted that the output thrust of the plunging wing in the adjustment stage does not necessarily increase as the frequency or amplitude increased as shown in Figure 8 (b) and Figure 10 (b). While either the frequency or amplitude is decreased, the output instantaneous thrusts reduce, as shown in Figure 7 (a) and Figure 9 (a).

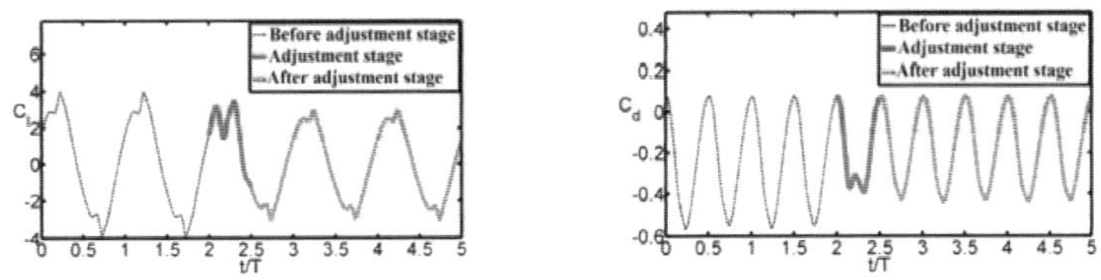

Fig.7 lift coefficient and drag coefficient for decreasing amplitude when the attach angle is from 0°

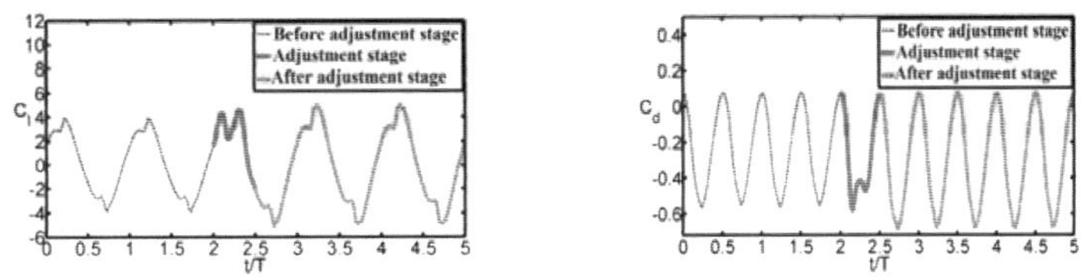

Fig.8 lift coefficient and drag coefficient for increasing amplitude when the attach angle is from 0°

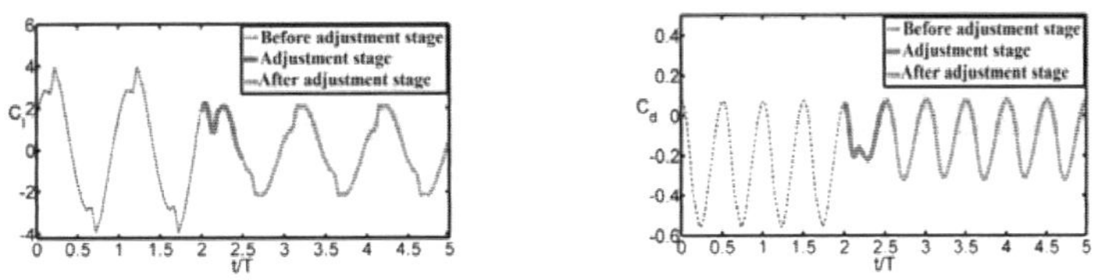

Fig.9 lift coefficient and drag coefficient for decreasing frequency when the attach angle is from 0°

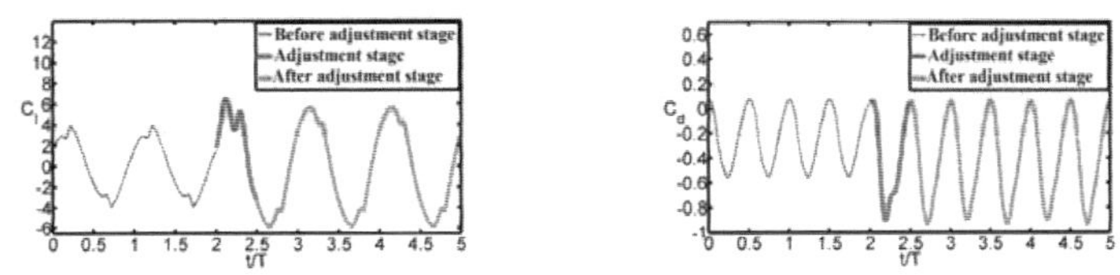

Fig.10 lift coefficient and drag coefficient for increasing frequency when the attach angle is from 0°

To further investigated the aerodynamic characteristics in the adjustment stage, the averaged values of thrust coefficient and vertical force coefficient at different attack angles is compared in Figure 11. As shown in Figure 11, the attack angles are range from -5°to 7.5°with an increment of 2.5°for each cases. It is seen that the averaged values of thrust coefficient decreases as the attack angle increased. The averaged vertical force coefficient monotonically increases with the attack angle.

IOP Conf. Series: Journal of Physics: Conf. Series **916** (2017) 012009 doi:10.1088/1742-6596/916/1/012009

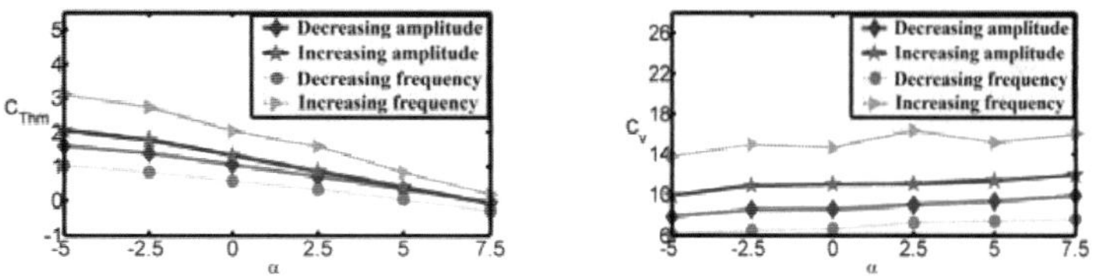

(a) Averaged thrust force coefficients (b) Averaged vertical force coefficients

Fig.11 Variations of averaged thrust force coefficients and averaged vertical force coefficients when the attach angle is from -5°to 7.5°.

The vorticity contours around the plunging airfoil for four cases are shown in Fig.12, respectively. In order to facilitate the quantified comparison on the wake structure, the vorticity contour of a complete period when the plunging motion is sinusoidal equation before adjustment stage is shown in Figure 12(a). The beginning of the adjustment stage is at t=0 instant. After t=0.5T instant, the adjustment stage is over and next new stage starts, as shown figure 12 (b) to (d). The vortex shedding plays an important role in the variation of aerodynamic forces. It is seen that the vortex shedding at the leading edge and trailing edge of the plunging airfoil are substantially influenced in the adjustment process. Especially, the shape of trailing edge vortexes is very different with the one of sinusoidal plunging motion, as shown in figure 12 (b) to (d). The trailing edge vortexes at the three adjustment cases are a little smaller and weaker than that of no adjustment case (e.g. Fig 10 at time instant of 0.5T marked with red circle). This variation corresponds to the decrease in the output thrust as shown in above figure 7 to figure 9.

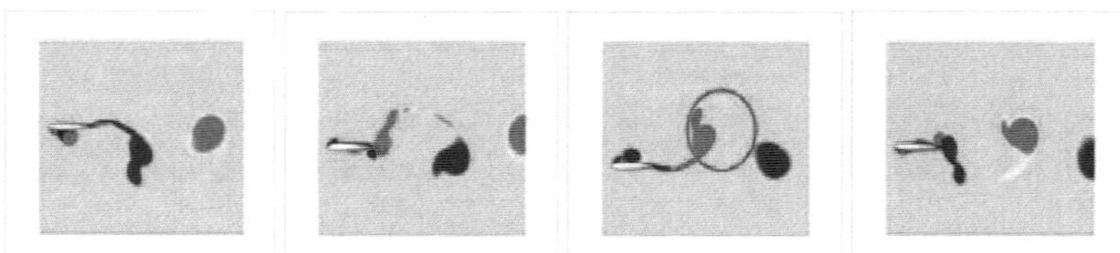

t=0 t=0.25T t=0.5T t=0.75T

(a) Before adjustment stage in the sinusoidal plunging motion

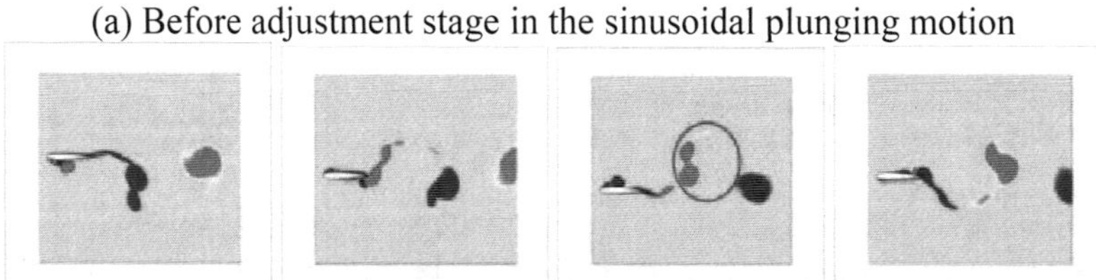

t=0 t=0.25T t=0.5T t=0.75T

(b) Decreasing amplitude

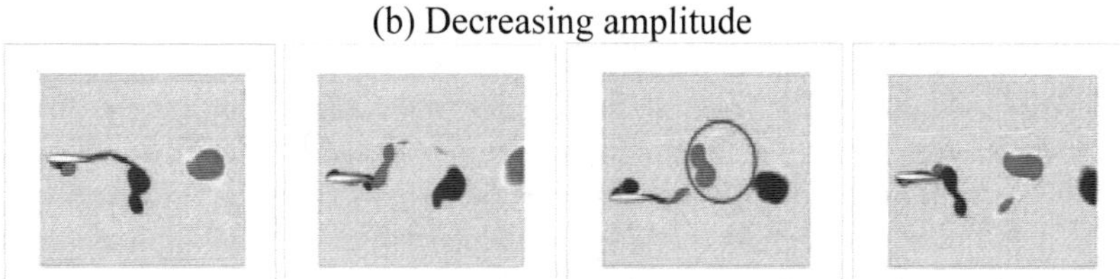

t=0 t=0.25T t=0.5T t=0.75T

(c) Increasing amplitude

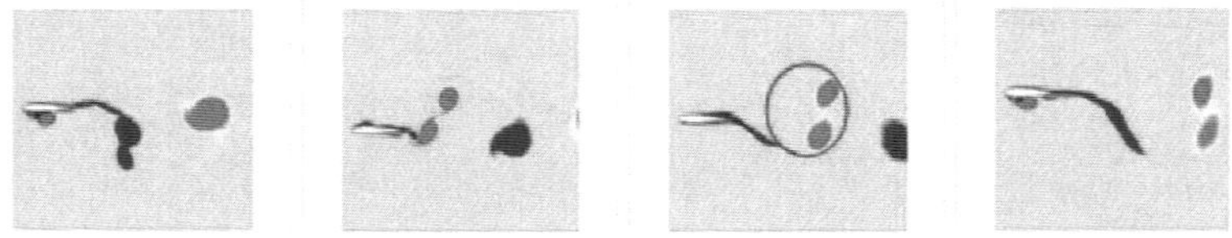

t=0 t=0.25T t=0.5T t=0.75T

(d) Decreasing frequency

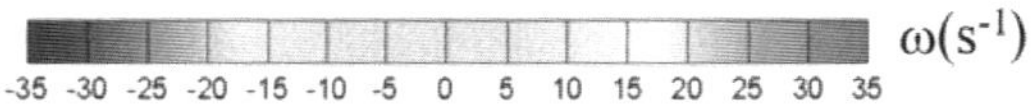

Fig.12 Vorticity contours of four cases in the adjustment stage, when the attach angle is from 0°. (blue: clockwise, corresponding to negative vorticity; red: counterclockwise, corresponding to positive vorticity).

To further investigate the effective output load and thrust efficiency of the plunging airfoil, the averaged consumption power coefficient, the ratio of mean lift to averaged consumption power, the thrust efficiency are calculated. The payload provided by the unit input power for four adjustment cases is characterized by $C_v/$ as shown in Figure 13. It can be observed that the values of $C_v/$ during the four adjustment cases are increasing slightly with the attack angles. Further, it is also found that the decreasing frequency case is easier to get higher output lift force with a lower input power and keep efficient flight than other three adjustment cases. In addition, the thrust efficiency at different kh_0 at same attack angle, is also depicted in Fig.14. As shown in Figure 14, the highest thrust efficiency appears at =-5°and the corresponding Strouhal number is 0.26.

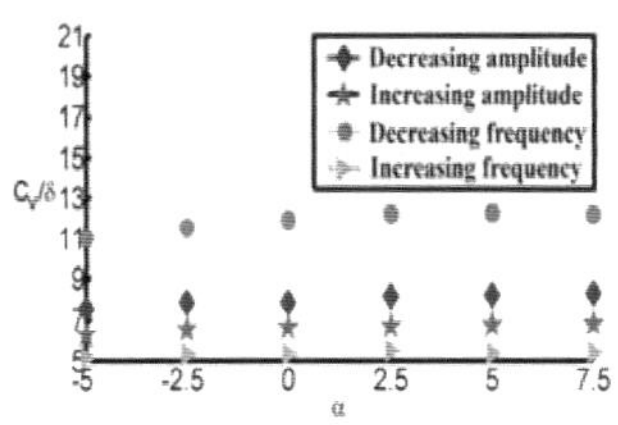

Fig.13 Variations of C_v/d with a for different adjustment processes.

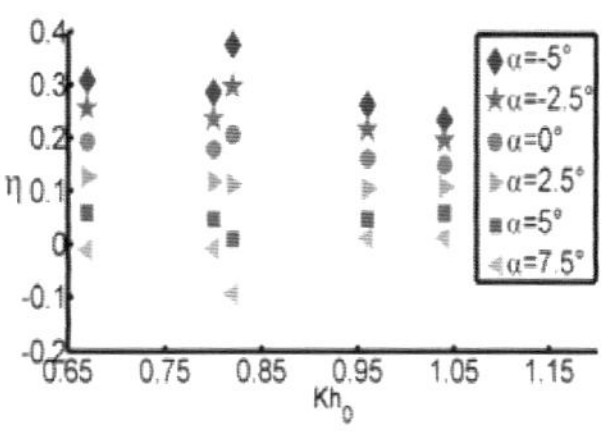

Fig.14 Variations of h with a at different kh_0.

4. Conclusion

A NAGA-II multi-objective optimization algorithm is used to establish the flapping movement models in adjustment stage. Four adjustment cases are systematically investigated. Then, a comprehensive numerical study on the aerodynamic performance of a plunging NACA0014 airfoil at Reynolds number 10^4 has been carried out by using finite volume method. The aerodynamic forces and corresponding wake structures of the airfoil are detail analyzed and the thrust efficiencies are presented. Current results show the increasement of plunging amplitude or frequency can enhance the output lift force in the adjustment stage during forward flight. The decrease of plunging amplitude or frequency will result in less lift force and less thrust force. However, the output thrust does not necessarily increase as the plunging amplitude increases in the adjustment stage. This study also found that the decreasing frequency adjustment case is much easier to get higher output lift force with a lower input power than others three adjustment cases. This interesting observation is very helpful for the further study on the aerodynamic performance of take-off stage and landing flight, and need be further studied. This study also indicates that t the manner using multi-objective optimization algorithm to generate a movement model in adjustment stage, to connect other two specific plunging motions, is a feasible and effective method. These findings can be used to guide further study the aerodynamic analysis on forward, take-off stage and landing flapping flight, and also provide some helpful information for improvement of control-abilities of artificial flapping wing air vehicles.

Acknowledgments
This study was supported by the National Youth Natural Science Foundation of China (Grant No. 51405279). The financial support is gratefully acknowledged.

References
[1] Wei-Fongh,T., Biology and physics of locust flight. IV. Notes on sensory mechanisms in locust flight. Bio.Sci.,239(1956)553-584.
[2] Zarnack,W., The effect of for-wing depressor activity on movement during locust flight. Bio.Cybernet.,59(1988)55-70.
[3] Tobalske,B. and K.Dial, Flight kinematics of black-billed magpies and pigeons over a wide range of speeds. J,Exp.Bio.,199(1996)263-280.
[4] Norberg,U.M.L and Y.Winter., Wing beat kinematics of a nectar-feeding bat, Glossophaga soricina, flying at different speeds and Strouhal numbers. J.Exp.Biol.,209(2006)3887-3897
[5] Schouveiler, L., F.S.Hover and M.S.Triantafyllou, Performance of flapping foil propulsion. Fluids and Struc.,20(2005)949-957.
[6] Tuncer, I.H and M.F.Platzer, Thrust generation due to airfoil flapping. AIAA J.,34(1996)324-331.
[7] Lewin, G.C. and H.Haj-Hariri, Modelling thrust generation of a two-dimensional heaving airfoil in a viscous flow. J.Fluid Mesh.,492(2003)339-362.
[8] Kalyanmoy Deb, Amrit Pratap, Sameer Agarwal, and T.Meyarivan. A Fast and Elitist Multiobjective Genetic Algorithm: NAGA-II, VOL.6 NO.2,2002,pp.182-197.
[9] K. Deb, Multiobjective Optimization Using Evolutionary Algorithms. Chichester, UK.Wiley,2001.
[10] Rhie, C.M. and W.L.Chow, Numerical study of the turbulent flow past airfoil with trailing edge separation. AIAA j.,21(1983)1525-1532.
[11] Versteeg, H.K. and W.Malalasekera, An Introduction to Computational Fluid Dynamics: The Finite Volume Method. Wiley, New York.
[12] Menter,F.R., Zonal two-equation k-w turbulence models for aerodynamic flows. AIAA 24th Fluid Dynamics Conference, Orlando, FL,USA.
[13] Issa, R.I., Solution of implicitly discretized fluid flow equations by operator-splitting. J. Comput. Phys.,62(1985)40-65.
[14] Issa, R.I., B.Ahamadi-Befrui,K.R.Beshay and A.D.Gosman, Solution of the implicitly discretized reading flow equation operator-splitting.J.Comput.Phys.,93(1991)388-410.
[15] Leonard,B.P., A stable and accurate convective modelling procedure based on quadratic upstream interpolation. Comput Meth Applied Mechan. Eng.,19(1979)59-98.
[16] Lesoinne, M and C. Farhat,. Geometric conservation laws for flow problems with moving boundaries and deformable meshes and their impact on aerodynamic computations. Comp Meth Applied Mechan Eng.,134(1996)71-90.
[17] Elhadi,E.E.,L.Xiaosong and W.Keqi, Numerical simulation of 2D separated flows using two different turbulence models. J. Applied Sci.,2(2002)1057-1062.
[18] Xingwei Zhang, Chaoying Zhou, et al. Numerical study on effect of leading-edge tubercles Aircraft Engineering and Aerospace Technology, Vol.85 No.4(2013)247-257.
[19] Xingwei Zhang, Chaoying Zhou, et al. Numerical Investigation on the Aerodynamic Characteristics of a Forward Flight Flapping Airfoil with Nonsymmetrical Plunging Motion.pp748-758,2011.

IOP Conf. Series: Journal of Physics: Conf. Series **916** (2017) 012010 doi:10.1088/1742-6596/916/1/012010

CFD simulation of rotor aerodynamic performance when using additional surface structure array

Bing Wang [1, 2], **Deyi Kong** [1,*]

[1]State Key Laboratory of Transducer Technology, Institute of Intelligent Machines, Chinese Academy of Sciences, Hefei 230031, China
[2]University of Science and Technology of China, Hefei 230026, China

*Corresponding author e-mail: kongdy@iim.ac.cn

Abstract. The present work analyses the aerodynamic performance of the rotor with additional surface structure array in an attempt to maximize its performance in hover flight. The unstructured grids and the Reynolds Average Navier-Stokes equations were used to calculate the performance of the prototype rotor and the rotor with additional surface structure array in the air. The computational fluid dynamics software FLUENT was used to simulate the thrust of the rotors. The results of the calculations are in reasonable agreement with experimental data, which shows that the calculation model used in this work is useful in simulating the performance of the rotor with additional surface structure array. With this theoretical calculation model, the thrusts of the rotors with arrays of surface structure in three different shapes were calculated. According to the simulation results and the experimental data, the rotor with triangle surface structure array has better aerodynamic performance than the other rotors. In contrast with the prototype rotor, the thrust of the rotor with triangle surface structure array increases by 5.2% at the operating rotating speed of 3000r/min, and the additional triangle surface structure array has almost no influence on the efficiency of the rotor.

1. Introduction

In the last few decades, the multi-rotor Micro-aerial vehicles (MAVs) have generated great interest in the rotor aerodynamics. Due to the unique hovering capability and a small rotor diameter without a separate anti-torque system, these kinds of vehicles are especially suited to perform missions considered tough, dirty, or dangerous [1, 2]. Although the development of the rotor aerodynamics of the micro-scale rotorcraft was hobbled by technology insufficiencies through most of the 20th century, the efforts were focused on the computational and experimental study to overcome the problem of poor aerodynamic performance resulted by the low Reynolds (often below 10^5 based on characteristic chord dimensions) for the MAVs, where the viscous effects tend to increase viscous losses and adversely affect lifting surface behavior [3]. As a result, the particular challenge of improving the aerodynamic performance lies in overcoming the various aerodynamic problems associated with low Reynolds number flight, which generally manifests as lower efficiencies and higher power consumption.

It has been studied by many researchers that good aerodynamic performance could be achieved through using surface structures. The riblet technology has been used on aircraft wings, which could provide a benefit to aerodynamic efficiency by reducing skin friction drag. The experiment by L.P.

Chamorro et al. [4] reported a reduction of roughly 6% in a turbine airfoil with riblets. The study by K.A. Seffen [5] shows that generating a surface texture in a shell of an airfoil section model can improve how the shell or structure interacts with the world around it, and the testing of similar specimens in a wind tunnel indicates that switching on the texture, at the correct wind speed, can reduce the drag significantly - as the dimples do on a golf ball. However, the additional surface structure array has never been tried on the rotor blade.

In this work, rotors with additional surface structure array are introduced to optimize the aerodynamic performance of the rotor. The computational fluid dynamics software Fluent is used to simulate the thrust of the prototype rotor and the rotor with additional surface structure array. Based on the accuracy of CFD results, surface structures in different shapes are being tried on the rotor blade. And the optimal rotor model is selected through CFD simulation. The purpose of this study is to propose a new design of the rotor blade used on the MAVs to improve the aerodynamic performance.

2. Numerical method

2.1. Parameters of the models

The prototype rotor in this paper is widely used on the MAVs, which is made of carbon fiber. The parameters of the prototype rotor are shown in Table 1.

Table 1. The parameters of the prototype rotor

Model	Rotor
Number of blades	2
Diameter	17 inch
Pitch	5.8 inch
Average blade chord	38 mm

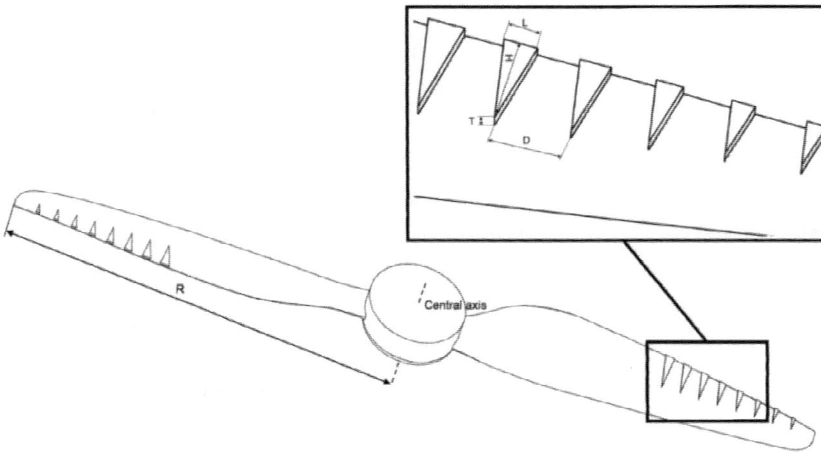

Figure 1. The rotor with additional surface structure array

In this study, the additional surface structure array used on the rotor blade is located on the specific area of lower surface close to the blade trailing edge. The parameters of the rotor with additional surface structure array are shown in Figure 1. The surface structure array on each blade consists of 8 triangle surface structures. The width of the triangle surface structure's edge near the blade trailing edge ranges from 2 mm to 5 mm (2 mm$\leq$ L $\leq$5 mm), and the height from 7 mm to 17 mm(7 mm$\leq$ H $\leq$17 mm). The thickness of all the surface structures is 1mm (T=1 mm). The spacing between the two adjacent surface structures is 10 mm (D=10 mm). The radial distance between the rotor center and the surface structure closest to the central axis is 134 mm (0.62R).

2.2. Calculating method and boundary conditions

Simple method and Reynolds Average Navier-Stokes k- ε viscous model are used to get the solution of RNS equation in this paper. The unstructured grids are used in the meshing of the rotor, for it can use optional shape elements as the control region, which have characteristics of well skin quality and self-adaptation [6, 7]. The mesh distribution of the rotor with additional surface structure array is shown in Figure 2. The Cartesian coordinate system is set up: X coordinate is the same with the blade's radius line; Z coordinate is the direction of outlet flow, which presents consistency with the rotated axis of the rotor; Y coordinate complies with the rule of right-handed.

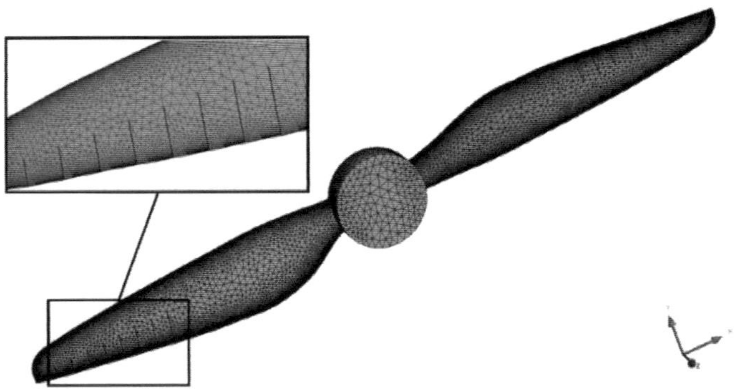

Figure 2. The mesh distribution of the rotor with additional surface structure array

This work takes the cylinder as the control volume. The mesh is the unstructured triangle elements. The positions of inlet and outlet are respectively two and three times of the rotor's diameter from the position of the rotor. The position of the outer boundary is two times of the rotor's diameter from the axis of the hub. The boundary conditions include velocity inlet, pressure outlet, and RNG k-ε viscous model. The thrusts of the rotors are calculated at the rotating speed of 2000r/min, 2200r/min, 2400r/min, 2600r/min, 2800r/min and 3000r/min, respectively.

3. Results and analysis

3.1. Preliminary calculation and experimental validation

This study has a preliminary research on the calculation of the rotor with additional surface structure array, and proposes a new way to improve the aerodynamic performance of rotors used on the MAVs.

With the potential theory, the lifting body moves in the flow field, it must produce a trailing vortex surface [8]. When calculating the aerodynamic performance of the rotor with additional surface structure array, the wake of this rotor is different from that of the prototype rotor, and it's difficult to predict the influence. It's believed that the more the real practical calculating model is used, the more precise pressure distribution of the rotors can be predicted.

This study has firstly calculated the thrusts of the prototype rotor and the rotor with additional triangle surface structure array in open air. In order to validate the results of the numerical simulation, experiments are carried out on the thrust test platform. The triangle surface structures are made through 3D printing, and fixed on the designed location of the rotor blade.

The thrusts of the rotor with triangle surface structure array and the prototype rotor are measured at different rotating speeds. The CFD calculation and the experiment results of the two rotors are shown in Figure 3.

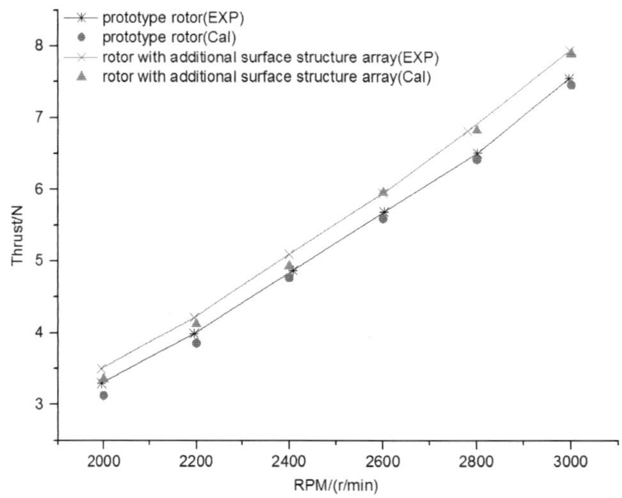

Figure 3. Thrust of the prototype rotor and the rotor with additional surface structure array

Obviously, the experimental data are in reasonable agreement with the numerical simulation results, and the calculation error is confined in the allowable range. It indicates that the thrust of the rotor with triangle surface structure array is greater than that of the prototype rotor at the same rotating speed. That is to say, the rotor with triangle surface structure exhibits better aerodynamic performance. Compared with the prototype rotor, the thrust of the rotor with triangle surface structure array increases by 5.2% at the operating rotating speed of 3000r/min.

According to the experiment results, the correlation between Power consumption and Thrust showed in Figure 4 can prove that additional surface structure array has almost no influence on the efficiency of the rotor. On the other hand, the results in Figure 3 show that the calculation model used in this paper is correct.

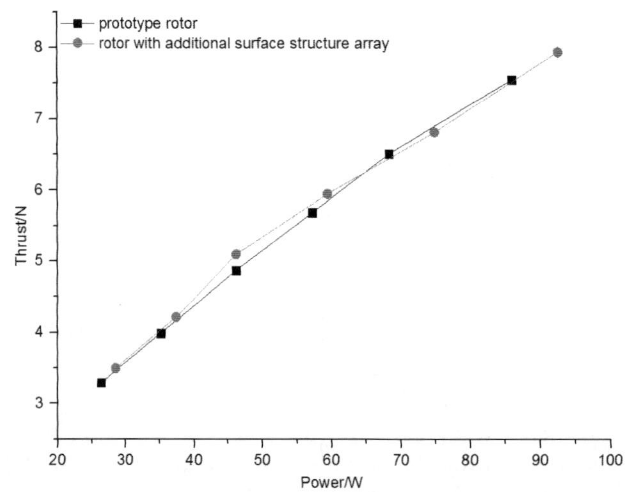

Figure 4. The correlation between power consumption and thrust

3.2. Simulation of rotors with arrays of surface structure in different shapes

Using this theoretical calculation model, the thrust of the rotor with arrays of surface structure in different shapes are calculated in this study. As shown in Figure 5, the shape of surface structure is changed into streamline and tetrahedron, respectively.

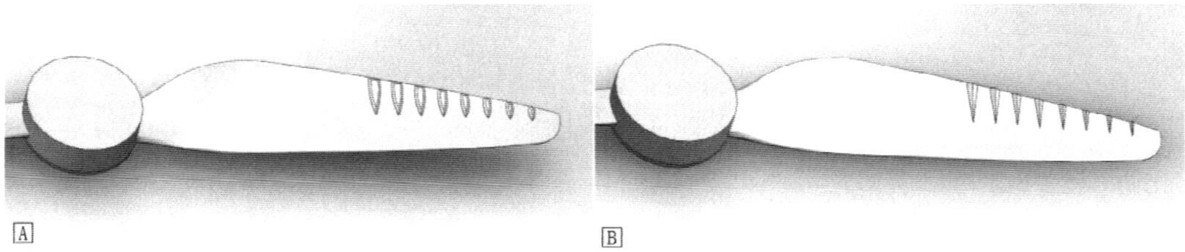

Figure 5. Rotors with arrays of surface structures in two different shapes, (A) the rotor blade with streamline surface structure array, (B) the rotor blade with tetrahedral surface structure array.

The simulation thrust of the rotor with arrays of surface structure in different shapes and the prototype rotor are shown in Figure 6.

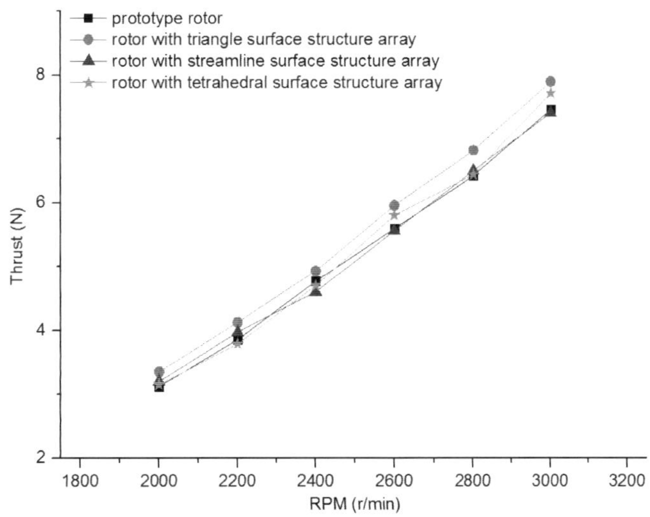

Figure 6. The simulation results of the rotor with surface structures in different shapes and the prototype rotor

According to the results of the simulation, it could be obviously observed that the rotor with triangle surface structure array has the prior on aerodynamic performance than the other rotors do. And the optimal rotor model is the rotor with triangle surface structure array. The pressure distributions of the rotors calculated in this paper are shown in Figure 7. It could be seen that the pressure of the rotor with triangle surface structure array is higher than that of the other rotors. This is in agreement with the experimental results. The pressure distributions show that the surface structures have the capability of improving the aerodynamic performance of rotors, and the rotor with triangle surface structures achieves the best performance among these rotors. The pressure distribution at the location where surface structures are fixed shows that the surface structures could improve the rotors' wake, and the airflow near the trailing edge of the modified rotors is better than that of the prototype rotor. It's supposed that the surface structure array could cause more turbulence near the trailing edge, which has a beneficial effect on pressure distribution. The surface structure array could make the pressure distribution at the trailing edge stronger, and the triangle surface structure array could make the pressure distribution in the most regions of the lower surface stronger. The results of simulations clearly indicate that the design of using additional surface structure array could achieve better aerodynamic performance.

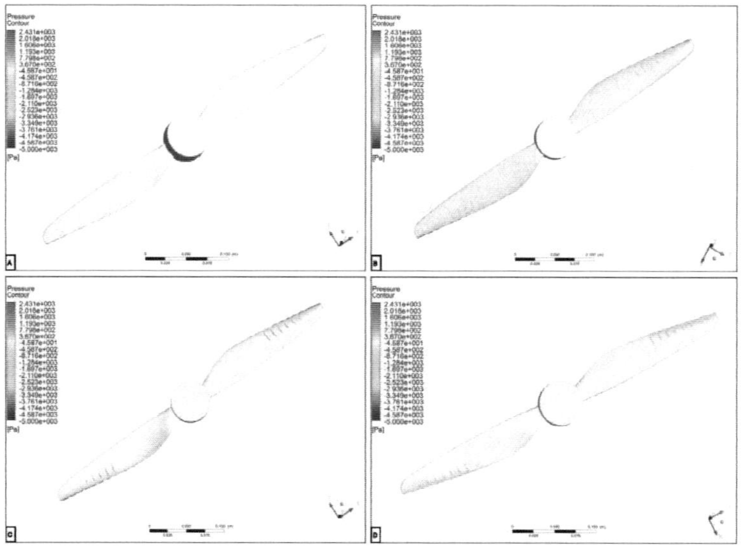

Figure 7. The pressure distribution of the rotors, (A) pressure distribution of the prototype rotor, (B) pressure distribution of the rotor with triangle surface structure array, (C) pressure distribution of the rotor with streamline surface structure array, (D) pressure distribution of the rotor with tetrahedral surface structure array.

4. Conclusion

MAVs are an emerging technology that can provide an inexpensive and expendable platform to a wide array of missions where larger vehicles are impractical to transport or operate. However, the capabilities of current rotor system fall short of various mission requirements due to its limitations arising from aerodynamic issues. This paper provides a new design of rotor with additional surface structure array.

The CFD software Fluent is used to simulate the thrusts of the prototype rotor and the rotor with triangle surface structure. Compared with the experimental data, the calculated results are in good agreement with the experiments and the errors are allowable. The results indicate that the calculation model used in this work is correct. Using this theoretical calculation model, the thrusts of the rotor with arrays of surface structure in different shapes are calculated.

The numerical simulation shows that the rotor with triangle surface structure array has better aerodynamic performance than the prototype rotor and the rotors with surface structures in streamline and tetrahedral shape. And the experimental data prove that the rotor with triangle surface structure array shows better aerodynamic performance than the prototype rotor. In contrast with the prototype rotor, the thrust of the rotor with triangle surface structure array increases by 5.2% at the operating rotating speed of 3000r/min. The additional triangle surface structure array has almost no influence on the efficiency of the rotor, for the power consumption of the rotor with triangle surface structure array is almost the same as that of the prototype rotor at the same thrust.

The method of using additional surface structure array on rotor blade is effective in improving the aerodynamic performance of rotors. In the future, the parameters of surface structure array could be optimized, including the spacing between two adjacent surface structures, the distribution position of the array and the three-dimensional size of surface structures.

Acknowledgements
This work was supported by the National Natural Science Foundation of China (No.11474291).

References

[1] T.E. Lee, Lee, Design and Performance of a Ducted Coaxial Rotor in Hover and Forward Flight, Dissertations & Theses - Gradworks (2010).
[2] B.R. Hein, I. Chopra, Hover performance of a micro air vehicle: rotors at low Reynolds number, J. Journal of the American Helicopter Society 52.3 (2007): 254-262.
[3] M.S. Genç, İ Karasu, H.H. Açıkel, An experimental study on aerodynamics of NACA2415 aerofoil at low Re numbers, J. Experimental Thermal & Fluid Science 39.5(2012):252-264.
[4] L.P. Chamorro, R.E.A. Arndt, F. Sotiropoulos, Drag reduction of large wind turbine blades through riblets: Evaluation of riblet geometry and application strategies, J. Renewable Energy 50.3(2013):1095-1105.
[5] K.A. Seffen, S.V. Stott, Surface Texturing Through Cylinder Buckling, J. Journal of Applied Mechanics 81.6(2014):061001.
[6] F.J. WANG, Calculated fluid mechanics analysis-CFD software principle and application, Tsinghua University Press, Beijing, 2004.
[7] Z.Z. HAN, J. WANG, X.P. LAN, FLUENT fluid engineering emulation calculation example and application, Press of Beijing University of Science and Technology, Beijing, 2004.
[8] P.K. Kundu, I.M. Cohen, D.R. Dowling, Fluid Mechanics, Fifth Edition, Academic Press, 2011.

Combustion and NO_x emissions in deep-air-staging combustion of char in a circulating fluidized bed

Zhiqiang Gong*, Zhentong Wang, Lei Wang and Aixun Du

College of Chemical Engineering, China University of Petroleum (East China), Qingdao, China

*Corresponding author e-mail: gzhiq@163.com

Abstract. Combustion and NO_x emissions in deep-air-staging (with higher level secondary air (SA) injection) combustion of char have been investigated in a CFB test rig. A good fluidized condition and uniform temperature distribution can be achieved with injection of higher level SA. NO_x emission decreases with injection of higher level SA and the reduction effect is more obvious at higher temperature. NO_x emission decreases with combustion temperature increasing for char combustion.

1. Introduction

Nitrogen oxides (NO_x) emitted from coal-fired power plant is a serious environmental issue for years worldwide. Currently, with the increasing deterioration of the environment, NO_x emission standards become much more stringent in many countries, for example, the limit allowed for power plants is 100 mg/m^3 at 6% O_2 from 2014 in China. NO_x reduction techniques from coal combustion process have attracted great attention.

Circulating fluidized bed (CFB) combustion technology has received great interest and made great progress over the past two decades [1-3]. Compared with traditional pulverized coal combustion, CFB has a great advantage in low NO_x emission. However, this advantage is weakened under stringent emission limit. NO_x emission from CFB combustion depends on many operating parameters, among which air stoichiometric ratio (λ) and secondary air (SA) ratio in air staging combustion are of great importance. Besides, some researchers have studied the effect of SA position [4-7] on NO_x emission. Wang [4] studied the impact of SA ratio and position on the NO and N_2O emission and first revealed the axial distribution of NO and N_2O in the furnace. Ersoy [5] have studied the influence of SA injection on the hydrodynamics of CFB test rig and found that a dense turbulent zone and a relatively dilute bed above the injection port were divided by the SA. Koksal [6] have investigated the effects of air staging on wall-to bed heat transfer and pointed that heat transfer depends strongly on the cross-sectional average suspension density for both SA and non-SA operation.

It has been proved that the air staging technique can significantly reduce NO_x emission but variations of SA position have little effect on further NO_x reducing. However, most researchers changed the SA position in the dense area of the riser, the study of effect of SA height along the whole furnace (deep-air-staging) on the hydrodynamics, combustion and NO_x emission in coal combustion is not sufficient. This paper focused on the effect of SA arrangements (heights of SA port from the distributor plate) along the whole furnace on the hydrodynamics, combustion and NO_x emission in char combustion in a CFB test rig. The goal of this paper is prepared for seeking super-low NO_x emission solutions.

2. Experimental section

2.1. *Fuel*

Coal char was made from slow pyrolysis of a sub-bituminous coal at relatively low temperature from 600 °C to 700 °C. After pyrolysis, char was sieved into different particle sizes. Proximate and ultimate analyses of char are listed in Table 1. Compared with the parent coal, the volatile content of char decreases significantly and the fixed carbon and ash contents increase obviously. The diameters of char are 0-4 mm with 50% cut mean diameter (d_{50}) of 2.90 mm.

Table 1. Proximate and ultimate analyses of char.

Items	Char
Proximate analysis (wt%)	
Moisture[a]	14.60
Volatile matter[b]	9.89
Fixed carbon[a]	66.72
Ash[a]	11.36
Ultimate analysis[a] (wt%)	
Carbon	68.31
Hydrogen	0.85
Oxygen	4.01
Nitrogen	0.58
Sulfur	0.30
Low heating value[a] (MJ/kg)	23.32

[a] As received.

[b] Dry and ash free basis.

2.2. *CFB test rig*

Experimental tests of char were carried out in a CFB test rig. A schematic diagram of the test rig is demonstrated in Fig. 1, which is composed of a CFB and an auxiliary system. The CFB furnace adopts heat insulation structure with a diameter of 430 mm and a height of 9 m. The recirculation loop is combined of a cyclone, a pneumatic distribution valve, an external heat exchanger (EHE), a loop seal, and two screw feeders. The preheated primary air (PA) is fed into the furnace at the bottom through the air distribution plate, and preheated SA is fed through the ports arranged at four different heights of 1.2 m, 2 m, 4 m and 6 m above the air distribution plate. There are two symmetrical SA ports at first and second level, respectively. Only one SA port is arranged at third and fourth level, respectively. Each SA port is designed to make the SA enter down into the furnace at an angle of 45°, which can make SA a long effective injection distance, pass through the center of horizontal plane [5], resulting in a good gas-solid mixing performance.

There are seven thermocouples and five pressure transmitters along the furnace. Three sampling ports are set: two are in the back pass duct for sampling fly ash and flue gas, respectively; and one is at the bottom slag port for sampling bottom ash. The gas is sampled, dried and filtered before entering individual online analyzers. Gasmet FTIR DX-4000 analyzer and KM9106 portable flue gas analyzer are used for analyzing flue gas. Oxygen content in flue gas is monitored online by a CY-IS-G zirconia oxygen analyzer in the back pass duct.

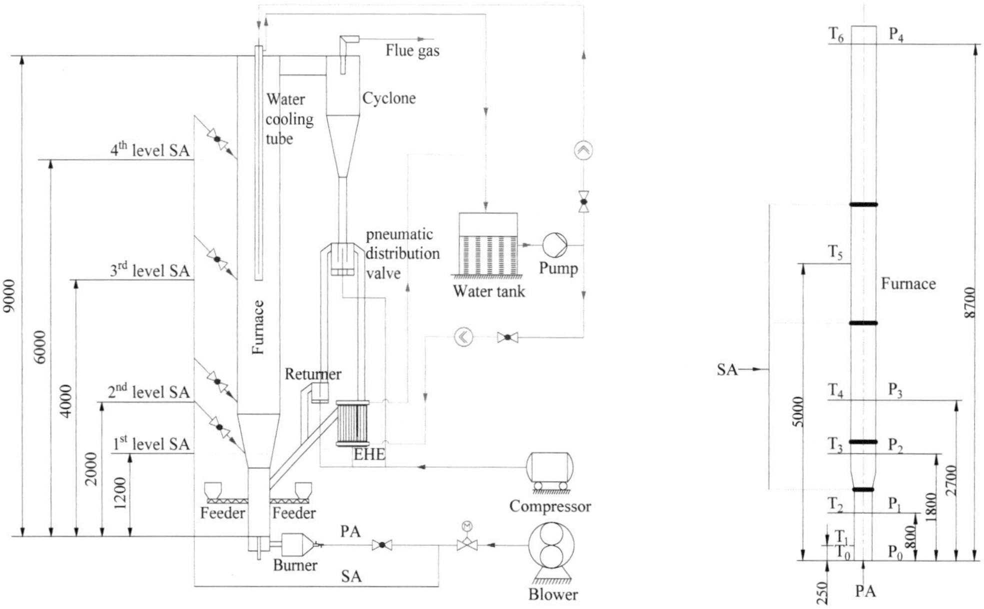

Figure 1. Schematic diagram of the CFB test rig.

Table 2. Operation conditions of the experiments.

Cases		Combustion temperature	λ	β	SA injection arrangement			
					1st level	2nd level	3rd level	4th level
1	1.1	872°C	1.15	52%	on	off	off	off
	1.2	874°C	1.18	52%	on	off	on	off
	1.3	873°C	1.20	50%	on	off	on	on
2	2.1	925°C	1.22	51%	on	off	off	off
	2.2	926°C	1.21	50%	on	off	on	off
	2.3	923°C	1.22	51%	on	off	off	on
3	3.1	936°C	1.22	56%	on	off	off	off
	3.2	948°C	1.30	49%	on	off	on	off
	3.3	943°C	1.18	53%	on	off	on	on

2.3. Experimental conditions

The experimental conditions are listed in Table 2. Each test lasted more than one hours stably. About 70 kg silicon sand with the particle size of 0-1 mm was added into the furnace before the experiments, establishing ash circulation and making the fuel to be fluidized well in the furnace. In the following discussion on the results, NO_x concentration has been normalized to flue gas with dry and oxygen concentration of 6%. Combustion temperature (represented by the highest temperature in the furnace for each case) was different for Cases 1, 2 and 3, 870°C, 920°C, and 940°C, respectively. Air stoichiometric ratio λ was maintained at 1.20, SA ratio β was maintained at 50% and the first level SA was injected and second SA was not injected for all experimental cases. There are three SA injection arrangements for Cases 1, 2 and 3. As mentioned above, the status of the first and second SA would not be changed, therefor, the variation for each case is the status of the third and fourth SA. Detailed information can be seen in Table 2. In order to develop and maintain a good fluidized state in the furnace, the PA and the first level SA were injected in all the experiments and the amount of PA and the first level SA should ensure the superficial gas velocity was larger than 3 m/s in the primary region, under this circumstance, the amount of third and fourth level SA was 20%-30% of the total amount of air.

3. Results and discussion

Pressure distribution and distribution of air stoichiometric and superficial gas velocity along the furnace for various SA injection arrangements are shown in Fig. 2 and 3, respectively.

Fig. 2 shows that as the higher level SA injects, there is an increased pressure in the dense region while the pressure in the upper region is almost the same. The higher pressure represents a larger solids holdup in the dense area as the injection of higher SA. With the third or fourth level SA injecting into the furnace, the amounts of PA and first level SA in the dense region decrease as the amount of the total air is constant for each case in Cases 1, 2 and 3, respectively. Therefore, solids carrying capacity in the dense region decreases, resulting in an increasing solids holdup and pressure in the dense region. The difference in pressure is decreasing at higher position as third or fourth level SA injecting. From Fig. 3, it can be seen that a good fluidized condition has been established in all cases, and the influence of higher level SA injection on solids holdup is not significant.

Fig. 3 shows the distribution of air stoichiometric λ and superficial gas velocity along the furnace for various SA injection arrangements. The air stoichiometric λ is about 1.20 in the exit of the furnace in all experiments. When the third and fourth level SA inject, λ is between 0.75 and 0.92 below 2 m and increases to a value between 0.92 and 1.03 below 4 m. While for cases of no injection of the third and fourth level SA, λ is larger or near 1.00 above 1.2 m (the height of the first level SA). Obviously, the reduction region is significantly lengthened in the furnace when higher SA injects, which can avoid much NO_x produced from oxidation of nitrogen-containing precursors in the primary region and strengthen the NO_x reduction reactions in the upper region.

Distribution of superficial gas velocity is uniform for cases of no injection of the third and fourth level SA. The lowest gas velocity appears at the lowest region in the furnace while it becomes larger when the first level SA injects and maintains the value to the upper region. For cases of injection of the third and fourth level SA, gas velocity firstly decreases as the sudden expansion of the cross area at the height of 1.2 m, as can be seen in Fig. 1 and then increases above 4 m or 6 m as the injection of the third or fourth level SA. The lowest gas velocity is 2.80 m/s in Case 3.3, which can develop the basic fluidized condition in the furnace.

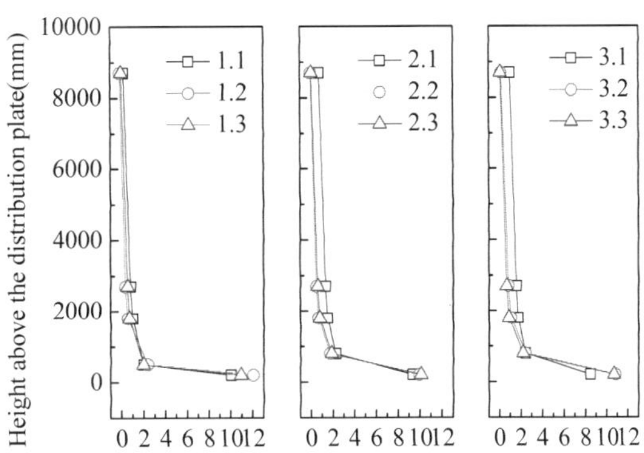

Figure 2. Pressure distribution along the furnace for various SA injection arrangements.

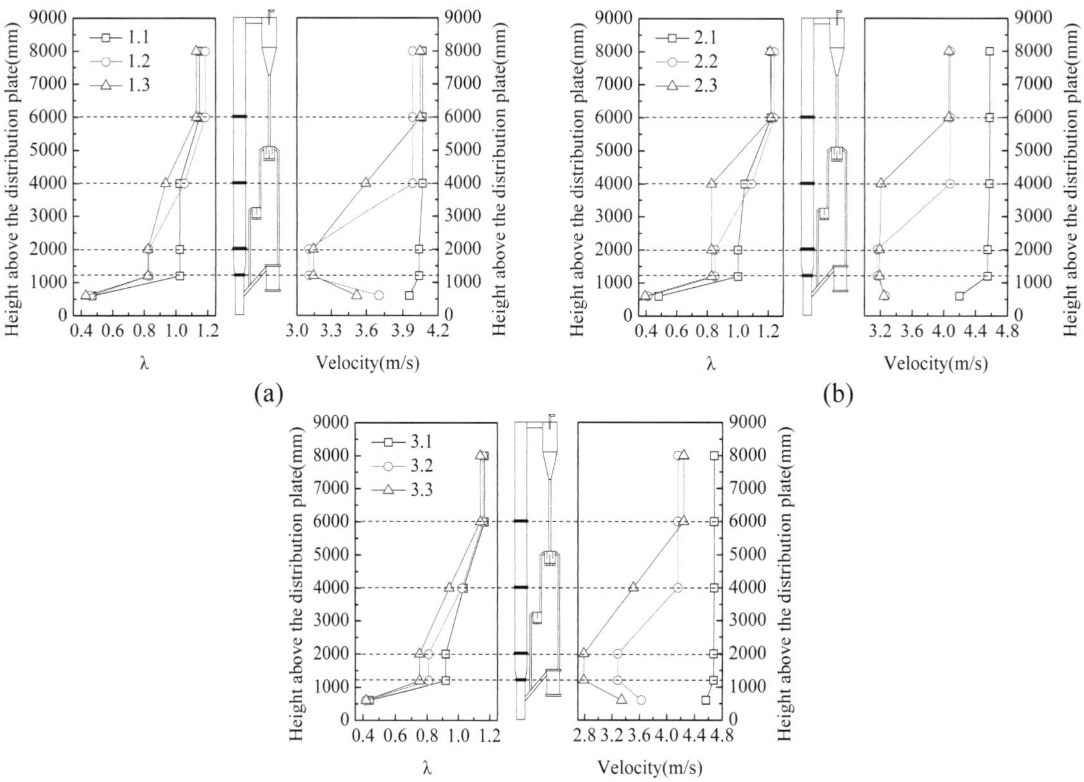

Figure. 3. Distribution of air stoichiometric and superficial gas velocity along the furnace for various SA injection arrangements.

Temperature distribution along the furnace for different SA injection arrangements are demonstrated in Fig. 4. Temperature in the furnace firstly increases and then decreases in all experiments. The temperature in primary region is lower because of the adding of cold fuel and returning of the circulating materials at a relatively lower temperature. The temperature in upper region decreases is due to the endotherm of water-cooling tube. The difference of the highest temperature and temperature in the upper region is within 30°C. The temperature distribution is relatively uniform in all experiments. The influence of injection of the third and fourth level SA on the temperature distribution is not significant.

From Fig. 5, it can be seen that the higher level SA injects, the lower NO_x emission in the flue gas. As mentioned above, the reduction region is significantly lengthened in the furnace when higher SA injects, which can avoid much NO_x produced from oxidation of nitrogen-containing precursors in the primary region and strengthen the NO_x reduction reactions in the upper region, resulting in lower NO_x emission. It also shows that NO_x reduction is more obvious at higher temperature due to the intensified NO_x reduction reactions at higher combustion temperature.

Another conclusion from Fig. 5 is that NO_x emission decreases with combustion temperature increasing. Generally, NO emission increases with the increase in the combustion temperature, which has been verified by many researchers [4,6,7-8]. However, the trend is opposite for char. Related work has been done and can be seen in the future.

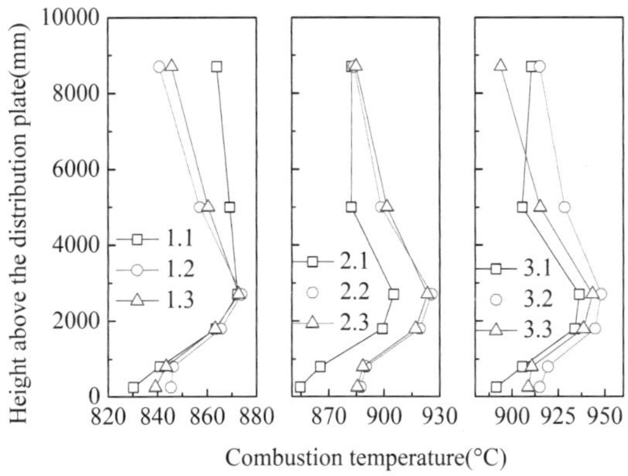

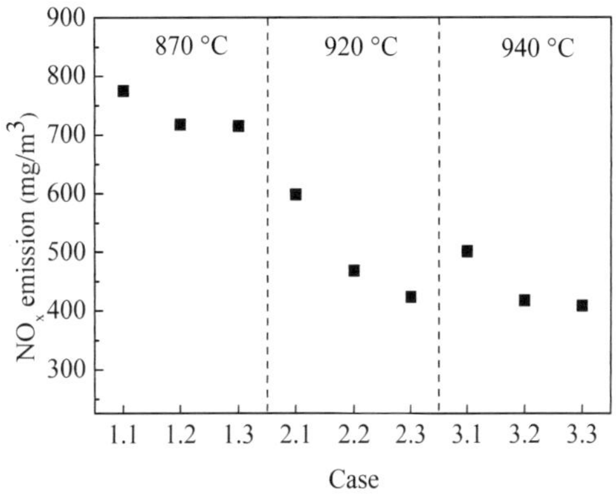

Figure. 4. Temperature distribution along the furnace. **Figure. 5.** NO_x emission in different cases.

4. Conclusion

A CFB test rig is used to study the combustion and NO_x emissions of char with higher secondary air (SA) injection. A good fluidized condition and uniform temperature distribution can be achieved with injection of higher level SA. NO_x emission decreases with injection of higher level SA and the reduction effect is more obvious at higher temperature. NO_x emission decreases with combustion temperature increasing for char combustion.

References

[1] J. M. Bursi, L. Lafanechere, L. Jestin. Basic design studies for a 600MWe CFB boiler (270b, 2*600°C), Proceedings of the 15th International, Conference on Fluidized Bed Combustion, Savannah, Georgia, 1999.

[2] Zhen Fan, Archie Roberson, Steve Goidich. Foster Wheeler. 800MWe Circulating Fluidized Bed Boiler with 1300°F Supercritical Steam, 33rd International Technical Conference on Coal Utilization and Fuel Systems, Clearwater, Florida, USA, 2008.

[3] Arto Hotta, Kari Kauppinen, Ari Kettunen. Foster Wheeler. Towards New Milestones in CFB Boiler Technology-CFB 800 MWe/New 460 MWe Super-Critical Plant with CFB Boiler in Lagisza-First Experience Update, Power-Gen International 2010, Amsterdam, The Netherlands, 2010.

[4] L.F. de Diego, C.A. Londono, X.S. Wang, Bernard M. Gibbs. Influence of operating parameters on NO_x and N_2O axial profiles in a circulating fluidized bed combustor, Fuel. 75 (1996) 971-978.

[5] Anders Lyngfelt, Lars-Erik Åmand, Lennart Gustavsson, Bo Leckner. Methods for reducing the emission of nitrous oxides from fluidized bed combustion, Energy conversion and management. 37 (1996) 1297-1302.

[6] A. Tourunen, J. Saastamoinen, H. Nevalainen. Experimental trends of NO in circulating fluidized bed combustion, Fuel. 88 (2009) 1333-1341.

[7] H. Kassman, M. Karlsson, Lars-Erik Åmand. Influence of air-staging on the concentration profiles of NH_3 and HCN in the combustion chamber of a CFB boiler burning coal, Fuel and Energy Abstracts. 41 (2000) 398.

[8] Carlos Lupiáñez, Luis I. Díez, Luis M. Romeo. Influence of gas-staging on pollutant emissions from fluidized bed oxy-firing, Chemical Engineering Journal. 256 (2014) 380-389.

Study on atomization characteristic of dual-orifice pressure-swirl injector

Fang Zhao*, Xianfeng Li, Zebin Ren, Yu Shi and Yang Yang

Facility Design and instrumentation institute, China Aerodynamics Research & Development Center, Mianyang, China

*Corresponding author e-mail: 1968683223@qq.com

Abstract. Fuel atomization greatly affects the combustion efficiency and uniformity of gas generator, therefore investigation on spray characteristics of injector is significative. In the present study, dual-orifice pressure-swirl injectors were designed and manufactured to meet the practical needs of mass flow rate and combustion performance of gas generator; water was used as the simulant medium for fuel, and Malvern optical instrument and SLA camera were applied to investigate the atomization characteristics of injectors. The results show that the mass flow rate, spray cone angle and Sauter Mean Diameter (SMD) all satisfy the design index request; the Sauter Mean Diameter decreases with the increase of injection pressure drop, but the decreasing tend becomes slow; when the spray develops to a certain position of injector exit, the Sauter Mean Diameter of axial sections keeps constantly.

1. Introduction

Gas generator is a kind of device which obtains high temperature gas by burning propellant. It is widely used in aerospace engineering, petroleum industry, automobile industry, ocean transportation and other fields, such as scramjet ground test heating system, negative pressure/vacuum pumping system, laser pressure recovery system and marine gas turbine [1-12], etc.

As one of the key components of gas generator, fuel injector greatly affects combustion performance of gas generator, such as combustion stability and combustion uniformity. Good atomization performance of fuel injector can reduce fuel droplet diameter, increase fuel evaporation rate and combustion rate, increase combustion efficiency, thus improving ignition and flame stability, and providing better outlet temperature field distribution.

At present, various kinds of injectors including straight flow injector and pressure-swirl injector have been applied to the gas generator [13, 14], and the latter is widely used because of its good atomization performance. However, lacking of perfect theoretical analysis model, the existing researches on the atomization performance of the pressure-swirl injector depend mainly on experiment means of measuring the spray cone angle and droplet size distribution parameters [15, 16], etc.

The dual-orifice pressure-swirl injector, which is the improved single-orifice pressure-swirl injector, has been a hot spot in the research of pressure-swirl injector. Its prominent advantage lies in wide range fuel adjustment, and obtaining good atomization quality under low flow rate condition, thereby meet the stable combustion and complete combustion for gas generator can be met with various working conditions [17].

According to the working characteristics of a certain type of gas generator, dual-orifice pressure-swirl injectors were designed and processed in this paper on the basis of comprehensive analysis of the advantages and disadvantages of various injector schemes. With the help of SLR camera, Malvern optical instrument, etc., related researches are carried out, and preliminary results are obtained.

2. Injector Type

The dual-orifice pressure-swirl injector works as follows, shown in Fig. 1: when air pressure is relatively low, atomization depends mainly on the fuel supply pressure; under the condition of higher air pressure, primary spray is realized by the action of fuel supply pressure, and then with the effect of the first stage swirler and secondary swirler, the secondary spray gets achieved; fuel spray is driven forward by the airflow generated by the two stage swirler, and fuel film forms on the inner surface of venture; at last, the two airflow produce shear breaking for fuel film at the venturi exit. Test conditions limited, single-phase with liquid spray test was carried out in this article.

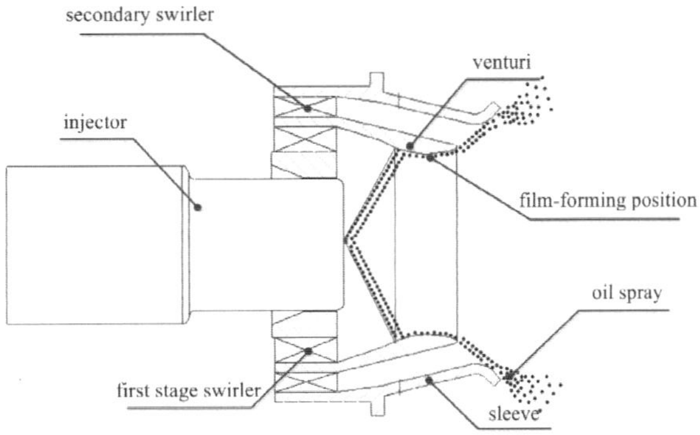

Figure 1. Atomization process of dual-orifice pressure-swirl injector with swirler

Design of the injector of this paper can be realized by two single fuel injector designs, and the designed dual-orifice pressure-swirl injector is shown in Fig. 2. The injector is mainly composed of shell, swirl core, main injection jet and secondary injection jet, etc. Among them, the injector shell is welded with the inlet pipe, connected with main injection jet by screw thread, thus forming the isolated channel between main circuit and secondary circuit; the function of swirl core is to generate tangential velocity for fuel of secondary circuit; the secondary injection jet is used to produce tangential velocity for main circuit and accelerate the injection channel fuel for secondary circuit; the main injection jet forms the fuel-accelerated injection channel for main circuit, and the equispaced orifices around are used to blow down the carbon deposition.

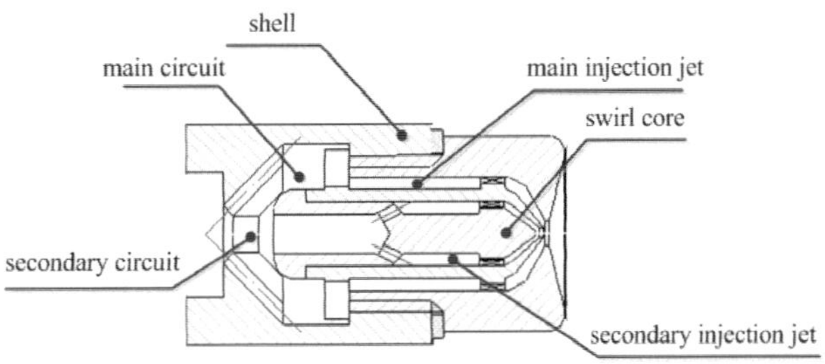

Figure 2. Structure diagram of dual-orifice pressure-swirl injector

3. Testing System

Due to the limitation of the test conditions, there was no back pressure on the downstream of the injector, which means the atomization test was carried out at atmospheric pressure. The main test system, including test objects, spray stand, supply system, measure & control system and measuring equipment (mainly including SLR camera, high precision electronic scale and Malvern optical instrument), and the schematic diagram and flow diagram of spray test are shown in Fig. 3, Fig. 4, respectively.

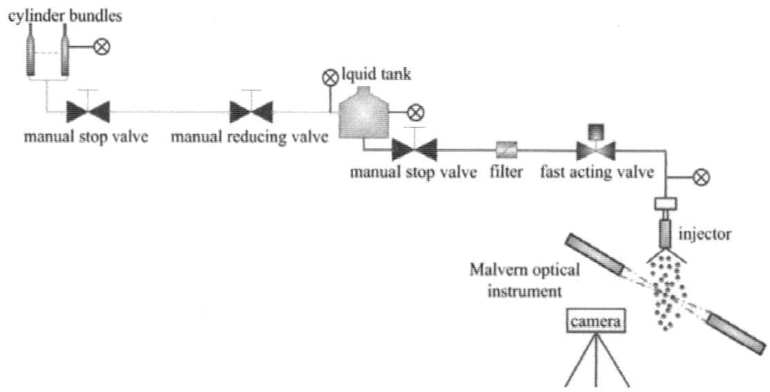

Figure 3. Schematic diagram of spray test platform

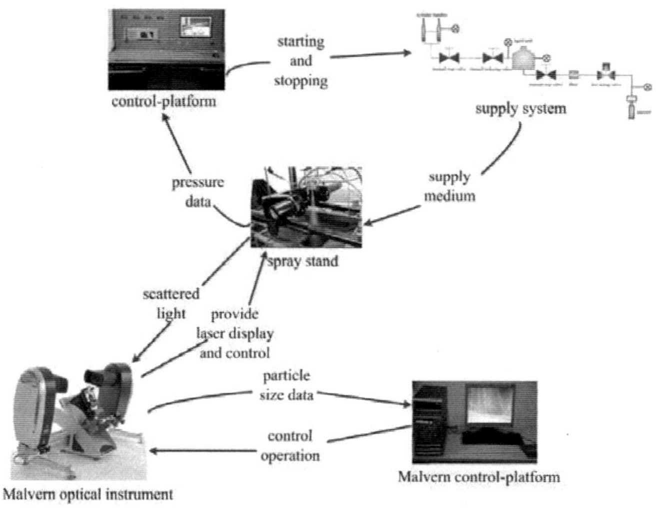

Figure 4. Flow diagram of spray test system

3.1. Supply, measure and control system

Considering the safety factors, water was used as the simulant medium for fuel. The supply system, which can smoothly supply simulant liquid and extrusion gas, mainly including:

Extrusion gas path: filter, manual stop valve, manual reducing valve, manual stop valve and fast acting valve were arranged in the pipeline.

Liquid tank: the tank top was equipped with refueling port, manual inlet valve, extrusion gas inlet, safety valve and manual gas release valve, and a liquid outlet was arranged in the bottom of the tank.

Liquid path: the discharge valve, manual stop valve, filter and fast acting valve were set up in turn.

Measure and control system were mainly responsible for spray operation sequence control, safety interlock, parameter measurement and data processing. The control objects included pressure relief valve, fast acting valve and other valves, and they were configured with signal feedback of opening and closing. In addition, measure and control system measured and recorded the pressure of each measuring point during the test. The pressure sensor adopted the pressure-resistance pressure transmitter of Mike sensor company, with the maximum response frequency of 30kHz and the measurement accuracy of 0.5% FS (full scale).

3.2. Measuring equipment

The measuring equipment used in the experiment mainly includes SLR camera, high precision electronic scale and Malvern optical instrument. Among them, the SLR camera was used to shoot the injector spray field and get the spray cone angle after post-proposing. The high precision electronic was applied to weigh mass flow rate, and the final result was the average of the 3 groups in the same working condition.

Malvern optical instrument was used for measuring the size distribution of spray and smoke in real time [18]. It provides the data required to fully understand spray and atomization processes. It will: measure across a wide size range (0.1 microns～2000 microns) without requiring constant optics changes; resolve rapid changes in droplet size over time, by measuring up to 10,000 measurements a second; deliver accurate, concentration-independent results using a patented multiple scattering analysis; characterize wide spray plumes without risking optical contamination; simply reveal the dynamic changes in spray particle size through its unique size history analysis software. It mainly consists of the following parts (as shown in Fig. 5): transmitter module, receiver module and optical bench, etc. Among them, the transmitter module which contains the collimated laser light source used to illuminate the spray during a measurement. The receiver module which can hold one of two lenses (300mm or 750mm) which focus any light scattered by the spray onto a series of detectors. These detectors accurately measure the intensity of light scattered by the spray droplets over a wide range of angles. The optical bench which ensures the transmitter and receiver are aligned. The length of this can be changed in order to fit with different applications, with the longest bench being 2.5m long.

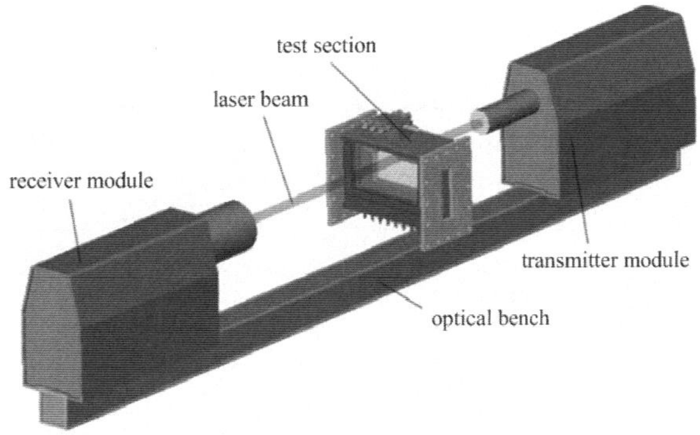

Figure 5. Component diagram of Malvern optical instrument

4. Experimental Results and Analysis

4.1. Discharge coefficient

Discharge coefficient refers to the ratio between actual mass flow rate and theoretical mass flow rate. The actual flow rate is obtained by experiments, and the theoretical flow rate can be calculated from continuity equation and Bernoulli equation. Calculation formula for liquid flow rate as shown below:

$$\dot{m} = C_A A \sqrt{2\rho \Delta P} \tag{1}$$

Where $\dot{m}$ refers to liquid flow rate; C_A is discharge coefficient; A is flow area; ρ refers to density of working fluid; ΔP is injection pressure drop. Therefore, discharge coefficient of liquid can be calculated from the following formula:

$$C_A = \dot{m} / (A\sqrt{2\rho\Delta P}) \tag{2}$$

The curve of injector flow rate with injection pressure drop is shown in Fig. 6. It can be seen that the flow rate and injection pressure drop basically change in exponential form, which is consistent with the flow rate calculation formula.

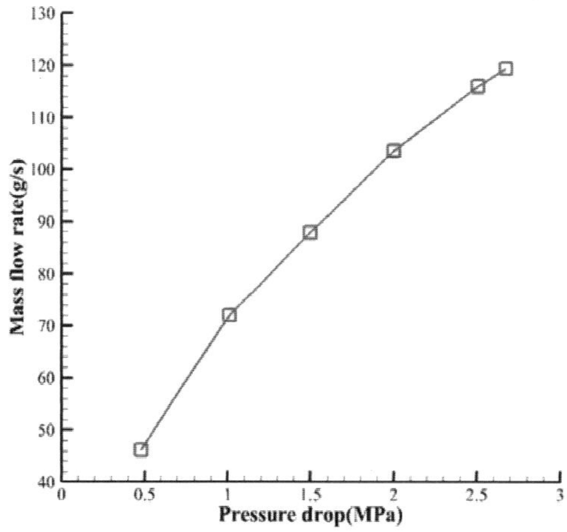

Figure 6. Mass flow rate with injection pressure drop

Fig. 7 shows the curve of discharge coefficient with the change of injection pressure drop. Discharge coefficient remained at around 0.22 within the pressure drop range in test. Due to the error caused by injector processing and flow rate measurement, discharge coefficient fluctuates within a narrow range. The overall trend shows that the discharge coefficient is less affected by injection pressure drop.

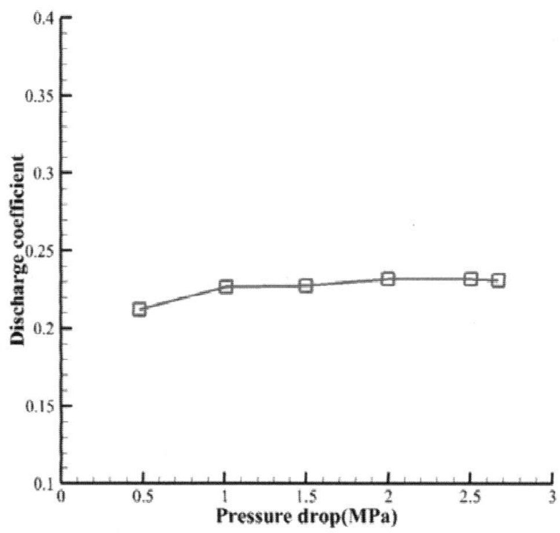

Figure 7. Discharge coefficient with pressure drop

The relationship between the mass flow rate and injection pressure drop is presented in Table 1, as well as error analysis. Mass flow rate is proportional to the 0.5 power of injection pressure drop (R-square refers to fitting parameter, the closer that this value is to 1, the better fitting effect is). This conclusion is basically consistent with the theoretical result of formula (1).

Table 1. Fitting function of mass flow rate with pressure drop

Fitted Formula	$\dot{m} = A \bullet (\Delta P)^B$	
Parameters	Fitted value	R-square
A	72.574	0.995
B	0.5	

4.2. Spray cone angle

Spray cone angle is an important factor affecting the atomization quality. Larger angle is advantageous to the gas-liquid mixing effect, promoting the primary and secondary spray, obtaining smaller and more uniform atomization particles, which benefits the gas generator combustor lean blowout. When the spray cone angle is too large, fuel tends to spray into the combustion liner wall in the combustion chamber, producing carbon deposition, which is not conducive to the combustion chamber life. So the proper spray cone angle is very important to the combustion organization of gas generator.

Spray cone angle is obtained by shooting spray field with SLR camera after post-proposing. Normal spray field at 2.5MPa is shown in Fig. 8. The spray is divided into two layers with obvious boundary, and the corresponding cone angles are 105.5 degrees and 70.9 degrees, respectively. This is consistent with design value (104 degrees and 70 degrees respectively).

However, due to reasons such as machining precision, individual injector appears splitting beam, inapparent boundary and disalignment between internal and external spray. Fig. 9 shows the abnormal spray field with splitting beam and inapparent boundary. The reason for splitting beam is that injector processing is not enough refined, thus burr exist on the injector inner surface. Because of the excessive thermal deformation brought by improper temperature control during welding, the gap between the main injection jet and the secondary main injection jet is too small, resulting in no layer-built.

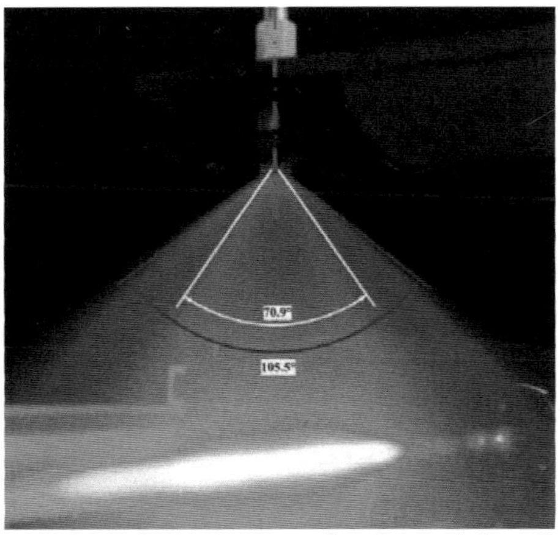

Figure 8. Typical photograph of atomizing field

Figure 9. Abnormal photograph of atomizing field

4.3. Sauter mean diameter (SMD)

Injector particle diameter usually refers to the SMD (Sauter mean diameter) for gas generator. Its physical meaning is the same size with the diameter of droplets instead of actual droplets, keeping the surface area to volume ratio the same as the real droplet group. SMD is an important index to evaluate the atomizing performance of injector, and the smaller the SMD value is, the better the atomizing performance of injector is.

SMD value with different injection pressure drop as shown in Fig. 10. SMD value decreases as the injection pressure drop increases, and the reason is that film thickness at injector exit decreases gradually as the injection pressure drop increases. Moreover, A. K. Jasuja [19] has obtained the relationship about $SMD \propto h^{0.4}$, and h refers to film thickness, hence the SMD value decreases gradually. However, this effect is obvious at the beginning, when the pressure drop increases to 1.5MPa, with the increase of injection pressure drop, the tendency to be gradual.

Z.T. Kang [13] showed that, with the increase of distance from injector exit, SMD value decreases gradually. The reasons are as follows: the rotating conical liquid sheet produced by the pressure-swirl injector, interaction with the ambient air, keeps growing and broken into liquid silk; then broken into droplets under the action of aerodynamic force; secondary spray of the large droplets achieved, resulting in small droplets. However, when the distance away from injector exit reaches to a certain value, which is to fully atomized, SMD value tends to the same order of magnitude, and differs very little, as shown in Fig. 11 (measuring height refers to the distance from injector exit to the laser axis center). With the increase of injection pressure drop, the tendency and value of SMD in each cross section show the same.

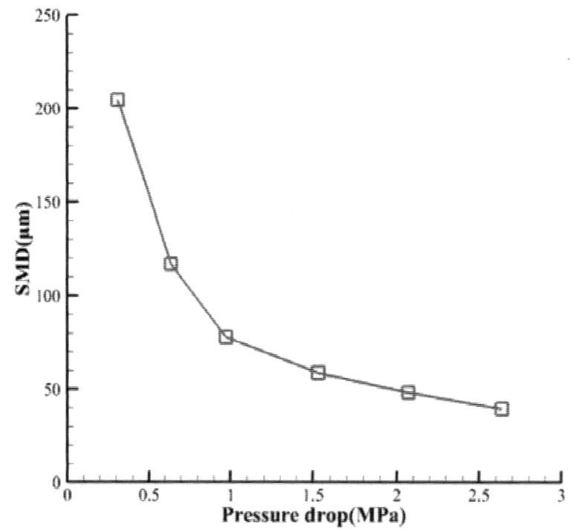

Figure 10. SMD with pressure drop

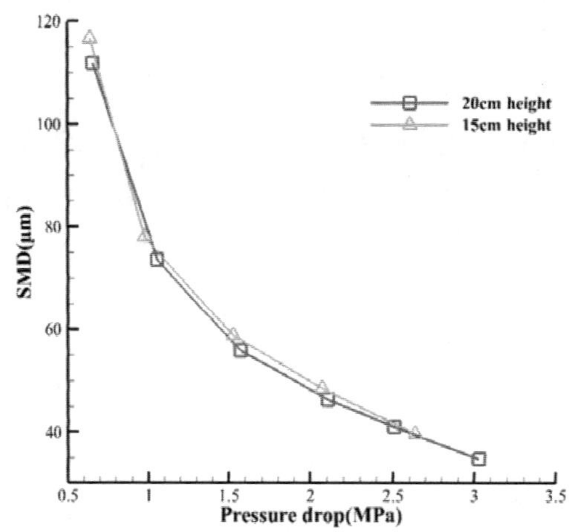

Figure 11. SMD with pressure drop of different cross-section

5. Conclusion

In order to ensure the uniformity and stability of the combustion of the air/ethanol gas generator, dual-orifice pressure-swirl injectors were designed in this paper. Spray characteristics of injectors were measured by SLR camera and Malvern optical instrument. The results are summarized as follows:

1. Mass flow rate of designed injector meets the target, and the spray field has obvious layer; moreover, the internal and external cone angle are consistent with the design goal.
2. Discharge coefficient of injector remains unchanged at different injection pressure drops.
3. SMD value decreases with the increase of injection pressure drop, and the trend of decrease is obvious at the beginning. However, when the injection pressure drops up to 1.5MPa, the tendency tends to be stable.
4. When fully atomized, the injector SMD value is essentially constant as the distance from the injector exit increases.

Acknowledgments

This work was financially supported by National 863 Plan of China.

References

[1] Connaughton, J.C. Application of a hydrazine gas gen-erator to vacuum ejector pumping of a chemical la-ser[C]. AIAA-77-0892, Presented at the AIAA/SAE 13th Propulsion Conference, July 11-13, 1977.

[2] F. Zhao. Numerical simulation and experimental research on direct-connected air heater of low total temperature[D]. Changsha: National University of Defense Technology, 2012.

[3] J.H. Feng. Numerical simulation and experimental numerical simulation and experimental research on high enthalpy and pressure air-heater [D]. Changsha: National University of Defense Technology, 2011.

[4] Z.Y. Cai, T. Cai, Y.H. Tan. Furnishing some theoretical basis for design of oxygen-rich gas generator [J]. Journal of Northwestern Polytechnical University, 2005, 23(5): 557-561.

[5] D.Y. Ma, G. Lu, X.P. Zhang, et al. Re-search on hot tests of LOX/methane gas generator [J]. Journal of Rocket Propulsion, 2013, 39(3): 21-26.

[6] J.H. Feng, C.B. Shen, F. Zhao. Analysis of liquid and gaseous oxygen influence on the combustion flow field of air heater [J]. Journal of National University of Defense Technology, 2012, 34(4): 43-48.

[7] S.T. Liu, X.W. Hu, Q. Liu, et al. The ex-perimental studies of gas generator based on nitrous oxide and ethanol [J]. Ship Science and Technology, 2008, 30 (6): 223-226.

[8] Q.L. Li, Q. Li, Z.G. Wang. Experimental research of the start-up process of gaseous oxygen/ethanol gas generator [J]. Journal of Rocket Pro-pulsion, 2010, 36(1): 13-18.

[9] Z.Q. Chen, D.X. Liao, Z.Z. Liu, et al. Experimental studies of gas generator based on hydro-gen peroxide and hypergolic ethanol [J]. High Power Laser and Particle Beams, 2007, 19(9): 1409-1412.

[10] S.Y. Jin, H.B. Xu, L. Ji. Technology research on air/jet fuel gas generator [J]. Journal of Rocket Propulsion, 2014, 40(3): 29-32.

[11] W.W. Xu. Study of high performance, high com-pression ratio pressure recovery system for chemical laser [D]. Changsha: National University of Defense Technology, 2003.

[12] W.L. Pan. New type of surface combatant and de-velopment of marine gas turbine[J]. Shipborne Weap-ons, 2010, 02: 47-72.

[13] Z.T. Kang. Research on the gas generating method based on the direct combustion of air and ker-osene [D]. Changsha: National University of Defense Technology, 2012.

[14] L. Yuan Research on changing operation combustion characteristics and combustion stabilities of hydro-gen/oxygen engine [D]. Changsha: National University of Defense Technology, 2011.

[15] D. Mondal, A. Datta, A. Sarkar. Prediction of drop size distribution in a spray from a pressure

IOP Conf. Series: Journal of Physics: Conf. Series **916** (2017) 012012 doi:10.1088/1742-6596/916/1/012012

swirl atomizer using maximum entropy formalism[J]. Journal of Me-chanical Engineering Science, 2003, 217: 831-838.
[16] D. Mondal, A. Datta, A. Sarkar. Droplet size and ve-locity distributions in a spray from a pressure swirl atomizer: application of maximum entropy formal-ism[J]. Journal of Mechanical Engineering Science, 2004, 218: 737-749.
[17] L.Y. Hou, X.C. Hou. Nozzle technical manual [M].Beijing: China Petrochemical Press, 2010.
[18] Information on http://www.malvern.co.uk
[19] Jasuja. A. K. Atomization of crude and residual fuel fuels [J]. Journal of Engineering for Gas Turbines & Power, 1979, 101(2): 250-258.

IOP Conf. Series: Journal of Physics: Conf. Series **916** (2017) 012013 doi:10.1088/1742-6596/916/1/012013

Performance Investigation on an Ultra-compact Interstage Turbine Burner with Trapped-vortex Slot Inlet

Hongtao Zhang *, **Guangqi Luo, Lei Guan, Jianchen Zeng**

Airforce Engineering University, Xi'an, China

*Corresponding author e-mail: nuaa202zht@163.com

Abstract. Ultra-Compact Combustor (UCC), which is one of mainstream design concepts of Interstage Turbine Burner (ITB), has the advantages of compact structure and high combustion efficiency. A design concept of an UCC with trapped-vortex slot inlet was proposed and numerical simulation of the stability, emissions, internal flow velocity and temperature distribution was carried out. The results indicated that the UCC with trapped-vortex slot inlet could enhance the mixing of combustion mixture and the mainstream airflow, improve the combustion efficiency, outlet temperature and the uniformity of outlet temperature field.

1. Introduction

Aviation gas turbine engine generally use after-burners which has the advantages of improved engine thrust, but with low combustion efficiency and higher fuel consumption rate. Increasing turbine inlet temperature will become the main method to increase thermal efficiency due to temperature ratio [1]. But the high temperature brings material, cooling and other aspects of the problem for the purpose of enhancing the performance of the engine. ITB realized the improvement of engine cycle work contribution to burning again through the channel between high and low pressure turbine blades [2,3].

There are two main design options for ITB at present. One is that a combustion chamber is arranged in the high and low pressure turbine transition section of the channel, the other one is that UCC scheme, whichadopts the addition of a circumferential cavity through around the low pressure turbine vane, for combustion of fuel mixed with air [4-6]. Mawid M A promoted the mixing of gas and mainstream in the circumferential cavity by setting a radial vane cavity (RVC) on the UCC guide vanes [7,8]. Because of the conventional UCC performance decreasing with the increase of the main channel size, a rectangular structure based on the principle of cavity vortex combustion is proposed [9].

The secondary flow is the source of high speed swirling flow in the circumferential cavity, which determines the centrifugal acceleration of the oil droplets, and has an important influence on the combustion performance of the UCC. Therefore, it is necessary to study inlet way of the secondary flow. Compared with the inlet way from hole inlet, slot inlet can make the secondary flow obtain a wider axial distribution when entering the circumferential cavity, so as to promote mixing fule-air in the circumferential cavity, and improve the combustion efficiency. The trapped-vortex combustion has the advantages of high combustion efficiency, low fuel-depleted boundary and low total pressure loss. On the basis of the inlet way of slot, a model of an UCC with trapped-vortex slot inlet is proposed in this paper, which has the trapped-vortex combustion and g-loading combustion advantages. By means of numerical simulation, the combustion performance and internal flow characteristics of an UCC with trapped-vortex slot inlet are analyzed.

2. Definition and verification of the mode of UCC

2.1. Definition of the mode of UCC

The verification model of UCC is established and used as the benchmark model (Model-0) [10]. In view of the fact that the benchmark model is a central symmetric structure and purpose of saving computing resources and speeding up the calculation, the 1/6 model is only simulated in this paper, as shown in Figure 1. In order to describe the flow field, a blade with one side of a radial slot is used as a guide vane.

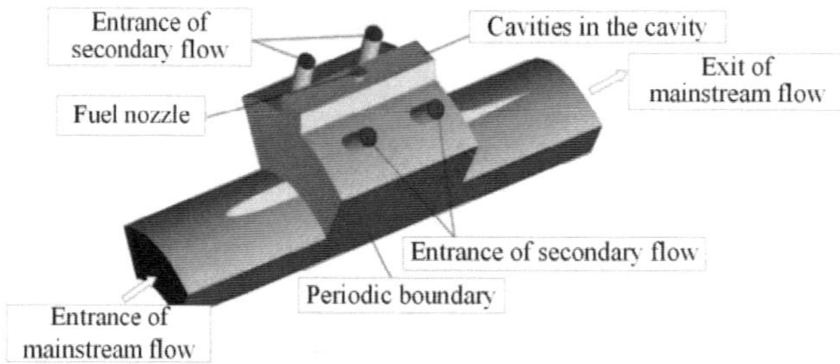

Figure 1. The configuration of Model-0

Three typical test cases are set up by UCC benchmark model [10]. Specific parameters are shown in Table 1. The turbulent chemical reactions in the baseline model are numerically simulated under the three conditions with CFD. The stability and emission parameters are obtained and compared with the experimental data of Reference 10. The accuracy of the calculation method is verified.

Table 1. Three typical test cases of UCC

Condition	*Main*(kg/min)	*Cavity*(kg/min)	*P*(kPa)	*dP/P*(%)	*T*(K)	$\Phi_{overall}$	Φ_{cav}
LMLP	2.200	0.484	284.1	4.7	528	0.294	1.620
LMMP	2.245	0.469	340.6	3.0	541	0.289	1.677
HMHP	4.914	0.960	408.2	8.2	550	0.147	0.899

2.2. Computational Grid

The computational domain is meshed by ANSYS, ICEM and CFD and the mesh type is structured grid, with a total number of 1 million 250 thousand cells, as shown in Figure 2. At the same time, the wall region is encrypted and the first layer mesh is controlled between 30~100 y^+, which satisfies the condition of non-equilibrium wall function.

Figure 2. Computational grid of Model-0

IOP Conf. Series: Journal of Physics: Conf. Series **916** (2017) 012013 doi:10.1088/1742-6596/916/1/012013

2.3. Boundary Condition

The convective boundary condition is adopted at the top wall of the circumferential cavity. The convection heat transfer coefficient is set at 8.1 W/(m2 • K) and the ambient temperature is set to 500K. The combined boundary conditions are applied to both sides of the circumferential cavity and the entrance and exit walls. The convection heat transfer coefficients are set at 8.1 W/(m^2 • K), 10.2 W/(m^2 • K) and 12.1 W/(m^2 • K) respectively. The ambient temperature is set at 300K and the emissivity of radiation heat transfer is set at 0.85. The other walls are subjected to adiabatic wall condition without sliding. In order to approximate the real distribution of oil droplets better, five cone sprays are set up to simulate the fuel injection. The parameters such as diameter, half cone angle and flow rate of oil droplets are shown in Table 2 and the injection speed is set at 30.5m/s.

Table 2. Five groups of conical spray parameter

Spray group	Half cone angle(°)	Droplet diameter(μm)	Percentage of flow(%)
1	40.00	40.00	10
2	30.00	42.50	25
3	35.00	45.00	30
4	32.00	47.50	25
5	38.00	50.00	10

2.4. Verification of the Calculation Result

The comparison between the numerical simulation results of the model, the experimental results and calculated results of the Reference 10 in the three test cases is shown in Table 3. It can be seen that the total pressure loss, carbon dioxide molar concentration and combustion efficiency calculated in this paper are greater than the experimental results. The molar concentration of carbon monoxide, oxygen and nitrogen oxides, and total outlet temperature are lower than the experimental data. However, the error between the parameters and the experimental data is kept within the acceptable range, and it can be well consistent with the law of each parameter changing with the working condition under the test condition.

Table 3. Comparisons between numerical simulation results of the model and the experimental and calculated results

Condition	Type	*dP/P(%)*	CO (ppm)	CO_2(%)	O_2(%)	NO_x(ppm)	T_{exit}(K)	η_b(%)
	A*	5.00	1089	3.20	16.40	37.30	1254.0	97.50
LMLP	B**	3.76	1377	3.69	14.99	38.76	1166.5	94.90
	C***	5.73	990	3.97	14.66	13.40	1204.5	98.81
	A*	3.30	1264	3.70	15.70	48.20	1256.0	97.70
LMMP	B**	2.74	1423	3.74	14.91	54.86	1167.1	95.95
	C***	4.40	1222	4.00	14.65	13.60	1230.2	98.57
	A*	8.30	478.0	1.60	18.70	15.60	931.0	98.00
HMHP	B**	6.39	287.2	1.78	18.16	15.93	871.5	96.49
	C***	8.90	304.0	2.04	17.80	3.80	915.3	99.30

*A** *Experiment in Reference*
*B*** *Model in Reference*
*C**** *Model established in this article*

3. Effects of Trapped-vortex Slot Inlet on the Performance of UCC

Figure 3 shows a geometry model of an UCC with trapped-vortex slot inlet (Model-1). On the basis of Model-0, the inlet way of secondary flow is changed from whole inlet to slot inlet. The normal angle of the slot inlet is set to 37 degrees to keep the inlet flow angle of secondary flow unchanged and the width of the slot inlet is set at 0.96 mm to keep the inlet flow speed constant. A circumferential inlet slot is arranged at the back of the circumferential cavity to promote the formation of the stationary vortex in the second half of the circumferential cavity. The other structures are the same as those of Model-0.

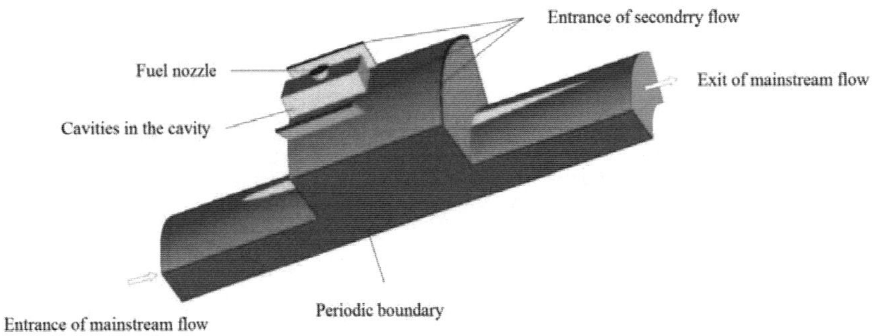

Figure 3. The configuration 0f Model-1

3.1. Stability and Emission

Table 4 shows the comparison of pressure loss, combustion efficiency, outlet temperature, component concentration and pollutant emission between Model-1 and Model-0 in three conditions.

Table 4. Performance parameters Contrast of Model-1 with Model-0

Condition	Model	dP/P (%)	CO_2 (%)	O_2 (%)	CO (ppm)	UHC (ppm)	NOx (ppm)	T_{exit} (K)	η_b (%)
LMLP	Model-0	5.73	3.97	14.66	990.3	110.80	13.4	1204.5	98.24
	Model-1	6.22	4.04	14.62	3.2	0.04	7.0	1242.0	99.99
LMMP	Model-0	4.40	4.00	14.65	1222.0	30.80	13.6	1230.2	98.57
	Model-1	5.35	4.01	14.66	16.2	0.04	11.6	1245.2	99.98
HMHP	Model-0	8.90	2.04	17.78	304.1	8.60	3.8	915.3	99.53
	Model-1	10.94	2.04	17.78	23.4	0.52	7.0	929.3	99.95

By comparing the parameters of Model-1 and Model-0, it can be seen that the CO and UHC molar concentrations of the Model-1 exit section are significantly lower than those of Model-0 but higher than it in total pressure loss. The reduction of combustible components CO and UHC means that combustion reactions in Model-1 are more complete than Model-0. Therefore, the combustion efficiency of Model-1 is obviously improved compared with that of Model-0 and remains above 99.95%. At the same time, the average temperature of the outlet section has also increased to a certain extent, the growth extent is above 14K. This is mainly due to introducing trapped-vortex in the circumferential cavity could enhance the mixing of combustion mixture with the mainstream air, promote combustion reactions in the comustor, thereby greatly reducing the concentration of combustible components and improving combustion efficiency. While enhancing the extent of mixing also brings a problem that increased the total pressure loss.

3.2. Velocity Field

The velocity distribution nephogram of Model-1 and Model-0 under LMLP conditions are shown in Figure 4. Both Model-1 and Model-0 have high velocity zones in the downstream channel near the circumferential cavity exit, but the radial and circumferential distrabution range of high-speed region of Model-1 is most extensive than Model-0, especially at the side of the blade basin. It illustrates that introducing the trapped-vortex in the circumferential cavity could drive radial transport of combustion products from the circumferential cavity into the main airflow and enhance fuel-air mixing, which is beneficial to improve the outlet temperature field of the combustor. Because the momentum exchange of micro fluid grop, the high-speed region of Model-0 will disappear rapidly. Although the high speed region of Model-1 is still shrinking, but the outlet section has not disappeared completely. At the same time, the velocity stratification of Model-1 in the downstream channel of the circumferential cavity is earlier than that of Model-0. The distribution of lower velocity region in the middle part of the channel is larger.

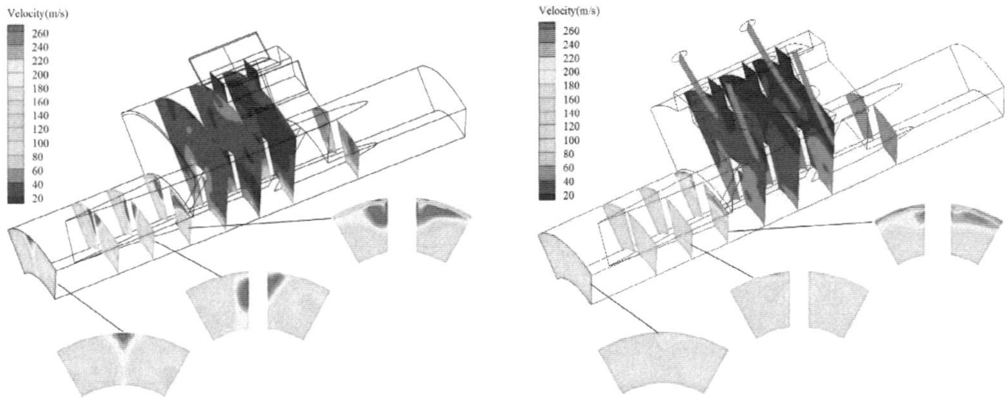

(a) Model-1 (b) Model-0

Figure 4. Velocity distribution nephogram of Model-1 and Model-0 under LMLP condition

3.3. Temperature Field

Temperature distribution nephogram of Model-1 and Model-0 under LMLP condition are shown in Figure 5. In Model-1, there is no high temperature region similar to Model-0 in the cavity-in-cavity, the distribution area of the high temperature region in the circumferential cavity is also greatly reduced, and the temperature in the latter half of the circumferential cavity is much higher than that in the first half. This is mainly due to the trapped-vortex of second half that enhancing mixed- combustion of the mainstream and the combustion products of first half. It is not only improving flow temperature of second half in the circumferential cavity, but also improving temperature distribution of the cavity-in-cavity and circumferential cavity. In the main-stream channel of downstream in the circumferential cavity, Model-1 only distributes a small high temperature area on both sides of the blade, decreases with the flow of the main-stream and disappears completely at the outlet section. Compared with Model-0 which has always a high temperature area at the top of the blade channel, Model-1 effectively improves the working environment of the outlet wall. At the same time, the lower temperature range distribution at the bottom of the main-stream channel of Model-1 is smaller than that of Model-0 and disappears rapidly with the flow of the main-stream. This is mainly due to radial transport capacity of the circumferential cavity combustion mixture was improved, high temperature combustion products can be further drive into the mainstream. Through the energy exchange, the temperature of the main-stream is lowered with the main-stream temperature raising, thus improving the overall temperature level of the downstream of the main-stream. In the outlet section, the top and bottom of Model-0 are also distributed high temperature zone and low temperature zone, but the temperature distribution of Model-1 is very uniform.

IOP Conf. Series: Journal of Physics: Conf. Series **916** (2017) 012013 doi:10.1088/1742-6596/916/1/012013

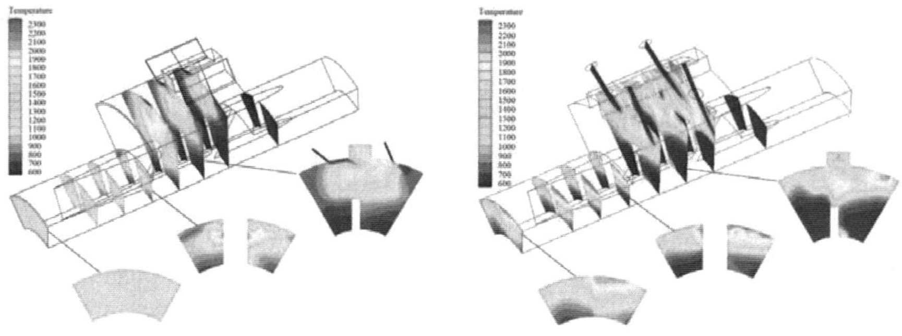

(a) Model-1 (b) Model-0

Figure 5. Temperature distribution nephogram of Model-1 and Model-0 under LMLP condition

The outlet radial temperature varying with the relative height of the guide vanes of Model-1 and Model-0 under three conditions are shown in Figure 6. The radial mean temperature of the bottom region of Model-1 is higher than that of Model-0, while the radial average temperature in the top region is lower than Model-0, and the range of temperature change is much lower than that of Model-0. At the same time, the radial mean temperature of Model-1 increases with the vane height and presents a parabola distribution at first and then decreases. The maximum of radial mean temperatures are at 70%, 60% and 55% height of vane, being closer to the ideal inlet temperature distribution of the turbine rotor. Therefore, compared with Model-0, the radial temperature distribution of the outlet section in Model-1 is greatly improved.

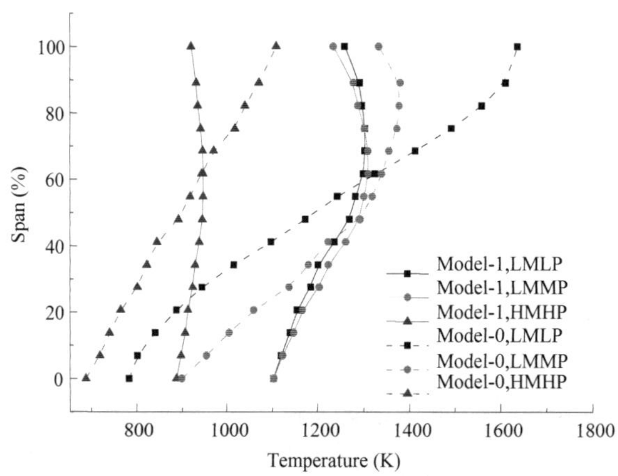

Figure 6. Exit radial temperature varying with the relative height

Table 5 shows the temperature distribution coefficient of outlet section of Model-1 and Model-0 under three kinds of working conditions. The outlet temperature distribution (OTDF) and radial temperature distribution (RTDF) of Model-1 are obviously smaller than that of Model-0 under three conditions. This is because of the presence of trapped-vortex in the circumferential cavity enhancing mixed-combustion of the combustion mixture and main stream, which improves the ability of driving high temperature combustion products from the circumferential cavity into the main airflow. The high temperature combustion products can exchange momentum and energy better with low temperature mainstream airflow in the downstream channel, so that the high temperature area and low temperature area of outlet section basically disappeared, improves the uniformity of the outlet temperature field of the combustor.

Table 5. Temperature distribution coefficients of outlet section of Model-1 and Model-0 under three conditions

Condition	Model	OTDF	RTDF
LMLP	Model-0	1.34	0.64
	Model-1	0.53	0.08
LMMP	Model-0	1.01	0.22
	Model-1	0.61	0.09
HMHP	Model-0	1.90	0.53
	Model-1	0.61	0.05

4. Conclusion

A geometric model of UCC with trapped-vortex slot inlet is established for numerical simulation about three-dimensional turbulent combustion reacting flow in the combustor. This article discusses the influence of inlet way of the secondary flow on the performance of UCC and the results are shown below:

(1) The model established accurately reflects the flow field parameters of UCC with good applicability and reference, which can be used for further investigations of UCC.

(2) Compared to the conventional UCC, an UCC with trapped-vortex slot inlet due to the inlet way of secondary flow by hole inlet to slot inlet and introducing trapped-vortex inside the circumferential cavity, enhanced the mixing of combustion mixture and mainstream airflow, could effectively improve the combustion efficiency and outlet temperature of the combustor, and improve the uniformity of the outlet temperature field, while the total pressure loss increased to a certain extent.

(3) The above-mentioned research conclusions provided valuable guidance and advice to the structure improvement of UCC under different condition.

References

[1] Shen Wei-dao, Jiang Zhi-Min, Tong Jun-geng, Engineering thermodynamics, Higher Education Press, (2000) 280-295.
[2] Sirignano W A, Liu F, Performance increases for gas~turbine engines through combustioninside the turbine, Journal of Propulsion and Power, 1999, 15 (1): 111-118.
[3] Shang Shou-tang, Cheng Ming, Liu Dian-chun, Research status and development trend of turbine interstage combustion chamber, Aeronautical Science and technology, 2011(4): 79-82.
[4] Lee A S, Singh R, Probert S D, Performance of two combustor turbofan engine under design and off design conditions, AIAA 2008-4901, 2008.
[5] Lee A S, Singh R, Probert S D, Two-Combustor Engine's Transient Performance, AIAA 2009-4837, 2009.
[6] Zelina J, Ehret J, Hancock R D, Ultra-compact combustion technology using high swirl for enhanced burning rate, AIAA 2002-3725, 2002.
[7] Mawid M A, Park T W, Thornburg H, et al, Numerical analysis of Inter-Turbine Burner (ITB) concepts for improved gas turbine engine performance, AIAA 2005-1162, 2005.
[8] M. A. Mawid, Thornburg H, Sekar B, Performance of an Inter-Turbine Burner (ITB) concept with three-different vane cavity shapes, AIAA 2006-4740, 2006.
[9] Sekar B, Thornburg H, Briones A M, Effect of Trapped Vortex Combustion with Radial Vane Cavity Arrangements on Predicted Inter-Turbine Burner Performance, AIAA 2009-4603, 2009.
[10] David S. Moenter, Design and numerical simulation of two-dimensional ultra-compact combustor model sections for experimental observation of cavity-vane flow interactions, Ohio： Air University, 2006.

A new fictitious domain approach for Stokes equation

Min Yang[1, 2,*]

[1]School of Aerospace Engineering, Nanjing University of Aeronautics and Astronautics, Nanjing, China
[2]School of Sciences, Hohai University, Nanjing, China

*Corresponding author e-mail: njhzm@sina.com

Abstract. The purpose of this paper is to present a new fictitious domain approach based on the Nietzsche's method combining with a penalty method for the Stokes equation. This method allows for an easy and flexible handling of the geometrical aspects. Stability and a priori error estimate are proved. Finally, a numerical experiment is provided to verify the theoretical findings.

1. Introduction

In many fields, such as fluid dynamics and elasticity, problems occur making it necessary to solve problem on domains being geometrically complex or time-dependent. As the generation of boundary-fitted meshes of good quality is a rather complex, often time consuming high costly, the fictitious domain method has been developed in order to overcome the meshing and re-meshing problem. One method for describing the essential Dirichlet boundary conditions, fitting into the context, is the Boundary penalty method in [1]. The real advantage of it is the combination with Nitsche's method in a fictitious domain in order to impose the essential boundary condition accurately in a weak sense. This approach was first proposed for an elliptic interface problem in [2] and later for a Stokes problem in [3]. In [4] and later in [5, 6] a stabilization of the classical Nitsche's method for the imposition of Dirichlet boundary conditions on a boundary not fitted to the mesh was considered for the Poisson problem and for the Stokes problem. In these methods the stabilization is applied in the boundary region and optimal convergence order and well-conditioned system matrices are ensured.

In this paper we propose a new fictitious domain approach for Stokes equations based on the Nitsche's method by using the lowest velocity-pressure projection penalty parameter. Rather than requiring calculation of higher order derivatives or edge based on data structures, we use a local projection to construct our scheme. We also show that the new method is inf-sup stable and optimally accurate. Moreover, its advantage is its ability to easily treat complex boundary conditions compared to existing ones. This method also can be of interest for computational domains having moving boundaries or boundaries with a complex geometry and various conditions on them. In this paper only Dirichlet and Neumann boundary conditions are considered. An extension to more complex boundary data is straightforward.

The paper is organized as follows. We describe the model problem which is represented by a linear scalar model problem with Neumann and Dirichlet boundary conditions. In Section 2 we describe the new method for the new model problem without any stabilization. In Section 3 we present new stabilization technique with the theoretical convergence analysis. We also give an error estimate in Section 4. Finally, some numerical experiments are presented in Section 5.

2. The model problem

In this paper we consider the following stationary Stokes equation in R^2:

$$\begin{cases} -\nabla\cdot(\gamma\nabla u)+\sigma u+\nabla p=f & in\quad \Omega \\ \nabla\cdot u=0 & in\quad \Omega \\ u=g_D & on\quad \Gamma_D \\ \gamma\partial_n u=g_N & on\quad \Gamma_N \end{cases} \tag{1}$$

Where Ω is an open, bounded subset of R^2, u denotes the average fluid velocity, p the pressure, γ the viscosity coefficient of the flow, σ the viscosity divided by the permeability, and f is a given external force. We assume that Ω has polygonal boundary $\partial\Omega$ and that the boundary is divided into two non-overlapping sets $\partial\Omega=\Gamma_D\bigcup\Gamma_N$. Then the weak formulation of (1) now takes the form: finding $(u,p)\in V(\Omega)\times Q(\Omega)$ such that

$$a(u,p;v,q)=F(v),\ \forall(v,q)\in V(\Omega)\times Q(\Omega)$$

Where $a(u,p;v,q)=(\gamma\nabla u,\nabla v)_\Omega+(\sigma u,v)_\Omega-(p,\nabla\cdot v)_\Omega+(\nabla\cdot u,q)_\Omega$;

$F(v)=(f,v)_\Omega+(g_N,v)_{\Gamma_N}$

$V(\Omega)=\{v\in(H_0^1(\Omega))^2|\ v|_{\Gamma_D}=g_D\ \};Q(\Omega)=L_0^2(\Omega)=\{q\in L^2(\Omega)|\ \int_\Omega qdx=0\ \}$

3. A stabilized formulation and its stability

We embed the domain Ω into the larger rectangular domain $\tilde{\Omega}$, where $\Omega\subset\tilde{\Omega}$. In a standard finite element method the mesh is fitted to the boundary or interpolates the boundary to some suitable order. Instead we propose to solve problem (1) approximately on a family of conforming triangulations T_h. Let T_h be a curvilinear triangulation of Ω, such that $\Omega\subset T_h$, but $T_h\not\subset\Omega$. For all triangles $K\in T_h$ holds $K\bigcap\Omega\neq\phi$ and we define the domain covered by T_h and $\Omega_T:=\bigcup_{K\in T_h}K$ and F_h denotes the set of interior faces of the triangles in T_h. Let h_K be the diameter of K which is less or equal to the grid-parameter h and h: $=\max_{K\in T_h}h_K>0$. By $G_h:=\{K\in T_h|K\bigcap\Gamma\neq\phi\ \}$ we denote the set of elements that are intersected by the boundary. For an element $K\in G_h$, let $\Gamma_K:=\Gamma\bigcap K$ be the part of Γ intersecting K. Similarly, $\Gamma_{D,K}:=\Gamma_D\bigcap K$ and $\Gamma_{N,K}:=\Gamma_N\bigcap K$. F_G Will denote the faces of triangles in G_h and not on the mesh boundary Γ. $F_G:=\{F\in F_h|\ F\subset\partial K,K\in G_h,F\bigcap\Gamma\neq F\ \}$. For each $F\in F_G$ we introduce the pair of element K and K' such that $F=K\bigcap K'$ and set $\hat{K}_F:=K\bigcup K'$. For details one can refer to [6]. Furthermore, the assumptions A1-A3 and definitions regarding the domain Ω and its boundary Γ are made as in [6]. Associated with T_h we have the finite element spaces

$$V_h=\{v_h\in(C^0(\tilde{\Omega}))^2:v_h|_K\in(P_1(K))^2,\forall K\in T_h\};$$
$$Q_h=\{q_h\in C^0(\tilde{\Omega}):q_h|_K\in P_1(K),\forall K\in T_h\ \}$$

Here $P_1(K)$ denotes the linear polynomials on T_h. Let J_l be the L^2-projection onto $P_1(\hat{K}_F)$ and let π_K be the L^2-projection onto $P_0(K)$ for each $K \in T_h$. Now we propose our stabilized Nietzsche method: finding $(u_h, p_h) \in V_h \times Q_h$ such that

$$A_h(u_h, p_h; v_h, q_h) + G_1(u_h, v_h) - G_2(p_h, q_h) = L_h(v_h), \quad \forall (v_h, q_h) \in V_h \times Q_h \tag{2}$$

Where $\lambda > 0$ is a penalty parameter and

$$A_h(u, p; v, q) = (\gamma \nabla u, \nabla v)_\Omega + (\sigma u, v)_\Omega - (p, \nabla \cdot v)_\Omega + (\nabla \cdot u, q)_\Omega - <\gamma \partial_n u, v>_{\Gamma_D} - <\gamma u, \partial_n v>_{\Gamma_D}$$
$$+ \sum_{\Gamma_{D,K}} \gamma \lambda h_K^{-1} <u, v>_{\Gamma_{D,K}} + <p, n \cdot v>_{\Gamma_D} - <n \cdot u, q>_{\Gamma_D}$$

$$L_h(v) = (f, v)_\Omega + (g_N, v)_{\Gamma_N} - (\gamma g_D, \partial_n v)_{\Gamma_D} + \sum_{\Gamma_{D,K}} \gamma \lambda h_K^{-1} < g_D, v >_{\Gamma_{D,K}} - < n \cdot g_D, q >_{\Gamma_D}$$

$$G_1(u, v) = \sum_{F \in F_G} \int_{\hat{K}_F} h_K^{-2} (u - J_l u)(v - J_l v) dx ; G_2(p, q) = \sum_{F \in F_G} \int_{\hat{K}_F} (p - \pi_K p)(q - \pi_K q) dx$$

On the discrete space space $W_h = V_h \times Q_h$ we define the following expression:

$$\|(v, q)\|_h^2 = \gamma |v|_{1,\Omega}^2 + \sigma \|v\|_{0,\Omega}^2 + \|q\|_{0,\Omega}^2 + \sum_{\Gamma_{D,K}} \gamma h_K^{-1} \|v\|_{0,\Gamma_{D,K}}^2 + \sum_F h_K^{-2} \|v - J_l v\|_{\hat{K}_F}^2 + \sum_F \|q - \pi_K q\|_{\hat{K}_F}^2$$

In order to analyze the method the following versions of trace/inverse inequalities are needed.

Lemma 3.1. (Trace inequality)There exists $C \in R$ such that for all $v_h \in V_h$ and $K \in T_h$ there holds

$$h^{\frac{1}{2}} \|v_h\|_{\Gamma_K} \le C(\|v_h\|_K + h_K \|\nabla v_h\|_K, \quad h^{1/2} \|\nabla v_h \cdot n_h\|_{\Gamma_K} \le C \|\nabla v_h\|_K,$$

Lemma 3.2. (Inverse inequality)There exists $C \in R$ such that for all $v_h \in V_h$ and $K \in T_h$ there holds

$$\|\nabla v_h\|_K \le C h^{-1} \|v_h\|_K,$$

Lemma 3.3. Let $p \in L^2(\tilde{\Omega})$ and $\tilde{\Omega} \backslash \Omega$ be sufficiently smooth. Then there exists a $w \in (H^1(\Omega))^2$ and constants $C_1, C_2 > 0$ independent of p, such that: $\|w\|_{1,\tilde{\Omega}} \le C_1 \|p\|_{0,\tilde{\Omega}}$; $\|w\|_{1,\Omega} \le C_2 \|p\|_{0,\Omega}$;

$$-(\nabla \cdot w, p)_\Omega = \|p\|_{0,\Omega}^2.$$

The proof can be found in [6].

Theorem 3.1. For all $(u_h, p_h) \in V_h \times Q_h$, there exists a constant β independent of h such that

$$\sup_{(v_h, p_h) \in V_h \times Q_h} \frac{A_h(u_h, p_h; v_h, q_h) + G_1(u_h, v_h) - G_2(p_h, q_h)}{\|(v_h, q_h)\|_h} \ge \beta \|(u_h, p_h)\|_h \tag{3}$$

Proof: For each $(u_h, p_h) \in V_h \times Q_h$, we start to construct a pair of (v_h, q_h) such that

$$A_h(u_h, p_h; v_h, q_h) + G_1(u_h, v_h) - G_2(p_h, q_h) \ge \beta \|(u_h, p_h)\|_h \|(v_h, q_h)\|_h$$

Firstly, setting $(v_h, q_h) = (u_h, -p_h)$, we obtain

$$A_h(u_h, p_h; u_h, -p_h) + G_1(u_h, u_h) - G_2(p_h, -p_h) = \gamma \|\nabla u_h\|_{0,\Omega}^2 + \sigma \|u_h\|_{0,\Omega}^2 - 2 < \gamma \partial_n u_h, u_h >_{\Gamma_D}$$
$$+ \sum_{\Gamma_{D,K}} \frac{\gamma \lambda}{h_K} \|u_h\|_{0,\Gamma_{D,K}}^2 + G_1(u_h, u_h) + G_2(p_h, p_h)$$

By applying the Cauchy-Schwarz, Young's inequality and using Lemma 3.1 we can get that

$$\left|2\gamma < \partial_n u_h, u_h >_{\Gamma_{D,K}}\right| \le 2\gamma C h_K^{-\frac{1}{2}} \|\nabla u_h\|_{0,K} \|u_h\|_{0,\Gamma_{D,K}} \le \frac{\gamma}{2}\|\nabla u_h\|_{0,K}^2 + \frac{2\gamma C^2}{h_K}\|u_h\|_{0,\Gamma_{D,K}}^2$$

Then we have with $\lambda \ge \lambda_0 \ge \frac{1}{2} + 2C^2$

$$A_h(u_h, p_h; u_h, -p_h) + G_1(u_h, u_h) - G_2(p_h, -p_h) \ge \frac{\gamma}{2}\|\nabla u_h\|_{0,\Omega}^2 + \sigma\|u_h\|_{0,\Omega}^2 + \sum_{\Gamma_{D,K}} \gamma(\frac{\lambda - 2C^2}{h_K})\|u_h\|_{0,\Gamma_{D,K}}^2$$

$$+ G_1(u_h, u_h) + G_2(p_h, p_h)$$

$$\ge \frac{\gamma}{2}\|\nabla u_h\|_{0,\Omega}^2 + \sigma\|u_h\|_{0,\Omega}^2 + \sum_{\Gamma_{D,K}} \frac{\gamma}{2h_K}\|u_h\|_{0,\Gamma_{D,K}}^2 + G_1(u_h, u_h) + G_2(p_h, p_h) \quad (4)$$

Secondly, as $p_h \in H^1(\tilde{\Omega})$, it exists a unique real constant c_0 with $p_h|_{\tilde{\Omega}} = p_{0,h} + c_0$, $p_{0,h} \in L_0^2(\Omega)$. From Lemma 3.3 it follows the existence of $w \in (H_0^1(\Omega))^2$ and constants such that

$$\|w\|_{1,\tilde{\Omega}} \le C_1\|p\|_{0,\tilde{\Omega}}; \qquad \|w\|_{1,\Omega} \le C_2\|p\|_{0,\Omega}; \qquad -(\nabla\cdot w, p)_\Omega = \|p\|_{0,\Omega}^2.$$

Let now $w_h := \Pi w \in V_h$, Π be L^2 interpolation operator, fulfilling the standard interpolation properties. Also we have $\|w - w_h\|_{0,\tilde{\Omega}} \le Ch\|w\|_{1,\tilde{\Omega}}$. Then there clearly exist positive constants $C_1', C_2' > 0$ with $\|w_h\|_{1,\tilde{\Omega}} \le C_1'\|p\|_{0,\tilde{\Omega}}$; $\|w_h\|_{1,\Omega} \le C_2'\|p\|_{0,\Omega}$;

Note that $-\int_\Omega p_h \nabla\cdot w_h dx = -\int_\Omega p_h \nabla\cdot(w_h - w)dx + \|p_h\|_{0,\Omega}^2$

Testing now with setting $(v_h, q_h) = (w_h, 0)$ in the second argument, we get

$$A_h(u_h, p_h; w_h, 0) + G_1(u_h, w_h) - G_2(p_h, 0) = \gamma(\nabla u_h, \nabla w_h)_\Omega + (\sigma u_h, w_h)_\Omega - (p_h, \nabla\cdot(w_h - w))_\Omega +$$

$$\|p_h\|_{0,\Omega}^2 - < \gamma\partial_n u_h, w_h - w >_{\Gamma_D} - < \gamma u_h, \partial_n w_h >_{\Gamma_D} + \sum_{\Gamma_{D,K}} \frac{\gamma\lambda}{h_K} < u_h, w_h - w >_{\Gamma_{D,K}} + < p_h, n\cdot(w_h - w) >_{\Gamma_D}$$

Using Schwarz inequality and Young's inequality, we have

$$\gamma(\nabla u_h, \nabla w_h)_\Omega \ge -\gamma\|\nabla u_h\|_{0,\Omega}\|\nabla w_h\|_{0,\Omega} \ge -C_3\gamma|u_h|_{1,\Omega}\|p_h\|_{0,\Omega} \ge -C_3(\frac{\gamma}{2\varepsilon}|u_h|_{1,\Omega}^2 + \frac{\varepsilon}{2}\|p_h\|_{0,\Omega}^2)$$

Similarly, $(\sigma u_h, w_h)_\Omega \ge -C_4(\frac{\gamma}{2\varepsilon}\|u_h\|_{0,\Omega}^2 + \frac{\varepsilon}{2}\|p_h\|_{0,\Omega}^2)$

Using intergration by parts, the interpolation estimate and the standard inverse inequality, we have

$$-\int_\Omega p_h \nabla\cdot(w_h - w)dx + \int_\Gamma p_h n\cdot(w_h - w)ds = \sum_K \int_K (w_h - w)\nabla p_h dx \ge -\sum_K \left|\int_K (w_h - w)\nabla p_h dx\right|$$

$$\ge -\sum_K \|w_h - w\|_{0,K}\|\nabla p_h\|_{0,K} \ge -C_5 h\|\nabla w\|_{0,\Omega} h^{-1}\|p_h\|_{0,\Omega} \ge -C_5\|p_h\|_{0,\Omega}^2$$

Then we focus on the boundary terms:

$$\sum_{\Gamma_{D,K}} \frac{\gamma\lambda}{h_K} < u_h, w_h - w >_{\Gamma_{D,K}} \ge -\sum_{\Gamma_{D,K}} \gamma\lambda h_K^{-\frac{1}{2}}\|u_h\|_{0,\Gamma_{D,K}} h_K^{-\frac{1}{2}}\|w_h - w\|_{0,\Gamma_{D,K}}$$

$$\ge -C_6(\sum_{\Gamma_{D,K}} \gamma h_K^{-1}\|u_h\|_{0,\Gamma_{D,K}}^2)^{\frac{1}{2}} (\sum_{\Gamma_{D,K}} h_K^{-1}\|w_h - w\|_{0,\Gamma_{D,K}}^2)^{\frac{1}{2}}$$

$$\geq -C_6(\sum_{\Gamma_{D,K}} \gamma h_K^{-1}\|u_h\|_{0,\Gamma_{D,K}}^2)^{\frac{1}{2}} (\sum_{\Gamma_{D,K}} h_K^{-2}\|w_h - w\|_{0,K}^2)^{\frac{1}{2}}$$

$$\geq -C_6(\sum_{\Gamma_{D,K}} \gamma h_K^{-1}\|u_h\|_{0,\Gamma_{D,K}}^2)^{\frac{1}{2}} (\sum_{K}\|w\|_{1,K}^2)^{\frac{1}{2}} \geq -C_6(\frac{1}{2\varepsilon}\sum_{\Gamma_{D,K}} \gamma h_K^{-1}\|u_h\|_{0,\Gamma_{D,K}}^2 + \frac{\varepsilon}{2}\|p_h\|_{0,\Omega}^2)$$

Similarly as the above, we can get

$$<\gamma\partial_n u_h, w_h - w>_{\Gamma_{D,K}} \geq -C_7(\frac{1}{2\varepsilon}\sum_{\Gamma_{D,K}} \gamma h_K^{-1}\|u_h\|_{0,\Gamma_{D,K}}^2 + \frac{\varepsilon}{2}\|p_h\|_{0,\Omega}^2)$$

$$<\gamma u_h, \partial_n w_h>_{\Gamma_{D,K}} \geq -C_8(\frac{1}{2\varepsilon}\sum_{\Gamma_{D,K}} \gamma h_K^{-1}\|u_h\|_{0,\Gamma_{D,K}}^2 + \frac{\varepsilon}{2}\|p_h\|_{0,\Omega}^2)$$

This yields the following inequality

$$A_h(u_h, p_h; w_h, 0) + G_1(u_h, w_h) - G_2(p_h, 0) \geq -\frac{C_3}{2\varepsilon}\gamma|u_h|_{1,\Omega}^2 - \frac{C_4}{2\varepsilon}\sigma\|u_h\|_{0,\Omega}^2$$

$$+(1 - C_5 - \frac{C_{10}}{2}\varepsilon)\|p_h\|_{0,\Omega}^2 - \frac{C_{11}}{2\varepsilon}\sum_{\Gamma_{D,K}} \gamma h_K^{-1}\|u_h\|_{0,\Gamma_{D,K}}^2$$

Choosing now $\varepsilon = \dfrac{1-2C_5}{C_{10}}$ and with an appropriate constant C_{13} then we can get that

$$A_h(u_h, p_h; w_h, 0) + G_1(u_h, w_h) - G_2(p_h, 0) \geq \frac{1}{2}\|p_h\|_{0,\Omega}^2 - \frac{C_{13}}{2}(\gamma|u_h|_{1,\Omega} + \sigma\|u_h\|_{0,\Omega}^2 + \sum_{\Gamma_{D,K}} \gamma h_K^{-1}\|u_h\|_{0,\Gamma_{D,K}}^2)$$

Finally, taking $(v_h, q_h) = (\delta u_h + w_h, -p_h)$ where $\delta \geq 0$ is a parameter

$$A_h(u_h, p_h; \delta u_h + w_h, -p_h) + G_1(u_h, \delta u_h + w_h) - G_2(p_h, -p_h)$$

$$\geq \frac{1}{2}\|p_h\|_{0,\Omega}^2 + (\frac{\delta}{2} - \frac{C_{13}}{2})\gamma|u_h|_{1,\Omega}^2 + (\delta - \frac{C_{13}}{2})\sigma\|u_h\|_{0,\Omega}^2 + \sum_{\Gamma_{D,K}}(\frac{\delta}{2} - \frac{C_{13}}{2})\gamma h_K^{-1}\|u_h\|_{0,\Gamma_{D,K}}^2$$

$$+\delta G_1(u_h, u_h) + \delta G_2(p_h, p_h)$$

Fixing the parameter δ by $\delta := \max\{1 + C_{13}, \frac{1}{2}\}$ yields

$$A_h(u_h, p_h; \delta u_h + w_h, -p_h) + G_1(u_h, \delta u_h + w_h) - G_2(p_h, -p_h) \geq \frac{1}{2}\|(u_h, p_h)\|_h^2 \tag{5}$$

On the other hand we have

$$\|(\delta u_h + w_h, -p_h)\|_h \leq \delta\|(u_h, p_h)\|_h + \|(w_h, 0)\|_h \leq (1+\delta)\|(u_h, p_h)\|_h$$

With (5) we can prove the stability estimate.

4. Error analysis

Lemma 4.1. Let $(u, p) \in V \times Q$ be the solution of (1) and $(u_h, p_h) \in V_h \times Q_h$ be the solution of (2), respectively, for all $(v_h, q_h) \in V_h \times Q_h$

$$A_h(E_2 u - u_h, E_1 p - p_h; v_h, q_h) = G_1(u_h, v_h) - G_2(p_h, q_h) \tag{6}$$

Where $E_i, i=1,2$ are two extension operators $E_1 : H^1(\Omega) \to H^1(\tilde{\Omega})$; $E_2 : H^2(\Omega) \to H^2(\tilde{\Omega})$ such that $(E_i v)|_\Omega = v$ and $\|E_i v\|_{H^k(\tilde{\Omega})} \le C\|v\|_{H^k(\Omega)}$;for all $v \in H^k(\Omega), i=1,2$

In order to give a priori error estimate we define:

$$e^h := (e_u, e_p) = (E_2 u - u_h, E_1 p - p_h) = (E_2 u - I^* u, E_1 p - J^* p) + (I^* u - u_h, J^* p - p_h)$$

Where $I^* := I_h E_2$ with I_h being the standard nodal interpolation and $J^* := J_h E_1$ with J_h the L^2-interpolation operator. For simply we set $\gamma = 1, \sigma = 1$.

Lemma 4.2. There exists a constant C>0 not depending on h such that

$$\|(E_2 u - I^* u, E_1 p - J^* p)\|_h \le Ch(\|u\|_{2,\Omega} + \|p\|_{1,\Omega}) \tag{7}$$

Proof: As $\|(E_2 u - I^* u, E_1 p - J^* p)\|_h = |E_2 u - I^* u|^2_{1,\Omega} + \|E_2 u - I^* u\|^2_{0,\Omega} + \|E_1 p - J^* p\|^2_{0,\Omega}$

$$+ \sum_{\Gamma_{D,K}} \gamma h_K^{-1} \|E_2 u - I^* u\|^2_{0,\Gamma_{D,K}} + \sum_F h_K^{-2} \|E_2 u - I^* u - J_l(E_2 u - I^* u)\|^2_{0,\hat{K}_F}$$

$$+ \sum_F \|E_1 p - J^* p - \pi_K(E_1 p - J^* p)\|^2_{0,\hat{K}_F}$$

The individual terms can be handled in the following way

$$\|E_2 u - I^* u\|^2_{1,\Omega} \le Ch^2 \|E_2 u\|^2_{2,\tilde{\Omega}} \le Ch^2 \|u\|^2_{2,\Omega}, \|E_1 p - J^* p\|^2_{0,\Omega} \le Ch^2 \|p\|^2_{1,\Omega}$$

$$h_K^{-2} \|E_2 u - I^* u - J_l(E_2 u - I^* u)\|^2_{0,\hat{K}_F} \le h_K^{-2} (\|E_2 u - J_l E_2 u\|^2_{0,\hat{K}_F} + \|I^* u - J_l I^* u\|^2_{0,\hat{K}_F})$$

$$\le Ch_K^2 (\|E_2 u\|^2_{2,\hat{K}_F} + \|I^* u\|^2_{2,\hat{K}_F}) \le Ch_K^2 \|u\|^2_{2,\hat{K}_F}$$

$$\|E_1 p - J^* p - \pi_K(E_1 p - J^* p)\|^2_{0,\hat{K}_F} \le \|E_1 p - \pi_K E_1 p\|^2_{0,\hat{K}_F} + \|J^* p - \pi_K J^* p\|^2_{0,\hat{K}_F}$$

$$\le Ch_K^2 (\|E_1 p\|^2_{1,\hat{K}_F} + \|J^* p\|^2_{1,\hat{K}_F}) \le Ch_K^2 \|p\|^2_{1,\hat{K}_F}$$

The boundary term is treated by using trace inequalities and inverse inequality

$$h_K^{-\frac{1}{2}} \|E_2 u - I^* u\|_{0,\Gamma_{D,K}} \le C(h_K^{-1} \|E_2 u - I^* u\|_{0,K} + |E_2 u - I^* u|_{1,K})$$

$$\le Ch_K \|E_2 u\|_{2,K} \le Ch_K \|u\|_{2,K}$$

Hence by squaring and summing together we can get that

$$\sum_{\Gamma_{D,K}} \gamma h_K^{-1} \|E_2 u - I^* u\|^2_{0,\Gamma_{D,K}} \le Ch^2 \|u\|^2_{2,\Omega}$$

Finally, with the above analysis we can get the result.

$$\|(E_2 u - I^* u, E_1 p - J^* p)\|_h \le Ch(\|u\|_{2,\Omega} + \|p\|_{1,\Omega})$$

Theorem 4.1. Let $(u, p) \in V \times Q$ be the solution of (1) and $(u_h, p_h) \in V_h \times Q_h$ be the solution of (2), Then there exists a constant C>0 not depending on h such that

$$\|(u - u_h, p - p_h)\|_h \le Ch(\|u\|_{2,\Omega} + \|p\|_{1,\Omega}) \tag{8}$$

Proof: Firstly, $\|(u - u_h, p - p_h)\|_h \le \|(E_2 u - u_h, E_1 p - p_h)\|_h$

$$\le \left\|(E_2u - I^*u, E_1p - J^*p)\right\|_h + \left\|(I^*u - u_h, J^*p - p_h)\right\|_h$$

Secondly, by applying Theorem3.1 and Lemma 4.1 we have

$$\left\|(I^*u - u_h, J^*p - p_h)\right\|_h$$

$$\le \beta \sup_{(v_h,q_h)\in V_h\times Q_h} \frac{A_h(I^*u - u_h, J^*p - p_h; v_h.q_h) + G_1(I^*u - u_h, v_h) - G_2(J^*p - p_h, q_h)}{\left\|(v_h, q_h)\right\|_h}$$

$$\le \beta \sup_{(v_h,q_h)\in V_h\times Q_h} \frac{A_h(I^*u - E_2u, J^*p - E_1p; v_h.q_h) + G_1(I^*u, v_h) - G_2(J^*p, q_h)}{\left\|(v_h, q_h)\right\|_h}$$

With the definition of $A_h(\cdot,\cdot)$ and $G_i(\cdot,\cdot), i = 1,2$ and the proof of Lemma 4.1 we have

$$A_h(I^*u - E_2u, J^*p - E_1p; v_h.q_h) \le Ch(\|u\|_{2,\Omega} + \|p\|_{1,\Omega})\|(v_h, q_h)\|_h$$

$$G_1(I^*u, v_h) - G_2(J^*p, q_h)$$

$$= G_1(I^*u - E_2u, v_h) + G_2(E_1p - J^*p, q_h) + G_1(E_2u, v_h) - G_2(E_1p, q_h)$$

$$\le Ch(\|u\|_{2,\Omega} + \|p\|_{1,\Omega})\|(v_h, q_h)\|_h$$

Finally, with Lemma 4.2 we can get the result.

5. A numerical experiment

Let Ω is a circle centered in (0, 0) with the radius 0.6 and $\gamma = 1, \sigma = 1$. $\tilde{\Omega} = [-1,1]\times[-1,1]$ is the unit square with pure Dirichlet boundary conditions computed from the known exact solution. The analytical solution of the problem is $u = -\nabla p = (\cos(x)\sinh(y), \sin(x)\cosh(y))$, $p = -\sin(x)\sinh(y) - (\cos(1)-1)(\cosh(1)-1)$. Since the pressure p is harmonic, the solution is independent of the viscosity. In the subsequent computations Nietzsche stability parameters are $\lambda = 0.1$. A direct solver based on the UMFPACK library has been employed for solving the resulting system. The results regarding the usual error analysis are showed in Table 1.

Table 1. Relative error-reduction for the velocity and pressure component

h	$\|e_u\|_{0,\Omega_h} / \|u\|_{0,\Omega_h}$	rate	$\|e_p\|_{0,\Omega_h} / \|p\|_{0,\Omega_h}$	rate
0.2022	4.63e-02	-	2.89e-01	-
0.1011	3.47e-03	1.24	7.76e-02	1.22
0.0505	4.15e-04	1.67	2.45e-02	1.56

6. Conclusion

We proposed a new stabilized fictitious domain method to solve the Stokes problem. The method is enforced based on the local projection and Nietzsche's method. We use the lowest equal order finite element spaces to approximate the velocity and the pressure. The optimal order error estimates are obtained. The numerical test also confirmed the theoretical analysis. In this paper only Dirichlet and Neumann boundary conditions are considered. An extension to more complex boundary data and other problem will also be one of our future subjects.

Acknowledgments
This work was supported by the Fundamental Research Funds for the Central Universities (Grants No.2013B07314)

References

[1] J.W.Barret, C.M.Elliott, Finite element approximation of the Dirichlet problem using the boundary penalty method, Numer.Math. 49(1986) 343-366.

[2] A.Hansbo, P.Hansbo, An unfitted finite element method based on Nitsche's method for elliptic interface problems, Comput.Methods Appl.Mech.Eng. 191(2002) 5537-5552.

[3] R.Becker, E.Burman, P.Hansbo, A Nitsche extended finite element method for incompressible elasticity with discontinuous modulus of elasticity, Comput.Mehtods Appl.Mech Eng. 198(2009)3352-3360.

[4] E.Burman, Ghost penalty, C.R.Math.Acad.Sci.Paris, 348(21-22)(2010) 1217-1220.

[5] E.Burman, P.Hansbo, Ficititious domain finite element methods using cut elements: II. A stabilized Nitsche method, Appl.Numer.Math. 62(2012) 328-341.

[6] E.Burman, P.Hansbo, Ficititious domain finite element methods using cut elements: III. A stabilized Nitsche method for Stokes'problem, ESAIM: Math. Model. Numer. Anal. 48 (2014) 859-874.

Study on the Reasonably Liquid Withdrawal Rate under Different Development Conditions of Water Flooding Reservoir

Chunju Li *

No.3 Oil Production Plant, Daqing Oilfield Limited Company, Daqing 163000, China

*Corresponding author e- mail: 2760913691@qq.com

Abstract. The reasonably liquid withdrawal rate was studied at the different geological characteristics and different development periods of water flooding reservoirs. Through the study we can see: (1) the reasonably liquid withdrawal rate of positive rhythm reservoir is less than that of inverted layered reservoir; (2) with the increase of permeability, the reasonable recovery rate of oil well is also increased; (3) with the increase of the permeability difference and the thickness difference, the reasonable recovery rate of the oil well is gradually reduced; (4) with the increase of water cut, the reasonable rate of liquid increased gradually, indicating that with the development of oil field, raise the amount of liquid production can be used as a measure to improve the recovery rate.

Keyword: Reasonably Liquid Withdrawal Rate; Reservoir Nature; Pattern Configuration

1. Introduction

Water flood development as a simple and low- cost way to add energy is used by most oil fields, it is very important to determine the reasonable production rate of oil wells[1- 2]. When the production rate of oil wells is too fast, it is easy to cause injecting water to form an inefficient cycle, which is not conducive to the improvement of oil recovery; when the flow rate is low, the development period is too long, the development efficiency is deteriorated, therefore, it is very important to carry out research on the reasonable production rate of oilfield [3- 4]. Due to the complex properties of reservoir property rhythm, permeability difference and thickness difference, the well pattern is different too, so the reasonable liquid production speed will change greatly [5- 6]. In this paper, the numerical simulation technique is used to study the reasonable production rate under different development conditions of water flooding reservoirs, and the reasonable extraction rate under different physical condition and development period is given.

2. Determination of liquid production rate of oil wells under different reservoir physical properties

2.1. Different rhythm

When the voidage replacement ratio is 1, use the five- point well pattern, study on liquid withdrawal rate of oil well for different rhyme reservoirs. The study found, The reasonable production rate of the positive rhythm reservoir is 0.15PV / a, while the reasonable production rate of the inverted layered reservoir is 0.20PV / a.

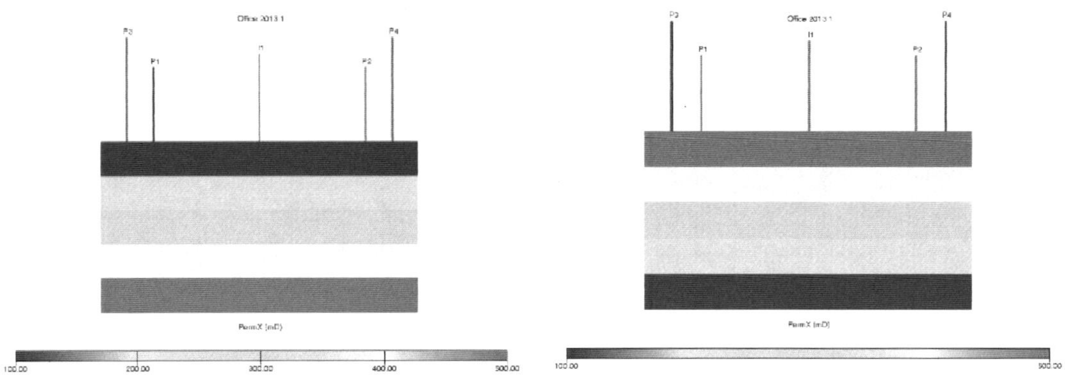

Figure 1. Permeability distribution of different rhythmic reservoirs

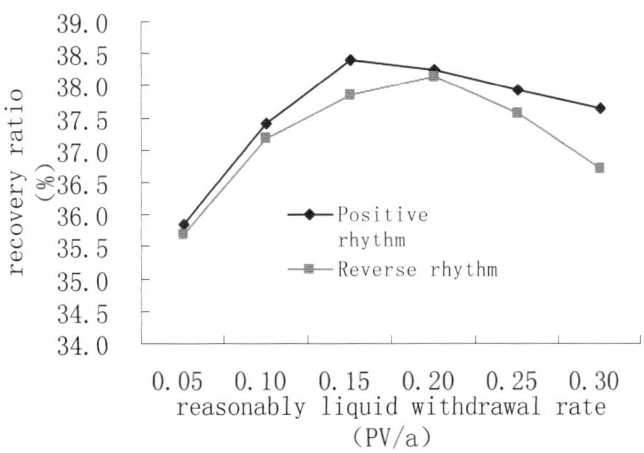

Figure 2. The curves of recovery rate with the reasonably velocityin different rhythm cases

2.2. Different permeability

The five- point ideal model is designed with permeability 100× $10^{-3}\mu m^2$、200× $10^{-3}\mu m^2$、300× $10^{-3}\mu m^2$、400 × $10^{-3}\mu m^2$ and 500 × $10^{-3}\mu m^2$ respectively. By numerical simulation, with the increase of permeability, reasonable liquid production rate is gradually increased, however, when the permeability is greater than 400× $10^{-3}\mu m^2$, the reasonable production rate is not increasing, 0.25PV / a.

IOP Conf. Series: Journal of Physics: Conf. Series **916** (2017) 012015 doi:10.1088/1742-6596/916/1/012015

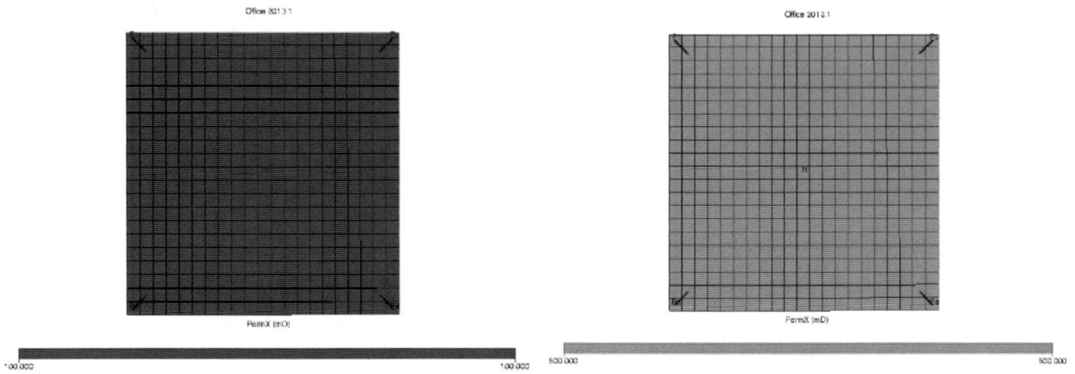

Figure 3. The 100× $\mathbf{10^{-3}}\boldsymbol{\mu m^2}$ and 500× $\mathbf{10^{-3}}\boldsymbol{\mu m^2}$distbution of permeability

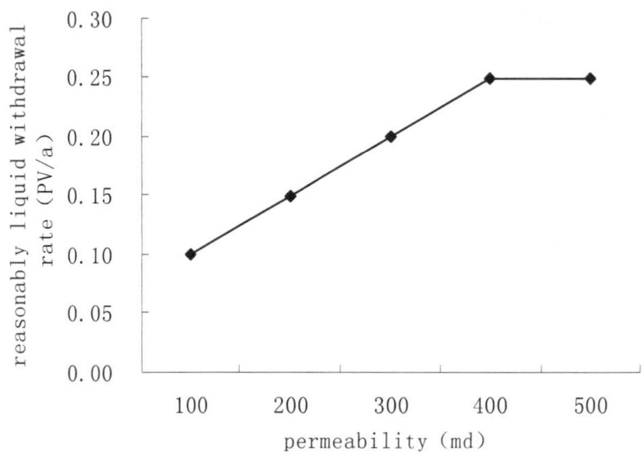

Figure 4. The curve of reasonably liquid withdrawal rate of difference permeability

2.3. Different permeability difference

The ideal model of the five- point method is designed with 1, 2, 3, 4 and 5, respectively. As can be seen from the calculation, with the increase of permeability difference, the reasonable rate of liquid production is gradually reduced, when the permeability difference is greater than 3, the reasonable production rate is stable, 0.05PV / a. Analysis of the reasons we can see, with the increase of permeability level, the higher rate of liquid production, a lot of water enters the high permeability horizon, resulting in low permeability layer is not used, affecting the region's oil recovery.

IOP Conf. Series: Journal of Physics: Conf. Series **916** (2017) 012015 doi:10.1088/1742-6596/916/1/012015

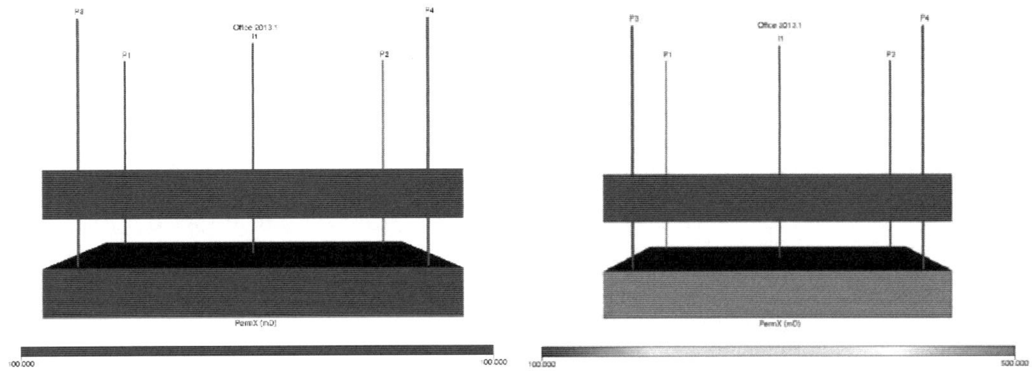

Figure 5. The permx distbution of permeability difference 1 and 5

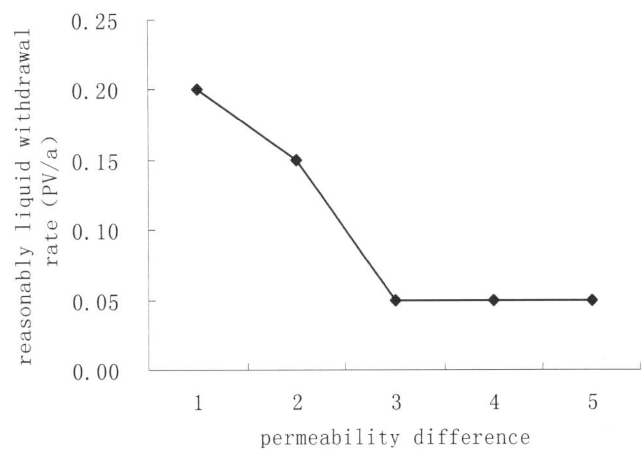

Figure 6. The curve of reasonably liquid withdrawal rate of Permeability difference 1 to 5

2.4. Different thickness differences

When the total thickness of the model is constant, the ideal model of the five - point method is designed with thickness difference of 1, 2, 3, 4 and 5. As can be seen from the calculation, as the thickness difference increases, the reasonable rate of liquid extraction decreases, when the permeability difference is greater than 5, the reasonable liquid production rate is 0.05PV/a.

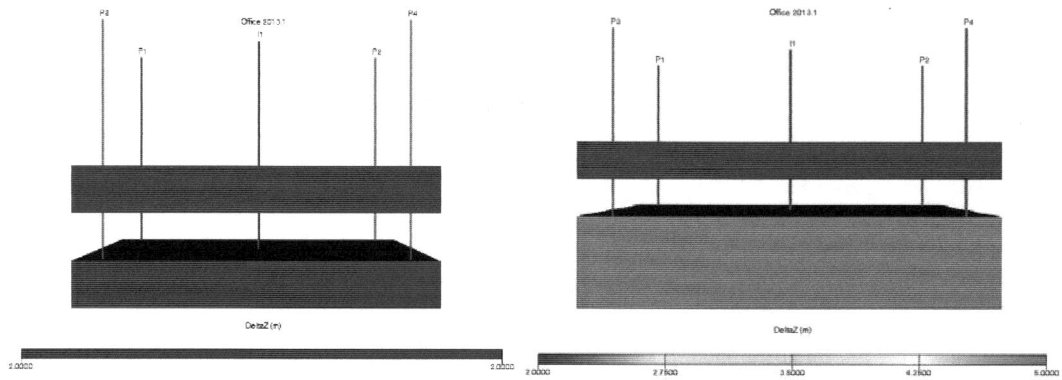

Figure 7. The DZ destruction of thickness difference 1 and 5

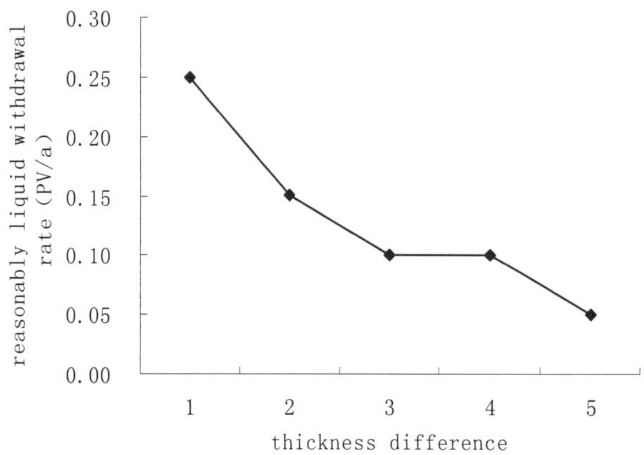

Figure 8. The curve of reasonably liquid withdrawal rate of Thickness difference 1 to 5

3. Determination of the Liquid Reasonable Production Rate in Different Water Cut

The initial production rate is 0.05PV/a, when the water content is 0%, 40%, 60% .80% and 95%, increase the amount of liquid production, The production rate was increased to 0.10 PV/a, 0.15 PV/a, 0.20 PV/a, 0.25 PV/a and 0.30 PV/a. According to the results, with the increase of water content, the reasonable rate of liquid production increased gradually, indicating that increasing the amount of liquid production can be used as an effective measure to improve the recovery rate

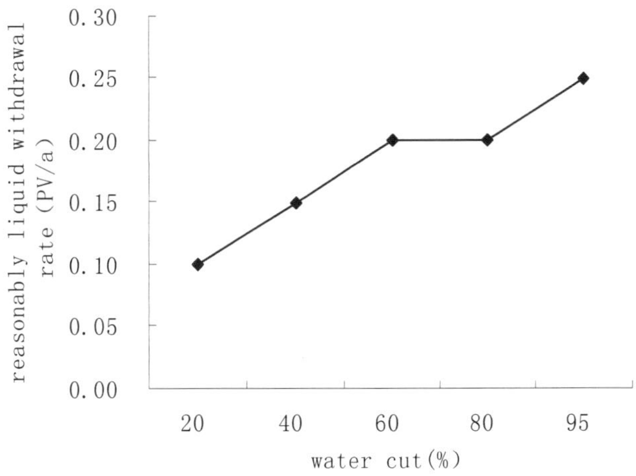

Figure.9 The curve of reasonably liquid withdrawal rate in different water cut

4 Conclusion

(1) For different rhythmic reservoirs, the reasonable production rate is different, and the rate of positive rhythm reservoir is less than that of inverted layered reservoir;

(2) With the increase of reservoir permeability, the reasonable velocity of oil well is gradually increased;

(3) With the increase of the permeability difference and the thickness difference, the reasonable recovery rate of the oil well is gradually reduced;

(4) With the increase of water cut, the reasonable rate of liquid extraction increased gradually, indicating that with the development of oil field, improving production can be used as efficient measure to improve the recovery rate.

References

[1] Guo Min. The Studies on Liquid Production of Subsequent Water Flooding of Polymer Flooding [D]. China University of Petroleum 2011.

[2] Huang Juanjuan. Factor Analysis and Evaluation System Research of Waterflood Status in High Water Cut Stage [D]. China University of Petroleum 2011.

[3] Mao Zhenqiang, Guo Jianwei, Dong Pingzhi. Study on Occurrence Mechanism and Prediction of Big Channels in Loose Cemented Sandstone Reservoir[J]. PGRE. 2011(01).

[4] Wang Xianrong, Feng Xubo, Duan Zhonghua. Development characteristics and countermeasures for stabilizing production of high water- cut reservoirs in Jinhu Sag [J]. Complex Hydrocarbon Reservoirs. 2012(03).

[5] LI Ai- fen,ZHANG Dong,YAO Jun. Physical Simulation of Water Flooding in Fractured- Vuggy Unit [J]. Journal of China University of Petroleum (Edition of Natural Sciences). 2012(02).

[6] DONG Xuemei, XU Huaimin, Madina Mawutihan. Well Infill Adjustment Test of Xishanyao Ultra- Low Permeability Reservoir in High Water Cut Stage in Wellblock Cai- 9 of Cainan Oilfield, Junggar Basin [J]. Xinjiang Petroleum Geology. 2014(05).

IOP Conf. Series: Journal of Physics: Conf. Series **916** (2017) 012016 doi:10.1088/1742-6596/916/1/012016

Experiments on the merged vortex formation in a flapping plate

Yong Huang[1], Jiaming Wu[1], Le Fang[2,*], Lixiang Wu[1] and Liang Wang[1]

[1]School of Energy and Power Engineering, Beihang University, Beijing, China
[2]LMP, Ecole Centrale Pékin, Beihang University, Beijing, China

*Corresponding author e-mail: le.fang@buaa.edu.cn

Abstract. A two-dimensional flapping device was designed by consisting of a front flapping flat plate and a rear fixed horizontal plate in a rectangular tunnel. The flow characteristics near the plate and the downstream were studied under different flapping parameters. A high-speed camera was used to obtain the flow images to analyse the shedding vortex formation. Results show that a merged vortex is generated when the plate flaps and, relatively, an inertial vortex is generated when it stops. The vortex merger is in agreement with the Kovaznay model after the non-dimensional time t/T of 0.3, implying its non-equilibrium properties during energy transfer. The length-scale ratio of the merged vortex to the neighbour existing minor vortex is around 2, providing an experimental verification for the simulations of Cardesa et al. [Science 2017]. In addition, the most clear merged vortex is formed when the plate flaps between 30 and 60 degree at a tip flapping linear velocity 3 times bigger than the inlet flow velocity.

1. Introduction

Since ancient times, humans are always eager to fly like a free bird. With the flapping wing flight becoming a technical hotspot in the aircraft development [1], there have been an increasing number of researches for investigating the flapping wing flights. In 1998, the first electrically powered palm-sized ornithopter "Microbar" was successfully developed, being capable of turning left or right and pitching up or down [2]. In 2008, Wood [3] developed a flapping-wing micro-robot, which can produce sufficient thrust to accelerate vertically. In 2012, a revolutionary design "Smart Bird", with an ingenious transmission mechanism and a precise control system, can fold and reverse the wings to have good aerodynamic performance [4].

Compared with the flapping-wing prototype, the theory of the flapping wing flight also has a rapid development because of the progress of experimental techniques and numerical simulations. Pomphrey et al. [5] have studied the mosquitoes flapping wing kinematics, arguing three key features: leading edge vortices, trailing-edge vortices caused by a form of wake capture at stroke reversal, and rotational drag. André and Jacquin [6] have presented 2D direct numerical simulations at Reynolds number Re = 1000. They analyzed three fundamental mechanisms, suction by the leading edge vortex, added mass reaction and wake capture. Chandra and Damodaran [7] have proved that the thrust is generated as a result of vortex asymmetry and the relative convection rates of positive and negative vortices are a key parameter involved in thrust generation. Taylor et al. [8] have confirmed that a general principle of oscillatory lift-based propulsion for flapping wing flight seems to be tuning cruise kinematics to optimize St Number.

Most of the existing studies indicate that flapping wing flight is based on the vortex formation and dissipation around the wing. But now it lacks of a convincing explanation for the vortex formation and dissipation of the flapping wing, which relates to complicated turbulence theory on vortex formation and dissipation. Indeed, visible observations of the Kolmogorov-Richardson energy cascade [9-11] have confused researchers for quite long time, until the very recent study by Cardoza et al. [12], which successfully simulated the process that energy is transfered from large to small vortices with accurate quantitative descriptions. However, to our knowledge, there is still no experiment for verifying this important numerical advance.

In this contribution, in order to clarify the vortex formation of the flapping wing flight with comparison to theoretical predictions, the question is simplified as the vortex formation of a flapping plate. A two-dimensional flapping device is designed to help to better understand the formation and underlying mechanisms of the vortices.

2. Experimental set-up and calibration

As shown in Fig.1, a two-dimensional flapping device is designed, consisting of a front flapping plate and a rear fixed horizontal plate in a Plexiglas rectangular tunnel. The flapping plate is driven by a stepper motor and the fixed plate is fixed by a square shaft. The flapping and fixed plates are located at 120 mm height in z direction. The rectangular Plexiglas tunnel is 50 mm width in y direction and 200 mm height in z direction. The flow images are obtained by an IDT's X-Stream VISION XS-3 high-speed camera whose full-angle pixel is 1280 × 1024, the maximum frequency is up to 10000Hz and the minimum exposure time is down to 1μs.

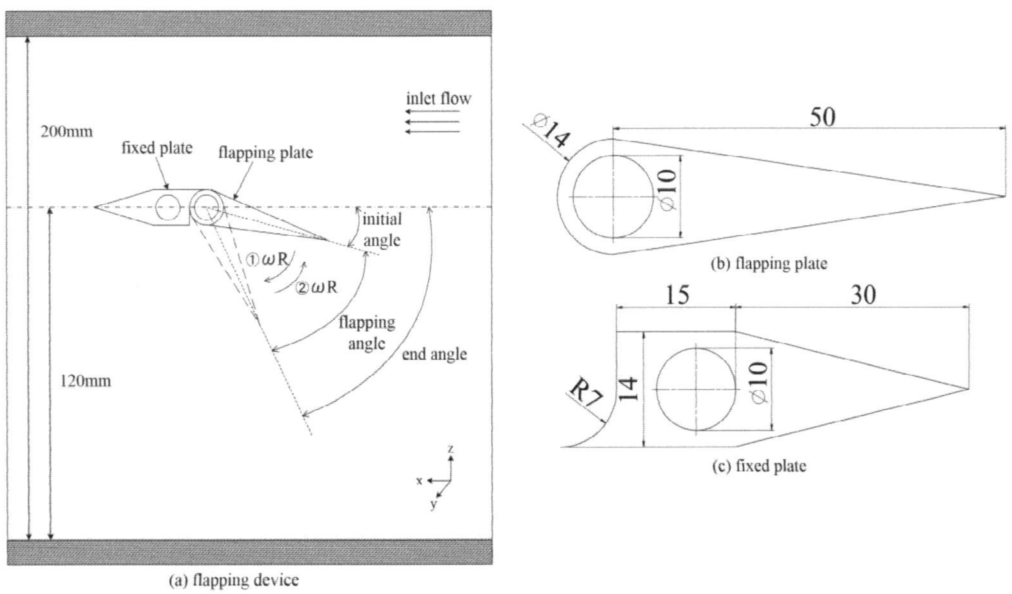

Figure 1. Experimental set-up

The flapping plate flaps clockwise from the initial angle to the end angle, and returns to its original position. An initial angle is defined as the angle at which the plate starts to flap. An end angle is defined as the angle at which the plate stops and changes its flapping direction. Thus, a flapping angle is defined as the difference between the initial and end angle. Regarding the x direction as 0 degree, three kinds of flapping angle are set, including the difference between 0 and 90 degree, between 30 and 90 degree and between 30 and 60 degree.

IOP Conf. Series: Journal of Physics: Conf. Series **916** (2017) 012016 doi:10.1088/1742-6596/916/1/012016

Because of the system response delay and the manufacturing error, the real tip flapping velocity of the plate is different from the set velocity. The real tip flapping linear velocity is measured by the high-speed camera. As shown in Fig.2, the scatter curves present the real velocity, while the line curves present the set velocity. The inlet flow velocity is about 0.25 m/s, so the velocity ratio φ of the flapping velocity ωR to the inlet flow velocity v_o , $\varphi = \omega R / v_o$ ranges from 3 to 4.6.

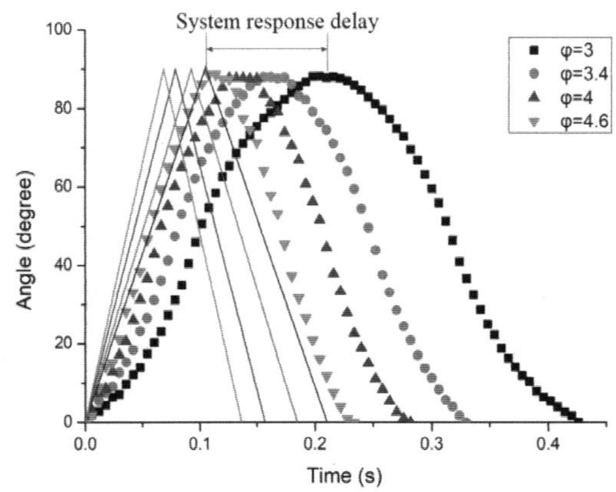

Figure 2. The calibration of the tip flapping linear velocity

3. Results and discussion

3.1. Shedding vortex formation

A merged vortex is defined as the vortex merged by several minor vortices. A minor vortex is defined as the existing small vortex, which is neighbor to the merged vortex and will be merged in the future. Fig.3 (a), (b) and (c) show that when the plate flaps from the initial angle, the streamline is bent and the minor vortices are generated and shed alternately on the tip of the plate. Subsequently, the minor vortices merge as a merged vortex. Larger than 60 degree, few minor vortex is generated while the merged vortex keeps expanding and eventually turn into the largest merged vortex. During the entire clockwise flap, the vortices are attached to the trailing edge of the plate.

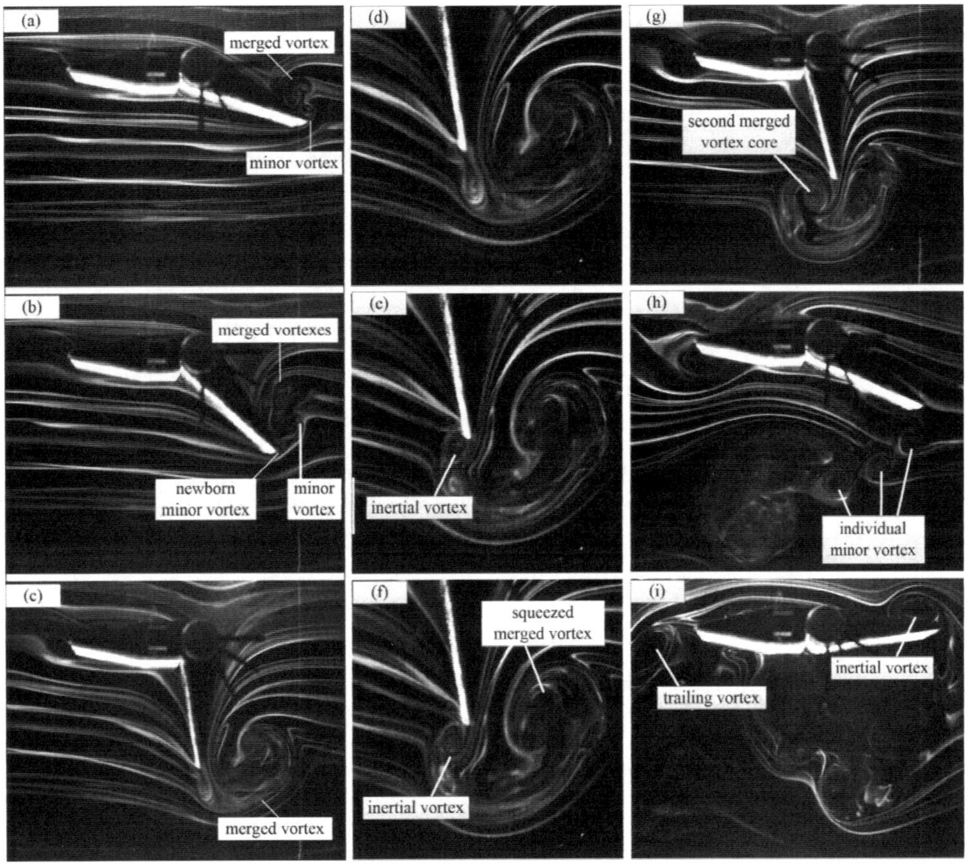

Figure 3. Vortex formation ($\varphi=4$ and between 0 and 90 degree)

During the short pause between the clockwise and counterclockwise flap, a reversely rotatory inertial vortex is generated in the other edge of the plate in Fig.3 (e). Meanwhile the merged vortex is squeezed, with the length-scale decreasing as shown in Fig.3 (f). The inertial vortex and the merged vortex seem to behave as a pseudo asymmetric vortex pair.

When the plate returns to the initial angle counterclockwise from the end angle, the pseudo asymmetric vortex pair shed from the tip of the plate. Then a minor vortex is generated and sheds into the neck of the pseudo asymmetric vortex pair quickly. This new minor vortex rotates fast as a vortex core, squeezing and breaking the inertia vortex and the merged vortex as shown in Fig.3 (g). The pseudo asymmetric vortex pair are destroyed totally and a second merged vortex is formed relatively. Compared with the minor vortices in the clockwise flap, some minor vortices in the counterclockwise flap not only shed individually without merger, but also depart from the plate. As presented in Fig.3 (h), these minor vortices appear in the area where the tip passes through and remain substantially in place for a short time like a string of pearl necklaces.

Fig.3 (i) indicates that once the plate returns to the initial angle, there will be a second inertial vortex on the tip. In addition, the air near the trailing of the fixed horizontal plate is oscillated continuously during the whole flap, thus a trailing vortex is regularly generated and sheds later.

It could be concluded that the flapping plate generates a merged vortex every direction flap. There is a possible reason that the flap creates a pressure distribution, which takes an important role in the vortex formation. The air is driven by the differential pressure to roll up, which develops as the merged vortex eventually. Due to the inertia, even though the plate stops, the upstream air still maintains its original movement, forms a velocity shear layer and then generates an inertial vortex.

3.2. Vortex merger

As mentioned above, the minor vortices merge as the large merged vortex. From the viewpoint of turbulent energy transfer, this inverse energy cascade is a typical two-dimensional phenomenon, which supports the quasi-two-dimensional design of the present experiment. In order to quantitatively describe this, we introduce the classical Kovaznay model, which provides an understanding of the local non-equilibrium time between scales [13, 14]. The relation between the non-equilibrium time t and the length-scale l during the vortex merger is

$$\log(l) \sim \frac{3}{2}\log(t) \tag{1}$$

Because most of the vortices are irregular, the length-scale l of a given vortex is measured as the average of the length of the vortex in the direction parallel to the plate and the length in the direction perpendicular to the plate. The non-dimensional length-scale l/L is defined as the ratio of the length-scale of the merged vortex to the length-scale of the final largest merged vortex. Time t is calculated from the plate start to flap. The non-dimensional time t/T is defined as the ratio of the present merger time to the time the merged vortex formation finally spends. Fig.4 (a) shows the comparison between Kovaznay model and our result. It is obvious that results are in quite good agreement with the model after a non-dimensional time t/T of 0.3 (corresponding to $\log(t/T) > -0.5$). To explain this fact, we recall that the non-equilibrium time describes a local time-scale that turbulence tends to equilibrium [15]. This then implies that vortex merger can be considered as a non-equilibrium phenomenon, which is also in agreement with the fact that energy backscatter is a typical non-equilibrium turbulence transfer [16]. For smaller time scales, this non-equilibrium scaling is not satisfied in Fig. 4(a), which might correspond to the experimental difficulty that small vortices are difficult to be accurately measured.

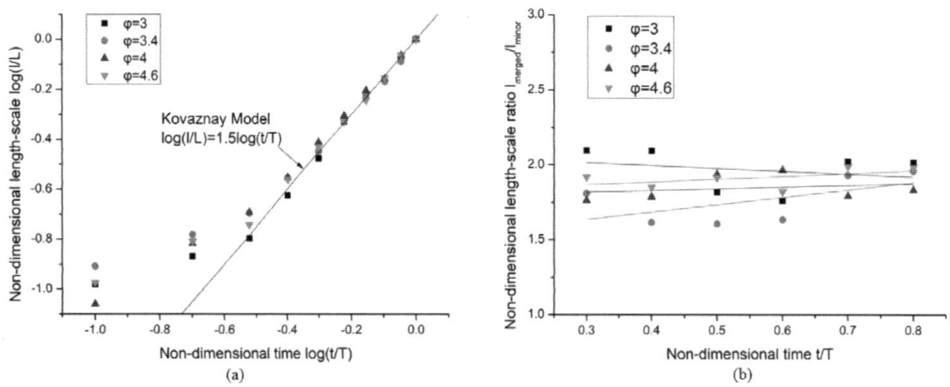

Figure 4. Change of the length-scale along the time in the vortex merger

Simulation results of Cardesa et al. [12] also suggest that in an inverse energy cascade, vortices of a given length-scale originate from the vortex of half their scale. A non-dimensional length-scale ratio is then defined as the ratio of the non-dimensional length-scale of the merged vortex to that of the minor vortex. In Fig.4 (b), the non-dimensional length-scale ratio is presented along the time. It is in general around the expected value 2, which provides an experimental verification for the simulation results of Cardesa et al.

3.3. *Influence of the flapping velocity and angle*

According to a previous report result [8], there is a suitable range of the flapping velocity for a given inlet flow velocity and a given plate size that makes the merged vortex structure most clear. Too high flapping velocity not only cannot effectively generate a merged vortex, but also brings a too strong perturbation to the flow field, causing a wide range of turbulence pulsations. In our experiment, the best velocity ratio φ of the flapping velocity to the inlet flow velocity varies from three kinds of flapping angle. Fig.5 indicates that it is about 4.6 for 90 degree, 3.4 for 60 degree and 3 for 30 degree. From the view of energy input, a bigger flapping angle involves a larger area of air, requiring a more increasing energy input, which refers to a much higher velocity.

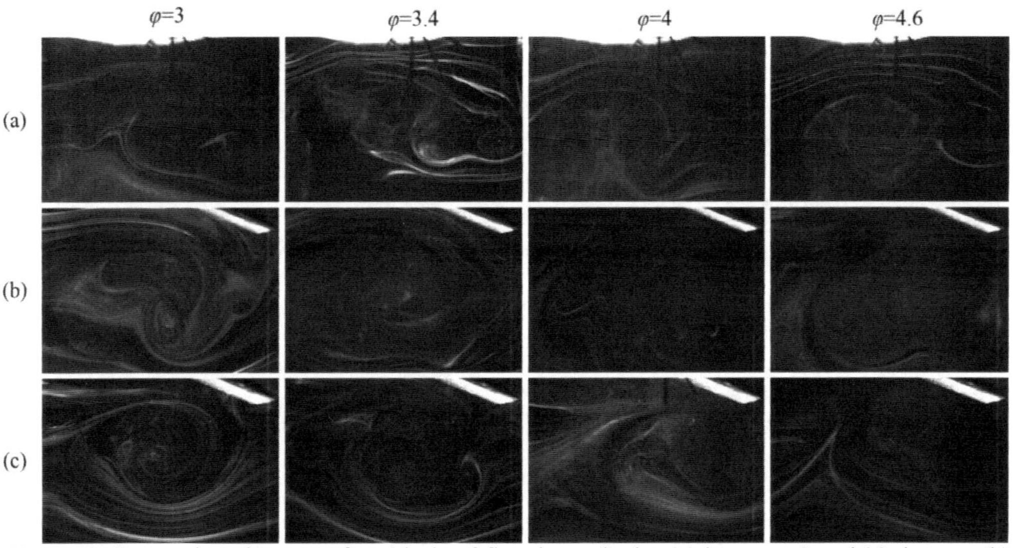

Figure 5. Comparison between four kinds of flapping velocity (a) between 0 and 90 degree; (b) between 30 and 90 degree and (c) between 30 and 60 degree.

If the plate flaps between 30 and 90 degree, a significant difference is that the second inertial vortex is hardly observed. Sometimes even the trailing vortex only has an unclear structure. The second inertial vortex and the trailing vortex seem to be inhibited by the plate, which stops at 30 degree as shown in Fig.6 (b).

If the flapping angle reduces to a small range between 30 and 60 degree, the merged vortex, involving 2 or 3 minor vortices, is quite fuzzy until the plate tends to stop at the end angle. Besides, an exciting phenomenon is that the second merged vortex shows an obviously clear vortex structure, involving just a few minor vortices as shown in Fig.6 (c). Confusingly, several trailing vortices appear again on the upper surface of the fixed plate instead of the trailing edge. Probably, those trailing vortices are not the trailing vortices mentioned in 90 degree flapping angle, but the minor vortices generated in clockwise flap. It is such a small angle that the vortices cannot cross the tip in a limited time.

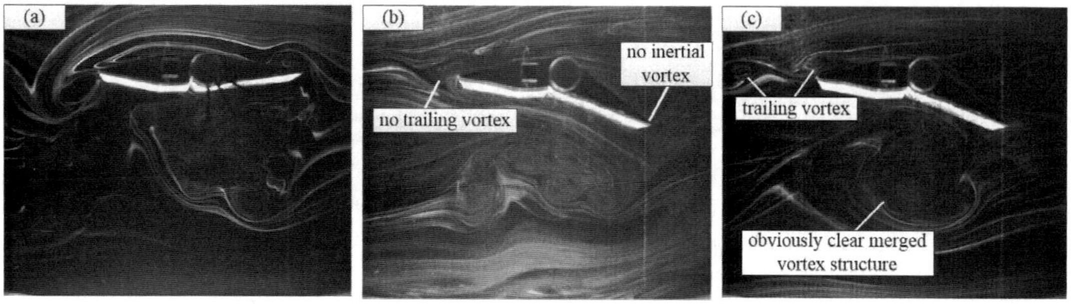

Figure 6. Comparison between three kinds of flapping angle at φ=3.4 (a) between 0 and 90 degree; (b) between 30 and 90 degree and (c) between 30 and 60 degree.

By adjusting the flapping angle, the inertial vortices and the trailing vortices can be inhibited. It may be explained by that the plate, which stops at 30 or 60 degree, can play a guiding role. The plate changes the air movement relative to the inlet flow from orthogonal to oblique, which weakens the velocity shear layer strength in some extent.

Conclusion

The present contribution investigates the merged vortex formation in a flapping plate, aiming at providing new understandings of the merged vortex. A two-dimensional flapping device is designed, which enables the plate to flap at a tip flapping velocity 3 to 4.6 times bigger than the inlet flow velocity. Main results from this work are listed as follows:

(1) The flapping plate generates practically a merged vortex, which merges with several minor vortices, by creating a low-pressure area upstream to drive the air to roll up and an inertial vortex when stops.

(2) The vortex merger follows the Kovaznay model after the non-dimensional time t/T of 0.3 and the non-dimensional length-scale ratio of the merged vortex to the minor vortex is around 2, implying that the vortex merger is a non-equilibrium phenomenon.

(3) The clearest merged vortex is formed when the plate flaps between 30 and 60 degree at a tip flapping velocity 3 times bigger than the inlet flow velocity, maintaining enough energy input and inhibiting the inertial vortex and the trailing vortex formation.

References

[1] J.M. McMichael, M.S. Francis, Micro air vehicles-toward a new dimension in flight, DARPA document. (1997).
[2] T.N. Pornsin-Sirirak, Y.C. Tai, C.M. Ho, et al., Microbat: A palm-sized electrically powered ornithopter, Proceedings of NASA/JPL Workshop on Biomorphic Robotics. 14-17 (2001).
[3] R.J. Wood, The first takeoff of a biologically inspired at-scale robotic insect, IEEE transactions on robotics., 24-2 (2008) 341-347.
[4] W. Send, M. Fischer, K. Jebens, et al., Artificial hinged-wing bird with active torsion and partially linear kinematics, 28th Congress of the International Council of the Aeronautical Sciences. (2012) 23-28.
[5] R.J. Bomphrey, T. Nakata, N. Phillips, et al., Smart wing rotation and trailing-edge vortices enable high frequency mosquito flight, Nature. 544-7648 (2017) 92-95.
[6] J.Y. Andro, L. Jacquin, Frequency effects on the aerodynamic mechanisms of a heaving airfoil in a forward flight configuration, Aerospace Science and Technology. 13-1 (2009) 71-80.
[7] D.D.J. Chandar, M. Damodaran, Computation of unsteady low Reynolds number free-flight aerodynamics of flapping wings, Journal of Aircraft. 47-1 (2010) 141.
[8] G.K. Taylor, R.L. Nudds, A.L.R. Thomas, Flying and swimming animals cruise at a Strouhal number tuned for high power efficiency, Nature. 425.6959 (2003) 707.

[9] A.N. Kolmogorov, The local structure of turbulence in incompressible viscous fluid for very large Reynolds numbers, Dokl. Akad. Nauk SSSR. 30.4 (1941) 299-303.
[10] A.N. Kolmogorov, A refinement of previous hypotheses concerning the local structure of turbulence in a viscous incompressible fluid at high Reynolds number, Journal of Fluid Mechanics. 13.1 (1962) 82-85.
[11] L.F. Richardson, Weather prediction by numerical process, Cambridge University Press, 1922.
[12] J.I. Cardesa, A.Vela-Martín, J. Jiménez, The turbulent cascade in five dimensions, Science. (2017) eaan7933.
[13] L.S.G. Kovaznay, Spectrum of locally isotropic turbulence, J. Aeronaut. 15 (1948) 745.
[14] L. Fang, H.K. Zhao, L.P. Lu, et al., Quantitative description of non-equilibrium turbulent phenomena in compressors, Aero. Sci. Tech. (2017) accepted.
[15] R. Rubinstein, T.T. Clark, 'Equilibrium' and 'non-equilibrium' turbulence, Theo. Appl. Mech. Lett. (2017) accepted.
[16] L. Fang, Y. Zhu, Y.W. Liu, et al., Spectral non-equilibrium property in homogeneous isotropic turbulence and its implication in subgrid-scale modeling, Phys. Lett. A 379 (2015) 2331-2336.

Erosion of water-based fracturing fluid containing particles in a sudden contraction of horizontal pipe

Jiarui Cheng [1], Yinping Cao[2,*], Yihua Dou [2] and Zhen Li [2]

[1]State Key Laboratory of Multiphase Flow in Power Engineering, Xi'an Jiaotong University, Xi'an, China.
[2]Department of Mechanical Engineering, Xi'an Shiyou University, Xi'an, China.

*Corresponding author e-mail: caoyinping029@163.com

Abstract. A lab experiment was carried out to study the effects of pipe flow rate, particle concentration and pipe inner diameter ratio on proppant erosion of the reducing wall in hydraulic fracturing. The results show that the erosion rate and erosion distribution are different not only in radial direction but also in circumferential direction of the sample. The upper part of sample always has a minimum erosion rate and erosion area. Besides, the erosion rate of reducing wall is most affected by fluid flow velocity, and the erosion area is most sensitive to the change in the diameter ratio. Meanwhile, the erosion rate of reducing wall in crosslinked fracturing fluid is mainly determined by the fluid flowing state due to the high viscosity of the liquid. In general, the increase in flow velocity and diameter ratio not only cause the expansion of erosion-affected flow region in sudden contraction section, but also lead to more particles impact the wall.

1. Introduction

During hydraulic fracturing, prop pants constantly impact the pipe wall and the inner surface of pipe sheet (that connects the two pipes), which causes material loss of the reducing wall, especially in the pipe at wellhead or in the central tube of the downhome tool. Excessive material loss may lead to tube thread tripping, frequent failure and loss of valuable production time in oil or gas well.

Experiment on particle erosion has been studied from many aspects, such as mixed media flows, particle sizes and contents, mechanical properties of the metal surface [1] and corrosive characteristics. And erosion tests of special structure, such as sudden contraction, have been studied extensively by scientists. Malta et al. [2] investigated erosion－corrosion behavior in sudden pipe constrictions and found that the erosion rate suddenly increases in pipe constriction section. Wong et al. [3] obtained material loss information in air-suspended sands flowing through aluminum pipe annular cavity. Lin et al. [4] also analyzed the effects of cavity height on cavity erosion under gas－solid flow. Their results showed that micro-erosion occurs on the top area of aft wall, and the height difference between cavity walls markedly affects the maximum erosion rate.

Different pipe flow velocities, particle concentrations and diameter ratios can cause the difference in fracturing fluid flow properties and particle erosion state on the reducing wall. Therefore, a laboratory experimental was used to study the difference in the erosion rate and area of reducing wall by changing fluid flow velocity, particle concentration, and pipe diameter ratio. Meanwhile, the erosion pattern of reducing wall under different flow conditions was discussed. Finally, the wall erosion rate under large fracturing displacement fracturing was predicted by erosion-velocity functional model.

2. Experimental method

2.1. Pipe flow system
In this work, erosion damage was studied in a flow loop with a sudden contraction test cell. A schematic of the flow loop and an image of the test system are shown in Figure 1. The experimental loop mainly comprised a liquid storage tank, a screw pump, a magnetic flow meter (8712HR, Rosemount. Co., America), and an erosion test section.

The test section consisted of a sudden contraction and two straight pipes (Figure 2(a)). The large-diameter pipe (D=50 mm) was connected to a small-diameter pipe (d=35 mm), and both pipes were made of acrylics. Two annular samples (super 13Cr tubing steel) were fixed by four screws on the reducing wall. The outer diameter of the sample was 50mm and the inner diameters were 35mm, 30mm and 25mm, respectively (i.e., d/D=γ=0.7, γ=0.6 and γ=0.5) (Figure 2(b)). The sample surface profiles were verified by H1200WIDE confocal scanning laser microscopy (Laertes. Co., Ltd., Japan).

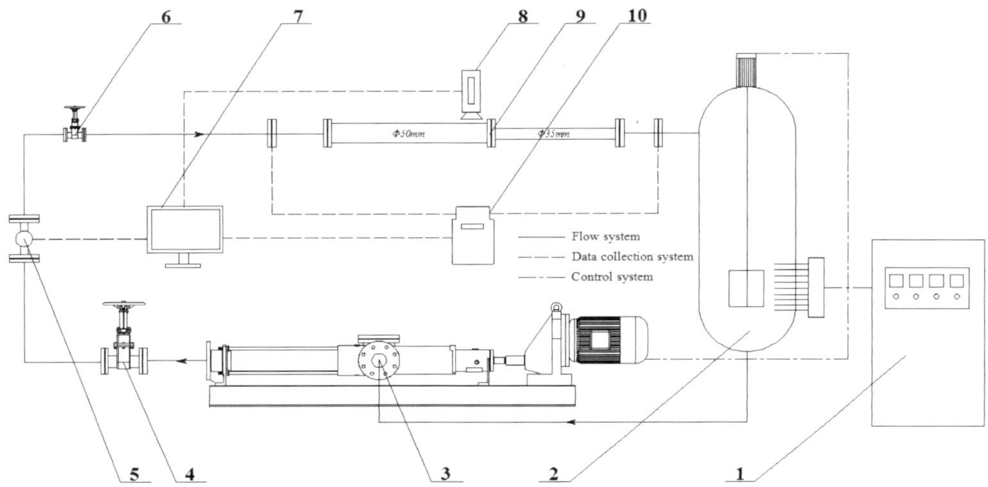

Figure 1. Schematic diagram and actual picture of experimental set-up:

(1)electrical control cabinet; (2)liquid storage tank containing electric heater ($2m^3$); (3)screw pump; (4)(6)cut off valve; (5)flow meter; (7)computer; (8)high-speed camera; (9)test section; (10)pressure transducer.

2.2. Experimental set-up

2.2.1 Solution: The base fluid used in this test added 0.4 wt% hydroxypropyl guar gum as the thickener and 0.4 wt% inorganic boron as the cross-linking agent. Before the experiment, the viscosities of base fluid and cross linked fluid were 27.42 mPa·s and 537.68 mPa·s, respectively. The viscosity of cross linked fluid at the end of the experiment was still higher than 150 mPa·s. Fracturing circular ceramists with a density of 1870 kg/m^3 and an average diameter of 0.6 mm were used as the erosion abrasive. Mass concentrations of the particles in suspension were 50 kg/m^3, 100kg/m^3 and 150 kg/m^3 (i.e., w=50 kg/m^3, w=100kg/m^3 and w=150 kg/m^3).

2.2.2 Fluid flow temperature and velocities: The liquid temperature was set to a fixed value of 50℃ with positive and negative fluctuations of 3℃ to simulate the actual pipe flow temperature. In order to avoid excessive temperature rise and to ensure adequate erosion time, the inlet flow velocities in this experiment were 1.5 m/s, 2 m/s, 2.5 m/s, 3 m/s, and 3.5 m/s.

2.2.3 The operating principle: Firstly, the liquid was set and heated to a predetermined temperature with stirring, followed by mixing the fluid and particles in the tank. Secondly, the mixture was stirred by the screw pump for 5 min in the bypass pipe. Finally, the cut-off valve was opened when the flow reached stability, and related data, including particle motion image, flow pressure, and temperature, were recorded. In order to prevent the particles from completely settling in the lower part of the pipe, each erosion test was limited to 1 h.

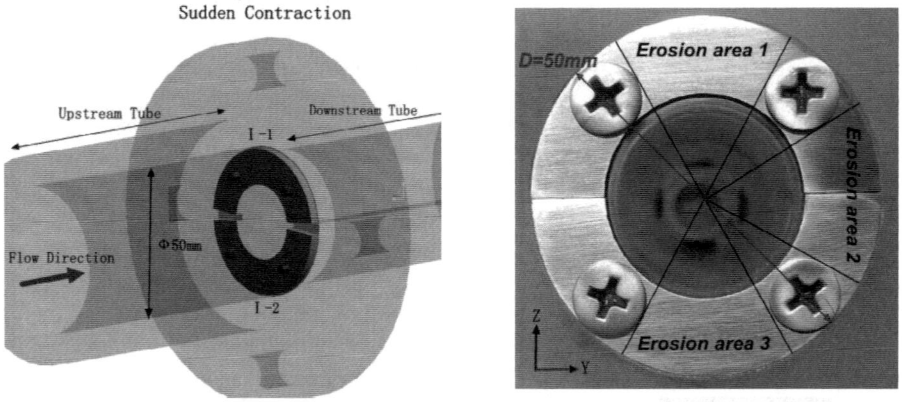

(a) Three-dimensional structure of test section (b) annular samples

Figure 2. Structure of the test section and the super 13Cr stainless steel samples.

3. Results

3.1. Erosion characteristics of reducing wall

As shown in Figure 3, the deepest erosion area is close to the inner edge of the sample, and Erosion area 1 has the minimum erosion depth. If we define the serious erosion area as an area where erosion depth is greater than 10% of maximum erosion depth, it will find that the serious erosion surface underwent multiple overlapping impacts with formation of craters, platelets, and extruding lips. And many scars caused by particle impingement were independently distributed on the outer circumference surface of sample. Although only 25% ~ 40% of total erosion areas are serious erosion areas, the surface of these areas are seriously damaged due to constant plastic deformation and cutting process acting by particle impacts. Therefore, the damage rate and width of each serious erosion area were measured as the important parameters.

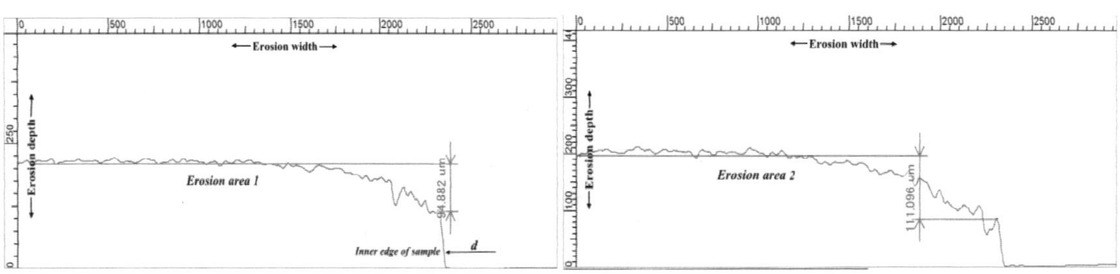

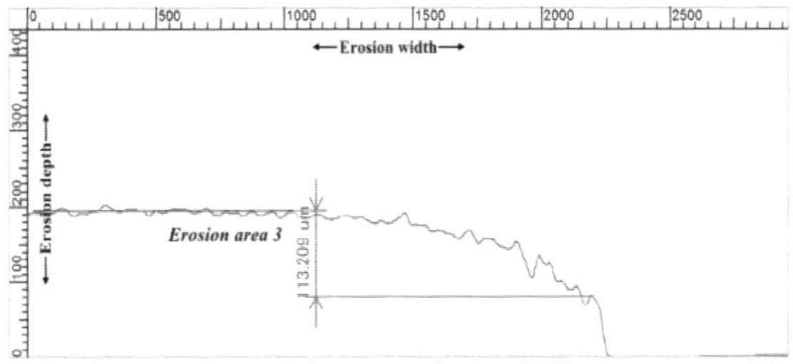

Figure 3. The profiles of the inner edge of the sample in different areas. The inlet flow velocity is 3 m/s.

3.2. Erosion rates of reducing wall

Figure 4(a) (b) (c) show the maximum erosion rates of each serious erosion area under different flow velocities, particle concentrations and diameter ratios. Except that the erosion rate increases with the increasing liquid flow velocity, the change of erosion rate with the particle concentration and diameter ratio is as follows:

3.2.1 Effect of particle concentration: When the particle concentration increases from 50 kg/m3 to 100 kg/m3, the growth of erosion rate is in the range of 23% ~ 31% (Figure 4(a)). And this growth is reduced to 18% with concentration of 150 kg/m3. The difference in erosion rate among different areas is large when w<100 kg/m3, and small at high particle concentration.

3.2.2 Effect of pipe diameter ratio: The smaller the diameter ratio, the greater the liquid flow velocity in the downstream tube, which may result in a greater particle impact velocity on the reducing wall. The growth of erosion rate varies in a wide range of 14% ~ 36% at flow velocity of 1.5m/s ~ 2.5m/s between γ=0.7 and γ=0.6 (Figure 4(a) and (b)), and this variation range decreases as the flow rate increases. When the diameter ratio is reduced to 0.5 (Figure 4(c)), the growth of erosion rate is less than 10% comparing to γ=0.6.

3.3. The widths of erosion area

As shown in Figure 4(d) (e) (f), the significant factor affecting the erosion area is the diameter ratio, followed by the concentration of particles. The widths of Erosion area 2 and 3 are similar and are much larger than the width of Erosion area 1. When the particle concentration increased by 50kg/m3, the growth of each width is almost less than 5% (Figure 4(d)), which indicates that the change in particle concentration has little effect on the erosion area. However, when the downstream pipe diameter is reduced, the width of erosion area increases significantly. Erosion width almost increased by 40% when diameter ratio is reduced from 0.7 to 0.6, which is maintained at a diameter ratio of 0.5.

IOP Conf. Series: Journal of Physics: Conf. Series **916** (2017) 012017 doi:10.1088/1742-6596/916/1/012017

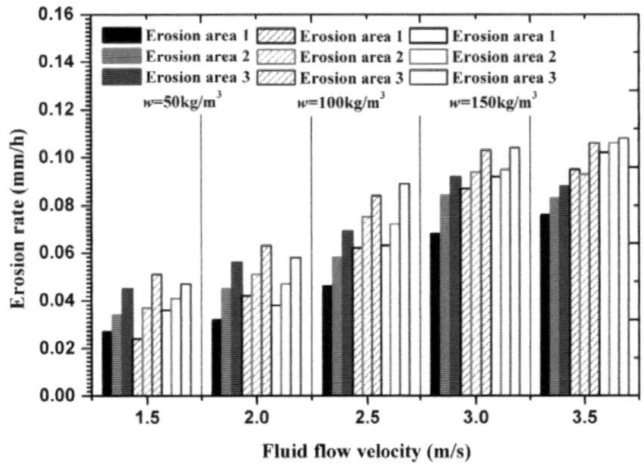

(a) $\gamma = 0.7$

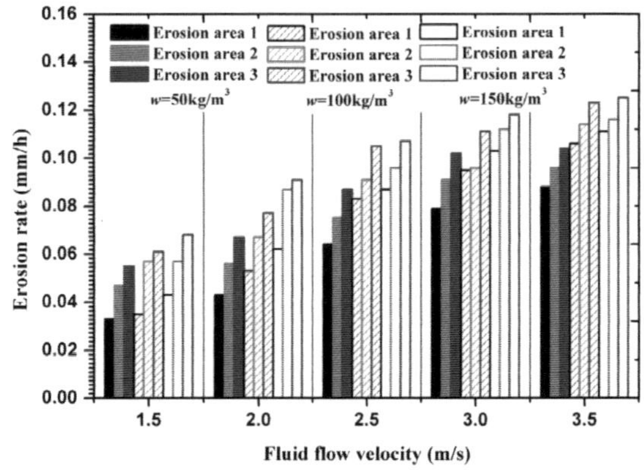

(b) $\gamma = 0.6$

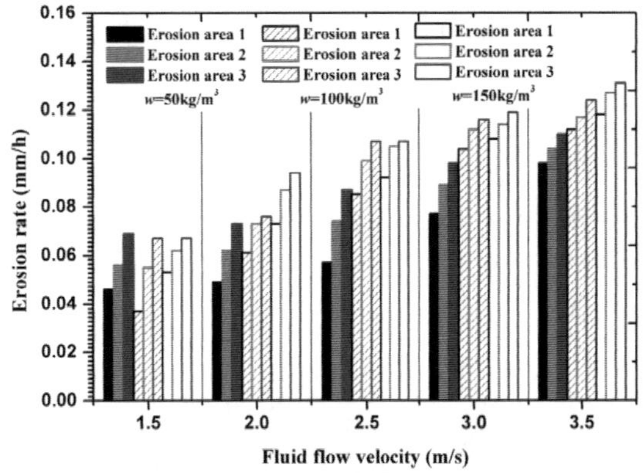

(c) $\gamma = 0.5$

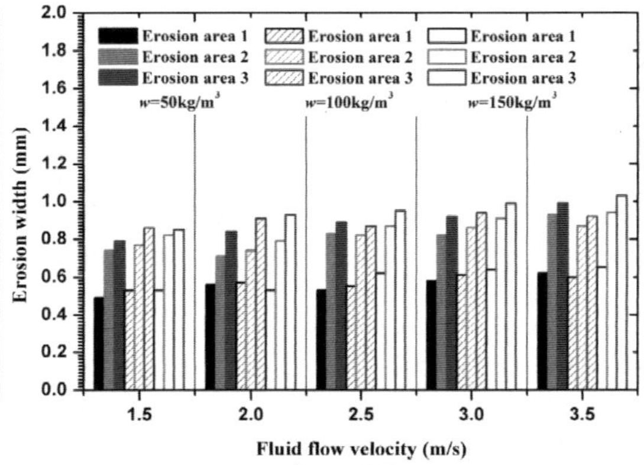

(d) $\gamma = 0.7$

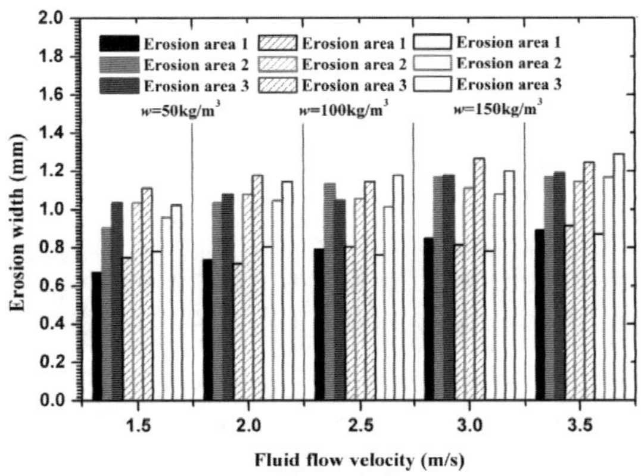

(e) $\gamma = 0.6$

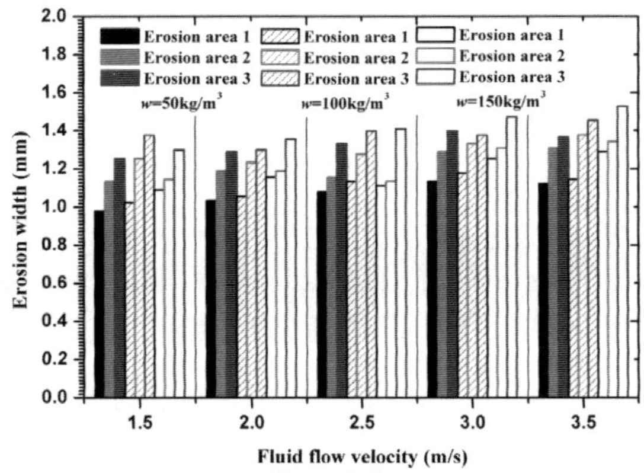

(f) $\gamma = 0.5$

Figure 4. Erosion rates and widths of reducing wall.

4. Discussion

In the process of changing the flow direction in the sudden contraction section, if the particles are out of the fluid control, they will directly impact on the wall due to inertia. In this case, the erosion area is mainly affected by inter-particle collisions and partial turbulence of liquid. And this erosion pattern can be called Particle Dominated Erosion. If the particles have good following properties, the erosion area and rate will change with the erosion-affected flow region. This erosion is called as Flow Dominated Erosion.

The characterization of particle following property mainly depends on the value of Stokes number, which can be expressed as $St=\rho p d p 2u/ (18\mu D)$ [5]. When $St>>1$, the fluid has little effect on motion of the particles. For $St<1$, particles follow fluid streamlines closely. When $St=1$, the erosion map of the reducing wall with different pipe inner diameters is shown in Figure 5. The area above each curve indicates the Flow Dominated Erosion, and the area under each curve represents Particle Dominated Erosion. Consequently, the particle erosion at the reducing wall in cross linked fracturing fluid (Figure 6(a)) is mainly determined by the flowing state due to the high viscosity of the liquid. And the erosion pattern in flow-back fluid (i.e., gel out fluid) can be treated as Particle Dominated Erosion (Figure 6(b)). Therefore, the flow velocity in the fracturing fluid injection process plays a crucial role for the reducing wall erosion.

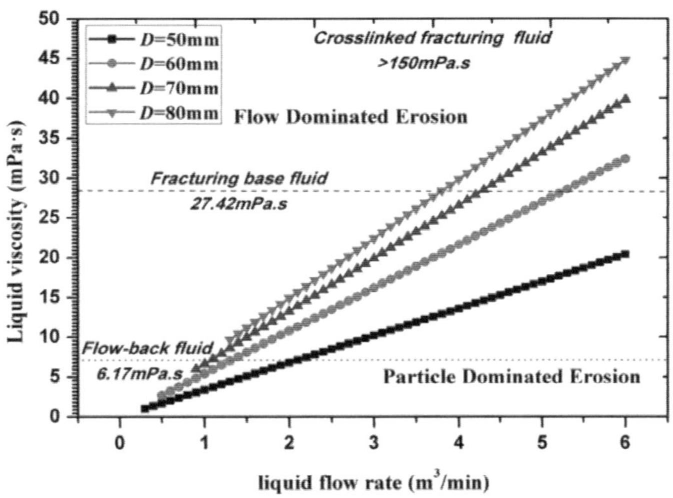

Figure 5. The erosion map of the reducing wall with different pipe inner diameters when $St=1$.

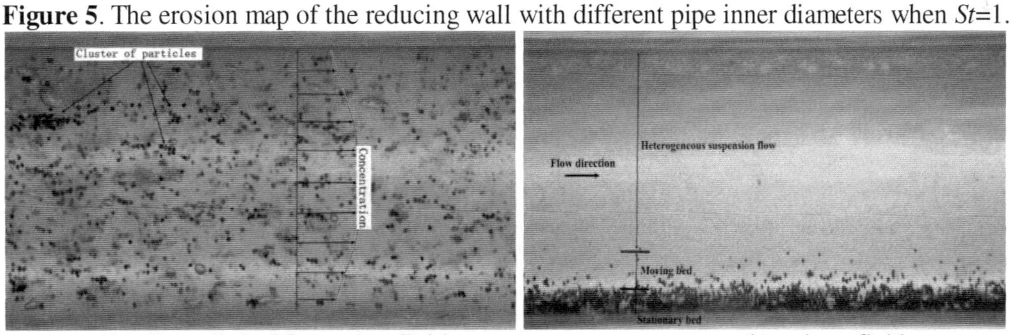

(a) Cross linked fracturing fluid; (b) gel out fluid.

Figure 6. Particle distribution in different fluids:

5. Conclusion

The conclusions derived from the present study can be summarized as follows:

(1) Both of the erosion widths and the erosion depths of each erosion area increase with increasing flow velocity, and the deepest erosion area is close to the inner edge of the sample. The width and the depth of Erosion area 1 are smaller than that of other erosion areas in each flow velocity.

(2) Erosion rate of reducing wall is most affected by fluid flow velocity, followed by particle concentration and diameter ratio. By contrast, the width of erosion area is most sensitive to the change in

the diameter ratio, then the particle concentration and flow velocity. The difference in erosion rate between Erosion area 1 and other areas is obvious at the low flow velocity than the high flow velocity. And the growth rate of erosion is slowing down with the increase of particle concentration and the decrease of diameter ratio.

(3) The particle erosion in different fluids can be divided into two erosion patterns based on particle following property. And the erosion rate of reducing wall in cross linked fracturing fluid is mainly determined by the flowing state due to the high viscosity of the liquid.

Acknowledgments

This work was supported by the National Natural Science Foundation of China (grant no. 51404198), and it was also performed by the group of profession and innovation for well testing integrity and safety of Xi'an Shiloh University.

References

[1] M.A. Habib., R. Ben-Mansour, H.M. Badr, M.E. Kabir, Erosion and penetration rates of a pipe protruded in a sudden contraction. Comput. Fluids. 37 (2008) 146-160.

[2] R. Malka, S. Nešić, D.A. Gulino, Erosion–corrosion and synergistic effects in disturbed liquid-particle flow. Wear. 262 (2007) 791-799.

[3] C.Y. Wong, C. Solnordal, A. Swallow, J. Wu, Experimental and computational modelling of solid particle erosion in a pipe annular cavity. Wear. 303 (2013) 109-129.

[4] Z. Lin., H. Xu, Y. Wang, Y. Li, B. Cui, X. Ruan, Experimental study of particle erosion in a cavity with a height difference between its walls. Powder. Technol. 286 (2015) 378-384.

[5] W.S. Peng, X.W. Gao, Numerical prediction of erosion distributions and solid particle trajectories in elbows for gas-solid flow. J. Nat. Gas Sci. Eng. 30 (2016) 455-470.

Numerical study of fairing installed between brackets based on CFD

Peng Xi[1], Ying Xiong[1,*] and Xin Tang[2]

[1]Naval University of Engineering, Wuhan, China
[2]Military Representative Office of the Navy Stationed in Wuhu, Wuhu, China

*Corresponding author e-mail: ying_xiong28@126.com

Abstract. In view of the low speed and instability of the flow between the two arms of the bracket in front of the propeller, the fairing is installed between the arms of the bracket taking example of compensating duct, in order to speed up the flow between the bracket arms and improve the flow quality. A four-propeller surface ship was studied and an integral mathematic model including hull, appendage and propellers was established. Using a RANS solver, its installation height, angle and airfoil is optimized. Then ship models with fairing and without fairing are calculated. The result shows that fairing improves propeller efficiency behind ship with 1.1% of the outer propeller and 1.6% of the inner propeller, which indicates that fairing helps improve the flow quality

Keywords. Fairing; propeller efficiency behind ship; four-propeller ship.

1. Introduction

Due to the fact of energy dilemma and poor environment, emission reduction, energy conservation and noise reduction of ships have become the research focus of the whole world. By installing new type of appendage, the uniformity of the wake flow around the stern could be improved and a favorable interference for the propellers could be aroused. In turn, a higher propulsive efficiency could be get. The compensating duct works notably in saving energy, and it has advantages of cheap cost and simple structures. It is widely researched home and abroad.

The compensating duct is a semicircle duct installed before the propellers, which can accelerate the velocity of the flow above the propeller, improve the uniformity of the flow around the stern and the propulsive efficiency of the propellers, and avoid blade tip cavitation erosion to some extent at the same time. In 1984, Germany Pro. Schneekluth.H firstly applied the compensating duct to 74000t "LudolfOldendorff" bulk carrier and worked well. Korkut[1] installed the same type of compensating duct to different boats and conducted contrastive analysis to the effect. The scaling effect to the energy saving effect of compensating duct was researched by Friesch[2]. MARIC[3] conducted 212 resistance tests and 475 self-propulsive tests to 22 ship models which showed that the energy saving effect of compensating duct differed from 5% to 11% due to different ship square coefficients. Wenhao Qian et al.[4] studied the measuring method of appended thrust of the compensating duct. Jieya Li et al. [5-6]verified it's saving energy and reducing vibration through experiment on compensating duct. Lijian Ou et al. [7] applied compensating duct in double-tail boat, which showed that compensating duct could improve the flow quality. Shaofeng Huang et al. [8] researched the saving energy effect of compensating duct in bulk cargo ship, and proved that the hull efficiency is a reasonable indicator in assessing its saving energy effect.

In this paper, the fairing is installed between the arms of the bracket taking example of compensating duct, in order to speed up the flow between the bracket arms and improve the flow quality. A four-propeller surface ship was studied and an integral mathematic model including hull, appendage and propellers was established. The fairing installation elements were optimized aiming at improving the flow quality and then its energy saving effect was estimated.

2. Mathematic model

2.1. Governing equations

Continuity equation is the concrete pattern of manifestation of mass conservation in the fluid.

$$\frac{\partial \rho}{\partial t}+\frac{\partial}{\partial x_i}(\rho u_i)=0 \tag{1}$$

Reynolds-Average Navier-Stokes (URANS) equations are represented as follows:

$$\frac{\partial}{\partial t}(\rho u_i)+\frac{\partial}{\partial x_j}(\rho u_i u_j)=-\frac{\partial p}{\partial x_i}+\frac{\partial}{\partial x_j}\left[\begin{array}{l}\mu_0(\frac{\partial u_i}{\partial x_j}+\frac{\partial u_j}{\partial x_i})-\\ \frac{2}{3}\mu_0\frac{\partial u_l}{\partial x_l}\delta_{ij}\end{array}\right]+\frac{\partial}{\partial x_j}(-\rho\overline{u_i' u_j'})+\rho f_i \tag{2}$$

Where ρ is the fluid density and p is the static pressure, f_i the mass force per unit mass, u_i 、 u_j the velocity components, $-\rho\overline{u_i' u_j'}$ the Reynolds stress representing turbulence effect.

Nowadays, turbulence energy equations and turbulence dissipation equations are basic method to reflect the turbulence characteristic. Added equations are needed to connect the pulse value and the time average in order to solve the increased Reynolds stress in N-S equations, which is called turbulence model. This paper adopted the SST $k-\omega$ turbulence model, which is developed from the standard $k-\omega$ turbulence model. This model owns high precision in the fluent simulation because it takes the advantage of the standard $k-\omega$ turbulence model, combines the cross-diffusion from equation ω, and takes the spread of turbulent shear stress into consideration.

2.2. Model and scheme

This paper studys a displacement surface ship with bulbous bow. The model is 8.56m at length and draft is 0.303m. There are four same propellers arranged at the stern, which are installed symmetrically. The propellers have five blades and changes in skew and trim. Fig.1 and Fig.2 show that the fairing is installed between brackets.

Figure 1. Fairing model from front view

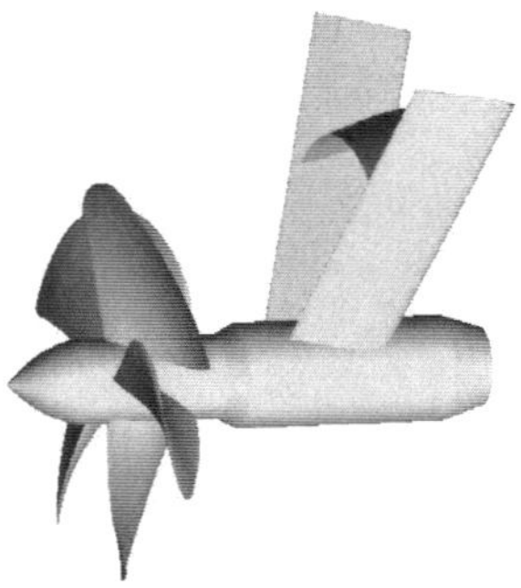

Figure 2. Fairing model from front side view

Fig.3 shows the factors of the fairing. The length of the fairing, which is 0.033m, is decided by the chord length of bracket cross section. In order to confirm the installation height and angle of the fairing, nine schemes are proposed. Table.1 shows the schemes with same airfoil and different heights and angles.

Table 1. Schemes of fairing with different height and angle

N umber	Height	Angle	Length/m	Aioli	Note
X1		10	0.033	65(1) -410	The front fairing and the back fairing in all schemes are the same.
X2	1.05D	15	0.033	65(1) -410	
X3		20	0.033	65(1) -410	
Y1		10	0.033	65(1) -410	
Y2	1.1D	15	0.033	65(1) -410	
Y3		20	0.033	65(1) -410	
Z1		10	0.033	65(1) -410	
Z2	1.15D	15	0.033	65(1) -410	
Z3		20	0.033	65(1) -410	

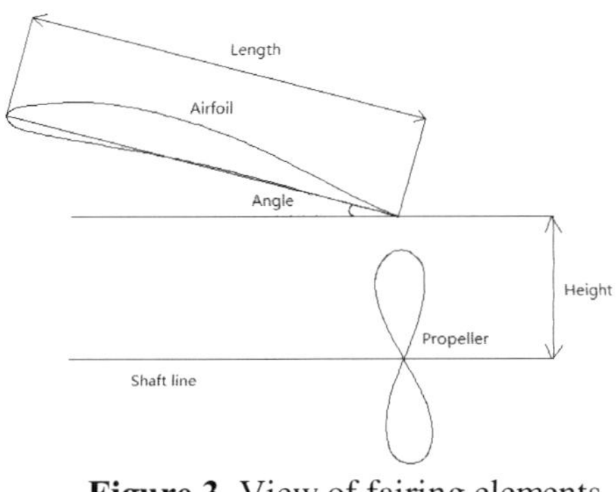

Figure 3. View of fairing elements

2.3. Mesh generation and boundary conditions

As the computational model is symmetrical, only starboard is considered in modeling and simulating. In order to eliminate the influence of the computational boundary on the simulated flow, the boundary of the computational flow field is set at a distance long enough away from the computational model. The inflow boundary, at a distance of one total model length upstream from the bow of the model, is imposed with the velocity-inlet boundary condition. The outflow boundary, at a distance twice the total model length downstream from the stern of the model, is imposed with the undisturbed pressure-outlet boundary condition. The outer boundary, at a distance of one total model length from the central axes of the model, is imposed with the undisturbed velocity boundary condition. The surface of the model is set as in no-slip boundary condition.

Due to the complexity of the model and the repeatability of the schemes, a multi-block grid consisting of four blocks is used, which is showed in Fig.4. Interface is used to connect each block. Block 1 includes the front bracket and its mesh is generated by structured grid. Block 2 includes the back bracket and its mesh is generated by structured grid. Block 3 includes the deadwood immediately and its mesh is generated by unstructured grid. The hull's mesh is generated by structured grid. The number of the mesh of block 1 and block 2 is 1.05 million, of block 3 is 4.3 million and the rest is 5.7 million. Fig.5 shows the mesh model.

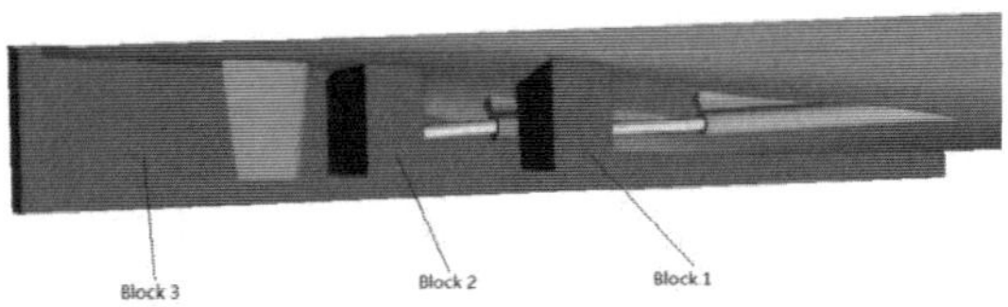

Figure 4. Blocks of the model

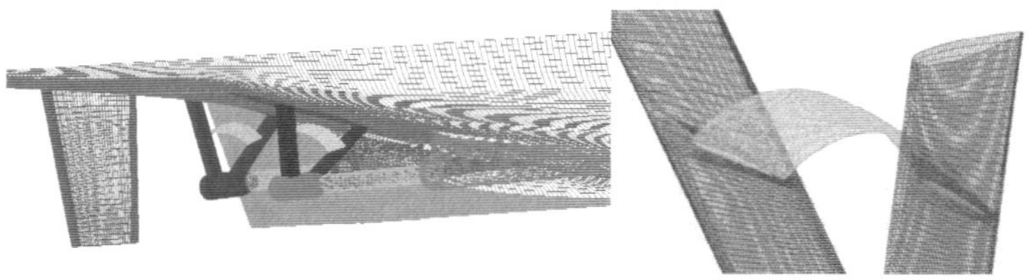

Figure 5. Stern mesh and fairing mesh

The fluid control equation and the turbulence equation are numerically solved by the finite volume method. The convection term is discretized by a second order upwind scheme, the diffusion term is discretized by the central difference scheme, the temporal term is discretized by a second order implicit scheme, and the pressure-velocity coupling equation is solved by the PISO method. The discretized algebraic equations are solved by the Gauss-Seidel iteration method.

3. Numerical computation

3.1. *Optimization of height and angle*

Nominal flow at the propeller of different schemes is obtained to determine preferred fairing installation height and angle. In order to show the improvement of flow quality accurately, uneven coefficient of each scheme is presented in table.2. Uneven coefficient comparison of all schemes is presented in fig.6.

Table 2. Uneven coefficient in different schemes

	The inner propeller				The outer propeller			
r/R	0.4	0.6	0.8	1.0	0.4	0.6	0.8	1.0
Original	0.4202	0.404	0.3715	0.3253	0.4196	0.3604	0.3322	0.3386
X1	0.4484	0.3987	0.3349	0.3257	0.4074	0.3421	0.3322	0.3833
X2	0.4422	0.3856	0.3176	0.3166	0.4046	0.3359	0.3302	0.3672
X3	0.4341	0.3645	0.3057	0.3025	0.3983	0.3252	0.3232	0.3731
Y1	0.4415	0.3988	0.3366	0.3083	0.4083	0.3423	0.334	0.3538
Y2	0.4383	0.3873	0.3187	0.3052	0.4054	0.338	0.3354	0.365
Y3	0.4328	0.3683	0.3053	0.3014	0.3979	0.3236	0.3201	0.3593
Z1	0.4515	0.4007	0.3348	0.3063	0.4119	0.3442	0.3331	0.3483
Z2	0.4395	0.3914	0.3166	0.3032	0.4043	0.3367	0.3295	0.3474
Z3	0.4379	0.3732	0.3005	0.3024	0.3841	0.3272	0.3249	0.3545

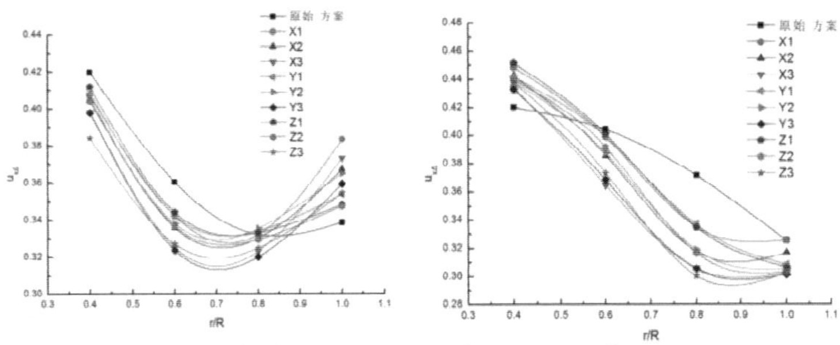

outer propeller inner propeller

Figure 6. Comparison of uneven coefficient on the propeller plane

Fig.6 shows that uneven coefficient on the outer propeller plane of each scheme is smaller than the original scheme's at r/R<0.8, bigger at blade tip. Comprehensive comparison of uneven coefficient on the outer propeller plane at different radius figures out that uneven coefficient reduction of scheme Z2 at r/R=0.4, 0.6, 0.8 is 3.62%, 6.56%, 0.83% respectively, and at blade tip the increase is 2.59%, which is the best among all schemes. Uneven coefficient on the inner propeller plane of each scheme is smaller than the original scheme's at all radius except at blade root. Scheme Y3's uneven coefficient reduction at r/R=0.6, 0.8, 1.0 is 8.83%, 17.81%, 7.37% respectively and increase at r/R=0.4 is 3.00%, which is the best among all schemes.

Installation elements are obtained from above, which is showed in table3.

Table 3. Installation elements of fairing

	height	angle/°
Front fairing	1.15D	15
Back fairing	1.1D	20

3.2. Optimization of airfoil

In this part, several airfoils such as NACA2410, 4412, 4415, 63(1) -410, 63(1) -412, 63(2) -615, 65(1) -206, 63-206 are researched to choose the best airfoil. Table4 shows the uneven coefficient of each scheme on the propeller plane. Uneven coefficient comparison of all schemes is presented in fig.7.

Table 4. Uneven coefficient in different schemes

	The inner propeller				The outer propeller			
r/R	0.4	0.6	0.8	1	0.4	0.6	0.8	1
Original	0.4202	0.404	0.3715	0.3253	0.4196	0.3604	0.3322	0.3386
2410	0.4409	0.3674	0.3161	0.3191	0.4326	0.3882	0.3706	0.3706
4412	0.4482	0.3762	0.3195	0.3181	0.4324	0.387	0.3877	0.4194
4415	0.4515	0.3831	0.3216	0.317	0.4331	0.3832	0.3753	0.409
63(1)-410	0.447	0.3743	0.3184	0.3199	0.4378	0.391	0.3917	0.414
63(1)-412	0.446	0.3724	0.3161	0.3204	0.434	0.388	0.3794	0.3961
63(2)-615	0.447	0.3752	0.3183	0.3183	0.4325	0.3878	0.3866	0.4268
65(1)-206	0.4392	0.3601	0.3133	0.3192	0.4405	0.3955	0.3881	0.4005
63-206	0.4323	0.3542	0.2977	0.3079	0.4089	0.3384	0.3335	0.3369

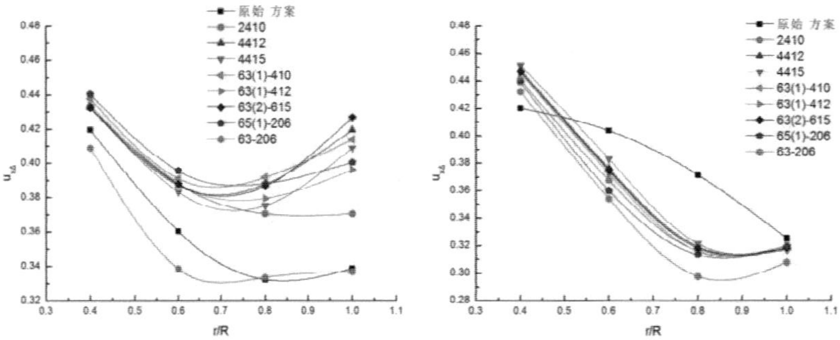

outer propeller inner propeller

Figure 7. Comparison of uneven coefficient on the propeller plane

Fig.7 shows that uneven coefficient on the outer propeller plane of each scheme is bigger than the original's except the scheme 63-206. Uneven coefficient on the inner propeller plane of each scheme is smaller than the original scheme's at all radius except at blade root. Among all schemes, uneven coefficient of scheme 63-206 is the smallest. Comprehensively, airfoil NACA63-206 is the best scheme. Uneven coefficient on the outer propeller plane reduces by 2.54%, 6.08% at r/R=0.4, 0.6, while stays unchanged at r/R=0.8, 1.0. Uneven coefficient reduction on the inner propeller plane at r/R=0.6, 0.8, 1.0 is 12.33%, 19.86%, 5.36% respectively and increase at r/R=0.4 is 2.9%.

3.3. Efficiency evaluation

In this part, a multi-block grid is used including five blocks. Block 1 and block 3 cover the brackets, including 1,100,000 structured grids. Block 2 and block 4 cover the rotating area, including 500,000 unstructured grids. Block 5 covers the dead wood, including 4,300,000 unstructured grids. The rotating area includes 700,000 unstructured grids, while the other area includes 5,700,000 structured grids. Fig.8 shows the block dividing. Table5 shows the comparison of the hydrodynamic coefficient.

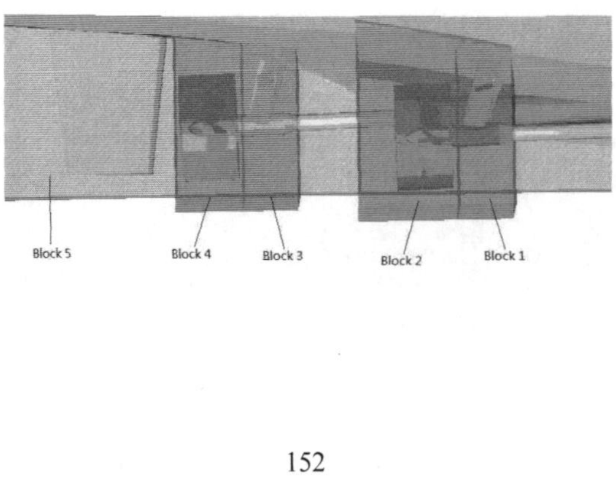

Figure8. Blocks of the model

Table 5.result comparison

Comparison of computational result	hydrodynamic coefficient					
	the outer propeller			the inner propeller		
	Kt	10Kq	η	Kt	10Kq	η
original scheme	0.1835	0.3898	0.6307	0.209	0.4423	0.633
fairing scheme	0.1807	0.3797	0.6374	0.2101	0.4377	0.6431
increase percentage/%			1.0678			1.5946

We can see from above that efficiency of the outer propeller and the inner propeller increases by 1.1% and 1.6% respectively after installing the fairing, which verifies the effect of improving the stern flow.

4. Conclusion

The research above tells us the installation elements of the fairing. The length of the fairing is 0.033m and the airfoil is NACA63-206. The height of the outer fairing and the inner fairing is 1.15D and 1.1D respectively, and the angle is 15° and 20° respectively. The fairing can improve the flow around the propeller and the efficiency is improved.

However, this paper just takes the flow and the efficiency into account, neglecting the resistance which is a negative effect of installing fairing. So next work should pay more attention to the comprehensive evaluation of the fairing.

References

[1] Korkute. A case study for the affect of a flow improvement device on ship powering characteristics[J]. Ocean Engineering, 2005, 33(2): 205-218.

[2] Friesch J, Ionannesn C. Propulsion optimization tests at high Reynolds numbers[J]. Sname Transactions, 1994(102):1-21.

[3] Qian Wenhao, Zhao Hanhun, Cai Yuejin, et al. Propeller inflow compensative simplified nozzle[J]. Ship Engineering, 1989, 6(3): 10-16.

[4] Qian Wenhao, Mo Cheng, Wei Yanqing, et al. Testing and study of the thrust of compensation duct[J]. Ship&Boat, 1997(2): 7-11.

[5] Li Jieya, Yu Ling, Yang Shaogang, et al. An experimental investigation on propeller inflow compensative nozzle for cargo ship[J]. Ship Engineering, 1996(3): 18-20.

[6] Li Jieya, Li Ganluo, Yu Ling. Damping effect of ship compensative nozzle[J]. Ship Building Technology, 1997(8): 14-16.

[7] Ou Lijian, Li Jieya, Wang Zhiyong, et al. A Study on the Application of the Compensating Duct on the Twin-Tail Fin Ship[J]. Journal of South China University of Technology(Natural Science Edition), 2000, 28(6): 49-52.

[8] Huang Shaofeng, Huang Guofu, Yang Yi. Numerical Prediction for Effectiveness of Wake Equalizing Duct[J]. Shipbuilding of China, 2012, 53(1): 7-12.

[9] Zhang Zhaoshun, Cui Guixiang, Xu Chunxiao. Turbulence Theory and Simulation[M]. Beijing: Tsinghua University Press, 2005.

[10] Chang Yi, Zhang Zhirong, Zhao Feng. Application of multi-block structured grid to computation of viscous flow around a ship with appendage[J]. Journal of Ship Mechanics, 2004, 8(1): 19-25.

Performance analysis of underwater pump for water-air dual-use engine

Jun Xia [a], **Yun Wang** [b], **Yu Chen** [c]

School of Aircraft Engineering, Nanchang Hangkong University, Nanchang, China.

[a]tomxj375@163.com
[b]wangyun66@126.com
[c]745317205@qq.com

Abstract. To make water-air dual-use engine work both in air and under water, the compressor of the engine should not only meet the requirements of air flight, but also must have the ability to work underwater. To verify the performance of the compressor when the water-air dual-use engine underwater propulsion mode, the underwater pumping water model of the air compressor is simulated by commercial CFD software, and the flow field analysis is carried out. The results show that conventional air compressors have a certain ability to work in the water environment, however, the blade has a great influence on the flow, and the compressor structure also affects the pump performance. Compressor can initially take into account the two modes of water and air. In order to obtain better performance, the structure of the compressor needs further improvement and optimization.

Key words: Water-air dual-use engine; Compressor; Underwater propulsion; Numerical Simulation; Flow field analysis.

1. Introduction

With the development of the war in the future, the all-dimensional weapon is more important [1], but, now the aircraft can only flight, the enemy region has the ability to survive the air is greatly reduced, If the aircraft can dive into the water after finished the mission, it can avoid tracking and search, then the survival ability can be greatly improved [2, 3]. So, the water-air aircraft are the product of such demands, the aircraft needed to achieve air fly and underwater propulsion, requiring a unique engine.

The water-air dual-use engine has two unique modes to work both in the air and the water. Among them, water-reactive metal materials are applied in the underwater working mode to achieve the underwater propulsion.The engine's thrust in air propulsion is from jet fuel combustion, which is similar to the current aero-engine [1]. Uunderwater propulsion is from high temperature gas which is produced by hydro-reactive metal reacting with water [4].

In order to explore the work condition of traditional aero-engine compressor under the underwater environment and evaluate the feasibility in compressive pump water performance of water-air dual-use engine in the underwater working mode, the CFD software has been applied in this paper to make the preliminary simulation analysis. Taking the example of the NASA STAGE35 compressor, this paper firstly makes analysis on its compression performance under the air environment to make comparison and verification, then making the compression work simulation of the compressive performance of the compressor under the water environment. Thus to analyse the compressive performance of the compressor under the water environment and provide the useful reference for the design of water-air dual-use engines.

2. Model Calculation And Analysis on the Gas Performance

2.1. Physical model
This paper uses NASA STAGE35 as calculation model. NASA STAGE35 is an axia-flow transonic compressor, which is one of four inlet stage for an advanced core compressor, designed by NASA Glenn research center in 1978. The rotor of NASA STAGE35 has 36 blades, the tip solidity is 1.3, the rotor tip speed is 455 meters per second, and the tip clearance is 0.2% of blade span. The stator has 46 blades, the tip solidity is 1.3, and the root gap is 0.5% of blade span [5, 6].

2.2. Model mesh generation and calculation
As the flow solver the well-known commercial software Numeral/ Fine turbo is used in present study. Auto grid/ IGG module is used to produce the stage three-dimension meshh (Han et al., 2010). The first layer of the wall grid is 0.003mm and the y+ is 1. After grid generation, the grid number of rotor is about 1.03 million, stator is about 1.18 million, and the total number of single channels is about 221 million [7] .As show in Fig1. FINE module is applied for flow field simulation, and Spalart-Allmaras model is applied for the turbulence model calculation, and the working fluid is ideal gas. The CFL number is 3. The boundary conditions as flows: the flow direction is axial, the compressor speed is 17188.7RPM / min, the total inlet pressure is 101325Pa, the total inlet temperature is 288.2K, and the inlet turbulent viscosity is 0.00005m2/s, also the average static pressure of the outlet is 130,000Pa. Mmeanwhile, considering the flow's periodicity and calculation, here only the single-channel zone is calculated and the rotation periodic boundary condition is applied.

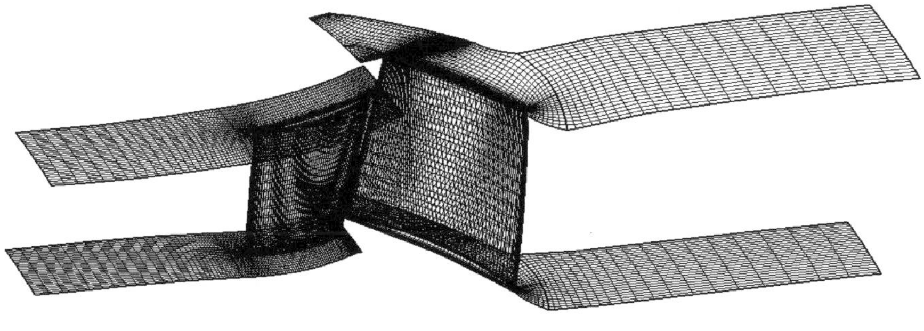

Figure 1. Grid of NASA Stage 35

2.3. Analysis on the calculation results
As the calculated results show, the flow rate is 20.59Kg / s, the pressure ratio is 1.79, the calculation error is within the allowable range with a small difference from the design condition, indicating the model and the meshing generation are correct.

3. Work simulation of compressors in underwater environment
When the compressor is in the water, the fluid is changed from air to water, to make the mesh more adaptable to fluid changes, the first layer of the wall grid modified to 0.00019mm. The rest conditions are not modified. Then, mmeshing again, as the compressor enters the water environment, the compressor operation should be readjusted. In the underwater environment, the compressor is similar

to a pump, the differential pressure and the efficiency have become significant underwater parameters for the compressor. The underwater calculation mode is the same as the air mode. However, the compressor speed is 1000RPM / min. The performance curve of this 'water pump' could be obtained by the change of the outlet mass flow. Preliminary consider the shallow water area, the boundary conditions as follows: the total inlet pressure is 110000Pa, the total inlet temperature is 293K, and the static pressure of the outlet is 130400 Pa.

3.1. Analysis of compressors' operating point performance &parameters of the rated water pressure in underwater working mode

According to the above boundary conditions, the compressor simulation &calculation are made in the underwater environment and the total pressure distribution map (as show in Fig. 2) of the compressor flow path in the underwater working is drew from the calculation results.

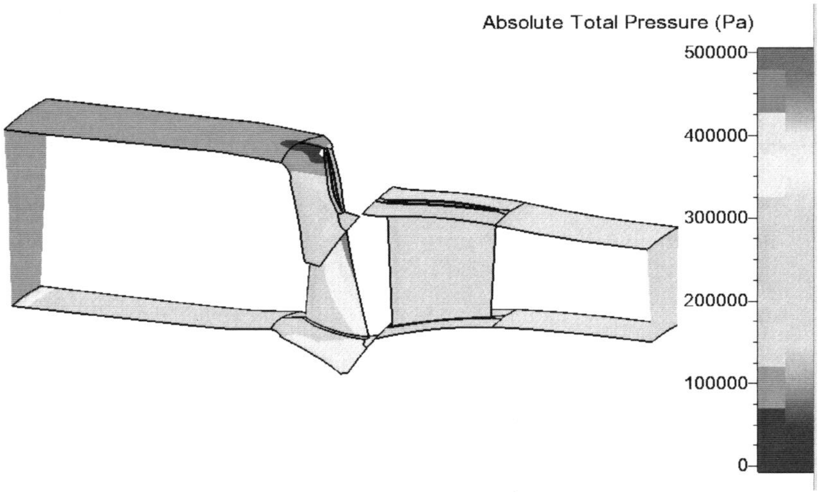

Figure 2. Total pressure distribution

The total pressure difference between the inlet and the outlet could be obtained from the total pressure distribution map, and this compressors operating point performance &parameter in underwater working mode is also obtained from this calculation result, respectively as: speed is 1000RPM/min, pressure differential is 1.72×10^{4}N, efficiency is 87%, and mass flow is 920kg / s. In the rated working state, its underwater working mode capability has initially met the underwater performance requirements, and further analysis is still needed.

3.2. Characteristic analysis of compressors' underwater operating mode

In order to understand compressors' underwater performance, the characteristic curve should be obtained. The compressor works similarly to the axial flow pump in the underwater, and external features of the pressure and efficiency need to be worked out [8, 9].

The efficiency of the pump can be calculated by the following equation:

$$\eta = \frac{gQ(P_{out} - P_{in})}{M\omega} \tag{1}$$

Where and M are mass flow and torque.

IOP Conf. Series: Journal of Physics: Conf. Series **916** (2017) 012019 doi:10.1088/1742-6596/916/1/012019

In order to calculate the characteristic curve for the underwater operating of the compressor, the outlet boundary condition should be changed as the mass flow outlet , flows of 840kg/s, 880kg/s, 920kg/s, 960kg/s, 1000kg/s, 1040kg/sshould be separately selected as the new series of outlet boundary conditions near the above-mentioned rated working point. The simulation results are plotted into mass flow-pressure differential curves (as show in Fig. 3) and flow-efficiency curve (as show in Fig. 4).

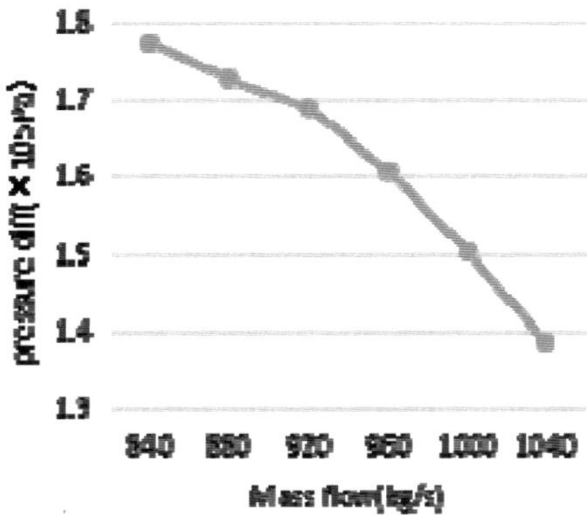

Figure 3. Flow-pressure differential curve

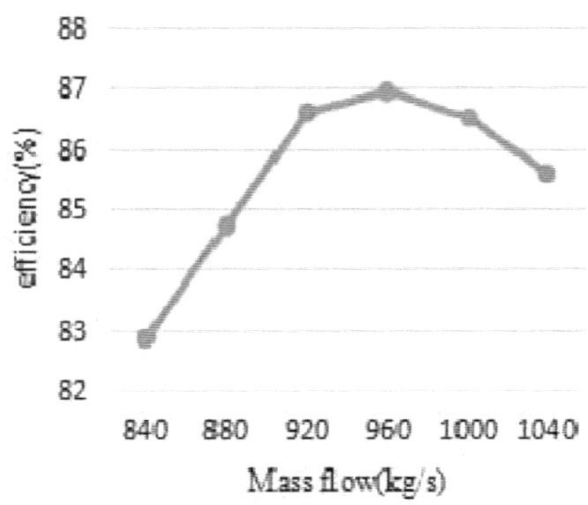

Figure 4. Flow-efficiency curve

The flow-efficiency curve graph shows that the underwater efficiency of the compressor has the tendency to rise first and then descend, with the maximum efficiency of 87% at 960kg / s. Ssimilarly, there's small changes of the efficiency near the flow. The compressor has the maximum underwater efficiency of 87%, has certain underwater work capabilities and meets the underwater work requirements for water-air dual-use engines. However, as the compressor is designed to compress the air, its underwater performance would be affected. Compared with the traditional pump, its efficiency is still relatively low, the working capacity is far less than the pump, and therefore the further analysis and optimization are needed.

3.3. Flow field analysis of compressors' underwater operating mode

In order to further improve the compressing efficiency, detailed analysis on the flow field should be made to provide reference for the subsequent impeller optimization. The underwater internal flow of the compressor mainly depends on its velocity distribution and pressure field. The working condition

IOP Conf. Series: Journal of Physics: Conf. Series **916** (2017) 012019 doi:10.1088/1742-6596/916/1/012019

with the maximum efficiency of 960kg/s is selected to verify the underwater working flow field of the compressor. The pressure surface and suction surface static pressure diagram of its rotor and stator are respectively given as below (the right sides of Fig. 5 and Fig. 8 are the leading edge of the blades, and the left sides of Fig. 6 and Fig. 7 are the leading edge of the blades). Meanwhile, surface S1 is made in the midpoint of the blade height to display its speed cloud picture (as show in Fig. 9). The speed cloud picture of the rotor is also provided (as show in Fig. 10).

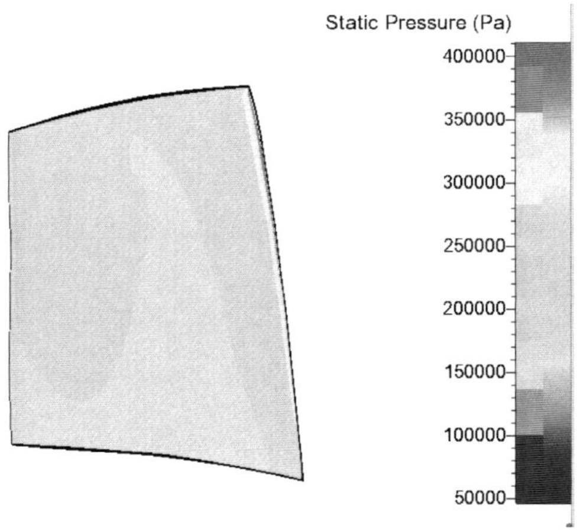

Figure 5. Rotor pressure surface

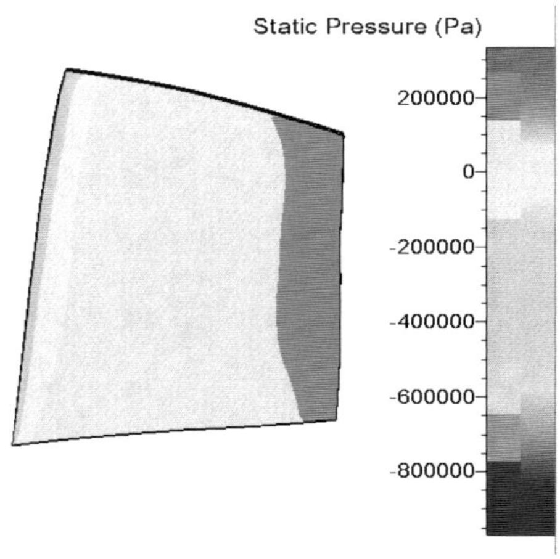

Figure 6. Rotor suction surface

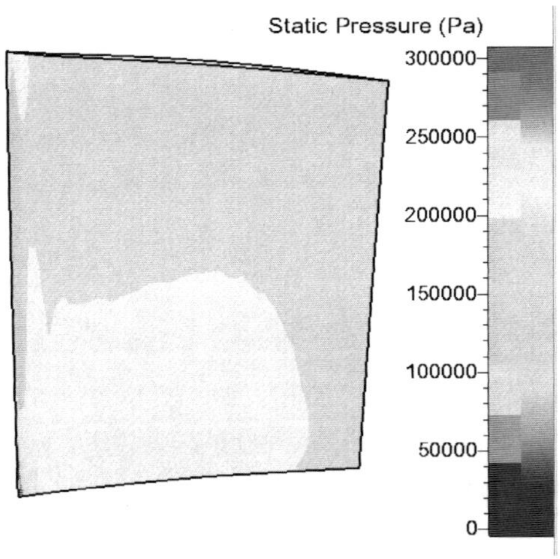

Figure 7. Stator pressure surface

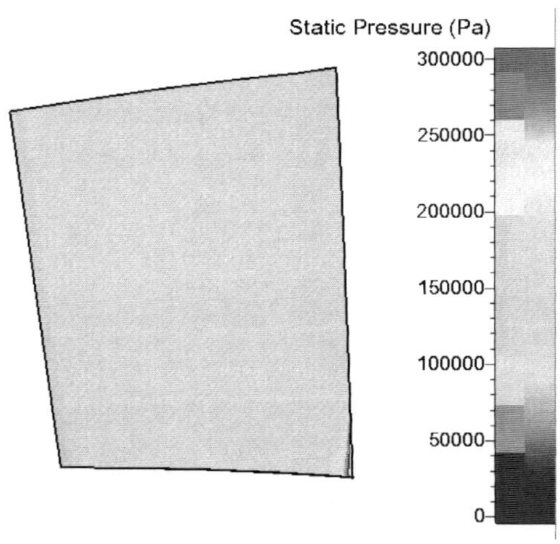

Figure 8. Stator suction surface

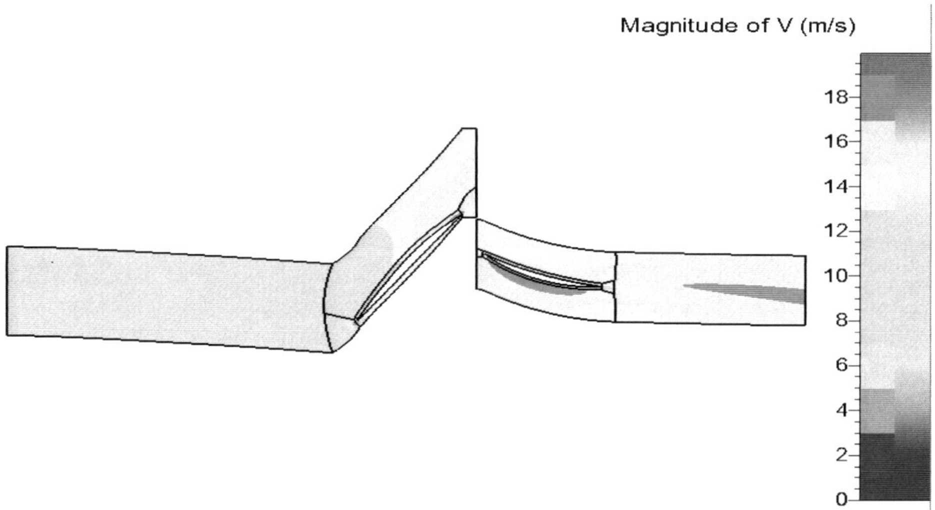

Figure 9. S1 surface speed

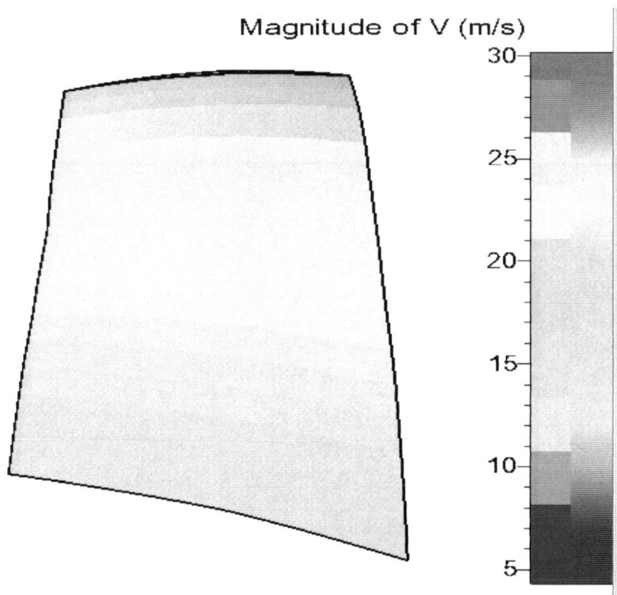

Figure 10. Rotor speed

As the pressure surface of the rotor indicates, a small range of high pressure zones is formed in the leading edge of the blades due to the impact of the water flow, meanwhile the pressure changes on its pressure surface are not uniform. On the suction surface, it can be seen that a small range of low pressure zones forms in its leading edge with even changes backwards, and no big changes in the water flow occur.

Low pressure zones form in the central region of the suction surface of the stator, as S1 section shows, the blade diversion has increased in the flow rate and decreased the pressure in this region, forming the low pressure zone in the central of the blade. Meanwhile, high pressure zones form in the upper end of the blade's leading edge as shown in Figure 8. Through the observation of the moving blade velocity distribution Figure 10, it could be seen that the water flow is high close to the tip of the moving blade, making more water flows strike on the tip of the stationary blade, thus to impede the water flow and increase the pressure, which also reduces the efficiency of the compressor. As the compressor channel is convergent, the reduced flow path would surely make the flow velocity increase in the case of constant flows, therefore a small range of low pressure zones from in the front edge of the stationary blade root as the flow velocity increases.

4. Conclusion

Through numerical simulation of water model of compressor underwater pump, we canconclude that:

(1) While the compressor operating underwater, the compressor speed is 1000RPM / min, and the outlet flow varies from 840kg / s to 1040kg / s, the compressor efficiency varies from 82.8% to 87%. Under this condition, the compressor has the ability to work under water.

(2) The air compressor model is used to simulate the underwater conditions, as the rotational speed of the compressor without any blade optimization is reduced to a certain extent, the performance of the compressor in the water has be reduced in a certain degree as shown in the cloud picture, the flow fluidity in the channel is not as smooth as in the air, especially the stator would impede the flow of the water and the compressor's work, while even failing to take into account of the operational capability in the underwater environment, the compressor to compress the air still have a certain degree of underwater work ability without modification and has the underwater work abilities for compressors. Meanwhile, to better meet the underwater working conditions, the compressor blades and even the layout still need to be improved. This paper also provides certain theoretical basis for the future optimization and improvement.

Acknowledgments

This work was supported by the National Natural Science Foundation of China NO.51266012

References

[1] W. Zhang, Aerodynamic design of impeller of centrifugal compressor for water-air engine. Nanchang, 2013.
[2] X.L.Zhong, the conceptual design and research on multi-bypass counter-rotating air-water engine. Nanchang, 2016.
[3] S.Zhu, The design and research of power system in Water-Air UAV. Nanchang, 2012.
[4] X. GAN, G. Q. HE and J. WANG, 2008, Experimental investigation on concept of water ram jet, Journal of Solid Rocket Technology, (01): 4-7.
[5] Z.H. Li, Y.M. Liu. Blade-end treatment for axial compressors based on optimization method. Energy, 2017, 126:
[6] W. J. Han, S. T. Wang and X. F. Wang, Design optimization of transonic compressor stage using CFD and response surface model. Journal of Harbin Institute of Technology, (01): 112-118. 2010.
[7] Information on https://wenku.baidu.com/view/9eb781e780eb6294dd886cdf.html.
[8] Q. Zhang, 2008, Improved Lifting Design Method and Internal Flow Field Calculation of Axial Flow Pump Impeller, Xihua University.
[9] F. Feng, X. Han and Y. Shi, 2009, Influence of impeller parameters on the performance of 020Q84 water jet axial flow pump, Ship Engineering, (01): 11-14.

IOP Conf. Series: Journal of Physics: Conf. Series **916** (2017) 012020 doi:10.1088/1742-6596/916/1/012020

Effect of Tip Clearance on Hydraulic Performance of Water-jet Pump

Duo Yang [a], Zuodong Huang, Ang Guo, Jiawei Xu and Lei Jiao *

Zhejiang University, Zhoushan 316021, China

*Corresponding author e-mail: jiaolei@zju.edu.cn, [a]21434070@zju.edu.cn

Abstract. The k-ω Shear Stress Transport (SST) turbulence model is adopted to study the hydrodynamic performance of the water-jet axial pump which is applied in Unmanned Surface Vehicle (USV). The numerical simulation of the whole passage of the water-jet pump is carried out for four different tip clearance with $\delta = 0.3$ mm, $\delta = 0.8$ mm, $\delta = 1$ mm and $\delta = 1.4$ mm. And the results, in term of external characteristics and internal flow field, show that, due to the tip leakage flow, the leakage vortex is formed behind the blade after the fluid flows over the impeller blades. Moreover, with the expansion of tip clearance, the head and efficiency of the water-jet pump will be affected, and this impact is weakened with the increase of the flow rate. Finally, the suggestion of structure optimization of water-jet pump can be made.

1. Introduction

USV is a kind of small-scale surface platform which can carry out multiple Tasks (e.g. Target detection) by itself and autonomous voyage. It can undertake terrain investigation, Search and Rescue Service, Surveillance and Reconnaissance etc. With the development of USV, the water-jet propulsion technology is gradually applied in the propelling of the USV. For example, "Jinghai No. 1" USV (designed by the Shanghai University Research Institute of USV Engineering and Donghai Navigation Safety Administration in 2013) and "Haiteng No. 1" USV (designed by Dalian Maritime University) both adopted the water jet propulsion technology [1]. In order to improve the speed and stability of USV, it is necessary to optimize the performance of the water-jet pump.

In the water-jet pump, when there is a tip clearance, under the action of the pressure differential, there will be a tip leakage flow in the tip clearance of the blade, resulting in the performance degradation of the water-jet pump. Therefore, tip clearance is one of the important factors that affect the performance of the water-jet axial flow pump.

Many studies have been engaged on the effect of tip clearance in recent years. Li Yaojun et al. [2] used the large eddy simulation method and the sliding mesh technique to simulate the axial flow pump with unsteady calculation, and analyzed the influence of the tip clearance on the pump external characteristics, the leakage vortex shape and the pressure fluctuation at the outer edge of blade. Zhang De-sheng et al. [3] carried out numerical analysis of an axial flow pump, found the cause and characteristic of the leakage vortex and the influencing factors of the tip leakage flow, which reveals the flow characteristics of the leakage vortex in the tip of the axial flow pump. Zhao Yu et al. [4] do experiments in a closed cavitation tunnel to observe the unsteady cavitating flows in a hydrofoil tip leakage region, during which high speed camera was used to capture the cavitation patterns. Then he proposed a cavitating vortex model to describe the development of vortex cavitation.

In addition, some scholars have brought above research methods in the field of structure optimization of the water-jet axial flow pump. HAN Ji-ang et al. [5] analyzed the cavitation characteristics of the water-jet pump and its affected process under different conditions of tip clearance. The study found that the critical net positive suction head (NPSH) of the pump and the cavitation area on the suction surface of the blade have a positive correlation with tip clearance. Wang Jianing [6] optimized the anti-cavitation performance of water-jet pump by analyzing the effect of tip clearance on the cavitation characteristics.

However, the previous studies on effect of tip clearance are mainly concentrated on the ordinary land axial pump, only a few studies focus on the effect of tip clearance in the field of water-jet pump. In other words, our knowledge of detailed characteristics and performance affected by tip clearance in water-jet pump is not widely studied. The aim of this paper is to explore the effect of tip clearance on the performance of water-jet pump and gain suggestions of structural optimization for water-jet pump used in USV by analyzing external characteristics, internal flow field under different tip clearance conditions.

2. Geometry and grid

2.1. Parameter selection and modeling of water-jet pump

The water-jet pump studied in this paper is designed for the propulsion unit of an USV. The USV is 7.5 m long and 2.4 m wide. Based on the basic design theory of pump [7] and power requirements of USV, the water-jet pump is designed. This paper selects flow (Q) 10 kg/s, speed (n) 1000 rpm and the head (H) 5 m as the design condition.

Determine the specific speed of the water-jet pump at design condition, then the number of blades and the hub ratio can be obtained. Based on the previous calculation, streamline method can help get the blade diameter and relevant parameters. The model of impeller and blades are shown in Figure 1, and the overall model is shown in Figure 2.

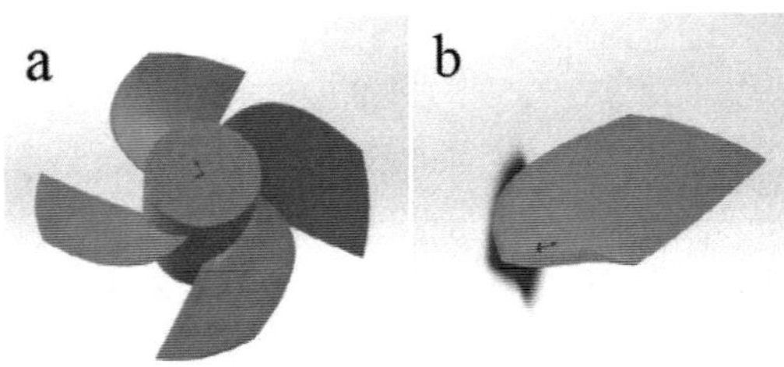

Figure 1. The model of impeller (a) and blade (b)

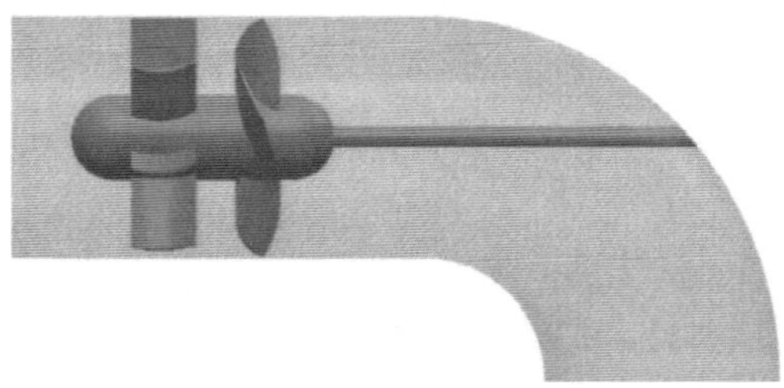

Figure 2. The overall model of water-jet pump

2.2. Grid and the test of grid independence

The computational domain is divided into four parts including inlet pipe, impeller with four blades, stator part with 7 guide vanes and outlet pipe before meshed. Then structural mesh methods is used to mesh the above parts respectively in the ANSYS ICEM CFD. O-type grid refinement are used at near wall of the blades and guide vanes for more accurate simulation of the internal flow of the pump. The number of grids in the blades and the guide vanes parts is 170W and 125W, respectively. The computational domain mesh is shown in Figure 3, in which Figure 3 (a) is the grid of full passage and Figure 3 (b) is the grid of the blades and hub. The enlarged view of local mesh at the surrounding area of the blades and guide vanes is enlarged as shown in Figure 4.

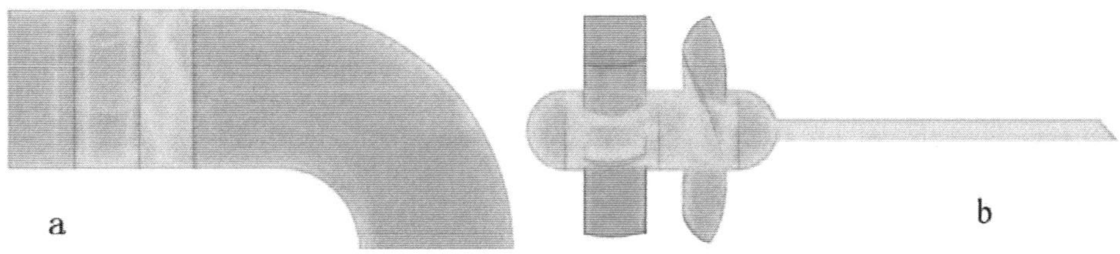

Figure 3. Computational domain mesh of (a) full passage and (b) blades and hub

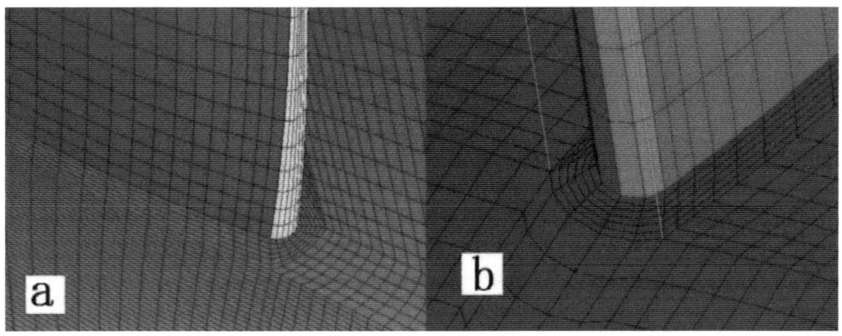

Figure 4. Enlarged view of mesh at the surrounding area of the (a) blades and (b) guide vanes

3 million, 5 million, 7 million, 8.5 million and 10 million grids are all tested in the simulation to find the best grid number, and make sure the accuracy and efficiency of the calculation. Compare the head in the results and verify grid independence. By comparison with the results of all above simulation, the grid of independence is verified. As shown in Figure 5, when the number of grids rises to 7 million, the change of the head is less than 1%, which means the number of grids has little influence on the result. So the final mesh consists of about 7 million grids.

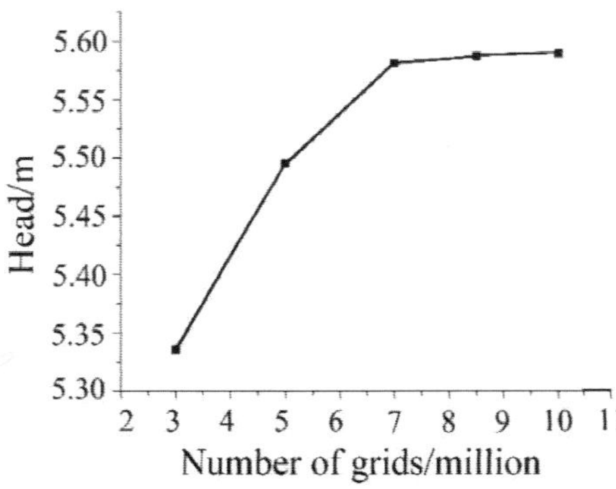

Figure 5. The test of grid independence

2.3. Boundary conditions

ANSYS CFX turbomachinery module was applied for the study. During the calculations the turbulence effects were modeled by k-ω SST turbulence model. The boundary condition is pressure inlet and mass flow rate outlet. The boundary of blades and the pump shell is no slip wall. For this study, the frozen rotor is employed for the interface between stationary and rotating components, which allows the flow to recirculate across the interface freely. The rotational speed (n) is 1000 rpm and the mass flow rate are 7.5 kg/s, 10 kg/s, 12.5 kg/s, 15 kg/s and 20 kg/s, respectively.

3. Results and discussion

In this paper, four different values of tip clearance (δ) were chosen among 0.1% ~ 0.5% diameter (D), which are 0.3 mm, 0.8 mm, 1 mm and 1.4 mm respectively. Under the above five conditions, calculate the external characteristic and the flow field Characteristics for the four models, and then comparative analyze.

3.1. External characteristics comparison

The change of head and efficiency with tip clearances of the water-jet pump is shown in Figure 6.

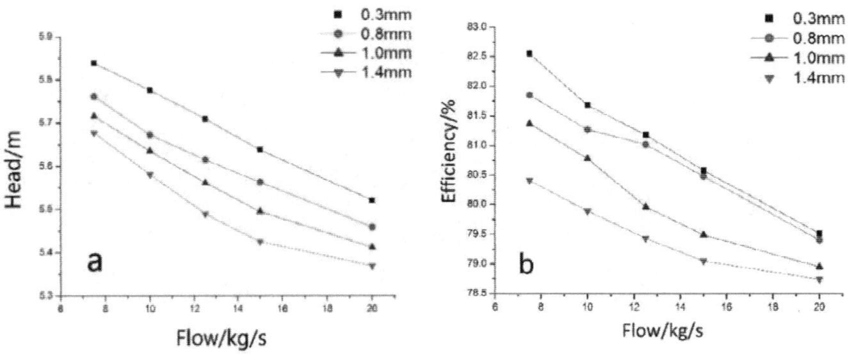

Figure 6. Change of head (a) and efficiency (b) with tip clearances

As can be seen from Figure 6 (a), when the tip clearance is constant, the head of the water-jet pump decreases as flow increases. Among these models, the head is the largest when the tip clearance is 0.3 mm, and reduced with the increase of the tip clearance.

As can be seen from Figure 6 (b), the efficiency of the water-jet pump decreases as the flow rate increases. When the mass flow rate is 7.5 kg/s, as the tip clearance increasing from 0.3 mm to 1.4 mm, the efficiency reduces by 1.2%; but when the mass flow rate is 20 kg/s, the efficiency reduces by only 0.5% as the tip clearance increasing from 0.3 mm to 1.4 mm.

In the case of the tip clearance are 1 mm and 1.4 mm, the efficiency respectively reduce by 1.4% and 1.0% when the mass flow rate increases from 7.5 kg/s to 12.5 kg/s. However, the efficiency reduce by 0.6% and 0.3% respectively when the mass flow rate increases from 15 kg/s 20 kg/s. This shows that when the tip clearance is large, the efficiency is greatly affected by the tip clearance at the low flow rate condition, and the effect is much smaller under the large flow condition.

Therefore, taking into account the optimization of efficiency and head, the tip clearance 0.3 mm is the best option for a water-jet pump.

3.2. Analysis of pressure nephogram on blades

The pressure distribution of blades as the tip clearance is 0.3 mm, 0.8 mm, 1 mm, 1.4 mm is shown in Figure 7, in which the left side is the pressure surface and the right side is the suction surface.

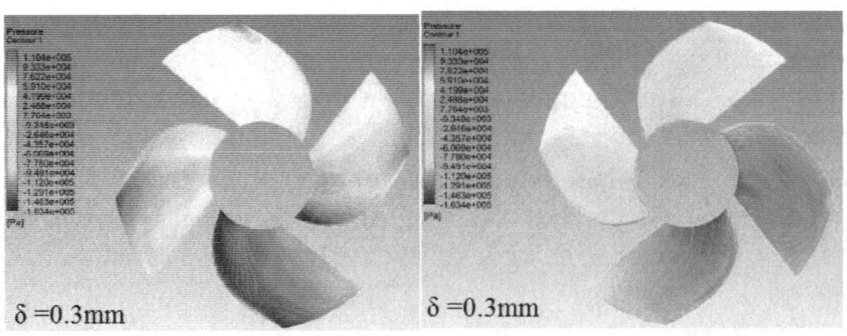

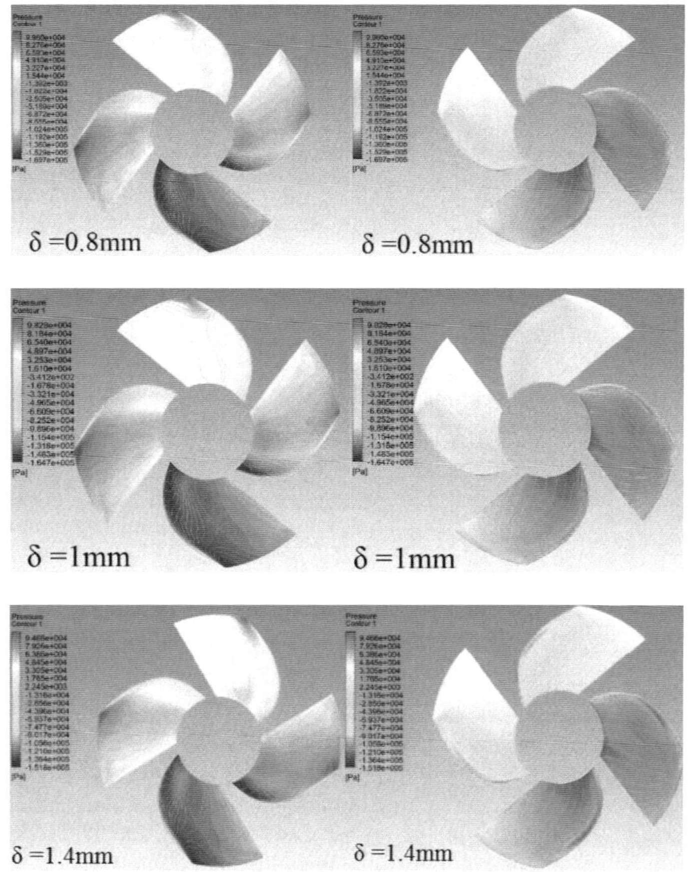

Figure 7. Pressure distribution of blades

As can be seen in Figure7, at about 40% ~ 80% radius (R) in the radial direction on the leading edge of the blades, there is a very small zone of low pressure. For the pressure surface, after the initial low pressure zone, the pressure on the blade increases rapidly along the chord (direction of the water flow), and the span is large. Then the pressure reaches the maximum at the middle of blades. This shows that the blade work increases as the area of blade contact with water increases. With the increase of the tip clearance, the area of the middle high pressure area increases, but the maximum pressure slightly reduces and the pressure gradient also reduces. It shows that, as the tip clearance increasing, the blades do less work on the flow, but the blades are uniformly forced.

In the four different models, the pressure gradient on the blade is basically along the direction of water flow. However, the pressure gradient distributes along the radial direction at the outer edge of the blade on suction surface, indicating that there is a radial flow near the outer edge of blade, named tip leakage flow. Tip leakage flow does not contribute to the increase in outlet pressure, which leads to the reduction of the head. This is one of the reasons why the tip clearance has an impact on performance of water-jet pump.

3.3. Analysis of the pressure cloud and velocity vector of the axial section at 50% chord length

The pressure distribution of the axial section at 50% chord length of the blade for three kinds of models with the tip clearance of 0.3 mm, 0.8 mm and 1.4 mm are shown in Figure 8, Figure 9 and Figure 10 respectively.

(a) Q=7.5kg/s (b) Q=10kg/s (c) Q=20kg/s

Figure 8. The pressure distribution on the axial section when δ=0.3 mm

(a) Q=7.5kg/s (b) Q=10kg/s (c) Q=20kg/s

Figure 9. The pressure distribution on the axial section when δ=0.8 mm

(a) Q=7.5kg/s (b) Q=10kg/s (c) Q=20kg/s

Figure 10. The pressure distribution on the axial section when δ=1.4 mm

The leakage vortex core and the pressure difference between the front and back of the blade can be clearly observed in Figure 8, Figure 9 and Figure 10. With the increase of the tip clearance, the pressure difference between the front and the back of the blade reduce, the area of leakage vortex becomes larger, pressure of the vortex center decreases as well. Although the leakage vortex core is not obvious in figures when tip clearance is 0.3 mm, there has been an obvious leakage vortex when tip clearance is 0.8 mm and 1.4 mm.

Under the low flow rate condition (a) and design condition (b), comparing with Figure 9, the pressure difference between the front and the back of black is higher, the size of leakage vortex is larger and the pressure in the leakage vortex core is higher in the Figure 10. So it can be seen, the tip clearance has a great influence on the pressure difference between the front and the back of the blade and leakage vortex.

Under the large rate condition (c), the pressure difference between the front and the back of blade, and the size and pressure of the leakage vortex in Figure 9 and Figure 10 are almost the same. This phenomenon shows that the influence of the tip clearance on tip leakage flow and leakage vortex is the biggest under low flow rate condition. Tip clearance has less effect on the internal flow field of water-jet pump with the increase of the flow rate.

(a) δ = 0.3 mm (b) δ = 0.8 mm (c) δ= 1.4 mm

Figure 11. Velocity vector on the axial section at 50% chord length

Figure 11 shows the velocity vectors on the axial section at 50% chord length of the blade for three models under design conditions. The tip leakage flow can be observed in this figure. Combining Figure 8, 9, 10, and 11, it can be seen that water flows into the tip clearance and form tip leakage flow due to the pressure difference between the front and back of blades. Then the main flow rolls up the tip leakage flow and then leakage vortex generates.

From Figure 11 (a) and Figure 8, it can be seen when tip clearance is 0.3 mm, there is little tip leakage flow or leakage vortex. It is because that the tip clearance is so small that the fluid at the clearance is greatly affected by the viscosity of the near wall, and the tip leakage flow has little influence on main flow which results in little leakage vortex. With the increase of the tip leakage, the tip leakage flow increases, the leakage vortex is more obvious, and the entrainment is graver. It can be seen from Figure 11 that, the leakage vortex has a great disturbance to the flow and thus produces a large flow loss, so that the pressure at the leakage vortex rapidly decreases and greatly reduces the work of the blade on the fluid.

It is concluded that, with the tip clearance of impeller blade increases, the tip leakage flow increases, thus, head and efficiency of water-jet pump will reduce.

4. Conclusion

To sum up, for different tip clearance, this paper identifies external characteristic curves and internal flow field distribution of water-jet pump which is used in USV. First, the water-jet pump is designed and modeled. Then structured grids were generated in model and O-type grid refinement is used at near wall of the blades and guide vanes. Numerical simulation is calculated in four different tip clearance (0.3 mm, 0.8 mm, 1 mm, 1.4 mm) by k-ω SST turbulence model. Then obtain the external characteristics and internal flow field. Through comparative analysis, conclusions are drawn as follows:

(1) The tip clearance has little effect on the pressure distribution of blade suction surface, while it has a great effect on pressure distribution of blade press surface.

(2) Through analysis of pressure distribution and velocity vector, it can be seen leakage vortex caused by tip leakage flow results in poor performance of the water-jet pump.

(3) The tip clearance will affect the head and efficiency of the water-jet pump, but this effect will be decreased as the flow rate increases.

From the above conclusion, in the design process of water-jet pump, it is suggested that tip clearance should be designed small enough under the engineering condition.

Acknowledgments

This work was financially supported by National Key Research and Development Plan (2017YFC0403802), National Science Foundation of China (51276158).

References

[1] Yue J, 2016, *China, Master Thesis, Dalian Maritime University,* Study on Modeling and Simulation of 6-DOF Motion of water-jet propelled Unmanned Surface Vehicle.

[2] Li Y, Shen J, Hong Y, Liu Z, 2014, *China, Transactions of the Chinese Society for Agricultural Machinery* ,Numerical Investigation of Pressure Fluctuations on Axial-flow Pump Blades Affected by Tip-gap Size, (**05**):59-64,58.

[3] Zhang D, Shao P, Shi D, Pan D, Wang H,2014, *China, Transactions of the Chinese Society for Agricultural Machinery* ,Numerical Simulation of Tip Leakage Vortex Hydrodynamics Characteristics in Axial Flow Pump, (**03**):72-76,82 .

[4] Zhao Y, Wang G, Huang B, Liu L,2016, *China, Journal of Ship Mechanics,* Investigation of vertical cavitating flows in tip leakage region of a hydrofoil, Journal of Ship Mechanics, (**11**):1304-1311 .

[5] Han J, Li P, Zhong J,2016, *China, Science Technology and Engineering*, Effect of Tip Clearance on Cavitation of Water-jet Axial-flow Pump, Science Technology and Engineering, (**33**):130-136.

[6] Wang J,2009, *China, Master Thesis, Harbin Engineering University* ,The Technology of Anti-Cavitation on Water-jet Pump and the Design on Optimization of Blade.

[7] Guan X,2011, *Modern Pumps Theory and Design* (*Beijing*: China Aerospace Press).

The Study on Flow Velocity Measurement of Antarctic Krill Trawl Model Experiment in North Bay of South China Sea

Shuai Chen, Lumin Wang, Hongliang Huang *, Xun Zhang

Key Laboratory of Oceanic and Polar Fisheries, Ministry of Agriculture; East China Sea Fisheries Research Institute, Chinese Academy of Fishery Sciences, Shanghai, 200090, China

*Corresponding author e-mail: yangpu79@hotmail.com

Abstract. From August 25 to 29, 2014, the project team carried out the experiment of Antarctic krill trawl in the Beihai Bay of the South China Sea. In order to understand the flow field of the network model in the course of the experiment, it is necessary to record the speed of the ship and to grasp the flow field of the ocean. Therefore, the ocean velocity is measured during the experiment. The flow rate in this experiment was measured using an acoustic Doppler flow meter (Vectoring Plus, Nortek, Norway). In order to compensate for the flow rate error caused by ship drift, the drift condition of the ship was also measured by the positioning device (Snapdragon MSM8274AB, Qualcomm, USA) used in the flow rate measurement. The results show that the actual velocity of the target sea area is in the range of 0.06-0.49 m / s and the direction is 216.17-351.70. And compared with the previous research, the influencing factors were analysed. This study proves that it is feasible to use point Doppler flow meter for velocity study in trawl model experiment.

1. Introduction

From August 25 to 29, 2014, the project team carried out the experiment of Antarctic krill trawl in the Beihai Bay of the South China Sea, and studied the resistance of the net, the expansion of the screen and the height of the network port. In order to understand the flow field of the network model in the course of the experiment, it is necessary to record the speed of the ship and to grasp the flow field of the ocean. Therefore, the ocean velocity is measured during the experiment.

2. Materials and methods

2.1. Flow rate measurement

The flow rate in this experiment was measured using an acoustic Doppler flowmeter (Vectrino Plus, Nortek, Norway) with a system sound of 10 MHz and an accuracy of ± 0.5% of the measured value, belonging to a high-precision three-dimensional instrument. It uses the principle of acoustic Doppler measurement to measure the flow rate with the Doppler shift of the acoustic wave in the flow field. The main components of the device 4-arm flat-panel sensor installed in the 40cm long rod-type fixed probe, there are four focus beam, the probe size is small, greatly reducing their impact on the flow field. No moving parts, no friction and hysteresis, measurement of induction time fast, you can 200 Hz sampling rate to measure the flow field velocity. The measurement data is read by the software (Vectoring + 1.14, Nortek), which monitors and records the change in flow rate over time.

In the measurement, the probe of the flow meter is fixed at 2 m below the water level of the ship side. The X axis is aligned with the direction of the bow, the Z axis is perpendicular and the Y axis is perpendicular to the X axis and the direction is perpendicular to the ship's side. Flow rate measurement data acquisition interface and data output as shown in Figure 1, a total of 20, from the data can be read in the sensor measurement data correlation coefficient were 98,97,97,96, in general screening correlation coefficient greater than 70 Data, this group of data quality is good. X, Y, Z in the direction of the flow rate were -0.147, -0.026, and 0.011. Data analysis, flow rate size, direction calculation and other basic statistical analysis and steady-state, correlation, spectrum analysis and other analysis in the flow rate analysis software (Explore Pro 1.58, Nortek AS, Norway).

According to the ocean hydrological observation GB, the flow rate measurement time is 3 min. The flow velocity in the experiment is based on the velocity in the three directions. According to the distribution of the flow direction in the whole measurement process, the direction of the flow velocity is the most instantaneous velocity direction, as shown in Fig. 2, when the flow velocity direction is -151 °. The direction of the direction of the bow is 0, the vertical ship is 90 ° outward, the stern direction is 180 °, and the vertical ship's side is -90 °.

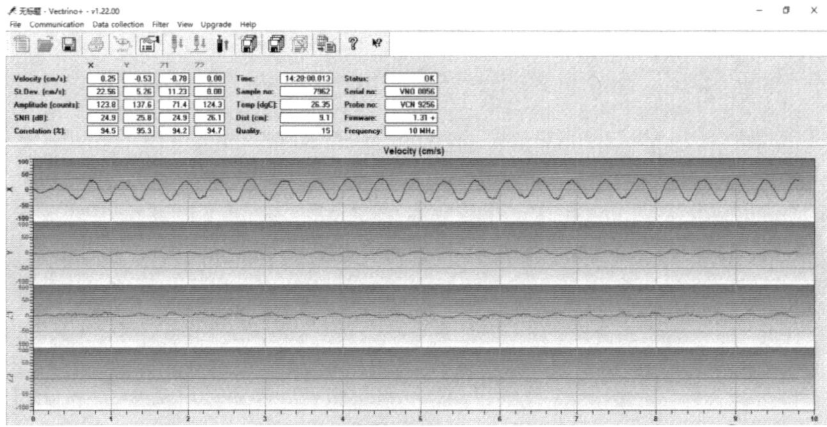

Figure 1. Flow rate measurement data acquisition interface and data output situation

16 01000011	-0.1470	0.1800	-0.0960	0.0000	163	177	177	177	23.4	23.4	23.4	23.4	99	99	99	99
17 01000011	-0.1310	0.1710	-0.1350	0.0000	167	163	183	168	23.6	22.7	23.7	22.9	97	99	99	99
18 01000011	-0.1540	0.1760	-0.1070	0.0000	152	150	185	176	22.8	21.9	23.8	23.3	98	99	99	99
19 01000011	-0.1430	0.1740	-0.1150	0.0000	169	181	182	175	23.7	23.6	23.6	23.3	99	99	99	99
20 01000011	-0.1370	0.1720	-0.1260	0.0000	168	181	175	178	23.7	23.6	23.3	23.4	98	99	99	99
21 01000011	-0.1370	0.1740	-0.1100	0.0000	154	178	178	174	22.9	23.4	23.4	23.2	98	99	99	99
22 01000011	-0.1230	0.1720	-0.1210	0.0000	149	176	181	177	22.6	23.3	23.6	23.4	94	99	99	99
23 01000011	-0.1290	0.1730	-0.1240	0.0000	156	165	176	178	23.0	22.8	23.3	23.4	97	99	99	99
24 01000011	-0.1570	0.1810	-0.0790	0.0000	172	176	179	169	23.9	23.3	23.5	23.0	96	99	99	99
25 01000011	-0.1410	0.1760	-0.1180	0.0000	173	172	181	159	23.9	23.1	23.6	22.4	99	99	99	99
26 01000011	-0.1310	0.1760	-0.1240	0.0000	158	176	173	152	23.1	23.3	23.2	22.1	99	99	99	99
27 01000011	-0.1310	0.1790	-0.1130	0.0000	160	180	177	167	23.3	23.5	23.4	22.9	98	99	99	99
28 01000011	-0.1230	0.1800	-0.1240	0.0000	157	173	171	166	23.1	23.2	23.1	22.8	99	99	99	99
29 01000011	-0.1250	0.1820	-0.1240	0.0000	162	181	179	181	23.4	23.6	23.5	23.6	98	99	100	99
30 01000011	-0.1190	0.1820	-0.1290	0.0000	177	178	177	175	24.1	23.4	23.4	23.3	99	99	99	99
31 01000011	-0.1190	0.1830	-0.1290	0.0000	171	183	183	171	23.8	23.7	23.7	23.1	99	99	99	99
32 01000011	-0.1170	0.1850	-0.1210	0.0000	140	182	180	178	22.1	23.6	23.5	23.4	99	100	99	99
33 01000011	-0.1250	0.1850	-0.1290	0.0000	161	186	172	174	23.3	23.8	23.1	23.2	99	100	99	99
34 01000011	-0.1290	0.1860	-0.1210	0.0000	168	185	176	169	23.7	23.8	23.3	23.0	99	99	99	99
35 01000011	-0.1170	0.1850	-0.1290	0.0000	160	182	178	182	23.3	23.6	23.4	23.6	99	100	99	99
36 01000011	-0.1140	0.1860	-0.1180	0.0000	163	170	178	175	23.4	23.0	23.4	23.3	99	99	99	99
37 01000011	-0.1190	0.1860	-0.1160	0.0000	163	181	178	182	23.4	23.6	23.4	23.6	99	99	99	100
38 01000011	-0.1150	0.1870	-0.1130	0.0000	174	182	175	175	24.0	23.6	23.3	23.3	99	99	99	99
39 01000011	-0.1210	0.1860	-0.1160	0.0000	174	183	174	161	24.0	23.7	23.2	22.6	99	100	99	99

Figure 2. Flow direction distribution in the measurement

2.2. Experimental ship motion measurement

According to the provisions of the National Standard Ocean Survey (Part 2: Ocean Hydrological Observations (GB / T 12763.2-2007) (hereinafter referred to as the national standard for marine hydrology observation), the ship needs to be anchored when measuring the ocean flow velocity, but this test Conditions cannot be related to the restrictions. In order to compensate for the flow rate error caused by ship drift, the drift condition of the ship was also measured by the positioning device (Snapdragon MSM8274AB, Qualcomm, USA) used in the flow rate measurement. The output of the positioning data is the NMEA0183 standard format developed by the National Ocean Electronics Association. It is a string of ASCII codes. Each set of data has two kinds of statements. GPGGA, GPRMC, for example, a set of data is as follows:

$ GPGGA, 15334, 1954.2826, N 10930.4979, E, 1,27,3, -10 M, 0000 * 3A
$ GPRMC, 15334, A, 1954.2826 N, 10930.4979, E, 0.6, 292.7, 290814, * 02

It can be seen that the location of the positioning equipment is single point positioning, the number of satellites is 27, the horizontal accuracy factor is 3, the experimental ship drift movement rate is 0.6 kn, the direction of motion is 292.7, the relevant data in the ocean data view application software ODV 4.5.6, AWI, German), the trajectory of the experimental ship drift is shown in Fig. 2. Generally speaking, the drift of the experimental ship can be considered as a uniform motion, and the whole of the drift trajectory can be obtained Drift displacement of the ship during flow measurement. Combined with the time of the whole process to obtain the speed of the ship drift.

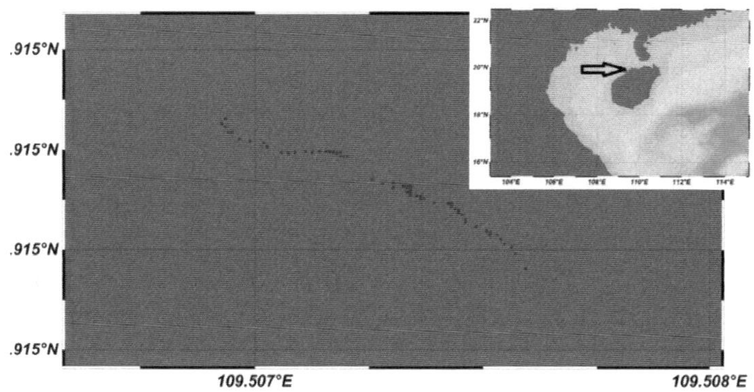

Figure 3. Trajectory of drift motion of experimental ship

2.3. Actual ocean velocity calculation

The actual velocity of the ship is superimposed on the velocity of the ship and the actual ocean velocity is obtained. According to the ocean hydrological observation of the national standard, continuous observation of the current at least once per hour observation. In order to maintain the consistency of the experimental conditions in the whole group of experiments, the flow rate measurement is carried out before the start of each experiment, and the actual ocean velocity is obtained as the ocean velocity in the experiment.

3. Results

The flow rate measurement results obtained according to the above method are shown in Table 1.

Table 1. Flow rate measurement results table

No.	Time	Flow (m/s)	Direction	Boat speed (m/s)	Boat speed Direction	Actual flow (m/s)	Actual flow Direction
1	08-28 10: 38	0.293	-151	0.37	295.3	0.18	243.98
2	08-29 09: 49	0.235	-111	0.42	305.8	0.40	272.64
3	08-29 15: 46	0.296	160	0.57*	316.3*	0.30	330.81
4	08-30 09: 31	0.117	-155	0.135	292.2	0.06	351.70
5	08-30 13: 45	0.192	-77	0.41	238.6	0.49	216.17

Note 1: the flow rate in the table refers to the measured flow rate, the speed refers to the process of measuring the speed of the ship drift, and the actual flow rate refers to the actual ocean flow rate.

Note 2: The direction of the flow direction is 0, the vertical ship's side is 90 ° outward, the stern direction is 180 °, and the vertical ship's side is -90 °.

Note 3: The flow rate unit in the table is m / s and the direction unit is °.

Note 4: The third group of experimental ship speed size and direction are instantaneous value, because this group of GPS data quality is poor.

4. Discussion

4.1. Related measurement results reference

For the experimental waters and roughly the same time, 2006-2007 Xiamen University, China Ocean University, South China Sea Fisheries Research Institute, South China Sea Institute of Ocean through 908 special (908-01-ST09) had Qiongzhou Strait west of the acoustic Doppler (ADCP, Acoustic Doppler Current Profilers) measured the flow rate [1], its position as shown in Figure 4 M3, and the experimental area is similar.

The site's velocity profile began on July 23, 2006, and finally on August 24 of August. The observed results show that the flow field of the site is stronger in August (Figure 3). During the period from 3 August to 21 August, the M3 was basically eastward, with the exception of August 5, 10, 11 and 14. In the August 21 to 24, M3 and turn to the west to the flow, which 21 and 24, the entire vertical water are west, but 22 and 23 of the West flow mainly in the upper 7 m of water.

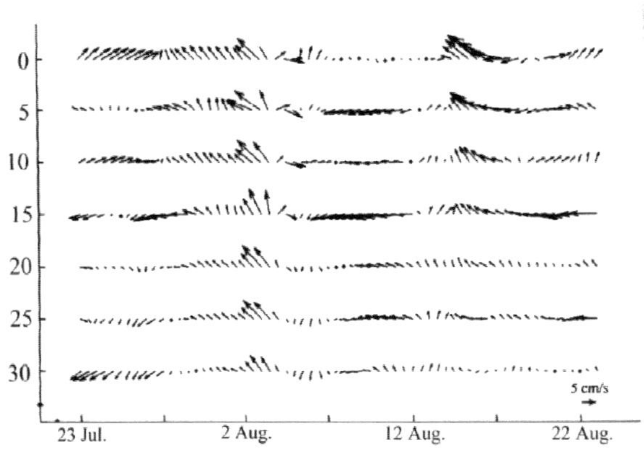

Figure 4. July 2006 - August 24 observed at the M3 station flow field [1]. Where y axis represents water depth, Arrows up and right respectively represent velocity towards north and east.

IOP Conf. Series: Journal of Physics: Conf. Series **916** (2017) 012021 doi:10.1088/1742-6596/916/1/012021

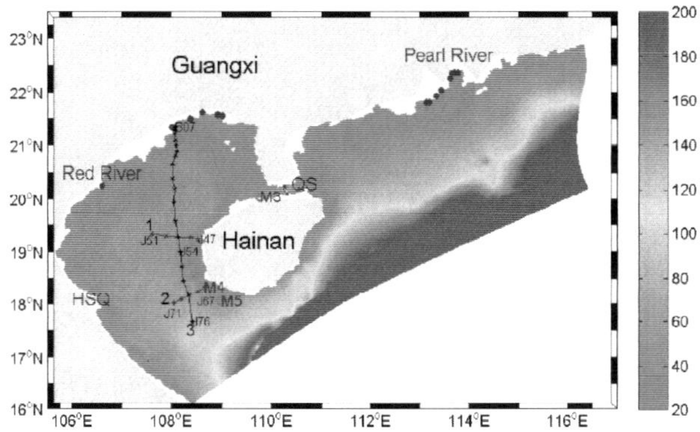

Figure 5. M3 station location diagram

4.2. A brief analysis of the influencing factors

In general, there are three factors, such as tidal and Qiongzhou Strait, which are affected by the velocity of the experimental sea. According to Li Shuhua et al. [2], the numerical simulation of the North Bay tidal wave system is carried out. The experimental waters may be normal tide Tidal opposite, for the irregular day trend; Qiongzhou Strait on the northern Gulf of water supply flux and vortices of the northern part of the northern part of the circulation also plays a key role; the other in the northeastern part of the northern part of the circulation and the monsoon higher correlation , There are significant seasonal changes in the characteristics of the North Bay is mainly controlled by the East Asian monsoon system, summer for the southwest wind, and winter for the northeast wind.

Acknowledgments

This work was financially supported by the National Science and Technology Pillar Program (2013BAD13B03), Project NO. 2016T02 Supported by Special Scientific Research Funds for Central Non-profit Institutes (East China Sea Fisheries Research Institute) .

References

[1] Zhang Guorong, Pan Weiran,etc, Characteristics of winter and spring water transport in the east and north of North Gulf in China, North Bay Ocean Science Research Proceedings (Part 2), Ocean Press,2009.

[2] Li Shuhua, Numerical Calculation of Tidal Current in North Gulf, Marine Science Bulletin,4(1985) :6-9.

Inverse Kinematics Solution and Verification of 4-DOF Hydraulic Manipulator

Zhiqiang Chao, Fei Wang *, Chuanqing Zhang, Huaying Li

Department of Mechanical Engineering, Academy of Armored Force Engineering, Beijing China

*Corresponding author e-mail: master809@163.com

Abstract. Aimed at four degree of freedom (DOF) hydraulic manipulator, an inverse kinematics solution is proposed from Cartesian space to drive space based on geometrical method. To the structural and diver characteristics of the manipulator, a forward kinematics is conducted by using D-H method. The position and orientation of manipulator's end-effector can be obtained under the kinematics constraint. By analyzing the structure, the solution of inverse kinematics of manipulator can be obtained, and the conversion between drive space and joint space can be got through the sport's mechanism kinematics. In order to meet the need of motion planning and control of the manipulator, the inverse kinematics and conversion are validated based on simulation.

1. Introduction

Hydraulic manipulator plays an important role in industry, agriculture and military because it has small size, high power, durability and large operation range. With hydraulic motor rotating and hydraulic cylinder expanding, the links can be moved. Further, hydraulic manipulator realize the position. Thus, in order to make sure the position of manipulator's end-effector in the Cartesian coordinate system, the conversion between Cartesian space and joint space and between joint space and drive space should be determined firstly.

Inverse kinematics of manipulator is solving the joint variables according to the known position and orientation of manipulator's end-effector. The result perhaps is no solution, unique solution or multiple solutions. The kinematics is used for analyzing, and the inverse kinematics is used for control of manipulator. At present, there are some common methods for inverse kinematics [1]: analytic method, geometric method and iterative method. Analytic method applies to multiple DOF manipulator, and its real time ability is poor because of a lot of inverse matrix multiplication in the calculation process. We always can't converge to correct solution by iterative method. Geometric method keeps the advantage of calculation being simple, and it applies to few DOF manipulator. Besides, there are some other method for inverse kinematics, such as dual matrix method, screw algebraic method [2], dual quaternion method, genetic algorithm [3] and intelligent hill climbing algorithm [4]. According to the characteristics of manipulator designed in this paper, the geometric method for inverse kinematics is adopted. In the control of hydraulic manipulator, the motion of every joint of manipulator can be acquired by inverse kinematics. The controller converts motion to control signal, and the servo valves are driven. So the hydraulic motor and cylinders can be controlled.

IOP Conf. Series: Journal of Physics: Conf. Series **916** (2017) 012022 doi:10.1088/1742-6596/916/1/012022

2. Model and forward kinematics of manipulator

Targeting 4 DOF hydraulic manipulator, the structure and coordinate system are shown in the figure 1. The four joint are all rotational. The $X_0Y_0Z_0$ is base coordinate system, the $X_1Y_1Z_1$, $X_2Y_2Z_2$, $X_3Y_3Z_3$ are link coordinate systems respectively, and $X_4Y_4Z_4$ is end-effector coordinate system. According to D-H parameters, the unique position and orientation of manipulator's end-effector in the base coordinate system can be acquired.

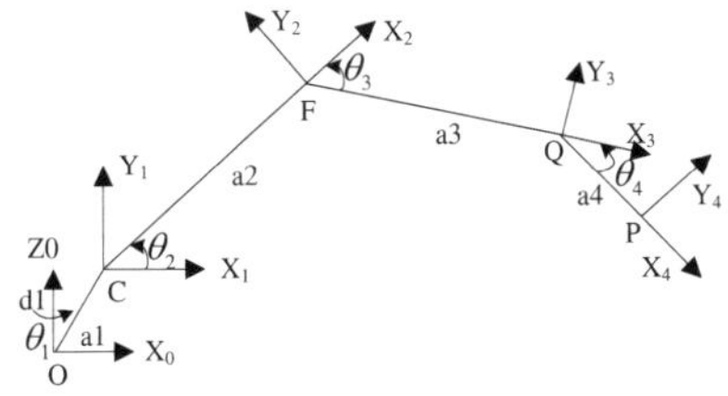

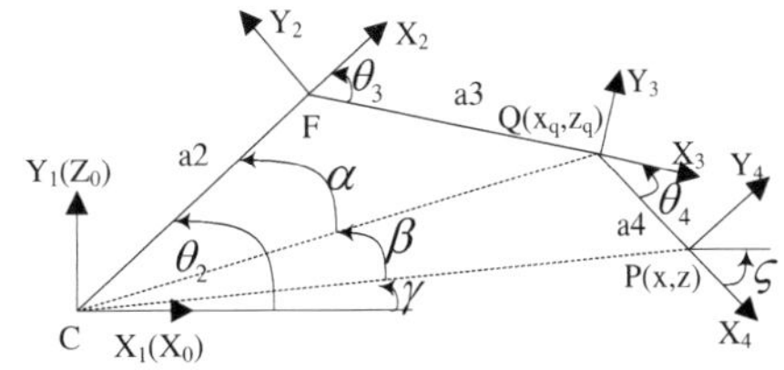

Figure 1. The simplified structure of manipulator **Figure 2.** The simplification for inverse kinematics

Table 1. D-H parameters of manipulator

i	α_{i-1}	a_{i-1}	d_i	θ_i
1	90	187.5mm	300mm	θ_1
2	0	L_2	0	θ_2
3	0	L_3	0	θ_3
4	0	L_4	0	θ_4

L1=515mm, L2=1100mm, L3=1350mm, L4=280mm

The mission is usually described in the Cartesian space, so the unique position and orientation can be determined in the Cartesian coordinate system. According to conversion matrix, the position and orientation of 4 DOF manipulator's end-effector can be described as

$$\begin{cases} X = c_1(a_4c_{234} + a_3c_{23} + a_2c_2 + a_1) \\ Y = s_1(a_4c_{234} + a_3c_{23} + a_2c_2 + a_1) \\ Z = a_4s_{234} + a_3s_{23} + a_2s_2 + d_1 \\ \xi = \theta_2 + \theta_3 + \theta_4 \end{cases} \quad (1)$$

3. Inverse kinematics

Compared to forward kinematics, inverse kinematics is more complicated. Its result perhaps is no solution or multiple solutions. Duffy proved that the result of inverse kinematics exists if three successive joints of manipulator are parallel [6]. According to this proof, the manipulator designed in this paper can acquired at most 16 groups solutions. The solution that rotates a minimum angle will be chosen. In this paper, the geometric method is chosen for inverse kinematics.

As shown in the figure 2, at first the rotation from joint 1 is not considered. So $a_1 = \theta_1 = d_1 = 0$, and point C is considered as point O. Taking the hinge point C of link 2 as the origin, the base coordinate system is built. And the position and orientation of manipulator's end-effector are $[x, 0, z, \varsigma]^T$, and $\varsigma = \theta_1 + \theta_2 + \theta_3$.

The angle γ shown in the figure 2 is solved firstly.

$$\gamma = \arctan\frac{z}{x} \tag{2}$$

The angle β shown in the figure 2 is solved secondly. According to coordinate value of point P, the coordinate value of point Q can be acquired by coordinate conversion. The equation ${}^{i}p = {}^{i}A_{i+1}{}^{i+1}p$ [7] has been known, so

$$\begin{bmatrix} x_q \\ y_q \\ z_q \\ 1 \end{bmatrix} = {}^{0}A_3 \begin{bmatrix} 0 \\ 0 \\ 0 \\ 1 \end{bmatrix}, \quad \begin{bmatrix} x \\ y \\ z \\ 1 \end{bmatrix} = {}^{0}A_4 \begin{bmatrix} 0 \\ 0 \\ 0 \\ 1 \end{bmatrix} \tag{3}$$

Because the rotation of joint 1 is not considered, ${}^{0}A_1$ become identity matrix. ${}^{0}A_3$ 和 ${}^{0}A_4$ are substituted to ${}^{0}p = {}^{0}A_4{}^{4}p$, so

$$\begin{bmatrix} x_q \\ y_q \\ z_q \\ 1 \end{bmatrix} = \begin{bmatrix} a_3c_{23} + a_2c_2 \\ 0 \\ a_3s_{23} + a_2s_2 \\ 1 \end{bmatrix} \quad \begin{bmatrix} x \\ y \\ z \\ 1 \end{bmatrix} = \begin{bmatrix} a_4c_{234} + a_3c_{23} + a_2c_2 \\ 0 \\ a_4s_{234} + a_3s_{23} + a_2s_2 \\ 1 \end{bmatrix} \tag{4}$$

The two equation are compared

$$\begin{bmatrix} x \\ y \\ z \\ 1 \end{bmatrix} = \begin{bmatrix} x_q \\ y_q \\ z_q \\ 1 \end{bmatrix} \begin{bmatrix} a_4c_{234} \\ 0 \\ a_4s_{234} \\ 0 \end{bmatrix} \tag{5}$$

We can know that

$$\begin{aligned} x_q &= x - a_4\cos\varsigma \\ y_q &= y \qquad\qquad \varsigma=\theta_1+\theta_3+\theta_4 \\ z_q &= z - a_4\sin\varsigma \end{aligned} \tag{6}$$

So far, the point Q can be acquired. In triangle PQC, according to cosine theorem, we can know

$$\beta = \arccos\frac{CQ^2 + CP^2 - a_4^{\,2}}{2CQ \cdot CP} \tag{7}$$

In triangle FQC, according to cosine theorem, we can know

$$\alpha = \arccos\frac{CQ^2 + a_2{}^2 - a_3{}^2}{2CQ \cdot a_2} \tag{8}$$

Based on equation (2)(7)(8)

$$\begin{aligned}
&\theta_2 = \alpha + \beta + \gamma \\
&\theta_3 = \pi - \arccos\frac{a_2^2 + a_3^2 - CQ^2}{2a_2a_3} \\
&\theta_4 = \varsigma - \theta_2 - \theta_3
\end{aligned} \tag{9}$$

Now the rotation of the first joint is considered, the orientation of manipulator' end-effector doesn't change, and the polygon CPQF consisting of link2, Link3, and link4 maintains the same shape. So $x = \sqrt{(x - a_1)^2 + y^2} and z = z - d_1$ are substituted to equation from (6) to (9). It means that the joint 1 take a counter rotating θ_1, and the origin Q is move back to O. So the line CP will project the X-axis just right. Finally

$$\theta_1 = \arctan\frac{y}{x}, \theta_2 = \alpha + \beta + \gamma, \theta_3 = \pi - \arccos\frac{a_2^2 + a_3^2 - CQ^2}{2a_2a_3}, \theta_4 = \varsigma - \theta_2 - \theta_3 \tag{10}$$

In this equation, the coordinate value is in the base coordinate system of manipulator. The CQ and CP can be expressed by equation(6) and point P coordinate.

$$CQ = \sqrt{(x_q - a_1)^2 + y^2 + (z_q - d_1)^2}, CP = \sqrt{(x - a_1)^2 + y^2 + (z - d_1)^2}$$

And α, β, γ respectively is

$$\alpha = \arccos\frac{CQ^2 + a_2{}^2 - a_3{}^2}{2CQ \cdot a_2}, \beta = \arccos\frac{CQ^2 + CP^2 - a_4{}^2}{2CQ \cdot CP}, \gamma = \arctan\frac{z - d_1}{\sqrt{(x - a_1)^2 + y^2}} \tag{11}$$

4. Conversion between joint space and drive space

Like the relationship between Cartesian space and joint space, the extension length of hydraulic cylinder can be acquired by manipulator joint angle. On the contrary, the manipulator joint angle can be acquired by extension length of hydraulic cylinder [8]. The space conversion depends on geometric structure of manipulator. Like inverse kinematics, geometric method can be used for solutions because of no coupling in these solution for each other.

In this paper, under the premise of ignoring joint 1, the relationship between joint angle and expansion length of hydraulic cylinder is built. The specific steps are:

4.1. The relationship between joint angle of link 2 and expansion length of hydraulic cylinder 1

The link 2 is manipulator's upper arm. It is driven by hydraulic cylinder 1. The relationship between joint θ_2 of link 2 and length S_1 of hydraulic cylinder can be acquired by simplified figure 3.

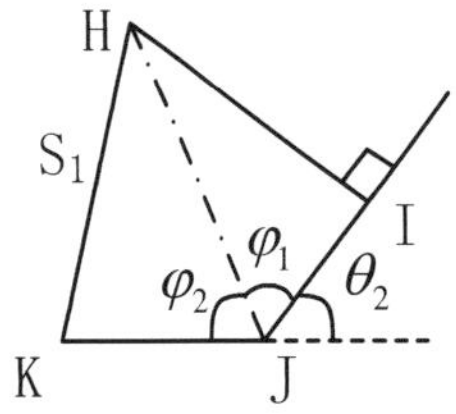

Figure 3. The mechanism sketch of link 2

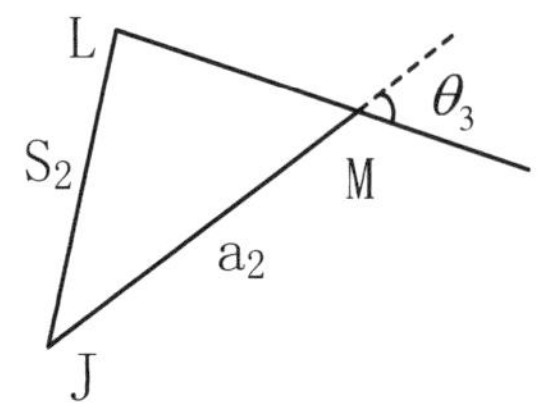

Figure 4. The mechanism sketch of link 3

S_1 is hydraulic cylinder length to be solved. JI is a part of link 2. HI ⊥ IJ. The lengths of IJ,KJ and HJ are known. Based on Cosine theorem,

$$\varphi_1 = \arccos\frac{JI^2 + JH^2 - HI^2}{2JI \cdot JH} \tag{12}$$

$$\varphi_2 = \arccos\frac{JK^2 + JH^2 - S_1^2}{2JK \cdot JH} \tag{13}$$

Because $\theta_2 = \pi - \varphi_1 - \varphi_2$, the relationship between θ_2 and S_1 is

$$\theta_2 = \pi - \arccos\frac{JI^2 + JH^2 - HI^2}{2JI \cdot JH} - \arccos\frac{JH^2 - JK^2 - S_1}{2JH \cdot JK} \tag{14}$$

4.2. The relationship between joint angle of link 3 and expansion length of hydraulic cylinder 2

The link 3 is manipulator's lower arm. It is driven by hydraulic cylinder 2. The relationship between joint θ_3 of link 3 and length S_2 of hydraulic cylinder can be acquired by simplified figure 4.

S_2 is hydraulic cylinder length to be solved. JM is the length a_2 of link 2. The length of LM is known. Based on Cosine theorem,

$$\theta_3 = \arccos\frac{LM^2 + a_2^2 - S_2}{2LM \cdot a_2} \tag{15}$$

4.3 The relationship between joint angle of link 4 and expansion length of hydraulic cylinder 3

The link 4 is the manipulator's wrist. It is driven by hydraulic cylinder 3. The relationship between joint θ_4 of link 4 and length S_3 of hydraulic cylinder can be acquired by simplified figure 5.

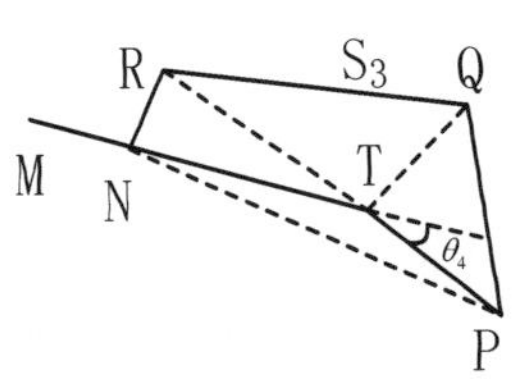

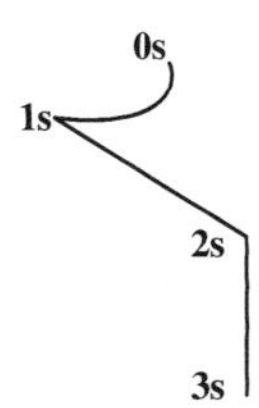

Figure 5. The mechanism sketch of link4 **Figure 6.** Desirable trajectory of manipulator's end-effector

S_3 is hydraulic cylinder length to be solved. The lengths of NR, QP, TP, NP, RT, TQ and NT are known. Because $\angle PTN + \angle NTR + \angle RTQ + \angle QTP = 2\pi$, we can acquire that

$$(\pi-\theta_4)+\angle NTR+\angle RTQ+\angle QTP=2\pi$$

Based on Cosine theorem,

$$\angle NTR=\arccos\frac{TR^2+TN^2-NR^2}{2TR\cdot TN} \tag{16}$$

$$\angle RTQ=\arccos\frac{TR^2+TQ^2-S_3^{\ 2}}{2TR\cdot TQ} \tag{17}$$

$$\angle PTQ=\arccos\frac{TP^2+TQ^2-QP^2}{2TP\cdot TQ} \tag{18}$$

According to equation(16)(17)(18), we can acquire

$$\theta_4=\arccos\frac{TR^2+TN^2-NR^2}{2TR\cdot TN}+\arccos\frac{TR^2+TQ^2-S_3^{\ 2}}{2TR\cdot TQ}+\arccos\frac{TP^2+TQ^2-QP^2}{2TP\cdot TQ}-\pi \tag{19}$$

5. Simulation and verification

The trajectory of manipulator's end-effector is known, the variable of joint and drive can be analyzed by MATLAB. If the trajectory has been given, the specific motion of manipulator's end-effector can be described as:

Begin-the first second: the only joint 1 rotates 25° counterclockwise;

The first second-the second seconds: uniform moves down in the straight line at a certain inclination angle;

The second second-the third second: uniform moves down in the straight line perpendicular to the ground.

Based on the inverse kinematics, the angle of joint space and length of hydraulic cylinder can be acquired in the figure 7 and figure 8.

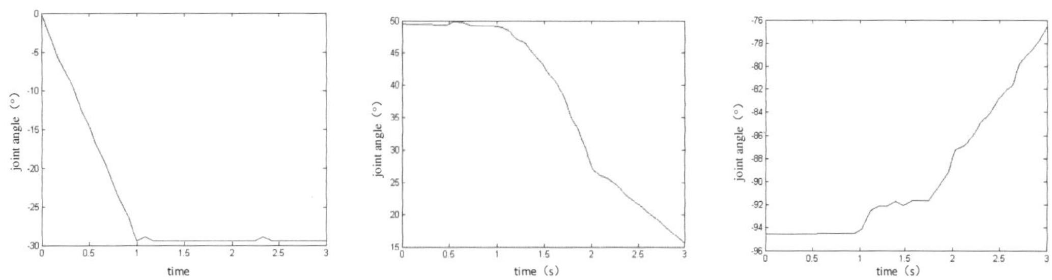

Figure 7(a). Angle curve of joint 1 **Figure 7(b).** Angle curve of joint 2 **Figure 7(c).** Angle curve of joint 3

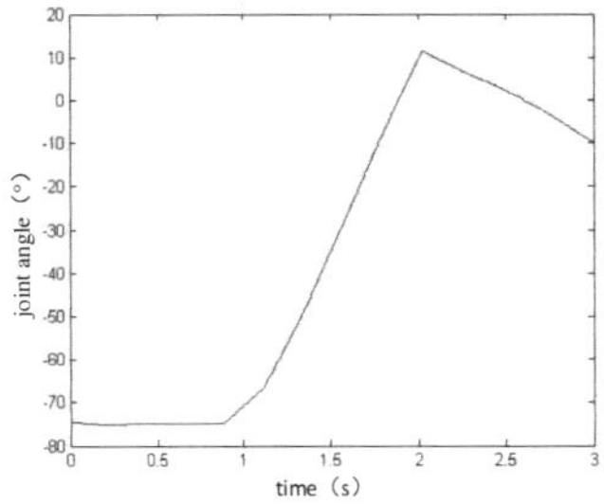

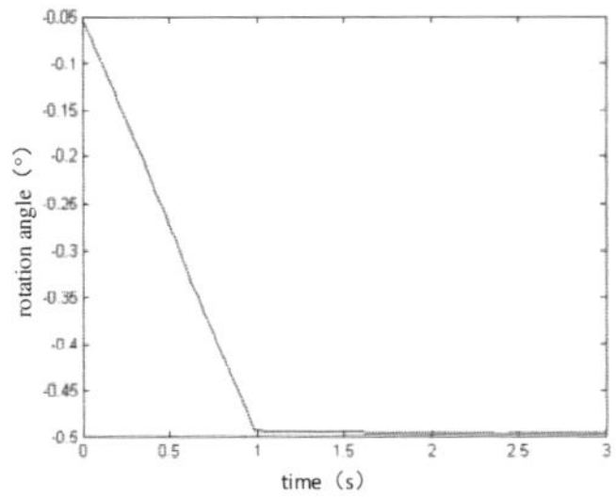

Figure 7(d). Angle curve of joint 4 **Figure 8(a).** Angle curve of motor

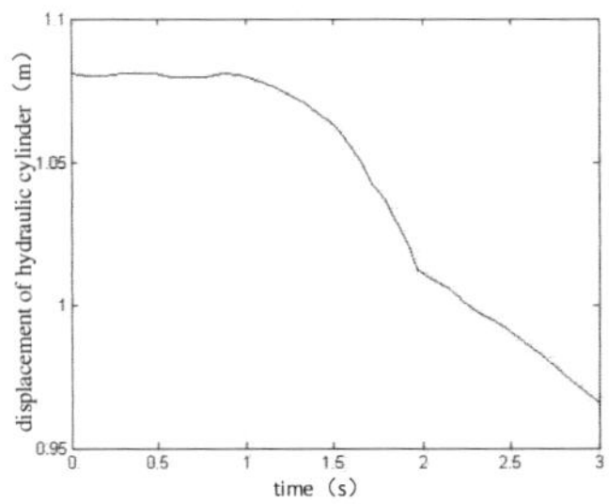

Figure 8(b). Displacement curve of cylinder 1

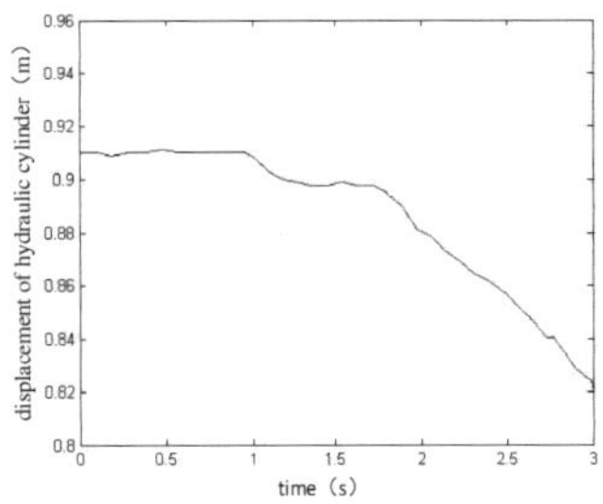

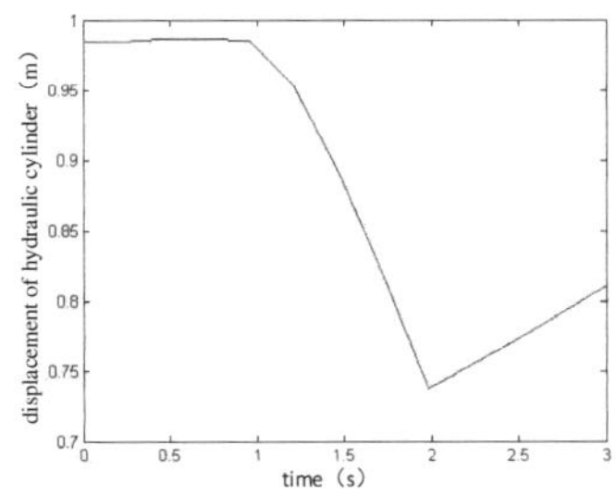

Figure 8(c). Displacement curve of cylinder 2 **Figure 8(d).** Displacement curve of cylinder 3

The simulation shows that the only joint 1 moves and joint 2,3,4 is stationary from begin to the first second. Then the manipulator's joints make a compound motion from the first second to the third second. It meets the given motion requirement.

6. Conclusion

The kinematics analysis is the basis on motion planning and control of manipulator. Taking 4 DOF manipulator as a study object in this paper, the forward kinematics can be acquired by D-H method. Based on the structure of the manipulator, the geometric method is chosen for the inverse kinematics. On this basis, the relationship between joint angle and expansion length of hydraulic cylinder can be expressed. Finally MATLAB validates the correctness. This paper provides a solving method for reverse conversion from Cartesian space to joint space to drive space. It provides theoretical basis for motion planning and control study of manipulator.

Acknowledgments

This work was financially supported by Military Research fund (No.2013BZ162).

References

[1] LV Gang, WANG Fu-dong, A new algorithm for inverse kinematics of manipulator. Journal of

Suzhou institute of silk textile technology,19(5),17-21(1999).
[2] ZHANG Zhi-wen, ZHANG Da-qing, SHI Sheng-xian, CHEN Yong, The analysis of excavator kinematics based on the exponential product formula. Modern machinery,(3),20-22(2005).
[3] WANG Li-quan, ZHANG Ming-jun, YUAN Zheng-you, ZHOU Yi, An algorithm for inverse problem of manipulator kinematics based on genetic algorithm. Journal of Harbin engineering university, (12),10-17(1998).
[4] HUANG Zhi-xiong, Research on the kinematics model and kinematic inverse method of the turbine blade field detection robot. Changsha: Central south university, (2002).
[5] GUO Xiu-li, Research on mechanism and control system of manipulator of logging harvester. Harbin: Northeast forestry university, (2011).
[6] JIANG Hong-chao, LIU Shi-rong, ZHANG Bo-tao, Inverse kinematics analysis for 6 degree of freedom modular manipulator. Journal of Zhejiang university(engineering-science) , 44(7), 1348-1354(2011)
[7] John J Craig, Introduction to robotics. Beijing: China machine press, (2006).
[8] ZHANG Da-qing, Research on motion control of manipulator of hydraulic excavator. Changsha: Central south university, (2006).

Numerical simulation of flow field and the study of the uniform distribution of fluid in uranium hydrometallurgy fixed bed based on CFD

Jian Deng, Zeyong Lei * and Lin Zhong

School of Mechanical Engineering, Nan Hua University, Hengyang, China

*Corresponding author e-mail: lzy1068@sina.com

Abstract. Uranium hydrometallurgy fixed bed is used to separate and extracting uranium compounds from the leaching of uranium ore. It is a very important equipment in the process of uranium purification. The distribution of the internal flow field of uranium hydrometallurgy fixed bed has great effect on the running efficiency of fixed bed. In this paper, on the basis of fluid mechanics, computational fluid dynamics software Fluent is used to numerical simulation for resin adsorption process in axial flow uranium hydrometallurgy, fixed bed that the diameter is 1600 mm and the height is 6800 mm and to research internal flow field distribution of the fixed bed. The results shows that the fluid distribution in the fixed bed is uneven in the process of adsorbent resin adsorption. The groove shunt filter plate at the exit is a powerful measure to realize the uniform distribution of fluid in the resin layer of fixed bed.

1. Introduction

Uranium hydrometallurgy fixed bed is a very important equipment used to separate and extracting uranium compounds from the leaching of uranium ore. And mainly used to complete the process of resin adsorption and leaching [1-2]. Because its structure is simple and easy to operate, it is widely used in our country's uranium mines [3]. In the process of resin adsorption, uranium leaching in the uranium ore enters the fixed bed from the top of the fixed bed, and the uranium compounds are adsorbed in the adsorbent resin through the ion exchange process of the resin bed. Then the adsorption tail is discharged from the bottom of the fixed bed, after the resin is saturated, the saturated resin is transported to the leaching fixed bed for uranium purification. During the adsorption of fixed bed, the fluid distribution in the bed directly affects the ion exchange process and the adsorption efficiency.

At present, the axial flow of fixed bed is mainly used in uranium hydrometallurgy resin adsorption process. The study of adsorption efficiency mainly concentrated in the resin and adsorption process. However, the study of fixed bed design and internal flow field is mainly dependent on experience and cold simulation [4-9].

In recent years, with the rapid development of computer technology and simulation methods, Computational Fluid Dynamics (CFD) provides a strong condition for the study of flow field in fixed bed. Through the calculation of fluid flow characteristics in the fixed bed to obtain the information of internal flow field, By which it can provides an important reference for the optimization of fixed bed. Zhong Siqing et c.[10] used CFX software to finish numerical simulation and experimental study of flow field in axial flow fixed bed with gas distributor and collector; Guo Xiangbo etc. [11]finished numerical calculation of flow field in the large fixed-bed reactor that the liquid entered from the bottom of the inlet distributor and went through the porous media; Calis etc. [12] finished the numerical simulation and experimental verification for the distributing of the velocity and pressure drop in the catalytic reactor.

In this paper, CFD Software is used to establish the hydrodynamic model of resin adsorption process and finish the numerical simulation of flow field in fixed-bed to get the distribution of fluid field in the process of resin adsorption. Then, the uniform distribution of internal flow field is studied, which is aimed at optimizing the structure of fixed-bed and improving the efficiency of uranium extraction.

2. Modeling

2.1. Physical modeling

Physical modeling of φ1600 Uranium hydrometallurgy fixed bed is shown in figure 1:

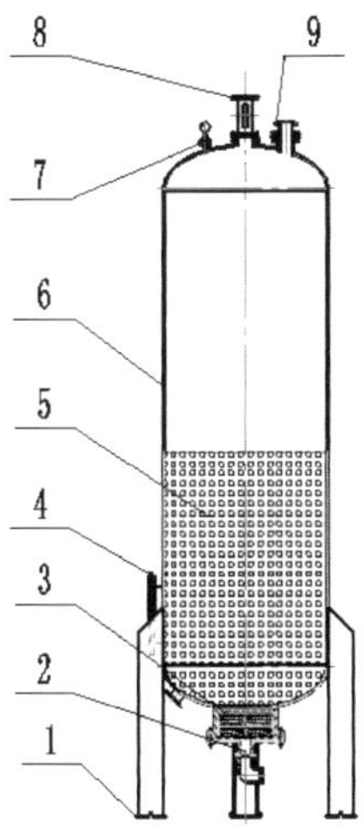

1.base 2. liquid outlet 3. filter 4. manhole 5. resin layer 6. tower body 7. pressure gauge 8. liquid inlet 9. spare port

Figure 1. φ1600 uranium hydrometallurgy fixed bed

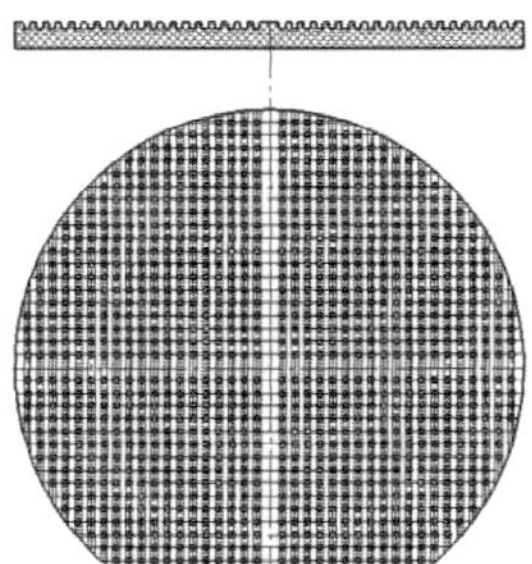

Figure 2. Grooved liquid shunt filter plate

The process of resin adsorption: The leaching of uranium ore enters the tower body (6) from the inlet (8),then go through the resin layer (5). The uranium compounds is adsorbed by the resin, and the tail liquid is discharge through the filter (3) and the liquid outlet (2).

Change the way of packing of resin inφ1600 uranium hydrometallurgy fixed bed. Instead of filter at the bottom of fixed bed, a grooved liquid shunt filter plate which can filter and prevent resin leakage is installed at junction between cylinder and bottom head. Shunt filter plate evenly open a diameter of 0.5 mm small round hole, the opening rate of entire shunt filter is 40%.As shown in figure 2.

2.2. *Mathematical modeling*

The fluid flow at the entrance of the fixed bed is calculated as turbulence. Taking into account the accuracy of fluid flow calculation and the computing capacity of the computer, the standard κ-ε turbulence model is used.

2.3. *Governing equations*

(1) Mass conservation equation

$$\frac{\partial(\rho)}{t}+div(\rho u)=0 \tag{1}$$

(2) Momentum conservation equation

$$\frac{\partial(\rho u_i)}{t}+div(\rho u u_i)=div(\mu grad u_i)-\frac{\partial p}{\partial x_i}+S_i \tag{2}$$

(3) Energy conservation equation

$$\frac{\partial(\rho T)}{t}+div(\rho u T)=div(\frac{k}{c_p}grad T)+S_T \tag{3}$$

(4) Turbulence control equation

$$k=\frac{\overline{u_i^{`}u_i^{`}}}{2}=\frac{1}{2}(\overline{u^{`2}}+\overline{v^{`2}}+\overline{w^{`2}}) \tag{4}$$

$$\varepsilon=\frac{\mu_t}{\rho}\overline{(\frac{\partial u_i^{`}}{\partial x_k})(\frac{\partial u_i^{`}}{\partial x_k})}\ \varepsilon=\frac{\mu_t}{\rho}\overline{(\frac{\partial u_i^{`}}{\partial x_k})(\frac{\partial u_i^{`}}{\partial x_k})} \tag{5}$$

2.4. *Mesh generation*

The two-dimensional geometric model is built and meshing of fixed bed is divided by GAMBIT software. Since the structure of fixed-bed is relatively complex, a quadrilateral unstructured grid is used, the grid nodes are 17980, and the grid cells are 35380. After the grid is divided, the grid model is imported into FLUNET for grid check and the result of check is qualified.

2.5. *Boundary conditions*

(1) Inlet boundary condition includes inlets of liquid. and the liquid inlet boundary condition is

$$V=0.56\ m/s \tag{6}$$

(2) The outlet is outflow boundaries.
(3) The no-slip boundary conditions is used on the wall and there is no velocity.

3. Results and analysis

3.1. The distribution of fluid field in the process of resin adsorption

Fluent software is used to numerical simulation for resin adsorption process in uranium hydrometallurgy, fixed bed .The diameter of inlet and outlet are 160mm, and the leaching of uranium ore enters the tower body from the top inlet of the fixed bed. The velocity is 0.56 m/s. Adsorbent resin is filled in the fixed bed, and the height of resin layer is 4300mm, the height of quartz sand filter layer is 200mm.The experiment shows that the porosity of the resin layer and the quartz sand filter layer is 0.4, and the simulation results are shown as follows:

(1) Velocity distribution

Figure 3 shows the internal velocity distribution of the resin adsorption process in φ 1600 uranium hydrometallurgy fixed bed. It can be seen from the figure that the leaching of uranium ore jet into the fixed bed at a high speed ,after reaching the resin layer, due to the resistance of the resin layer, the liquid flow to both sides of the upper part of the bed to form the phenomenon of backmixing. At the same time the velocity distribution of liquid in the resin layer is not uniform, especially in the vicinity of the outlet, the flow velocity at the central axis and the flow velocity near the wall are quite different. In order to understand the distribution of velocity in the resin layer for more clarity, a radial velocity distribution curve of 300mm, 1800mm and 3800mm from the top of the resin bed is built, as shown in figure 4.

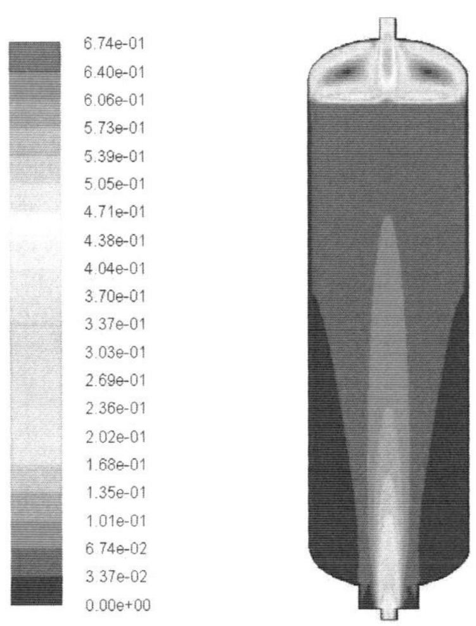

Figure 3. Nephogram of velocity distribution

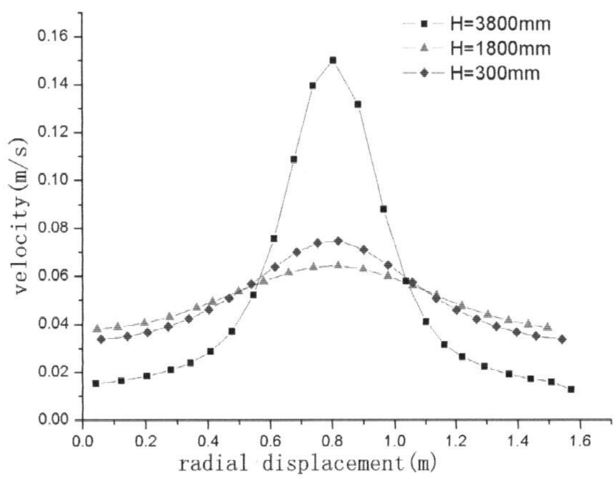

Figure 4 . Curve diagram of the radial velocity

It can be seen from figure 4, after the liquid enters the resin layer from the inlet, the velocity of the liquid in the resin bed is greatly reduced due to the resistance of the resin. In addition, the velocity distribution in the resin layer is inverted V- From the central axis to both sides of the decline, near the outlet, the speed difference between the central axis and the near wall is greater, which will lead to resin adsorption process, resin is not full aturation, so the adsorption efficiency will be reduced.

(2) Axial pressure distribution

Figure 5 shows the axial pressure distribution of the resin adsorption process inφ1600 uranium hydrometallurgy fixed bed. It can be seen from the figure that the pressure distribution in the fixed bed is not uniform and the distribution of the pressure drop from top to bottom is close to 0.4MPa, which is basically consistent with the field test data. In order to quantify the axial pressure, the axial pressure distribution curve in the resin bed is made from the top of the resin bed to the bottom about 4300 mm. It can be seen from the figure 6, The pressure in the resin bed is parabolic along the axis and the pressure loss is about 0.25 MPa.

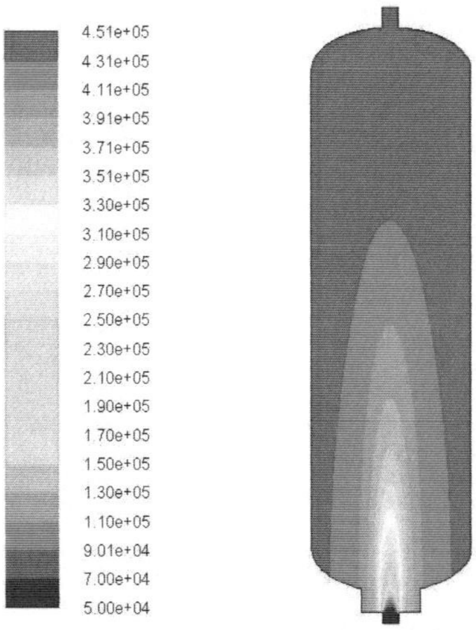

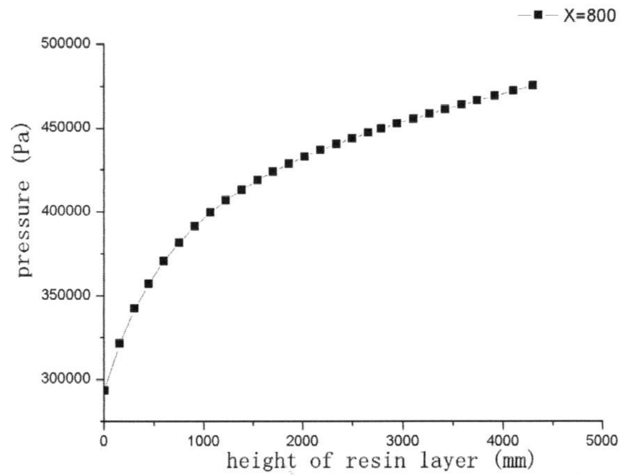

Figure 5. Nephogram of pressure distribution **Figure 6.** Curve diagram of the axial pressure

3.2. Influence of grooved liquid shunt filter plate on flow field distribution in fixed bed

(1) Velocity distribution

Figure 7 shows the internal velocity distribution of the resin adsorption process in φ1600 uranium hydrometallurgy fixed bed with a grooved liquid shunt filter plate. It can be seen from the Figure that the leaching of uranium ore jet into the fixed bed at a high speed ,after reaching the resin layer, due to the resistance of the resin layer, the liquid flow to both sides of the upper part of the bed to form the phenomenon of backmixing. Due to the addition of filter plate at the bottom fixed bed, the flow rate in the resin is evenly distributed and only a local convergence flow is formed at the outlet. The design of the outlet shunt filter plate provides a good improvement to the fluid distribution in the fixed bed. For a clearer understanding of the improved effect, a contrast curve of radial velocity is shown in figure 8

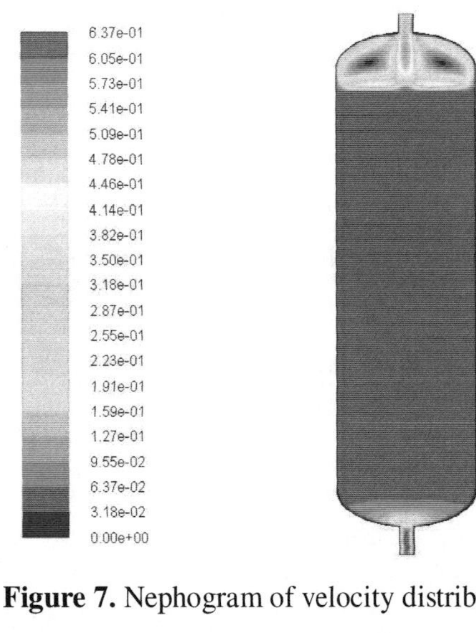

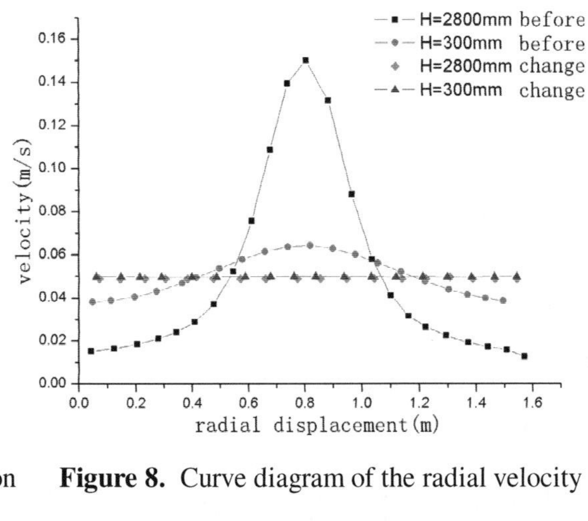

Figure 7. Nephogram of velocity distribution **Figure 8.** Curve diagram of the radial velocity

It can be seen from figure 7 that after installing a grooved liquid shunt filter plate at junction between cylinder and bottom head instead of filter at the bottom of fixed bed, the flow rate in the fixed bed resin is substantially linearly distributed in the radial direction. Consequently, the uniform distribution of the fluid in the fixed bed was realized.

(2) Pressure distribution

Figure 9 shows the axial pressure distribution of the resin adsorption process inφ1600 uranium hydrometallurgy fixed bed with a grooved liquid shunt filter plate. It can be seen from the figure, after changing the structure of filter, pressure changes have been improved in the fixed bed, and the pressure is hierarchical along the axial direction. We can see that the maximum pressure is about 0.09MPa which appears at the top of the fixed bed, and the pressure loss is 0.04 MPa, which is much better than before. The pressure distribution curve in the resin bed was made along the central axis at 4300 mm from the top of the resin bed, as shown in figure 10

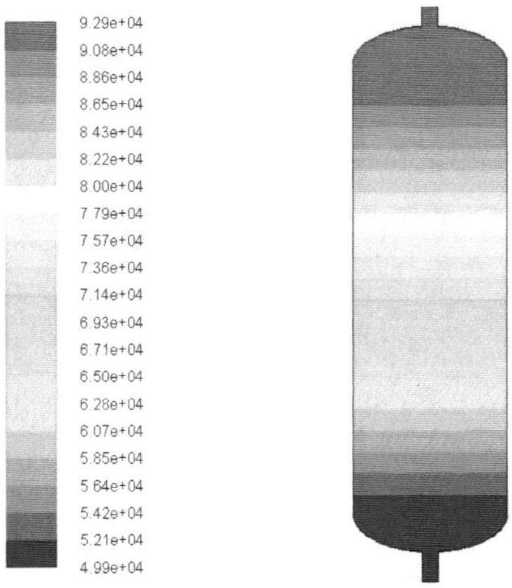

Figure 9.Nephogram of pressure distribution

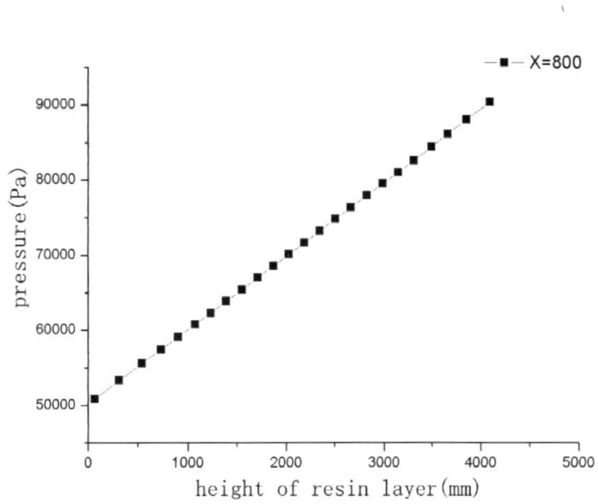

Figure 10.Curve diagram of the axial pressure

It can be seen from the figure, the axial pressure in the resin bed is linearly distributed after changing, and the pressure loss in the resin layer is almost one order of magnitude smaller than before.

From the comparative analysis, we can know that the application of the grooved filter plate has a great influence on the distribution of the flow field in uranium hydrometallurgy fixed bed, and it is an important means and effective measure to realize the uniform distribution of the flow field in the fixed bed.

4. Conclusion

In this paper, computational fluid dynamics software is used to numerical simulate for resin adsorption process in uranium hydrometallurgy, fixed bed that the diameter is 1600 mm. Based on the analysis of velocity distribution and pressure distribution, the following conclusions are drawn:

(1) During the resin adsorption process in axial flow uranium hydrometallurgy, fixed bed, the fluid distribution of the resin bed in fixed bed is not uniform, resulting that adsorption resin can not be fully saturated, and there is a "false saturation", so adsorption efficiency is reduced. At the same time, the pressure distribution in the fixed bed is decreasing from top to bottom, and the pressure loss is close to 0.4 MPa. The pressure in the resin bed is distributed along the axis to the parabola and the pressure loss is about 0.25 MPa.

(2)The design of grooved liquid shunt filter plate of uranium hydrometallurgy fixed bed joints between cylinder and bottom head is a powerful method to realize the uniform distribution of liquid velocity in the resin bed. At the same time, the design of grooved liquid shunt filter plate reducs the pressure loss in the fixed bed.

References

[1] Zeyong Lei, Jian Deng, Lin Lei,et al. Summary of new type of Φ1600 high efficient uranium

hydrometallurgy fixed bed ion-exchange device[J].Journal of University of South China(Science and technology),2014,(03):56-59.

[2] PingruZhong, Zeyong Lei, Yijun Zeng. Design and application of ZLX-1600 ion exchange column for uranium recovery[J].Uranium Mining and Metallurgy, 2006,25(3):144-146.

[3] Qingtao Han, Liangren Zhao, Chunlei Zhang, et al.Ion-exchange Tower Using in Uranium Hydrometallurgy[J].Hydrometallurgy of China, 2008,(03):176-180.

[4] Chenzhan Fu, Zeyong Lei. Development of device for adsorbing and eluting uranium at in-situleaching mine[J].Uranium Mining and Metallurgy,2004,23(1)：44～48.

[5] Yongqiang Wang, LijianZhi, WeibinWang,et al. Improvement of ion exchange technology in one uranium mine[J].Uranium Mining and Metallurgy,2015,(04):266-269.

[6] Bin Xu, YilingZhong, HuiminXiu, et al. Research on Recovery of Uranium from a Leaching Solution by Ion-exchange Resins[J]. Guangdong Chemical Industry,2016,(19):65-67.

[7] Changfu Wang, Zhirong Liu, Kai Zeng. Research progress on resin in adsorption of uranium[J].Hydrometallurgy of China, 2015,(04):259-263.

[8] Genfu Xu. Investigation of resin adsorption uranium mechanism from one uranium mine of CO_2+O_2 in situ leaching solution[J]. Uranium Mining and Metallurgy,2015,(01):8-11.

[9] Naizhong Liu, Xilong Li, Jianbu Wang, et al.The distributed ISL mining uranium and its key equipment[J].Uranium Mining and Metallurgy,2015,(02):61-67.

[10] SiqingZhong, Qingling Chen, Zhiqiang Chen, et al. Numerical simulation and experimental study of flow field in axial flow fixed bed [J]. Journal of Chemical Industry and Engineering (China), 2005,(04):638-642.

[11] XiangboGuo, JinWang.Numerical Simulation of Fluid Flow in Large Scale Axial Fixed Bed reactor[J]. Petrochemical Technology,2007,(07):705-711.

[12] Calis H P A, Nijenhuis J, Pakert B C. CFD Modeling and Experimental Validation of Pressure Drop and Flow Profile in a Novel Structured Catalytic Reactor Packing. Chem Eng Sci,2001, 56(4):1713-1720.

Construction of Fluid - solid Coupling Model with Improved Richards - BP & Its Engineering Application

Chengyu Xie[1], Nan Jia[2,*], Dongping Shi[1], Hao Lu[1]

[1]School of Environment and Resources, University Of Xiang tan 411105, China
[2]Institute of public safety, Tsinghua University, 100062, China

*Corresponding author e-mail: 575580701@qq.com

Abstract. In order to study the slurry diffusion law during grouting, Richards unsaturated-saturated model was introduced, the definition of the grouting model is clear, the Richards model control equation was established, And the BP neural network was introduced, the improved fluid-solid coupling model was constructed, Through the use of saturated - unsaturated seepage flow model, As well as the overflow boundary iterative solution of the mixed boundary conditions, the free surface is calculated. Engineering practice for an example, with the aid of multi - field coupling analysis software, the diffusion law of slurry was simulated numerically. The results show that the slurry diffusion rule is affected by grouting material, initial pressure and other factors. When the slurry starts, it flows in the cracks along the upper side of the grouting hole, when the pressure gradient is reduced to the critical pressure, that is, to the lower side of the flow, when the slurry diffusion stability, and ultimately its shape like an 8. The slurry is spread evenly from the overall point of view, from the grouting mouth toward the surrounding evenly spread, it gradually reaches saturation by non-saturation, and it is not a purely saturated flow, when the slurry spread and reach a saturated state, the diffusion time is the engineering grouting time.

Keywords: Richards-BP model, fluid-solid coupling, diffusion law, model construction.

1. Introduction

The grouting technology improves the physical and mechanical properties of the goaf through the replacement, filling, extrusion, etc., thereby improving its safety. At present, the choice of grouting parameters was usually based on engineering experience, the grouting design and process not only by the geological environment, but also by the grouting works and its comprehensive impact of the object. There were still many problems in the grouting was not resolved, Such as the diffusion radius of the slurry, grouting time, grouting pressure and so on. When the site information was incomplete, or the promotion of new technology and materials cannot guarantee the grouting effect, this will be the safety of the project will be reduced. In this context, many scholars from the theoretical and experimental aspects of grouting design and technology were studied [1~5]. But this series of studies did not systematically carry out the analysis of the overall grouting effect of the corresponding model was presented [6].

IOP Conf. Series: Journal of Physics: Conf. Series **916** (2017) 012024 doi:10.1088/1742-6596/916/1/012024

There were many factors that affect the effect of grouting. The research of theory and experiment cannot carry on the effective analysis to all aspects of grouting, the effective simulation of this study could be better applied to the engineering design, So the mathematical model had been used for reference, Through the appropriate improvement, and then the simulation of grouting effect was of great significance[7~8].

For the infiltration grouting process, grouting and anti-seepage and strengthening of the composition and structure of the individual and unpredictable were considered, The improved Richards-BP coupled model was constructed by numerical simulation of multi-field coupling. The Richards equation was introduced into the fluid flow model in the multi-field coupling software, The diffusion of the slurry, the main influencing factors and the diffusion effect were analyzed, Which provides an effective basis for grouting reinforcement design and grouting effect.

2. Improved Richards-BP model

2.1. Model building

The saturated - unsaturated seepage flow model of Richards equation was used to calculate the diffusion law of free surface. Richards equation could not only intuitively express the grouting effect, But also indirectly indicate the rationality of the various aspects of the grouting design [9].

Through the Richards equation control model was established, set the effective saturation and relative permeability expressed as a function of the matrix suction saturation zone for the positive pressure zone, the permeability coefficient was saturated permeability coefficient, the unsaturated region was a negative pressure zone, the infiltration coefficient was about zero [10]. The seepage field of grouting slurry was analyzed and studied, and then by means of multi-field coupling software for numerical simulation of effective slurry saturation - unsaturated simulation is obtained, to intuitively analyze the slurry diffusion law.

2.2. Richards grouting model control equation

Since the pressure in the seepage field was equal to the atmospheric pressure, the model control equation is set [11]

$$[C+S_eS]\frac{\partial H_p}{\partial t}+\nabla\cdot[-K\nabla(H_p+D)]=0 \qquad (1)$$

In the formula, pressure head Hp(m) as the dependent variable, C was the volume ratio of humidity (m-1), Se was effective saturation, S was the pulp coefficient (m-1), t is time,K was the permeability coefficient m/s, D (such as x,y or z) represents the coordinates of the vertical direction (m). The equation does not show the volume fraction of water θ, But it was a value that depends on HP, because C, Se and K With changes in HP and θ changes.

Volume ratio timing could be expressed as

$$C=\partial\theta/\partial H_p \qquad (2)$$

In the formula (2), the relationship between the slurry filling volume and the grouting pressure was described [12].

In order to simulate the slurry quality factor, it could be used more than the storage option.

$$S=\rho_f g(\chi_p+\theta\chi_f) \qquad (3)$$

In the formula (3), ρf indicates fluid density kg/m3, g was gravity acceleration,χp andχf represents the compression rate of solid particles and fluid (m·s2/kg). Formula such as van Genuchn, Brooks and

Corey, it could be described C, Se, K and θ With Hp change in the relationship,the Prerequisite wereθs andθr, as well as some other geomaterials α, n, m and l as known. In the van Genuchten equation, assuming that when the pressure of the fluid was equal to atmospheric pressure, soil was saturated, which was Hp=0, and according to the method of Brook and Corey, According to the intake air, Can distinguish the saturation of porous media (Hp>-1/α) and unsaturation status (Hp<-1/α), which was

$$\theta = \begin{cases} \theta_r + S_e(\theta_s - \theta_r) .... H_p < -\frac{1}{\alpha} \\ \theta_s .... -\frac{1}{\alpha} \le H_p \end{cases} \tag{4}$$

$$S_e = \begin{cases} \frac{1}{\left|\alpha H_p\right|^n} .... H_p < -\frac{1}{\alpha} \\ 1 .... -\frac{1}{\alpha} \le H_p \end{cases} \tag{5}$$

$$C = \begin{cases} \frac{-n}{H_p}(\theta_s - \theta_r)\frac{1}{\left|\alpha H_p\right|^n} .... H_p < -\frac{1}{\alpha} \\ 0 .... -\frac{1}{\alpha} \le H_p \end{cases} \tag{6}$$

$$K = \begin{cases} K_s S_e^{\frac{2}{n}+l+2} .... H_p < -\frac{1}{\alpha} \\ K_s .... -\frac{1}{\alpha} \le H_p \end{cases} \tag{7}$$

2.3. Improved Richards-BP Neural Network
First, using the Richards model described above, the trend of slurry diffusion in the collapse zone was predicted, Then, the absolute error of the Richards model was processed by BP neural network model, That is, the absolute error predicted by the Richards model was taken as the input and output value of the BP neural network model, The BP neural network model was used to correct the prediction error, and the model prediction error of Richards-BP was obtained, to improve the prediction accuracy of the model[13].

2.4. Coupling calculation
Since there was an iterative coupling process in the above calculation, the number of unsaturation layers N and the layer thickness ΔzN were constantly changing, and therefore need to use the saturated band calculation of the head results, so that the mesh of the non-saturated band was updated. On the other hand, when running the groundwater module [14], the upper boundary condition was the bottom flux qN, N + 1, to ensure the overall water balance, the water level should be equal to the free porosity in the unsaturated zone during the period, which was

$$\mu = \theta\, sN - \theta\, N \tag{8}$$

In the formula, θsN and θN are the saturated volume water content and volume water content of the last layer of the unsaturated zone, respectively.

In order to get the answer to this question, Initial conditions and boundary conditions could be set. Initially, the sample had the same pressure on the head HP0. The boundary condition was (n was the unit vector perpendicular to the boundary)

Boundary conditions	$n[-K\nabla(H_p + D)]$	$n[-K\nabla(H_p + D)]$	$H_p = H_{p0}$	$H_p = H_{p0}$
Initial value	$\partial\Omega$Sides	$\partial\Omega$Rings	$\partial\Omega$Base	$\partial\Omega$Surface

2.5. Model build process and calculation parameters

In order to study the law of slurry diffusion, combined with multi-field coupling software, and the numerical model of the calculation parameters in Table 1.

As the diffusion law by many factors, under certain environmental conditions set, The effect of the subsequent slurry material and the initial pressure on the diffusion law was studied. Based on experience, respectively, the slurry water-cement ratio was set to 0.8: 1 and 1:1, the grouting pressure was set to 0.5 to 1 MPa [15].

Table 1 Numerical simulation calculation parameters

Slurry properties						Model parameter setting		
Total drilling volume(m)	Grouting volume(m^3)	Use cement(t)	Water glass(kg)	Grouting pressure(Mpa)	Water-cement ratio	Saturated slurry volume fraction	Residual liquid volume fraction	Constitutive constants
440.5m	72	41.5	20	0.5-1	0.8:1,1:1	0.43	0.045	14.5

In the saturated stream, the permeability will change as the fluid flows through the porous medium which was filling the voids. Using the Richards equation to simulate flow, the sloughness of the voids could be considered by using the van Genuchten and Brooks-Corey formulas.

3. Engineering Applications

3.1. Engineering examples

A lead-zinc mine in a stope, due to the mine was not timely and mined-out area exposure time was too long, resulting in two groups of filling body collapse and the top ore body fall, in the mining process, due to the mined-out area and the surrounding rock of the plastic zone to expand and extend to the adjacent stope lead to through and a large range of crust damage. Combined with on-site investigation, the surrounding rock was very unstable, the collapse area must be filled with grouting. According to engineering practice, cement slurry grouting method was selected to fill the collapse area to fill grouting. The grouting process must involve the slurry in the filling body voids and cracks in the flow problem, it was possible to simulate the flow of the slurry in the filling body by means of numerical analysis.

3.2. Hole Mesh Model for Rectangular Grouting

Select a section of the collapse area to set the grouting hole, In the Richards model, the pressure head was uniform, and all the vertical slices were the same, it can select a 2D cross section, the establishment of the direction of the direction of Y = 636m, 100m to 220m down hole collapse model. In order to reduce the amount of computation, the free-split triangular mesh was selected, set the maximum unit size to 0.025, the results were shown in Fig1, a total of four grouting holes were arranged.

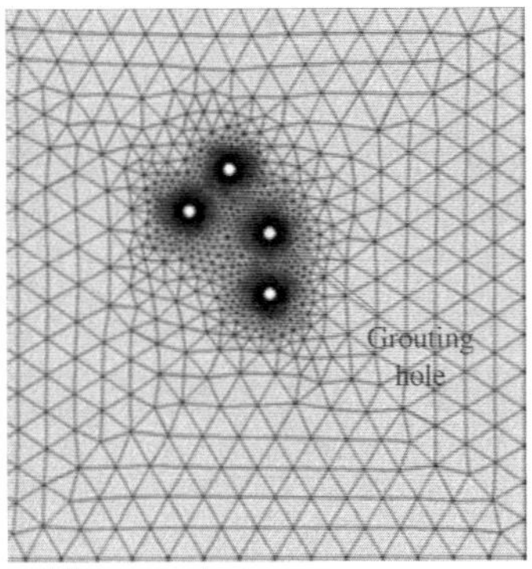

Fig.1 Model grid

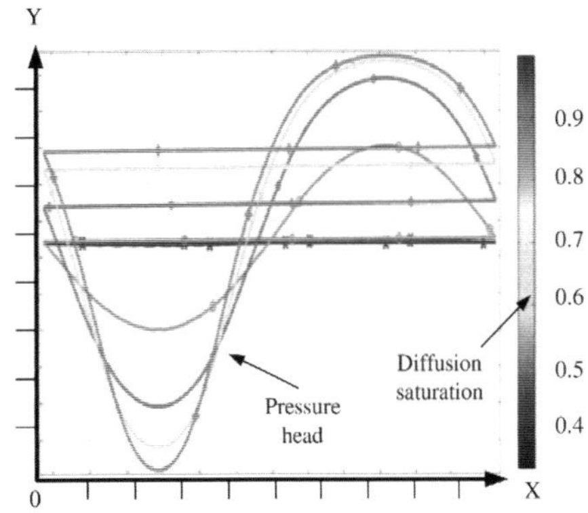

Fig.2 Effective saturation around the grout

3.3. Analysis of Basic Seepage Regularity of Grouting

The parameters of the material are converted into variable parameters in the Richards model, The results of the diffusion calculation of the four grouting holes are simulated, the distribution of effective saturation, pressure head (contour) and flow rate (arrow) is shown. The diffusion saturation in the filling body is in the order of blue, yellow and red, respectively, from low to high, it can be seen that the flow around the slurry hole has undergone significant changes, While in other parts of the change is basically uniform, as shown in picture 2, slurry diffusion is not spread evenly around, But the degree of slurry saturation will continue to change with the continuous injection of the slurry.

Figure 2 for the different time points out of the pulp around the slurry of the effective saturation of the one-dimensional map, X is the distance to the grouting hole, and y represents the effective saturation of the position. The angular coordinates (+ 180 °, 0 °) and (-180 °, 0 °) correspond to the upper and lower parts of the boundary, it shows the upper and lower sides of the pulp on the effective saturation, the effective saturation is 0s, 50s, 100s, 300s, 600s, 900s, respectively, from the middle to the sides (in the order of dark blue, green, red, light blue, rose). From Figure 2, ① the initial value of the effective saturation is a horizontal line, ② the influence of the initial pressure on the upper rock mass is larger than that of the lower rock mass, that is, the initial pressure is affected by the initial pressure. With the passage of time, the slurry will be affected by the initial pressure will become smaller and smaller, by the impact of gravity becomes larger, the effective saturation of the lower part of the region also began to increase, the final out of the mouth will be fully saturated. ③ the saturation of the slurry increases with time. ④ slurry diffusion is along a certain path, it constitutes an irregular closure curve, and it can be calculated by calculating the trajectory path perimeter to the theoretical diffusion radius, it can be seen that when other grouting conditions are constant, a certain initial pressure of the grouting directly affects the effect of grouting diffusion is set so that the grouting spread evenly.

3.4. With the law of time spread

On the basis of the above model, set the relevant parameters in the grouting design, change the time step, Set the time step are 0s, 50s, 100s, 150s, Thus, the slurry diffusion is studied in chronological order. Through the multi-field coupling software to simulate as shown in Figure 3.

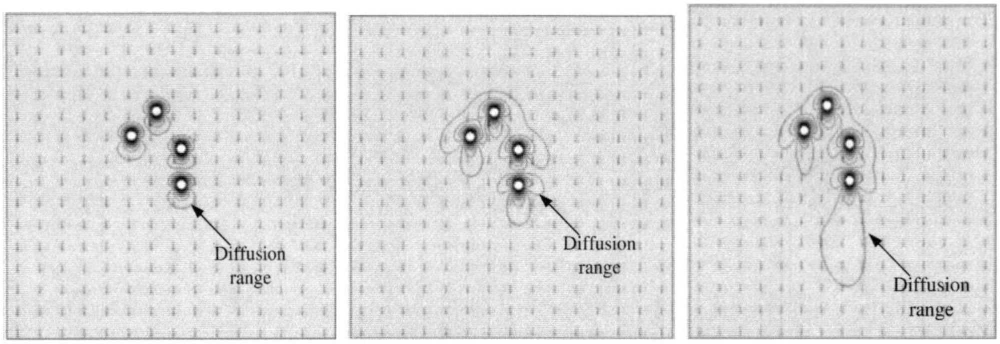

Fig.3 The law of the diffusion of the slurry over time

By comparing the diffusion maps of different times, ① the upper and lower sides of the slurry spread the same degree of beginning, but as time changes, when the slurry away from the grouting hole, which is affected by gravity, the pressure gradient is reduced to the critical pressure, the initial pressure is weakened and the diffusion on the upper side of the slurry is not uniformly diffused, that is, the initial pressure of grouting needs to be adjusted. ② Slurry in a certain grouting pressure gradient, it will flow along the chamber or fissure diffusion, Slurry density, viscosity increases, low flow rate, this may cause it to not spread completely, and the slurry may stop flowing.

Through the above verification, according to the demand for grouting numerical simulation, Such as the progress of the project tight, the material water-cement ratio that changes the material density and viscosity in the model is adjusted, the best water-cement ratio in the same area is obtained, this can effectively shorten the duration, but also to make the slurry will not premature condensation and spread in place.

4. Conclusion

(1) Richards unsaturated-saturated model is introduced, the definition of the grouting model is defined, and the Richards model control equation is established, introduced into the BP neural network, and the improved fluid-solid coupling model is constructed. Taking the engineering practice as an example, the numerical simulation of the slurry diffusion process is carried out on the basis of setting the grouting parameters.

(2) Through the slurry diffusion analysis can be obtained, the slurry began to flow along the upper part of the grouting hole flow diffusion, when it is away from the grouting hole, the pressure gradient will be reduced to critical pressure, this time, the slurry by gravity, It will flow to the bottom, and ultimately it gradually reached a uniform state, when the slurry tends to stabilize the diffusion, the shape of the class "8" shape.

(3) Slurry diffusion from the whole to be evenly dispersed, from the grouting port to its saturation around the uniform diffusion, and gradually from the saturation to saturation, it is not purely saturated flow. When the slurry time to promote the map did not show significant changes, That is, the diffusion of the slurry reaches saturation, and the diffusion time at this time is the grouting time of the project.

References

[1] Yang Zhiquan, HOU Kepeng, CHENG Yong et.al. Study of Column-hemispherical penetration grouting , Mechanism Based On Power-Law Fluid. 2014(S2) 3840-3846.

[2] Zhang Qingsong, Li Peng, ZHANG Xiao , et al. Model test of grouting strengthening mechanism for fault gouge of tunnel, Chinese Journal of Rock Mechanics and Engineering, 2015(05) 924-934.

[3] Yang Zhiquan, Niu Xiangdong, Hou Kepeng, et al. Study on diffuson parameters of bingham fluid base on Column-hemispherical penetration grouting, Journal of Sichuan

university(Engineering science edition). 2015 (S2) 47-53.

[4] LI Peng ,ZHANG Qing-song, ZHANG Xiao, et al. Analysis of fracture grouting mechanism based on model test, Rock and Soil Mechanics, 2014(11) 3221-3230.

[5] LIU Jian,LIU Rentai,ZHANG Xiao, et al. Diffusion law model test and numerical simulation of cement fracture grouting. Chinese Journal of Rock Mechanics and Engineering, 2012 (12) 2445-2452.

[6] Li Shucai, Zhangxiao, zhang Qingsong, et al. Research on mechanism of grout diffusion of dynamic grouting and plugging method in water inrush of underground engineering, Chinese Journal of Rock Mechanics and Engineering. 2012 (12) 2445-2452.

[7] Ruan Wenjun. Spreading model of grouting in rock mass fissures based on time-dependent behavior of viscosity of cement-based grouts, Chinese Journal of Rock Mechanics and Engineering, 2005 (15) 2709-2714.

[8] Ruan Wenjun. Research on diffusion of grouting and basic properties of grouts, Chinese Journal of Geotechnical Engineering, 2005 (01) 69-73.

[9] ZHANG Qingsong,ZHANG Lianzhen,zhang xiao, et al.grouting diffusion in a horizontal crack considering temporal and spatial variation of viscosity, Chinese Journal of Rock Mechanics and Engineering. 2015(06) 1198-1210.

[10] Luo Zhou-quan, Xie Cheng-yu, zhou Ji-ming, et al. Numerical Analysis of Stability for Mined-out Area in Multi-field Coupling, Journal of Central South University, 2015 22(02) 669-675.

[11] G.S. Zhao,G.Q. Zhou,G.R. Zhong,F.P. Zhu,H.C. Liang. Analysis of stratum grouting influence on shaft lining stress with the methods of simulation and in site measurements, Procedia Earth and Planetary Science, 2009 (1) 497-502.

[12] Hideki Shimada,Akihiro Hamanaka,Takashi Sasaoka,Kikuo Matsui. Behaviour of grouting material used for floor reinforcement in underground mines, International Journal of Mining, Reclamation and Environment, 2014 28(2) 133-148.

[13] Min Zhang,Xing-hua Wang,You Wang. Numerical evaluation of uplifting effect for upper structure by grouting, Journal of Central South University, 2012 19(2) 553-561.

[14] Zha Yuanyuan, Zhu Yan, Yang Jinzhong. A regional coupled groundwater model for unsaturated-saturated flow based on modified integrated Richards equation, Journal of Sichuan university(Engineering science edition), 2013(01) 107-115.

[15] Tian Fuqiang, Hu Heping. Numerical model of Richard's equation based on an ordinary differential equation solver, Tsinghua University (Sci& Tech) , 2007 (06) 785-788.

IOP Conf. Series: Journal of Physics: Conf. Series **916** (2017) 012025 doi:10.1088/1742-6596/916/1/012025

Numerical analysis of sliding bearing dynamic characteristics based on CFD

Kun Wang *, Jinwu Wu, Guoyang Zhao

School of Aircraft Engineering, Nanchang Hangkong University, Nanchang, China

*Corresponding author e-mail: shijianwangkun@163.com

Abstract. According to the characteristics of the flow field of the sliding bearing, the CFD numerical calculation model of the sliding bearing was established. The CFD method was used to simulate the flow field of the sliding bearing, and the pressure distribution of the sliding bearing flow field was obtained. Based on this, the stiffness and the damping coefficient of the dynamic characteristics were obtained by the dynamic grid technique. The influence of the rotational speed on the dynamic characteristic coefficient was further studied. The results show that the difference between the stiffness coefficient and the damping coefficient is less than 5%, and the accuracy of the model is verified. The absolute value of the oil film stiffness increases non-linearly with the increase of the rotational speed, and the influence of the rotational speed on the damping coefficient is small.

1. Introduction

The rotor-support system is at the core of the entire rotor system, and sliding bearings are the most important part of the system, and its operating conditions will directly affect the performance of the entire mechanical system [1]. The sliding bearing has the characteristics of simple structure, strong carrying capacity, and high rotation precision and long life. It is widely used in steam turbine, internal combustion engine, compressor, large motor and blower [2]. Important, the dynamic characteristics of sliding bearings are mainly described by the stiffness coefficient and the damping coefficient, and the dynamic characteristic coefficient is directly related to the stability of the rotor and the flow field.

The Reynolds equation is the theoretical basis for studying the sliding bearing. The basic way to calculate the bearing characteristics is to solve the corresponding Reynolds equation for different sliding bearings. Chen Boxian, Su Hong et al given the oil film in the three-dimensional stability of the two main ways to get the corresponding three-dimensional stability parameters from the theoretical analysis of three-dimensional oil film work conditions [3-4]. Yang Jinfu et al used numerical methods to study the journal radial offset and skew when the dynamic characteristics of sliding bearings, derived power coefficient matrix expression, and discussed the physical meaning of the elements [5]. Swift, H, W, GUO Zeng lin et al.conducted theoretical and experimental studies on tilted tile bearings from different perspectives [6-7]. It is gradually unable to meet the increasing demand for industrial technology and is gradually replaced by the higher accuracy CFD method.

CFD technology has been widely used in aviation, aerospace, automobile manufacturing and other high-end areas, CFD technology using Nervier-Stokes (N-S) equation and the establishment of appropriate models to solve the bearing incompressible oil film, and dynamic solution is usually used rotor micro-eddy method [8]. CFD technology has the advantage of being able to calculate the complex mathematical model to solve the complete N-S equation, the oil film pressure and other parameters of the specific distribution for the bearing selection, design, etc. to provide important data basis. The effects of aspect ratio, density, dynamic viscosity and rotational speed on the dimensionless damping coefficient were analyzed by fluid calculation software. The results show that the CFD method has been studied in these aspects of sliding bearing. A certain superiority, we can see that CFD technology for a variety of flow field calculation has a high degree of fit [9-11].

2. Model establishment and parameters

In order to make the calculation result accurate, this paper uses the model in [12], its geometric dimensions are shown in Table 1, and Figure 1 shows the axial structure of the cylindrical plain bearing.

Table 1. Calculates the geometry of the model.

Bearing radius (R/mm)	Rotor radius (r/mm)	Bearing width (l/mm)	Eccentricity ratio (ε)	Eccentricity angle (θ)
25	24.95	25	0.5	49.8°

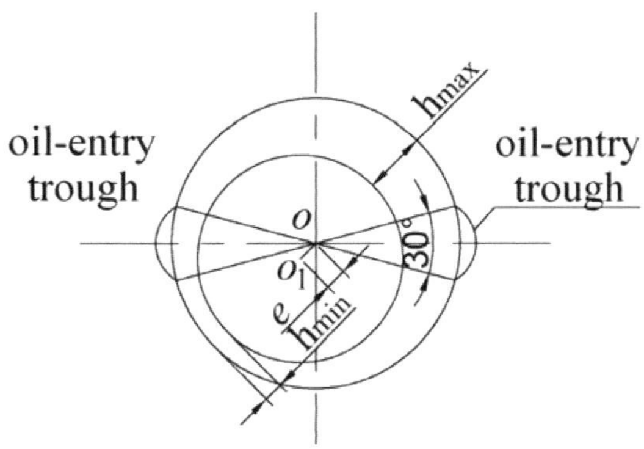

Figure 1. The model of the axial structure of the model.

The model of sliding bearing pressure is established by using fluent tool. For simplicity of comparison, the structural parameters of the model are the same as those in [12]. The model of ordinary cylindrical bearing is taken as the object of study. Bearing diameter D=50mm, bearing width B=25mm, speed n=9550r/min, relative clearance Ψ=0.175%, Eccentricity ratio ε=0.5, bearing level in the surface of the two into the oil tank. Lubricants are selected according to Zhang Zhiming's "hydrodynamic lubrication theory of plain bearings". The lubricating oil grades selected in this calculation are HU-20, dynamic viscosity μ=0.0125Pa.s, average operating temperature 50 ℃, in order to simplify the neglect of viscosity With the temperature change. Boundary conditions: set the oil pressure P=103000Pa, the oil pressure P=0Pa, from both sides of the oil tank into the oil, bearing both ends of the oil, journal speed n=9550r/min, the other part of the solid wall boundary. Figure 2 shows the resulting grid structure. The iterative residual graph shows the convergence, save the calculation result, and post-process the calculation result. Figure 3 shows the eccentricity of 0.5, speed 9550rpm, oil pressure 103000Pa under the conditions of the sliding bearing pressure.

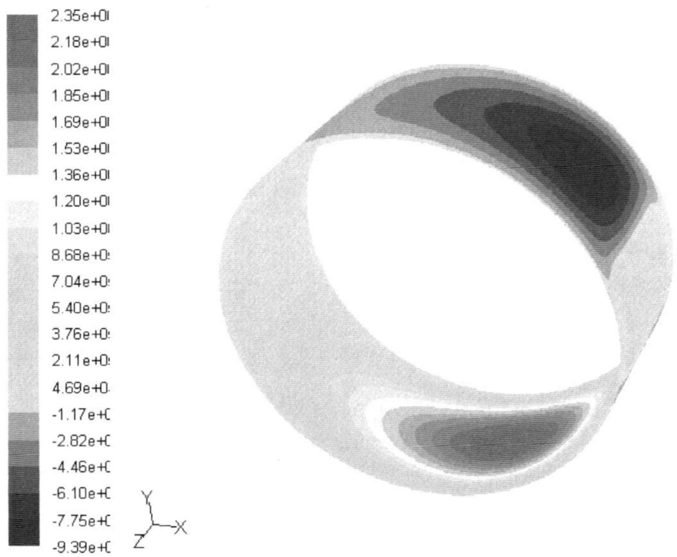

Figure 2. Sliding bearing grid structure.

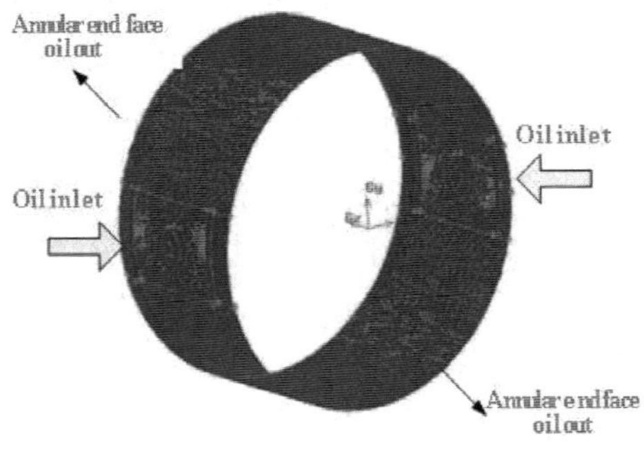

Figure 3. Pressure profile.

From this figure, it is clear that the pressure distribution of the oil film in the sliding bearing is clearly seen. The space within the bearing can be divided into convergent oil wedge and divergent oil wedge. The maximum pressure is generated in the convergent oil wedge region, which is 2.347MPa and the load capacity is 1256N. The calculated results in the literature [12] are 2.352MPa maximum pressure and 1160N load. The results show that result error is small and the calculation result is accurate. See Table 2 for specific data.

Table 2. Comparison of data and foreign language data.

	Maximum pressure (MPa)	Load capacity (N)	Sump angle (deg)
Foreign language data	2.352	1160	49.8
Data for this article	2.347	1256	49.8

3. Dynamic characteristic coefficient calculation

Sliding bearings in normal work, the static balance of the journal position is O0(x0, y0), the oil film on the journal of the force Fx0 and Fy0. When the journal moves slightly in the vicinity of its static equilibrium position Δe, its position is O(x, y) when it deviates from the static equilibrium position, and the force applied to the journal is Fx and Fy. Then the change in oil film force can be expressed as:

$$\begin{cases} \Delta F_x = k_{xx} \Delta x \\ \Delta F_y = k_{yx} \Delta x \end{cases} \tag{1}$$

$$\begin{cases} F_{x_0} - F_x = k_{xx}x + k_{xy}y + c_{xx}\dot{x} + c_{xy}\dot{y} \\ F_{y_0} - F_y = k_{yx}x + k_{yy}y + c_{yx}\dot{x} + c_{yy}\dot{y} \end{cases} \tag{2}$$

Where kxy and cxy represent the stiffness and damping coefficients of the bearing, the first subscript indicates the direction of the force increment, and the second subscript represents the direction of the displacement increment. x, y, $\dot{x}$, $\dot{y}$ Respectively, that the journal in the horizontal and vertical direction of the small displacement and speed. So that the crest in the horizontal direction produces a small displacement disturbance (Δx, 0), then, y=0, then each equation has only one stiffness unknown. Calculate the difference between the force of the oil film force in the horizontal and vertical directions before and after the migration, ie, ΔFx and ΔFy, and substituting the equation (2) to obtain the stiffness coefficients kxx and kyx of the oil film. Similarly, so that the journal in the vertical direction of a small displacement (0, Δy), It can get the oil film stiffness coefficient kxy and kyy.

3.1. Calculation of stiffness coefficient

The stiffness coefficients kxx and kyx of the oil film can be obtained by calculating the difference between the forces in the horizontal and vertical directions before and after the offset, ie, ΔFx and ΔFy. Similarly, the journal in the vertical direction of a small displacement, it can get the oil film stiffness coefficient kxy and kyy. Calculate the journal to produce a slight offset x/h0 and y/h0 (0.0005mm, 0.00025mm, and 0.0001mm), calculate the change of oil film force in two states, and obtain the stiffness coefficient of oil film. Figure 4 the pressure distribution of the oil film under different horizontal disturbances is shown.

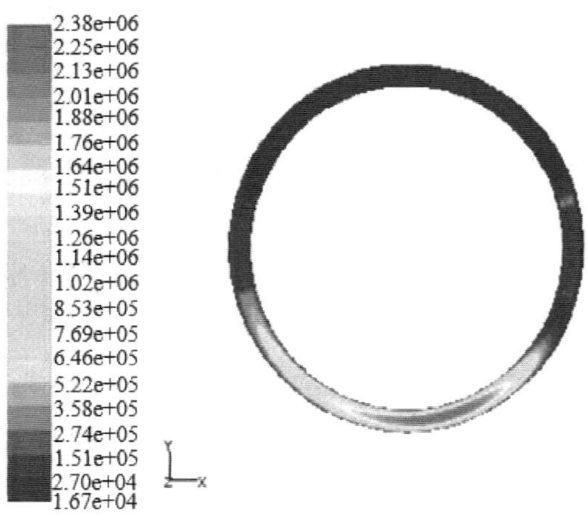

(a) x/h_0=0.01when the pressure distribution

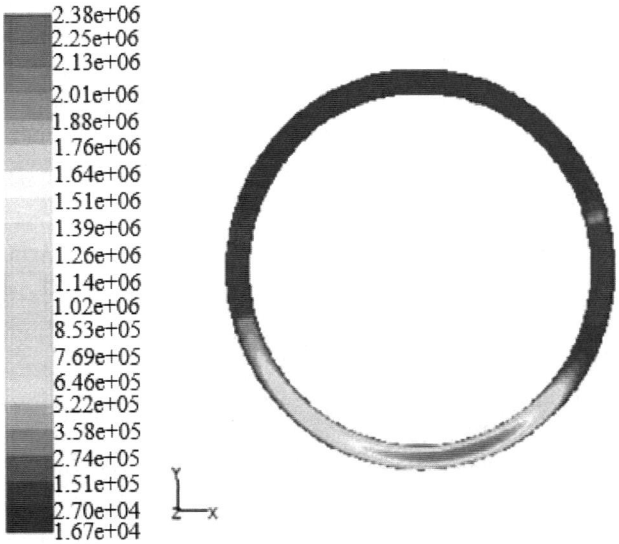

(b) x/h_0=0.002 when the pressure distribution

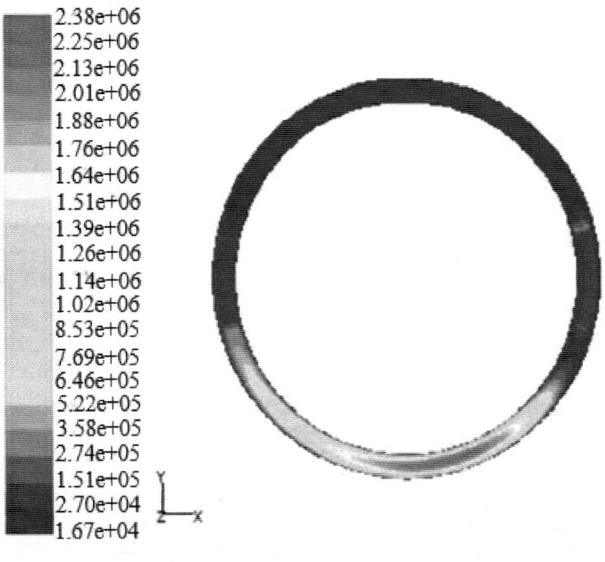

(c) x/h_0=0.005 when the pressure distribution.

Figure 4. Oil pressure distribution under different horizontal disturbances.

In Table 3 the CFD single-phase flow model in different disturbance position under the journal in the horizontal and vertical force is shown. It can be seen from the table, when the journal to the x-axis positive direction, the horizontal direction of the oil film force is less than the equilibrium position, and in the vertical direction is greater than the equilibrium position, with the journal to the x-axis direction of the increase, The force in the horizontal direction is gradual decline and the vertical direction increases gradually. When the journal is shifted in the negative direction of the y axis, the oil film has more force on the journal in both directions than the equilibrium position and it increases linear relationship as the offset increasing.

Table 3. CFD single-phase flow model in different disturbance Position horizontal and vertical direction of the oil film on the journal (N).

	Disturbance displacement	F_x	F_y	Disturbance displacement	F_x	F_y
CFD single phase flow model	x/h_0=0	40.0	1172.2	y/h_0=0	40.0	1172.2
	x/h_0=0.01	58.5	1217.1	y/h_0 =-0.01	54.4	1200.5
	x/h_0=0.002	42.5	1180.9	y/h_0 =-0.002	44.3	1178.2
	x/h_0=0.005	47.0	1192.8	y/h_0 =-0.005	48.8	1186.3

Table 4. Stiffness Coefficient of Oil Film for Different Models (106N/m).

Calculation method	k_{xx}	k_{xy}	k_{yx}	k_{yy}
VT-FAST	40.0	-19.4	87.2	59.1
CFX-TASCflow($x/h_0=y/h_0$ =0.01)	41.2	-21.9	88.0	60.0
CFX-TASCflow($x/h_0=y/h_0$ =0.005)	41.1	-22.1	88.1	58.0
CFX-TASCflow($x/h_0=y/h_0$=0.002)	38.9	-21.3	85.0	55.0
FLUENT Single phase flow ($x/h_0=y/h_0$=0.01)	43	-22.8	88.0	59.3
FLUENT Single phase flow ($x/h_0=y/h_0$=0.005)	40	-23.8	87.8	56.4
FLUENT Single phase flow $x/h_0=y/h_0$=0.002)	37	-23	86	54

In Table 4 the results of CFD single-phase flow model and other models are shown. It can be seen that the stiffness coefficient of the single-phase flow model calculated by fluent software is less than 5% compared with the CFX software.

3.2. Damping coefficient calculation

So that the crest in the horizontal direction of the small disturbance speed, then the oil film force in the vertical direction of the disturbance rate and the two directions of the disturbance displacement is zero, type (2) can be simplified as:

$$\begin{cases} \Delta F_x = c_{xx} \Delta \dot{x} \\ \Delta F_y = c_{yx} \Delta \dot{x} \end{cases} \tag{3}$$

The stiffness coefficients cxx and cyx of the oil film can be obtained by calculating the difference between the forces in the horizontal and vertical directions (ΔFx and ΔFy), in the two states cxx and cyx.

The damping coefficients cxx and cyx of the oil film can be obtained by calculating the difference between the forces in the horizontal and vertical directions, ie, ΔFx and ΔFy, in the two states. The pressure distribution of the oil film at four time is shown during the movement of the journal from the eccentric position to the static equilibrium position at a disturbance speed of -0.0005m/s in Figure 5.

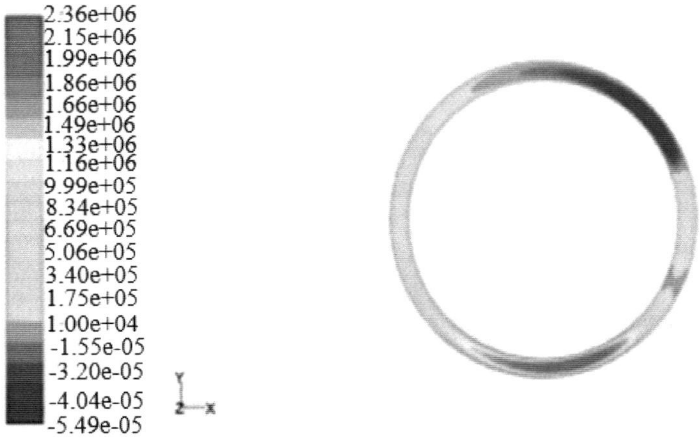

(a) $5e^{-5}$ time pressure distribution within the oil film

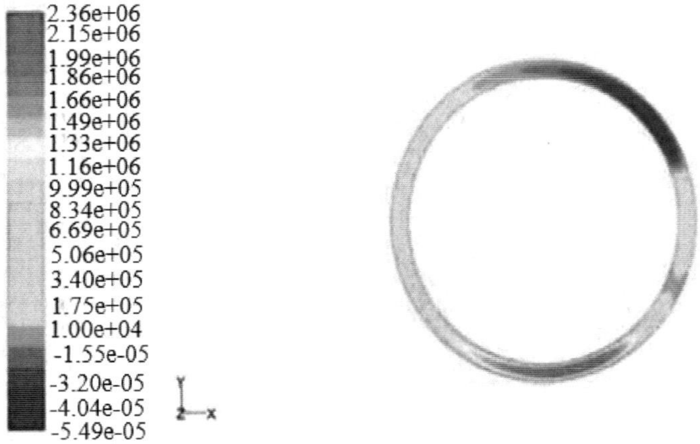

(b) $1e^{-4}$ time pressure distribution within the oil film

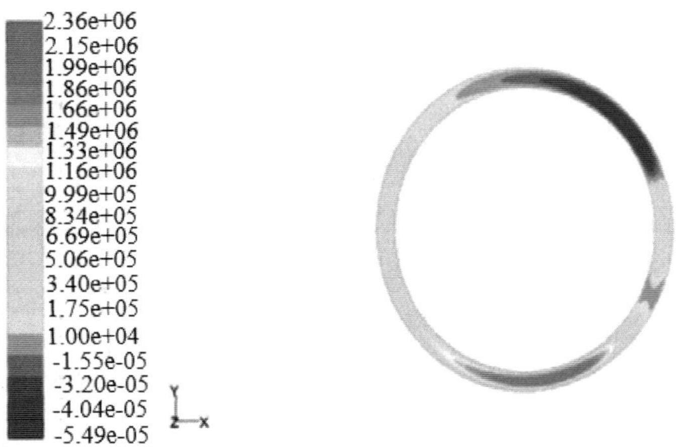

(c) $1e^{-4}$ time pressure distribution within the oil film

Figure 5. Oil pressure distribution at different times at the horizontal disturbance velocity.

Table 5. CFD single-phase flow model at different disturbance Speed horizontal and vertical direction of the oil film on the journal (N).

Disturbance speed	F_x	F_y	Disturbance speed	F_x	F_y
$\dot{x}/Nh_0$ =0	40.0	1172.2	$\dot{y}/Nh_0$ =0	40.0	1172.2
$\dot{x}/Nh_0$ =-0.01	72.1	1143.4	$\dot{y}/Nh_0$ =0.01	73.0	1155.6
$\dot{x}/Nh_0$ =-0.005	56.7	1157.4	$\dot{y}/Nh_0$ =0.005	54.6	1158.4
$\dot{x}/Nh_0$ =-0.002	46.5	1165.9	$\dot{y}/Nh_0$ =0.002	47.6	1163.4

The above table data into the formula (3) to obtain the oil film damping value, Table 4.5 shows the different models of oil film damping coefficient of the calculation results. This article references [12]. Use $\dot{x}/Nh_0 = \dot{y}/Nh_0$ =0.005 corresponding data for follow-up comparison.

Table 6. Oil film damping coefficient (104Ns/m).

Calculation method	c_{xx}	c_{xy}	c_{yx}	c_{yy}
VT-FAST	5.75	4.93	5.41	16.7
CFX-TASCflow($\dot{x}/Nh_0 = \dot{y}/Nh_0$ =0.01)	6.92	5.90	6.60	18.2
CFX-TASCflow($\dot{x}/Nh_0 = \dot{y}/Nh_0$ =0.005)	7.16	5.92	6.40	18.4
CFX-TASCflow($\dot{x}/Nh_0 = \dot{y}/Nh_0$ =0.002)	6.86	5.82	6.50	18.0
FLUENT Single phase flow ($\dot{x}/Nh_0 = \dot{y}/Nh_0$ =0.01)	6.8	5.8	6.5	18.3
FLUENT Single phase flow ($\dot{x}/Nh_0 = \dot{y}/Nh_0$ =0.005)	7.1	5.67	6.2	18.5
FLUENT Single phase flow ($\dot{x}/Nh_0 = \dot{y}/Nh_0$ =0.002)	6.75	6.0	6.3	18.6

Table 6 shows the results of CFD single-phase flow model and other model damping. It can be seen that the damping coefficient of the single-phase flow model calculated by fluent software is not much different from that calculated by CFX software. All the results are less than 5 %.

3.3. Damping coefficient calculation

Figure 6, It shows the CFD model of sliding bearings in the bearing dynamic characteristics of the impact. In the vicinity of the working speed, the absolute value of the oil film stiffness increases with the speed increase in Fig. 6 (a). It can be seen that the rotational speed has little effect on the oil film damping near the working speed in Fig. 6 (b).

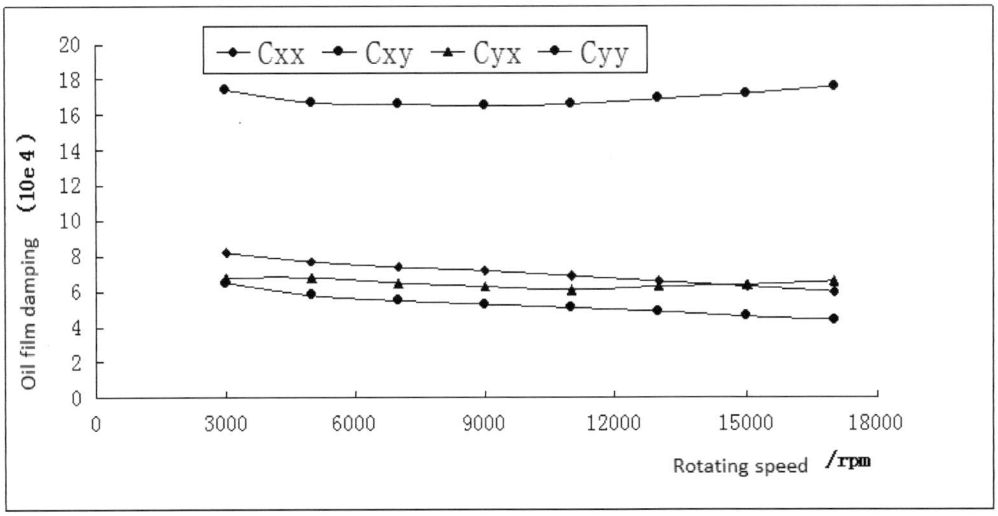

(a)The Influence of Speed on the Rigidity Coefficient of Sliding Bearing

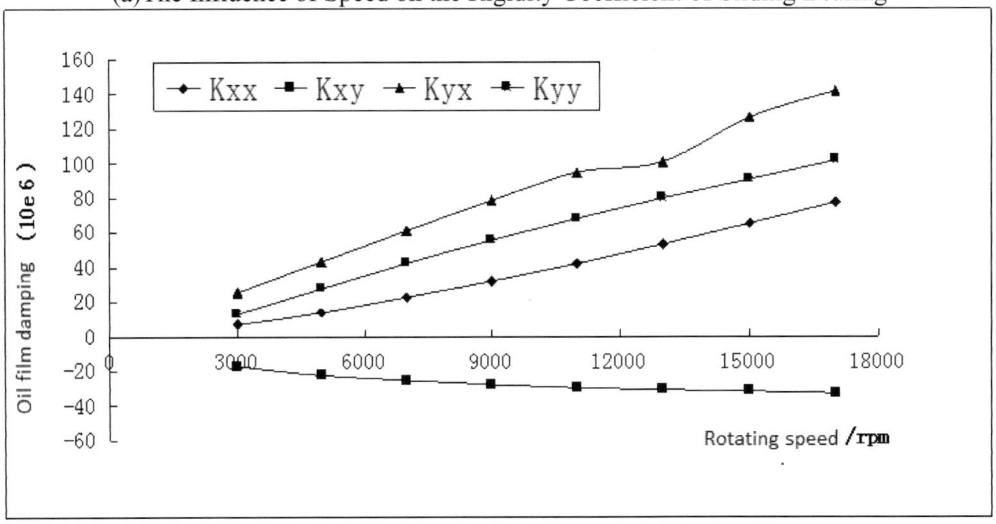

(b) Influence of Speed on Damping Coefficient of Sliding Bearing

Figure 6. Influence of Speed on Dynamic Characteristics of Sliding Bearing

4. Conclusion

(1) CFD method can directly solve the N-S equation, more accurately reflect the bearing dynamic characteristics, and be used for complex sliding bearing model to solve, this paper established a sliding bearing CFD solution model and the relevant literature results are less than 5% error, The accuracy of the method is verified, which provides the theoretical basis for the design of sliding bearing.

(2) The calculation results of the dynamic characteristics of the sliding bearing show that the calculated results of CFD and the related articles CFX are similar, which is different from the calculation results of the related articles Reynolds equation. This is because the CFD and CFX methods directly solve N-S equation, Reynolds equation method simplifies N-S equation, so the error is large.

(3) The results show that the oil film stiffness coefficient increases with the increase of the rotational speed, and the rotational speed has little effect on the damping coefficient.

Acknowledgments
This work was financially supported by Jiangxi Provincial Department of Education Science and Technology Research Key Project (DA201706712) fund.

References

[1] Zhang Zhiming. Hydrodynamic Lubrication Theory of Sliding Bearings[M]. Higher Education Pr-ess, 1986.
[2] Huang Ou, Yang Hong, Peng Zeying. Development Orientation for Domestic Ultra-Supercritical S-team Turbine[J]. THERMAL TURBINE, 2004, 33(1): 1-7.
[3] Chen Boxian, Qiu Zugan, Zhang Huisheng. Fluid lubrication theory and its application[M]. China Machine Press, 1991.
[4] Su Hong, Wang Xiaojing, Zhang Zhiming. A Comparison between Two Boundary Conditions of J-ournal Bearings[J]. LUBRICATION ENGINEERING, 2002 (5): 3-4.
[5] Yang Jinfu, Liu Zhansheng, Xie Yongbo, et al. A dynamic boundary condition study on doing Re-ynolds equation[J]. HEILONGJIANG ELECTRIC POWER, 2004, 26(5): 346-349.
[6] Swift H W. THE STABILITY OF LUBRICATING FILMS IN JOURNAL BEARINGS.(INCLU-DES APPENDIX)[C]//Minutes of the Proceedings of the Institution of Civil Engineers. Thomas T-elford-ICE Virtual Library, 1932, 233(1932): 267-288.
[7] Guo Z, Hirano T, Kirk R G. Application of CFD Analysis for Rotating Machinery: Part 1-Hydrod-ynamic, Hydrostatic Bearings and Squeeze Film Damper[C]//ASME Turbo Expo 2003, collocated with the 2003 International Joint Power Generation Conference. American Society of Mechanical Engineers, 2003: 651-659.
[8] Guo Shenglin, Bu Yan. A Novel Structure of Hydrostatic Mixed Radial Sliding Bearing[J]. MEC-HANICAL DESIGN, 1996 (8): 11-13.
[9] GUO Weiqi, LIU Guiping, MAO Wengui. Stability analysis and optimization for a high-speed built-in motorized hydrostatic spindle system[J]. Journal of Vibration and Shock, 2015, 34(6): 101-105.
[10] Huang Yacheng, Yao Tingqiang, Fu Yuanning. Analysis of Influence of Different Parameters on Dynamic Characteristics of Sliding Bearing[J]. Journal of Kunming University of Science and T-echnology (Natural Science Edition), 2015, 40(2).
[11] Zhang Aiping, Xie Meina, Lin Shengqiang. Analysis on Two Phase Flow of Oil Film Characteris-tics of Elliptical Bearing[J]. Lubrication Engineering, 2015, 40(3): 64-68.
[12] Tower B. First report on friction experiments[J]. Proceedings of the institution of mechanical eng-ineers, 1883, 34(1): 632-659.

Analysis of the cylinder's movement characteristics after entering water based on CFD

Xianlong Liu

CSIC 750 Test Range, Kunming, China

*Corresponding author e-mail: liuxianlong@yeah.net

Abstract. It's a variable speed motion after the cylinder vertical entry the water. Using dynamic mesh is mostly unstructured grid, and the calculation results are not ideal and consume huge computing resources. CFD method is used to calculate the resistance of the cylinder at different velocities. Cubic spline interpolation method is used to obtain the resistance at fixed speeds. The finite difference method is used to solve the motion equation, and the acceleration, velocity, displacement and other physical quantities are obtained after the cylinder enters the water.

1. Introduction

Buoys, beacons and other objects are dropped into the water by helicopters, and their velocity variation in the water is non-uniform. At present, the two-dimensional problem can be directly calculated by dynamic mesh method, and even simulating its rotation process. However, the calculation using unstructured grid, the credibility of the results is doubtful. For the three-dimensional case, the dynamic mesh method consumes more computer resources, and time-consuming, and the results are not necessarily right.

This study focuses on the movement of the cylinder vertical entry into the water, without considering the rotation of the cylinder. In order to taking advantages of CFD, using CFD method to calculate the cylinder resistance at fixed speed, and then using interpolation method to calculate the resistance at different speeds.

2. Principle of curve interpolation

Curve interpolation usually has exponential function interpolation, Lagrange interpolation, Cubic spline interpolation, B spline curve, NURBS curve and so on.

2.1. Cubic spline curve interpolation

There, $a = x_0 < x_1 < \cdots < x_n = b$, if function $y(x)$ satisfying two condition:

(a) $y(x)$ is Cubic Polynomial on each interval $[x_{i-1}, x_i], i = 1, 2, \cdots, n$;

(b) $y(x)$ is twice continuously differentiable function on interval $[a, b]$.

Then, $y(x)$ is Cubic spline on interval $[a, b]$. There, $x_i \, (i = 0, 1, \cdots, n)$ is the node. The interpolation points are shown in Table 1.

Table 1. Interpolation points

x	x_0	x_1	$\cdots$	x_n
y	y_0	y_1	$\cdots$	y_n

Cubic spline interpolation function:

$$s(x)=\frac{h_k+2(x-x_k)}{h_k^3}(x-x_{k+1})^2y_k+\frac{h_k-2(x-x_{k+1})}{h_k^3}(x-x_k)^2y_{k+1} +\frac{(x-x_k)(x-x_{k+1})^2}{h_k^2}m_k+\frac{(x-x_{k+1})(x-x_k)^2}{h_k^2}m_{k+1} \tag{1}$$

In equality (1),

$$h_k=x_{k+1}-x_k\,(k=0,1,\cdots,n-1) \tag{2}$$

$$\lambda_k m_{k-1}+2m_k+\mu_k m_{k+1}=g_k\,(k=1,2,\cdots,n-1) \tag{3}$$

Where,

$$\lambda_k=\frac{h_k}{h_k+h_{k-1}},\ \mu_k=\frac{h_{k-1}}{h_k+h_{k-1}},\ g_k=3(\mu_k\frac{y_{k+1}-y_k}{h_k}+\lambda_k\frac{y_k-y_{k-1}}{h_{k-1}})(k=1,2,\cdots,n-1)$$

3. Basic theory of fluid mechanics

3.1. Control equation

The continuity equation and momentum equation for incompressible viscous fluid are:

$$\frac{\partial\overline{u_i}}{\partial x_i}=0 \tag{4}$$

$$\rho\frac{\partial\overline{u_i}}{\partial t}+\rho\partial\overline{u_j}\frac{\partial\overline{u_i}}{\partial x_i}=\rho\overline{F_i}-\frac{\partial\overline{p}}{\partial x_i}+\frac{\partial}{\partial x_j}(\mu\frac{\partial\overline{u_i}}{\partial x_j}-\rho u_i^{'}u_j^{'}) \tag{5}$$

Where, ρ is density, μ is Viscosity Coefficient, $\overline{p}$ is average pressure, $\overline{F_i}$ is external force, $\overline{u_i}$ is mean velocity, $u_i^{'}$ is fluctuating velocity, $-\rho u_i^{'}u_j^{'}$ is reynolds stress.

3.2. Turbulence model

In order to make the equation closed, a new turbulent model equation must be introduced to link the fluctuating values with the time average in the stress terms. There, we chose the $RNGk-\varepsilon$ equtions. The transport equations of turbulent kinetic energy and turbulent fluctuation intensity in the $RNGk-\varepsilon$ equation are:

$$\frac{\partial}{\partial t}(\rho k)+\frac{\partial}{\partial x_i}(\rho k u_i)=\frac{\partial}{\partial x_j}[\alpha_k \mu_{eff}\frac{\partial k}{\partial x_j}]+G_k-\rho\varepsilon+S_k \tag{6}$$

$$\frac{\partial}{\partial t}(\rho\varepsilon)+\frac{\partial}{\partial x_i}(\rho\varepsilon u_i)=\frac{\partial}{\partial x_j}[\alpha_k \mu_{eff}\frac{\partial \varepsilon}{\partial x_j}]+G_{1\varepsilon}\frac{\varepsilon}{k}G_k-C_{2\varepsilon}\rho\frac{\varepsilon^2}{k}-R_\varepsilon+S_\varepsilon \tag{7}$$

Where, $\mu_{eff}=\mu+\mu_t$, $\mu_t=\rho C\mu\frac{k^2}{\varepsilon}$, $G_k=-\rho\overline{u_i^{'}u_j^{'}}\frac{\partial u_j^{'}}{\partial x_i}$, $R_\varepsilon=\frac{C_\mu\rho\eta^3(1-\frac{\eta}{\eta_0})}{1+\beta\eta^3}\frac{\varepsilon^2}{k}$, $\eta\equiv\frac{Sk}{\varepsilon}$,

$S=\sqrt{2S_{ij}S_{ij}}$, $S_{ij}=\frac{1}{2}(\frac{\partial u_i}{\partial x_j}+\frac{\partial u_j}{\partial x_i})$, S_k and S_ε are user-defined source terms.

Constant, $G_{1\varepsilon}=1.42, G_{2\varepsilon}=1.68, C_\mu=0.0845, \sigma_k=1.0, \sigma_\varepsilon=1.3, \eta_0=4.38, \beta=0.012$

4. CFD Simulation

The dimension of Cylinder is 200mm, the distance is 1000mm(Fig. 1).

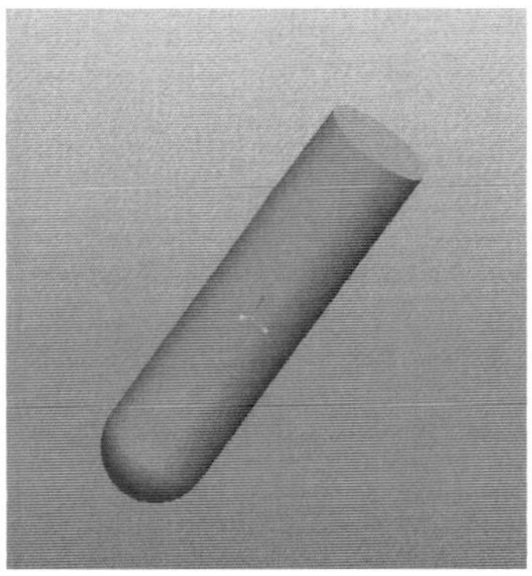

Figure 1. Cylinder

In order to obtain high-quality meshes, structured grids are used for the entire computational domain. The governing equations and turbulence models are discretized by the finite volume method and solved by the coupled implicit solver.

The pressure - velocity coupling iterative algorithm using SIMPLEC algorithm, pressure by the standard discretization, momentum, turbulent kinetic energy and turbulent dissipation rate by two order upwind scheme, relaxation factors are the default values. Wall function method is adopted in the near wall region.

Finally, the resistance values of the rotating body at different speeds are shown in tab. 2.

Table 2. Resistance of cylinder at different speeds

V(m/s)	0	0.5	1	2	3	4	5
R(N)	0	2.3	9.5	36.3	80.5	140.9	217.3

5. Calculation and analysis

When the cylinder moves in water, its force analysis can be seen in Figure 2. The gyroscopic motion and current action of cylinder after being thrown into the water by a flying vehicle are not considered.

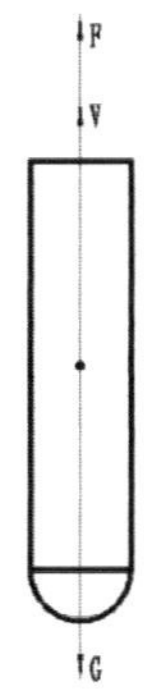

Figure 2. Force analysis of cylinder

The moment of motion of cylinder in water, its equation of motion is:

$$G - V - F = ma \tag{8}$$

The resistance F varies with the speed, so the differential equations can not be solved by theoretical methods, only numerical methods can be used. Changing eqution (8) to the following form:

$$G - V - F = m\frac{V_{t+\Delta t} - V_t}{\Delta t} \tag{9}$$

$V_{t=0}$ has been known, we can use cubic spline curve interpolation to calculate $V_{t+\Delta t}$ .Taking $V_{t=0} = 0$ as an example, numerical calculation when Δt equal to 0.05s, 0.1s,0.2s and 0.5s(Table 3.).

Table 3. Velocity, Resistance etc. at different Δt (t=5s)

V(m/s)	R(N)	a(m/s^2)	s(m)	Δt (s)
1.60	23.37	0.02	5.89	0.01
1.60	23.39	0.02	5.90	0.05
1.60	23.42	0.02	5.93	0.1
1.60	23.48	0.02	5.97	0.2
1.61	23.67	0.01	6.10	0.5

In order to ensure the efficiency of calculation and the accuracy of the result, $\Delta t = 0.05$ s, and the calculation results are shown in Table 4.

In table 4, the acceleration is 0 when the time is 10s, and the speed is stable at 1.62*m/s*. That means the cylinder's external force is 0 when the time after 10s. So, the cylinder will maintain uniform rectilinear motion.

Using this method, the time, velocity, acceleration and displacement of the cylinder can be calculated very quickly after it enter the water.

Table 4. Velocity, Resistance etc. at different time($\Delta t = 0.05$ s)

T(s)	V(m/s)	R(N)	a(m/s^2)	s(m)
1	0.72	4.00	0.68	0.39
2	1.23	13.81	0.34	1.41
3	1.47	19.62	0.15	2.79
4	1.56	22.33	0.06	4.32
5	1.60	23.39	0.02	5.90
6	1.61	23.78	0.01	7.51
7	1.62	23.92	0.003	9.13
8	1.62	23.97	0.001	10.75
9	1.62	23.99	0	12.37
10	1.62	24.00	0	13.99
11	1.62	24.00	0	15.62

6. Conclusion

Using CFD, cubic spline interpolation and finite difference method can effectively calculate the speed, acceleration and displacement etc. of the cylinder when it enter into water. Comparing with the dynamic mesh method, this method with less resource intensive, fast calculating. The method is suitable for analyzing the motion performance and trajectory analysis of heavy objects after entering water.

References

[1] Seddon C, Moatamedi M. Review of water entry with applications to aerospace structures [J]. International Journal of Impact Engineering, 2016, 32(7):1045—1067.

[2] Gu H B,Qian L,Causon D M,et al. Numerical simulation of water impact of solid bodies with vertical and oblique entries [J]. Ocean engineering,2013, 75(5):128-137.

[3] Tanvir Mehedi Sayeed, Heather Peng, Brian Veitch. Experimental investigation of slamming loads on a wedge [R]. The International Conference on Marine Technology. 2010:107-112.

[4] Tanvir Mehedi Sayeed, Heather Peng, Brian Veitch. Experimental investigation of slamming loads on a wedge [R]. The International Conference on Marine Technology. 2010:107-112.

[5] Tveitnes T, Fairlie-Clarke A. C, Varyani K. An experimental investigation into the constant velocity water entry of wedge-shaped sections [J]. Ocean Engineering. 2008, 35:1463-1478.

[6] Wang Yonghu, Shi Xiuhua, Review on research and development of water-entry impact problem. Explosion and Shock Waves, 2008, 28(3): 276-282.

[7] Arai M, Cheng L Y, Inoue Y. A computing method for the analysis of water impact of arbitrary shaped bodies[J]. Journal of the Society of Naval Architects of Japan,1995,176:233-240.

[8] Hadzica I, Hennigb J, PERICM, et al.Computation of flow-induced motion of floating bodies[J]. Applied Mathematical Modelling,2005,29(12):1196-1210.

[9] Qiu Haiqiang, Yuan Xulong, Wang Yadong, et al. Simulation on impact load and cavity shape in high speed vertical water entry for an axisymmetric body [J]. Torpedo Technology, 2013, 21(3): 161—164.

[10] Seddon C M, Moatamedi M. Review of water entry with applications to aerospace structures [J]. International Journal of Impact Engineering. 2006, 32:1045-1067.

[11] Carcaterraa A, Ciappi E. Hydrodynamic shock of elastic structures impacting on the water: theory and experiments [J]. Journal of Sound and Vibration. 2004, 271: 411-439.

[12] Zhao R, Faltinsen O M. Water-entry of two-dimensional bodies[J]. Journal of Fluid Mechanics, 1993, 246: 593-612.

An improved temperature compensation circuit for SLD light source of fiber-optic gyroscope

Xian Chen, Jianhua Yang*, Yilan Zhou and Xiaowu Shu

College of Optical Science and Engineering, Zhejiang University, Hangzhou, China

*Corresponding author e-mail: xxianhua@zju.edu.cn

Abstract. The super luminescent diode light source is commonly used in Fiber-optic gyroscope. The wavelength temperature stability is one of its key parameters. In this work, simulation model of temperature field of the SLD light source is established, and the temperature distribution of SLD internal structure is simulated and researched. A temperature compensation scheme is designed according to the simulation result and experiment is done to confirm the specific parameter of compensation circuit. Then the designed compensation scheme is used in temperature control circuit of SLD, and the performance of compensated SLD light source is tested. An improvement of 62% is demonstrated compared with uncompensated SLD.

1. Introduction

Fiber-optic gyroscope (FOG) is a kind of angular velocity sensor based on the Signac effect. It has shown great advantages such as small volume, light weight, wide range of accuracy, and no need for moving parts [1], and has been extensively applied in applications such as aviation, spaceflight, and navigation [2, 3].

There are two kinds of light sources commonly used in recent FOG technology, the super luminescent diode (SLD) and the aamplified spontaneous emission (ASE). The SLD light source shows various advantages such as small size, low cost and broad spectrum [4], and is commonly used in the low and medium precision FOG. Compared with SLD light source, the ASE light source shows better wavelength temperature stability [5, 6, 7], which makes it suitable for applications such as high precision FOG [8, 9].

Recently, as more and more attentions has been paid to space application of FOG, it becomes highly demanded to improve source system of the spatial FOG [10]. Since the light power and wave length of ASE light source will be unstable in space due to the cosmic radiation [11, 12], SLD light source is a common scheme for special FOG system. While the wavelength temperature stability of the sources remains problem to be solved for high precision FOG [13].

In this work, the cause factor of wavelength temperature instability in SLD is analysed and a simulation model of temperature field is established. Based on the simulation result, an improved SLD temperature compensation circuit has been proposed and demonstrated in experiment. An improvement of 62% is demonstrated.

IOP Conf. Series: Journal of Physics: Conf. Series **916** (2017) 012027 doi:10.1088/1742-6596/916/1/012027

2. Analysis of traditional temperature control scheme of SLD

In the construction of SLD light source, negative temperature coefficient (NTC) thermistor and thermoelectric cooler (TEC) are used to achieve temperature feedback control of tube core. Fig.1 shows the typical SLD structure [14]. TEC works as semiconductor cooler and nickel heat sink is put on its upper surface for heat conduction. The tube core is soldered on the heat sink and the NTC thermistor is soldered closed to it. The tube core behaves like a heat source in thermal analysis, and the thermistor beside is a heat sensor of tube core.

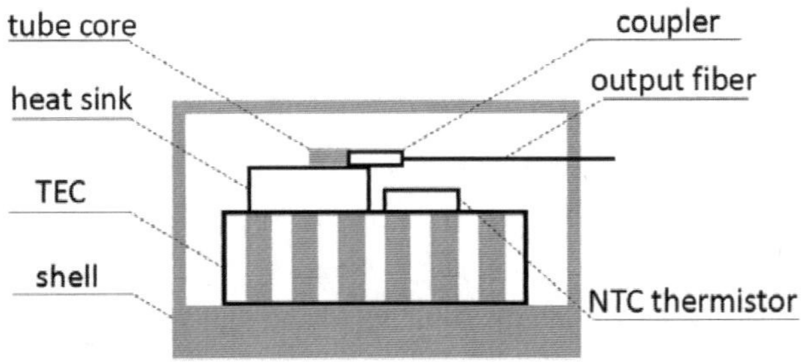

Figure 1. The typical construction of SLD light source

Special temperature control circuit is designed to keep SLD in a certain temperature. Fig.2 shows a traditional temperature control circuit of SLD light source [15]. R_0 , R_1 R_2 are precision resistance, and the value of R_1 is equal to R_2 . R_T Represents the NTC thermistor in SLD. U_1 is instrument op-amp which produces control signal of TEC. The feedback circuit works so that the value of R_T is controlled to be equal to R_0 , so that the working temperature of SLD is controlled by the value of R_T . In this way, the setting temperature of SLD keeps stable during working.

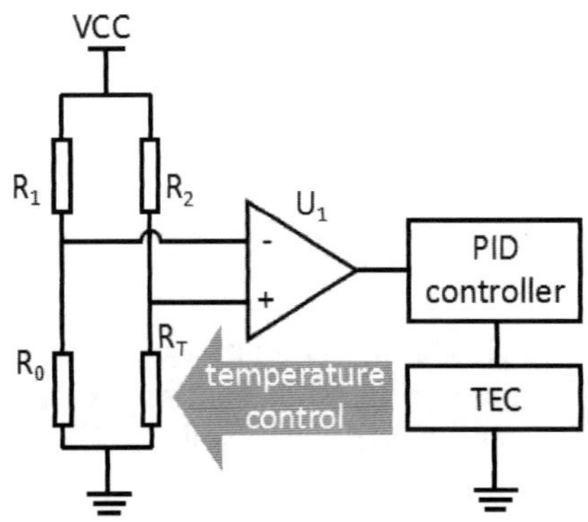

Figure 2. Traditional temperature control circuit of SLD light source

While the structure shown above shows great disadvantages, it still has drawback. When the circuit works, it is the temperature of NTC thermistor that has been controlled instead of tube core temperature. Since there is a small spatial distance from NTC thermistor to tube core, so there is a slight temperature difference between them. Although temperature of thermistor has been seriously controlled, slight temperature fluctuation still exists in tube core when external temperature changes. As the temperature characteristic of tube core is about $500\,ppm/°C$, the slight fluctuation in tube core leads to wavelength instability of SLD light source. In fiber-optic gyroscope, the mean wavelength of light source, $\overline{\lambda}$ is related to the scale factor of FOG output, K_S which can be described as

$$K_S = \frac{\Omega}{\phi_S} = \frac{\bar{\lambda} \cdot c}{8\pi S} \tag{1}$$

Where ϕ_S is the Sagnac phase shift, Ω is the angular velocity of rotation, c is light velocity and S is the total area of the closed light path. The mean wavelength of light source is proportional to the scale factor. If the mean wavelength is unstable, then the FOG output is unstable too. In such situations, the wavelength temperature instability problem greatly restricts SLD's applications in high precision FOG.

3. Temperature characteristic of SLD with traditional control scheme

The temperature distribution of SLD is researched in order to solve the temperature instability problem discussed above. Simulation model of temperature field is established according to the actual working status of SLD light source.

Heat has three transfer ways including radiation, convection and conduction. In SLD device, heat exchange happens in a small close space, so the heat mainly transfers through metal structures of SLD. This transfer way is called heat conduction. Heat conduction equation describes the temperature distribution inside solid structure as below:

$$\rho c \frac{\partial T}{\partial t} - \frac{\partial}{\partial x}\left(\kappa_x \frac{\partial T}{\partial x}\right) - \frac{\partial}{\partial y}\left(\kappa_y \frac{\partial T}{\partial y}\right) - \frac{\partial}{\partial z}\left(\kappa_z \frac{\partial T}{\partial z}\right) = \rho Q \tag{2}$$

Where T denotes temperature distribution, t denotes the time, ρ denotes the density, κ_x, κ_x, κ_x respectively denote heat conductivity coefficient through x, y, z direction, and Q denotes heat source inside solid structure. As the SLD works in steady state, transient response of Eq(2) can be ignored, then neat conduction equation of steady state can be described as.

$$\frac{\partial}{\partial x}\left(\kappa_x \frac{\partial T}{\partial x}\right) + \frac{\partial}{\partial y}\left(\kappa_y \frac{\partial T}{\partial y}\right) + \frac{\partial}{\partial z}\left(\kappa_z \frac{\partial T}{\partial z}\right) + \rho Q = 0 \tag{3}$$

The SLD works in a good heat dissipation condition, so the heat dissipation surfaces can keep the same temperature as outside environment. On purpose of simplify the boundary conditions during temperature field simulation, the convective boundary, radiation boundary and conduction boundary can all be replaced by a unified temperature boundary, which can be described as

$$T(M)\big|_{M \in S_1} = T_e \tag{4}$$

Where S_1 denotes the heat dissipation surfaces of SLD structure, and T_e denotes the environment temperature.

Then the heat simulation is carried out after temperature and structure model has been established. During the simulation, the temperature of NTC thermistor, T_{NTC} keeps $25°C$ within the interaction of tube core and TEC, and the temperature distribution has been calculated when environment temperature changes from $-40°C$ to $70°C$. Fig.3 shows the structure model and an example of thermal field distribution of SLD light source. In this example, the environment temperature was $10°C$, and NTC thermistor temperature was controlled in $25°C$. The tube core temperature was calculated to be $28.7°C$.

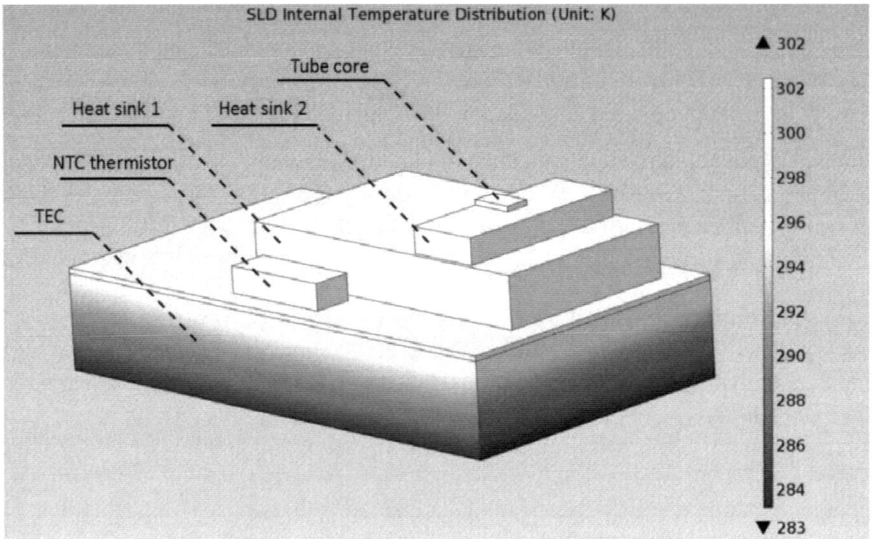

Figure 3. An example of simulation result of the temperature distribution inside SLD. The environment temperature is $10°C$, and NTC thermistor temperature is controlled in $25°C$. The shell and optical structure has been hidden in the graph.

From the result of thermal field distribution, a temperature difference of about $4°C$ can be demonstrated between the tube core and NTC thermistor. And this temperature difference can be influenced by the environment temperature. As a result, although the temperature of NTC thermistor, T_{th} ,is controlled to keep on $25°C$, the tube core temperature, T_{core}, still changes with the environment temperature, T_e. The simulation result about the relationship of T_{core}, T_{th} and T_e is shown in Fig.4.

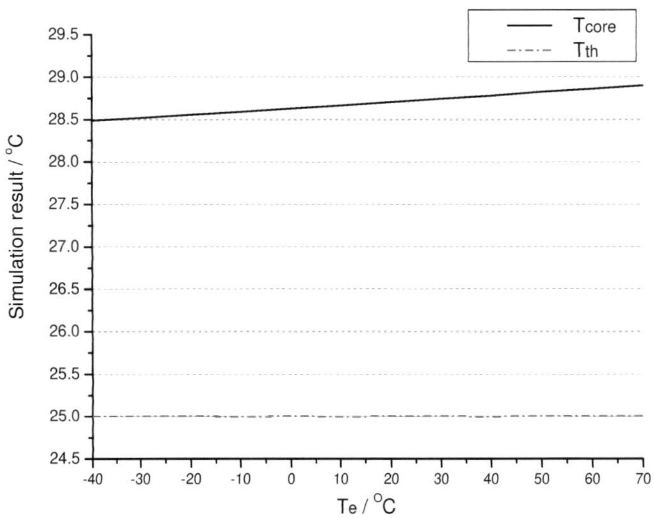

Figure 4. The simulation result as T_{th} is controlled to $25°C$

According to simulation result, there is a linear relationship between T_{core} and T_e. Experiment about the wavelength temperature instability has been done to demonstrate the simulation result. The mean wavelength of a SLD product has been measured by optical spectrum analyser with environment temperature changing from $-40°C$ to $70°C$. The test result is shown in Fig.5. The mean wavelength has changed about 300ppm in a linear fashion. Since the mean wavelength of SLD is almost linear to the temperature of the tube core, a conclusion can be gotten that tube core temperature is linear to environment temperature which is also demonstrated by simulation result mentioned above.

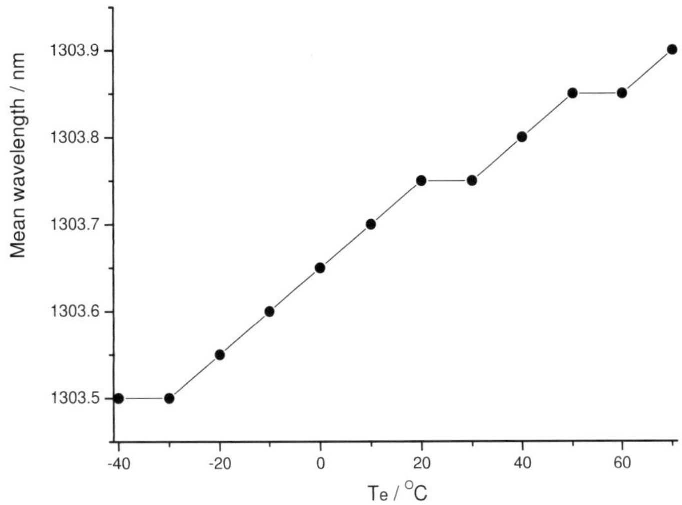

Figure 5. Test result of wavelength temperature instability without compensation

4. Temperature compensation scheme of SLD

In traditional temperature control scheme, the temperature of NTC thermistor keeps in a fixed value. This leads the tube core temperature to change linearly with environment temperature. In order to keep the tube core temperature stable, the temperature of NTC thermistor should also change with the environment temperature. Follow this approach, a temperature compensation scheme can be developed.

Another simulation has been taken to study the change of the thermistor temperature when tube core temperature is fixed. The temperature of tube core, T_{core} keeps on $25°C$ and the environment temperature, T_e changes from $-40°C$ to $70°C$. The simulation result is shown in Fig.6. In order to keep the tube core temperature stable in $25°C$, the temperature of NTC thermistor, T_{th} should change from $21.5°C$ to $21.1°C$ linearly.

IOP Conf. Series: Journal of Physics: Conf. Series **916** (2017) 012027 doi:10.1088/1742-6596/916/1/012027

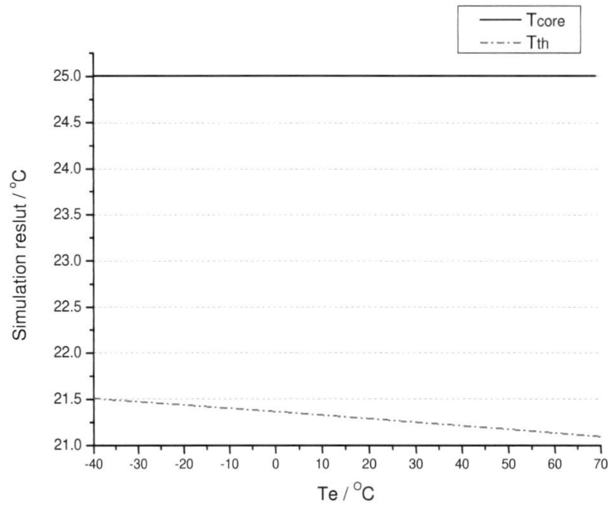

Figure 6. The simulation result when T_{core} is controlled to keep on $25°C$

Experiment has been designed according to the simulation result in order to find the appropriate compensation resistance. Fig.7 shows the experiment scheme. An 300Ω adjustable resistor, R_{adj}, was cascaded to R_0 on the basis of circuit shown in Fig.2. The control circuit and SLD light source were placed in a temperature control box, but R_{adj} was put outside so that its resistance can be adjusted by experimenter. As circuit works, the value of NTC thermistor inside SLD, R_T, can be determined by

$$R_T = R_{adj} + R_0 \tag{5}$$

So the controlled temperature of NTC thermistor can be adjusted by R_{adj}. The output of SLD light source was connected to an optical spectrum analyzer to monitor the mean wavelength of SLD.

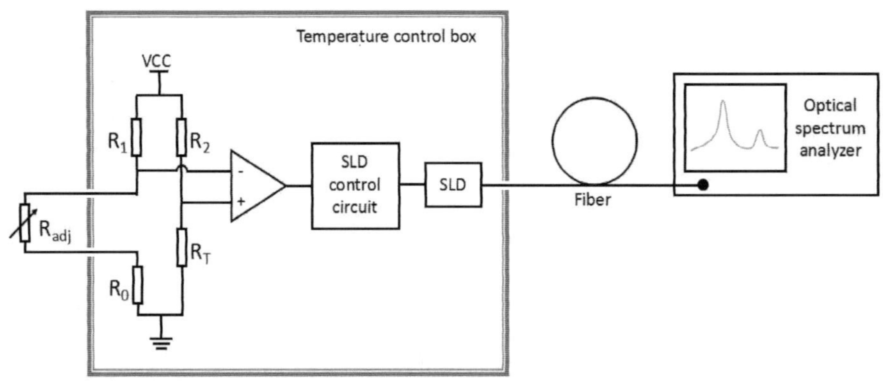

Figure 7. Experiment scheme to test out the appropriate compensation resistance

The temperature of the box, T_e changes from -40℃ to 70℃ in steps of 10℃. The mean wavelength at $-40°C$ was recorded as the original value. At each temperature spot, R_{adj} was slightly adjusted until the mean wavelength reached the original value. The values of R_{adj} with different temperature were recorded and the relationship curve between R_{adj} and T_e can be drawn as Fig.8.

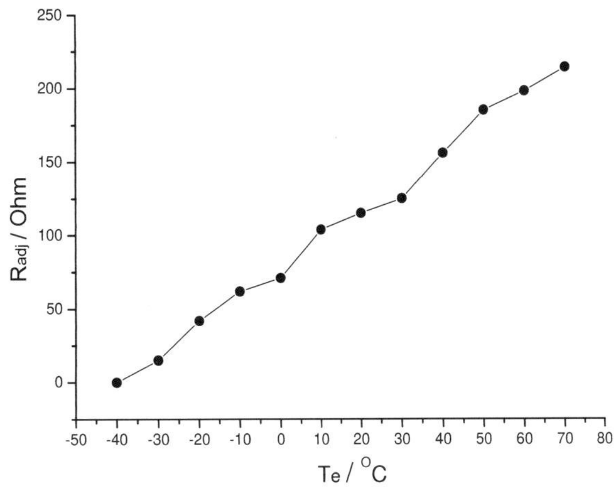

Figure 8. Test result of the compensation resistance

As shown in Fig.8, there is a linear relation between the resistance and the temperature, which means that a compensation resistor can be added to traditional SLD control circuit in order to get better wavelength temperature stability.

A positive temperature coefficient (PTC) thermistor, R_{PTC} was chosen as the compensation resistor according to Fig.8. Its resistance was 450Ω at 25°C, and changed from 330Ω to 530Ω linearly as temperature raised from -40°C to 70°C. The compensated SLD control circuit was shown in Fig.9. R_1, R_2 was 10$K\Omega$ high precision resistor. R_b Was 9.8$K\Omega$ high precision resistor. R_T Was the NTC thermistor inside SLD. As the circuit working, R_T can be determined by

$$R_T = R_{PTC} + R_b \tag{6}$$

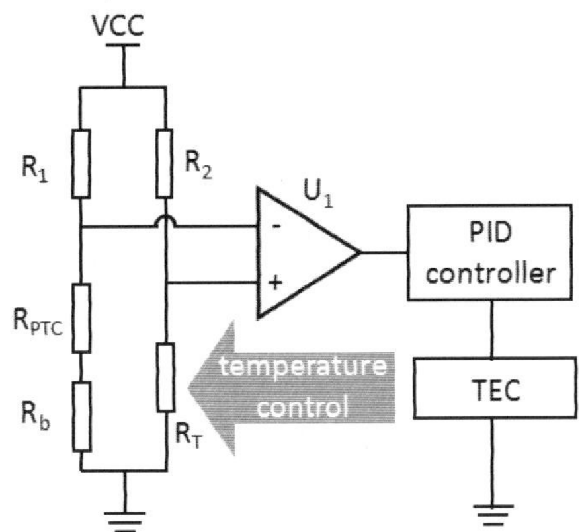

Figure 9. The compensated SLD temperature control circuit

IOP Conf. Series: Journal of Physics: Conf. Series **916** (2017) 012027 doi:10.1088/1742-6596/916/1/012027

The PTC thermistor, R_{PTC} provided temperature compensation for R_T . This compensation value made the temperature of NTC thermistor, T_{th} adjust slightly in the similar way as Fig.6. In this way, temperature of tube core, T_{core} became more stable when environment temperature changed, so wavelength temperature stability of SLD has been improved.

5. Temperature stability test of temperature compensated SLD

The temperature compensation scheme above provides a method to improve wavelength temperature stability of SLD light source. Temperature stability test has been done to verify the compensation effect of designed circuit. The SLD light source was controlled and driven by compensated circuit shown in Fig.9, and its wavelength temperature stability has been tested. Fig.10 shows the test result. The mean wavelength drift value was about 115ppm when environment changed form $-40°C$ to $70°C$. Compared with the temperature stability of uncompensated SLD light source in Fig.5, evident improvement of 62% was demonstrated.

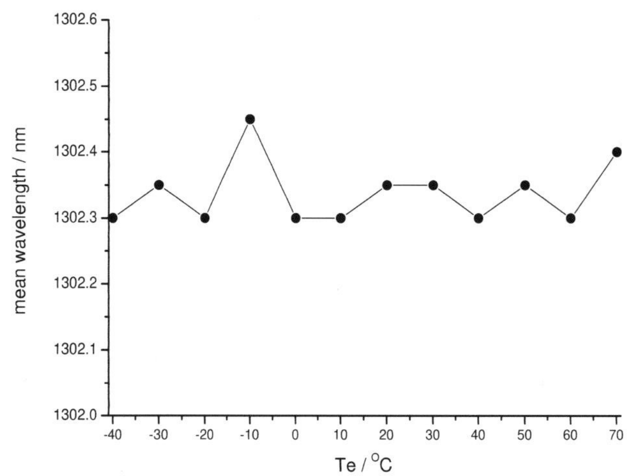

Figure 10. Test result of wavelength temperature stability after compensation

While as shown in Figure 10, small wavelength drift still exists after the SLD has been compensated. This is because this compensation scheme is not perfect and can be further improved. Firstly, the non-ideality of the PTC compensation thermistor confines the compensation effect. Secondly, operating environment of SLD provides random influence such as circuit noise and mechanical vibration. Thirdly, the test accuracy of optical spectrum analyzer is 0.05nm, which may amplify the rise and fall in test value. What is more, this compensation only aims at steady state of SLD, and the compensation effect can be further improved if transient characteristic of SLD has been studied and compensated.

6. Conclusion

The wavelength temperature stability of SLD light source is an important problem in the application of high precision FOG. Simulation and experiment has been done to study the temperature distribution of SLD internal structure, and a temperature compensation scheme has been designed. In this temperature compensation scheme, a specific compensation thermistor has been used to provide temperature compensation in the TEC control circuit of SLD. The compensation thermistor features with a certain resistance-temperature curve and can be used to change the setting temperature of SLD with environment temperature. In this way, the temperature of tube core can be more stable. This temperature compensation circuit has been applied and tested, and a wavelength temperature stability improvement of 62% was demonstrated in the compensated SLD light source.

IOP Conf. Series: Journal of Physics: Conf. Series **916** (2017) 012027 doi:10.1088/1742-6596/916/1/012027

References

[1] G.C Zhang. The Principle and Technologies of Fiber-Optic Gyroscope. (National Defense Industry Press, 2008)
[2] G.A. Sanders, S.J. Sanders, L.K. Strandjord, et al. Fiber optic gyroscope development at Honeywell. Spie Comm. + Sci. Sens. Imaging, 9852(2016):985207.
[3] S. Minakuchi, T. Sanada, N. Takeda, et al. Thermal strain in lightweight composite fiber-Optic gyroscope for space application. J. Lightwave Technol., 33.12(2015):2658-2662.
[4] X. Li, Z. Shao, J. Lu, et al. Study of high performance Fiber Optic Gyroscope based on SLD light source[C]. Control and Decision Conference. IEEE, 2013:1459-1463.
[5] W. Berglund, B. Ellerbusch, A. Gopinath, et al. A.S.E. alumina waveguide source at 1530 nm[C]. Lasers and Electro-Optics. IEEE, 2000:311-312.
[6] F. Hui, M.C. Li, L. Ma, et al. Double-pass backward ASE light source for three-axis FOG. Journal of Chinese Inertial Technology, 2016, 24(3):372-376.
[7] S.W. Lloyd, S. Fan, M.J. Digonnet, Experimental Observation of Low Noise and Low Drift in a Laser-Driven Fiber Optic Gyroscope. Journal of Lightwave Technology, 2013, 31(13):2079-2085.
[8] J. Chamoun, M.J. Digonnet. Aircraft-navigation-grade laser-driven FOG with Gaussian-noise phase modulation. Optics Letters, 2017, 42(8):1600-1603.
[9] Y. Yuan, J. Zhou, R.Z. Jiang. Optimization on Erbium-doped fiber source for high-precision optic gyroscope. Optical Instruments, 2015, 37(1):14-18.
[10] J. Nazir, T. Vivek, T.Jaisingh. Temperature stabilization in fibre optic gyroscopes for high altitude aircraft. Optik - International Journal for Light and Electron Optics, 2016, 127(20):9701-9710.
[11] H.C. Lefèvre. The fiber-optic gyroscope: Challenges to become the ultimate rotation-sensing technology[J]. Optical Fiber Technology, 2013, 19(6):828-832.
[12] W. Wang, X.F. Wang, J. Li, et al. Experiment on performance of erbium-doped fiber source for high performance fiber-optic gyroscope in a space irradiation environment. Infrared and Laser Engineering, 2012, 41(7):1826-1830.
[13] X.X. Wang, Y.N. Zhang, L.Y. Wei, et al. Wavelength change mechanism of light source SLD when starting up in fiber optic gyro. Infrared and Laser Engineering, 2012, 41(7):1826-1830.
[14] Z.T. Gong, X.W. Shu, X.D. Mou, et al. Analysis on the temperature fluctuating of FOG's source, Acta Photonica Sinica, 2005, 34(12):1910-1913.
[15] H. Cao, Y.F. Yang, S.B. Liu, et al. Temperature control system for SLD optical source of FOCS. Infrared and Laser Engineering, 2014, 43(3):920-926.

Numerical Study on Fire characteristics in Force-Ventilated Compartment with different air inlet locations

Jiaqing Zhang[1,*], Bosi Zhang[2], Kun Ji[1], Bo Yan[1], Minghao Fan[1], Wei Li[1] and Qiang Li[3]

[1]State Grid Anhui Electric Power Research Institute, Hefei, China
[2]China University of Labor Relations, Beijing, China
[3]Chinese People's Armed Police Academy, Langfang, China

*Corresponding author e-mail: dkyzjq@163.com

Abstract. The effects of air inlet locations on fire characteristics in force-ventilated compartment were investigated by computational fluid dynamics (CFD) simulations. The flame behaviour, the oxygen concentration above fire, the distribution of soot concentration and velocity of smoke in the closed compartment were discussed. The results show that the flame behaviour was mainly affected by the inflow air flow, and the flame inclined to the opened inlet in one-inlet scenario. The flame was found inclined to the exhaust vent in two-inlet scenarios. The difference between the soot concentrations in upper and lower space of the compartment was smaller in one-inlet scenarios compared with that in two-inlet scenarios, and the soot concentration was more uniform in the scenarios with higher air inlet elevation. The velocity of smoke on the side of inlet was larger than the other side in one-inlet scenario, which might cause stronger entrainment on the side with air inlet. In the two-inlet scenarios, the velocities of smoke on the sides of two air inlets were almost the same.

1. Introduction

For the compartments in ships, nuclear power stations and underground spaces, it is hard to acquire the natural ventilation, so the forced ventilation is usually used. The forced ventilation may still operate when the fire happens in these compartments. The fire characteristics in force-ventilated compartment are different from those in closed compartment and compartment with natural ventilation, and it deserves attentions.

There were already many studies focused on the forced ventilation enclosure fires [1-3]. Twenty seven experiments were conducted by Alvares et al. [4], and the results showed that the complete combustion could be achieved in forced ventilation enclosure fires with the oxygen concentration more than two times of the concentration with the stoichiometric ratio. Foote et al. [5] established a method for predicting the average temperature rise in forced ventilation enclosure fires, and the critical ventilation rate for avoiding flashover was also provided. Coutin et al [6] explored the electrical cabinet fires in closed and mechanically ventilated compartments, and established a method for achieving the heat release rate of fires with complex combustibles. To sum up, the effects of inlet locations on fire characteristics in force-ventilated compartment has not received sufficient attentions.

We have investigated the effects of air inlet configurations on forced ventilation enclosure fires experimentally [7]. The mass loss rate, the gas temperature rise and the oxygen concentrations in compartment have been presented. In this study, more parameters were provided to clarify this kind of enclosure fire by numerical simulation. The flame behaviours in different viewpoints, the oxygen concentration above fire source, the distribution of soot concentration and the vector velocity field of smoke in force-ventilated compartment with different air inlet locations were discussed in this study.

2. Numerical Simulations and Scenarios

The fire characteristics in forced ventilation compartment fires were investigated by using computational fluid dynamics (CFD) modelling. The software used in this study was Fire Dynamics Simulator (FDS), which was established by the National Institute of Standards and Technology. The FDS had been widely used for exploring many problems on fire science, and the results have been showed acceptable [8-10].

The dimensions of the closed compartment were 3.6 m (L) ×3.4 m (W) ×2.05 m (H). There were two air inlets and one exhaust vent in the compartment. The elevations of air inlet were changed in the simulations, and only one air inlet opened in the one-inlet scenarios. In all simulations, the elevation of the exhaust vent was 2.0m above the floor. The ventilation rates for both the air inlet system and the smoke exhaust system were 1500m^3/h, which is about 60 air change rates per hour. The ventilation rates of supply and exhaust ventilation systems were balanced in the simulation.

The fire source was located in the center of the force-ventilated compartment. The height of the fire source was 0.25 m above floor, and the area of fire source was 0.04 m^2. The HRR of fire was about 130 kW in all simulations. There were two viewpoints in the simulation model used for investigating the fire characteristic in the closed compartment. The view point 1 was set from the exhaust vent to the air inlets, and the view point 2 was set from one inlet to the other one.

There were four simulations conducted in this study. The scenario 1and 3 was one-inlet scenario with the air inlet elevations of 0.25 m and 1.65 m, respectively. In scenario 2 and 4, both the two inlets were opened, and the above two air inlet elevations were considered.

3. Results and Discussions

3.1. Flame behaviours

The flame behaviors of the forced ventilation enclosure fires with different inlet locations were investigated first. The flame behaviors from view point 1 and 2 in force-ventilated compartment with the inlet elevation of 0.25 m are shown in Figure 1 to Figure 4. In one-inlet scenario, the left inlet was opened as shown in Figure 1 and 2. From Figure 1, the flame inclined to the opened air inlet from viewpoint 1 in the experiments with one air inlet, especially at the middle and late stages of fire. It might due to the effect of inlet airflow on entrainment of the fire plume. The incline of flame was also found in scenario 3, in which there was also one air inlet opened and the air inlet was located at 1.65 m. In the one-inlet scenarios, the entrainment of fire plume on the side of air inlet was stronger than that on the other side, which might cause flame inclining to the air inlet. This hypothesis is also supported by the vector velocity field in the force-ventilated compartment, which discussed in below. The same phenomenon was also found in the previous experimental study [7]. In the scenarios with two opened air inlets, the incline of flame to air inlet was not observed, as shown in Figure 3. From viewpoint 1, the flame was almost in the middle of the two inlets during the whole simulation.

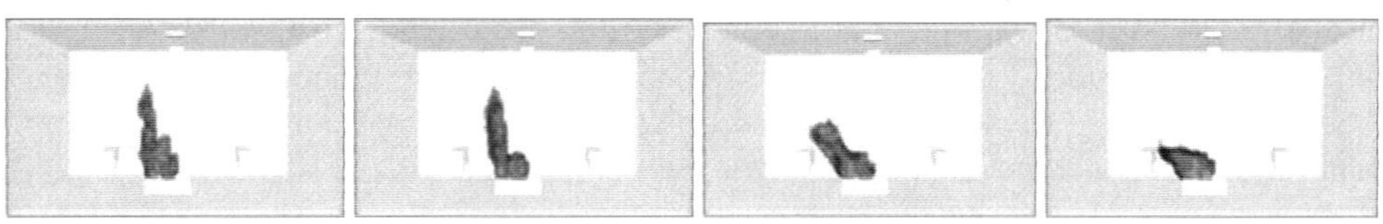

(a) t=30 s (b) t=60 s (c) t=150 s (d) t=300 s

Figure 1. Flame behaviours in scenario 1 from the View Point 1

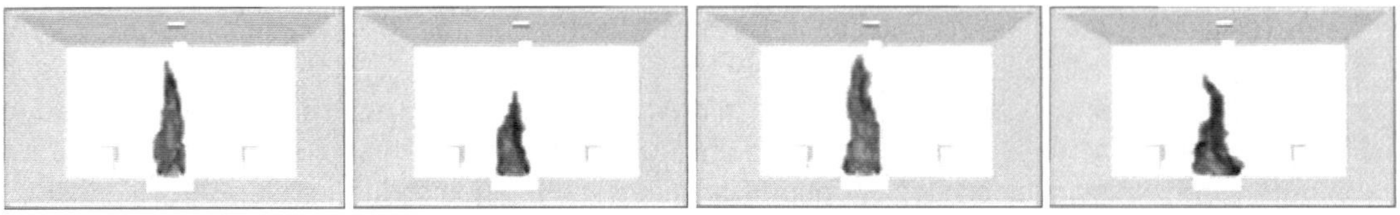

(a) t=30 s (b) t=60 s (c) t=150 s (d) t=300 s

Figure 2. Flame behaviours in scenario 2 from the View Point 1

Through the numerical simulation, it found that not only the air inlet, but also the exhaust vent might affect the flame behavior. The effects of exhaust flow on flame behavior were analyzed through the images from viewpoint 2. The flame behavior in scenario 1 was shown in Figure 3. It found that the effects of exhaust flow on flame behavior were not significant in the one air inlet scenario with the inlet elevation of 0.25 m. In scenario 1, the flame almost kept straightly from viewpoint 2 during the whole simulation duration. In scenario 2, in which both the two air inlets opened, the flame inclined to the exhaust vent obviously from the viewpoint 2, as shown in Figure 4. The incline of flame to the exhaust vent was due to the influences of the two air inlets and the exhaust vent on the entrainment of fire plume. Since the two opened air inlets were located at different sides of the fire source, the effects of the two air inlets might be counteracted, and the effect of exhaust vent was dominated. The inclining of flame to the exhaust vent was also appeared in the two-inlet scenario with the inlet elevation of 1.65 m, as shown in Figure 5.

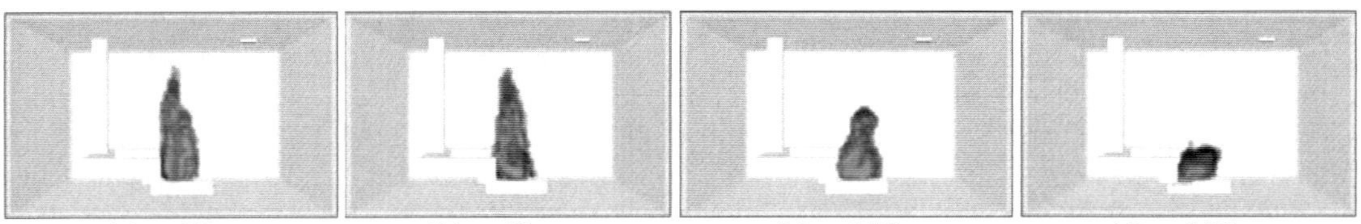

(a) t=30 s (b) t=60 s (c) t=150 s (d) t=300 s

Figure 3. Flame behaviours in scenario 1 from the View Point 2

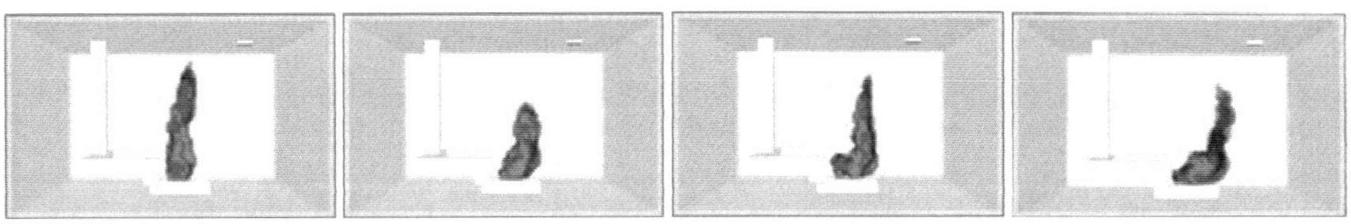

(a) t=30 s (b) t=60 s (c) t=150 s (d) t=300 s

Figure 4. Flame behaviours in scenario 2 from the View Point 2

3.2. Gas Concentrations

The oxygen concentrations above the fire source were measured in the numerical simulation. Due to the effects of combustion and ventilation, the oxygen concentration curves showed significant shock in all scenarios. The oxygen concentrations above the fire source in scenario 1 and scenario 2 are shown in Figure 5 to investigate the effects of air inlet number on oxygen concentration in forced ventilation enclosure fires. The oxygen concentration of two-inlet scenario was lower than that of the one-inlet scenario basically. The effects of air inlet elevation on oxygen concentration were illustrated in Figure 6 with the scenario 2 and scenario 4. It observed that the oxygen concentration of scenario 4 was slightly higher than that of scenario 2.

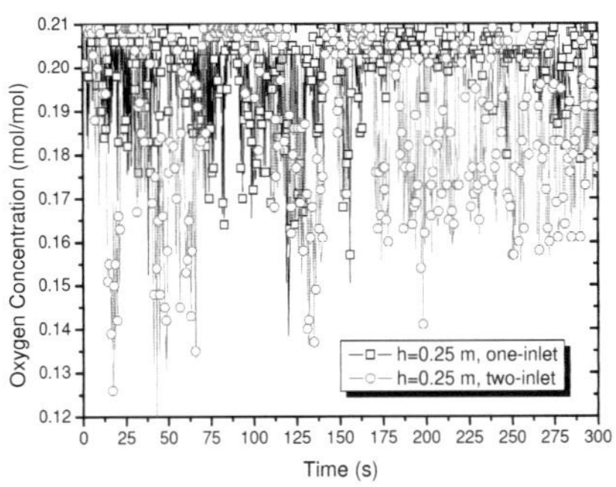

(a)

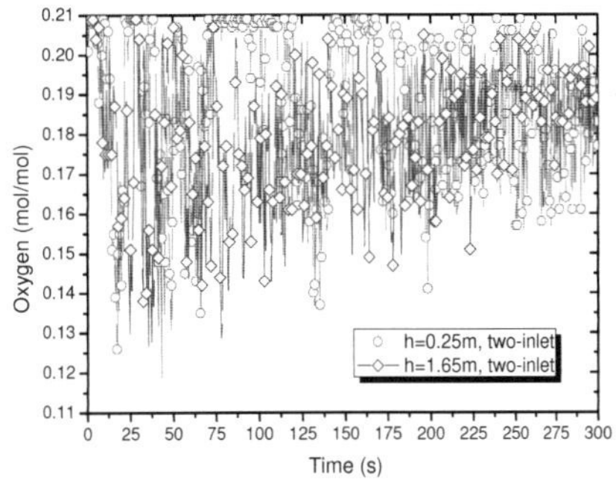

(b)

Figure 5. Oxygen concentrations above fire base: (a) effect of inlet number; (b) effect of inlet height

The distributions of soot concentrations in force-ventilated compartment at 150 s in scenarios 1-4 are shown in Figure 6. The soot concentration in the compartment was affected by the generation of soot due to combustion and the elimination of soot due to ventilation. Since the heat release rate of fire and the ventilation rate were accordant in all simulations, the maximum soot concentrations in the compartment were almost the same in different scenarios. The distributions of soot concentration were significant different in the scenarios with different air inlet numbers and elevations. Comparted with Figure 6 (a) and (b), the difference between the soot concentrations in upper and lower spaces of the compartment was more obviously in one-inlet scenario than that in two-inlet scenario, and the similar result was found in scenario 3 and 4 as shown in Figure 6 (c) and (d). From Figure 8, it also observed that the soot concentration in the compartment was more homogeneous in the scenario with the inlet elevation of 1.65 m than that in the scenario with the inlet elevation of 0.25 m.

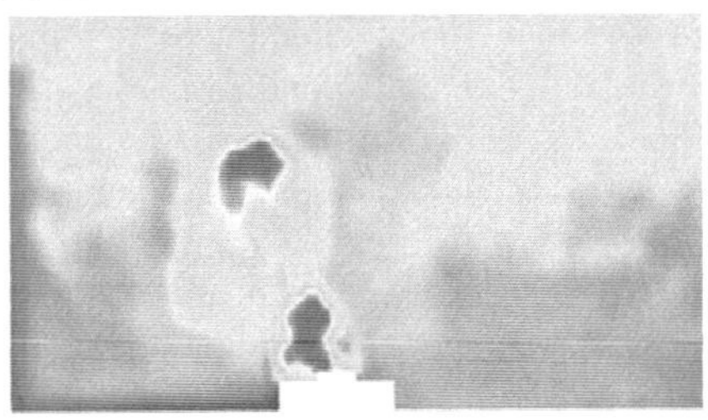

(a) Scenario 1

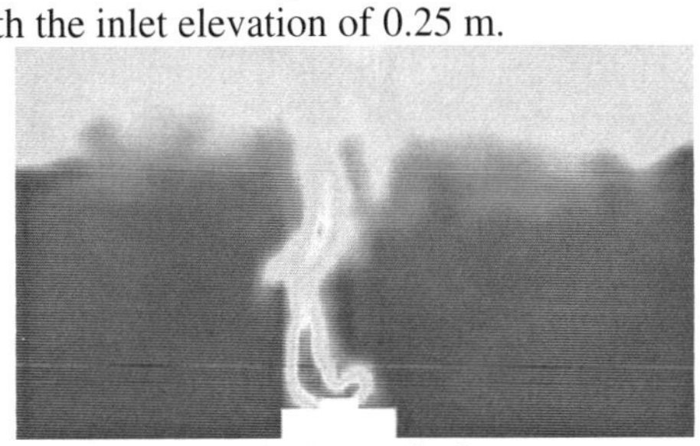

(b) Scenario 2

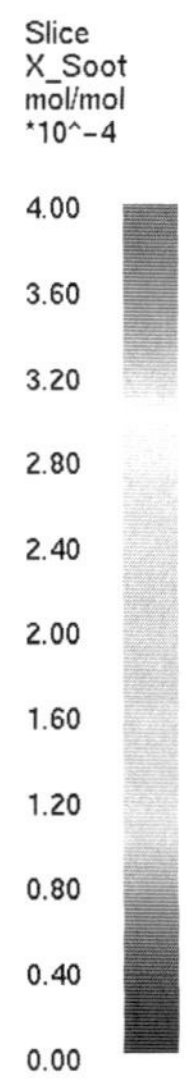

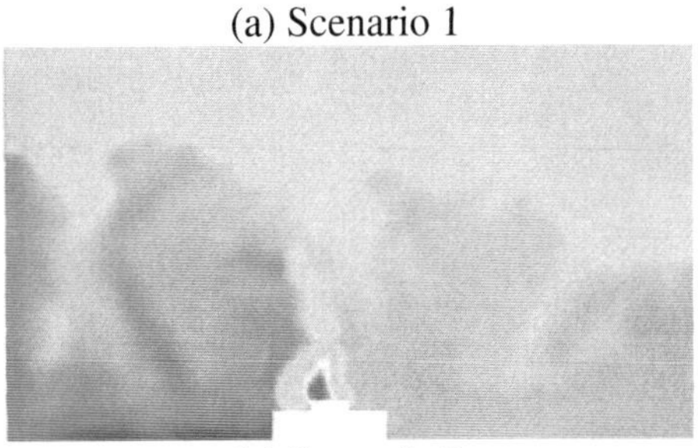

(c) Scenario 3

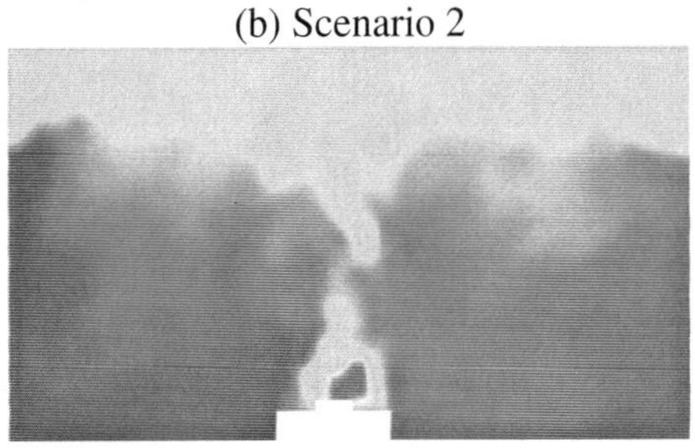

(d) Scenario 4

Figure 6. Soot concentration distribution in different scenarios at 150 s

3.3. Vector Velocity of Smoke

In the above discussion, the air inlet location showed obvious influence on the fire characteristic of the forced ventilation enclosure fires, which might relate to the entrainment of fire plume affected by the air flow. For further investigate the effect of air flow on the entrainment of fire plume, the vector velocity of smoke in the force-ventilated compartment are illustrated in Figure 7 and 8. From Figure 7,

IOP Conf. Series: Journal of Physics: Conf. Series **916** (2017) 012028 doi:10.1088/1742-6596/916/1/012028

the velocities of smoke in one-inlet scenarios were larger than those in two-inlet scenarios. This was because that the inflow air velocity in one-inlet scenario was twice larger than that in the two-inlet scenario. In scenario 1 and scenario 3, the velocity of smoke on the side of opened air inlet was obvious larger than the other side. It indicates that the entrainment of fire plume is stronger on the opened air inlet side than the other side in one-inlet scenarios, which may be the cause of inclining flame in these scenarios. In two-inlet scenarios, the velocities of smoke in the two sides of flame were almost the same from viewpoint 1. It means that the entrainments on the two sides were balanced, so the flame was stayed in the middle of the two inlets from viewpoint 1.

To sum up, some conclusions derived in the previous experimental research [7] were confirmed by numerical simulation, and the natural of the forced ventilation enclosure fire was further revealed through the simulations to some extent in the present study.

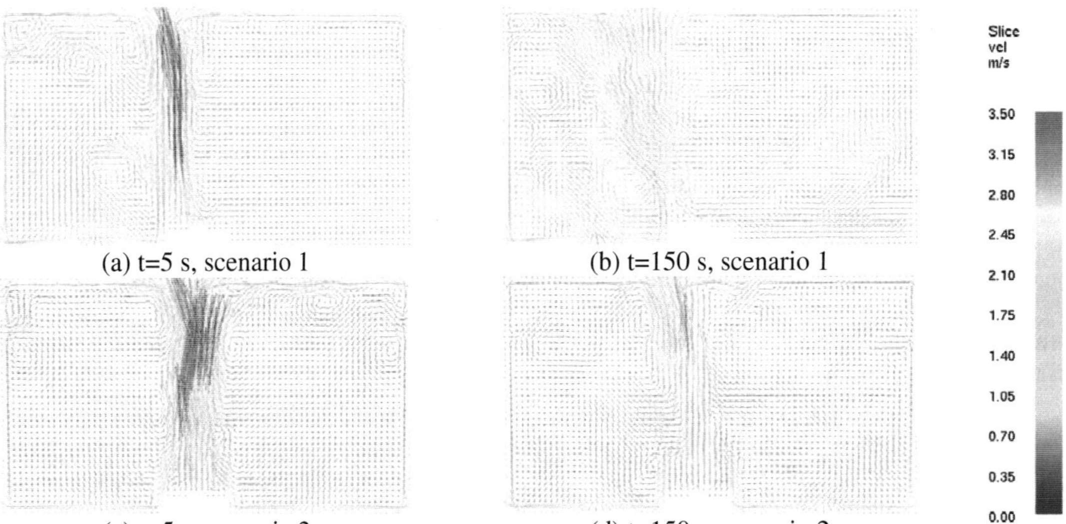

(a) t=5 s, scenario 1 (b) t=150 s, scenario 1

(c) t=5 s, scenario 2 (d) t=150 s, scenario 2

Figure 7. Vector velocity of smoke in scenarios with inlet elevation of 0.25 m

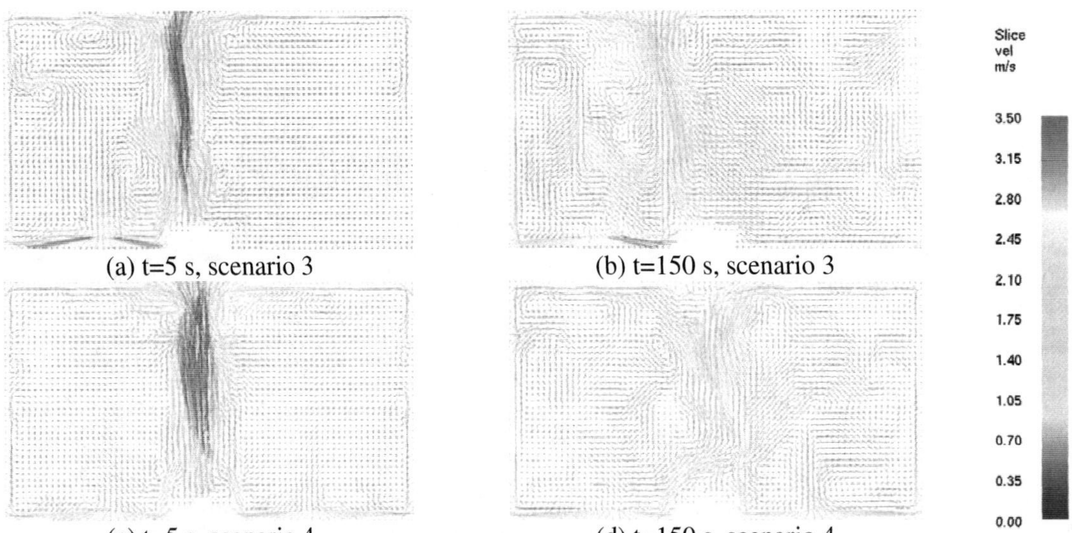

(a) t=5 s, scenario 3 (b) t=150 s, scenario 3

(c) t=5 s, scenario 4 (d) t=150 s, scenario 4

Figure 8. Vector velocity of smoke in scenarios with inlet elevation of 1.65 m

4. Conclusion

The fire characteristics in force-ventilated compartment with different air inlet locations were investigated by the CFD simulation. The fire behaviour, the oxygen concentration above the fire source, the distribution of soot concentration and vector velocity of smoke in the force-ventilated compartment were discussed. The results showed that both the inlet and exhaust vents affected the fire behaviour in the force-ventilated compartment. In the one-inlet scenarios, the flame inclined to the opened inlet and the effect of exhaust vent was not obvious. In the two-inlet scenarios, the effect of exhaust vent on flame behaviour was dominant, and the flames inclined to exhaust vent. The soot concentration distributions in one-inlet scenarios were more uniform than those in two-inlet scenarios, and difference between the soot concentrations in upper and lower spaces of compartment was smaller in the simulation with higher air inlet. The velocity of smoke in the side of opened air inlet was larger than that in the other side in one-inlet scenarios, and this might be the reason of the flame incline to the air inlet under this condition. The velocities of smoke on the two side of flame from viewpoint 1 in the two-inlet scenarios were almost the same.

Acknowledgments

This work was financially supported by Anhui Provincial Natural Science Foundation (Grant No. 1408085MKL94), Science and Technology Project of State Grid Corporation of China (Grant No. 52120517000X) and Hebei Provincial Natural Science Foundation (Grant No. E2017507012)

References

[1] Pretrel H, Le Saux W, Audouin L, Pressure variations induced by a pool fire in a well-confined and force-ventilated compartment. Fire Saf J. 2 (2012) 1–24.

[2] Pre´ trel H, Such J, Effect of ventilation procedures on the behaviour of a fire compartment scenario. Nucl Eng Des 235 (2005) 2155–2169.

[3] Chow WK, Chan W (1993) Experimental studies on forced-ventilated fires. Fire Sci Technol, 13(1993) 1_71–1_87.

[4] N. J. Alvares, K. L. Foote, P. J. Pagni, Forced Ventilated Enclosure Fires, Combustion Science & Technology. 39 (1984) 55-81

[5] K Foote, P Pagni, N Alvares, Temperature Correlations For Forced-ventilated Compartment Fires, Fire Saf Sci, 1 (1986) 139-148.

[6] Coutin, M Plumecocq, W Melis, S Audouin, L, Energy balance in a confined fire compartment to assess the heat release rate of an electrical cabinet fire, Fire Saf J, 52 (2012) 34-45

[7] B.S. Zhang, J.Q. Zhang, X.M. Wang, et al, Effects of Air Inlet Configuration on Forced-Ventilation Enclosure Fires on a Naval Ship, Fire Tech, 52 (2016) 547–562.

[8] ND Pope and CG Bailey, Quantitative comparison of FDS and parametric fire curves with post-flashover compartment fire test data, Fire Saf J, 41 (2016) 99-110.

[9] Y.M. Ferng, C.H. Liu, Investigating the burning characteristics of electric cables used in the nuclear power plant by way of 3-D transient FDS code, Nucl Eng Des. 241 (2011) 88-94.

[10] Qiang Liang, Yangfeng Li, Junmei Li, Hui Xu, Kaiyuan Li, Numerical studies on the smoke control by water mist screens with transverse ventilation in tunnel fires, Tunn Undergr Sp Tech, 64 (2017) 177-183.

The effect of high voltage pulsed electric field on water molecular

Xuejie Fan, Yaxiang Bai*, Ziying Ren

School of Science, Dalian Ocean University, Heishijiao Street, Dalian, China

*Corresponding author e-mail: Byx0671@163.com

Abstract. In order to study the mechanism of high voltage pulsed electric field pre-treatment on the food drying technology. In this paper, water was treated with high pulse electric field (HPEF) in different frequency, and different voltage, then, the viscosity coefficient and the surface tension coefficient of the water were measured. The results showed that indicated that the viscosity coefficient and the surface tension coefficient of the treated water can be decreased, and while HPEF pre-treatment was applied for 22.5kV at a frequency of 50Hz and 70 Hz, the surface tension and the viscosity coefficient of the pre-treatment treatment were reduced 13.1% and 7.5%, respectively.

1. Introduction

Fresh fruits and vegetables have a moisture content ranging from 75 to 90%. They are highly heat sensitive and are apt to be highly perishable. Drying is defined as the reduction of moisture from the products and is an efficient and cheap method for the preservation of fruits and vegetables. Hot air drying is one of the most commonly used procedures in dried food with main disadvantages being prolonged duration and high energy consumption. Therefore it is necessary to explore low power consumption processing for hot air drying.

Dry pretreatment technology is utilized to treat food through some physical methods before drying it. Pretreatment can not only improve the quality of dried food but also increase drying rate and thus reduce energy consumption. For example, during the process of drying and dehydration for fruits and vegetables, ultrasound pretreatment certain intensity will enhance their drying rate and thus reduce the drying energy consumption [1-3].

In recent years, a novel pretreatment technique, high voltage pulsed electric field drying pretreatment technology has been developed only recently, Angers Bach et al. showed an increase in permeability of potato tissue by high intensity electrical field pulse treatment, which resulted in improved mass transfer during fluidized bed drying [4]. Ade-Omowaye et al. found that high voltage pulsed electric field pretreatment could increase the cell membrane permeability and improve the drying rate [5]. Gachovska found that pulsed electric field treatment could increase the drying rate of carrots and enhance the quality of dried carrots [6]. Electrically assisted drying would reduce the processing time, operating temperature, and energy consumption [7]. Other papers reported similar results that pulsed electric field pretreated samples could significantly improve the drying rate [8-9].

But, it is not a profoundly explored method of drying pretreatment, and, its mechanism has not been accurately determined, therefore, study the mechanisms of high voltage pulsed electric field pretreatment is one of the important theoretical basis for the technology industrialization application use.

Material drying is a process in which the free water penetrates and diffuses from material interior to exterior under the effects of differential concentrations and temperatures in and outside the material. In addition to differential concentrations and temperatures, the speed of material drying process, namely the drying rate, is also affected by the structure of hydronic itself. Generally, the liquid water is composed by single hydronic and hydronic cluster, hydronic combined by hydrogen bond (figure 1). Such cluster structure of water is the reason of its high viscosity, surface tension, melting point, boiling point and heat capacity. However, current analysis on liquid hydronic cluster structure only focuses on its viscosity and surface tension etc. Therefore, this thesis plans to identify the effect of high voltage pulse electric field on hydronic cluster structure by measuring the viscosity and surface tension of hydronic after it being processed by high voltage pulse electric field.

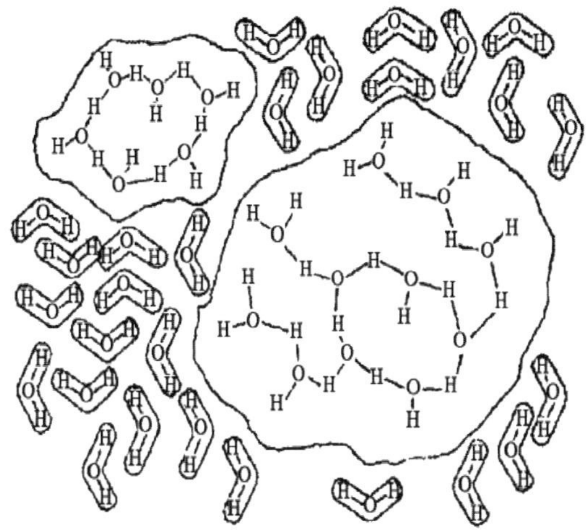

Figure 1. Schematic diagram of the structure of the water molecule

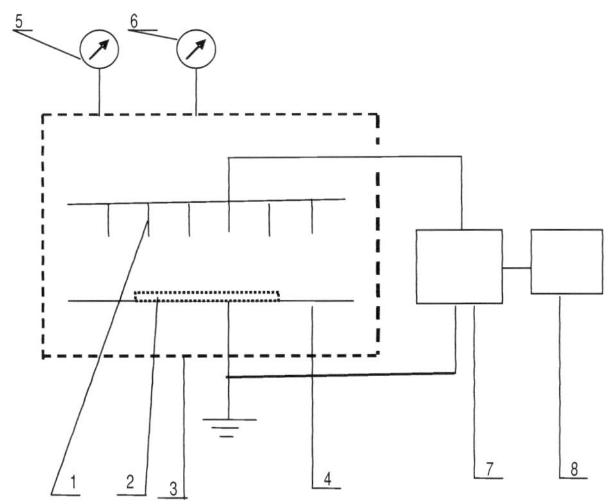

Figure 2. Schematic of the high pulse electric field experimental equipment.

2. Experimental procedure

2.1. Experimental equipment

Figure 2 shows a schematic diagram of the high pulse electric field setup developed for the experimental work. It consists of a vertically mounted electrode with multiple sharp pointed needles projected to a fixed horizontal grounded metallic plate on which the tap water to be pretreated is placed.

2.2. Experimental methods

The tap water were divided into 28 groups with similar weights. One was the control group and the others were experimental groups. The experimental groups were subjected to high-voltage pulsed electric field pretreatment with different voltage (18, 22.5, 27 kV), different frequency (50, 70, 90 Hz) and different time (2, 5, 8 min), respectively. Then, The Surface tension and Viscosity coefficient of all samples were measured.

2.3. Surface tension coefficient measurement

Align the transverse line on glass tube to the image in index mirror and scribed line in index mirror to ensure the spring bottom position is fixed, elongation of spring can be measured by the vernier (that is, the difference between the two readings before and after elongation).

According to Hooke's law, within the elastic limit, elongation of the springs proportional to the applied force F, i.e., measure stiffness coefficient of spring k.

Hang U wire on small hook under the index bar, turn the lifting button, make three wires aligned, and record the Vernier initial value.

Immerge U wire into water in breaker, rotate screw under platform to lower down the platform slowly. Because of surface tension is applied on the wire, the horizontal scribed line on the index bar drops accordingly, to adjust the lifting button, make three wires aligned. Lower the platform a bit and repeat above adjustment, until the platform lowers a little, the wire comes out the liquid surface. Write down the meter ruler readings indicated by Vernier zero line, can get spring elongation, and then calculate the emerged surface tension coefficient.

$$\sigma = \frac{k(s - s_0)}{2(l + d)} \tag{1}$$

Where l and d are the width and diameter of the U wire (cm) σ is the surface tension coefficient.

2.4. Viscosity coefficient measurement

The water of the control group and experimental group were load in a glass cylinder of 1000 ml, respectively, and the measuring cylinder should be placed directly in the cylinder, so that the ball of wax would fall down the center line of the cylinder.

Measure the distance between the upper and lower line AB with a meter stick. The diameter d was measured by a helix micrometer.

Using a pair of tweezers to put the ball into the measuring cylinder, the ball is measured with a stopwatch to measure the constant velocity of the ball. The time t is required by the upper and lower scales of A and B, and the coefficient of the viscosity of the water is eta, which is the viscosity coefficient of the water.

$$\eta = \frac{1}{18}\frac{(\rho - \rho_0)gd^2}{v_0} \tag{2}$$

Where ρ_0 and ρ are the density of water and the ball of wax ($g \cdot cm^{-3}$), and d is the diameter of the ball of wax (cm) sis

2.5. Effect of high voltage pulsed electric field on the surface tension coefficient of tap water

The influence of high voltage pulsed electric field on the surface tension coefficient of tap water was shown in Figure 3-5, respectively. It can be seen from the graphs that the surface tension coefficient of tap water treated by high voltage pulsed electric field was lower than that of the control. Under a 27 kV at a frequency of 50 Hz for 5min, the surface tension coefficient of treated tap water is 13.1% lower compared with that of the control group.

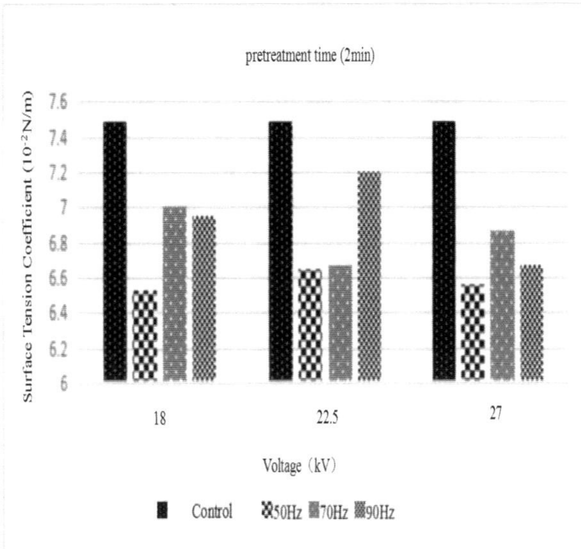

Figure 3. Effect of high voltage pulsed electric field on the surface tension coefficient of tap water

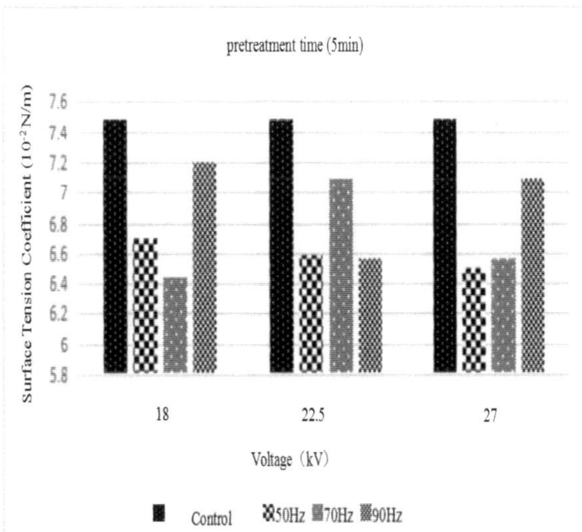

Figure 4. Effect of high voltage pulsed electric field on the surface tension coefficient of tap water.

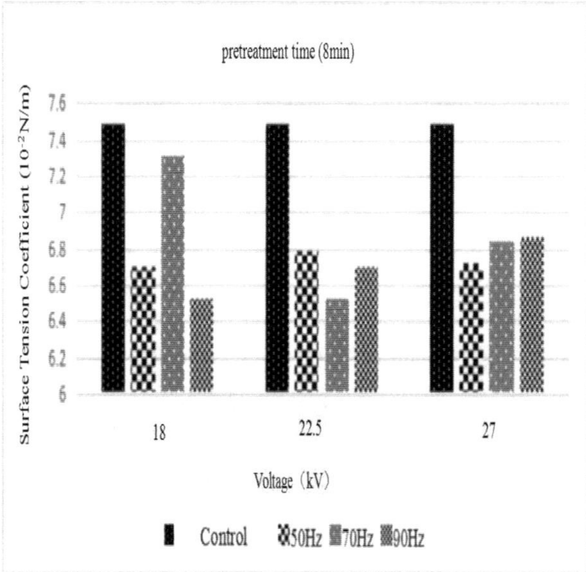

Figure 5. Effect of high voltage pulsed electric field on the surface tension coefficient of tap water.

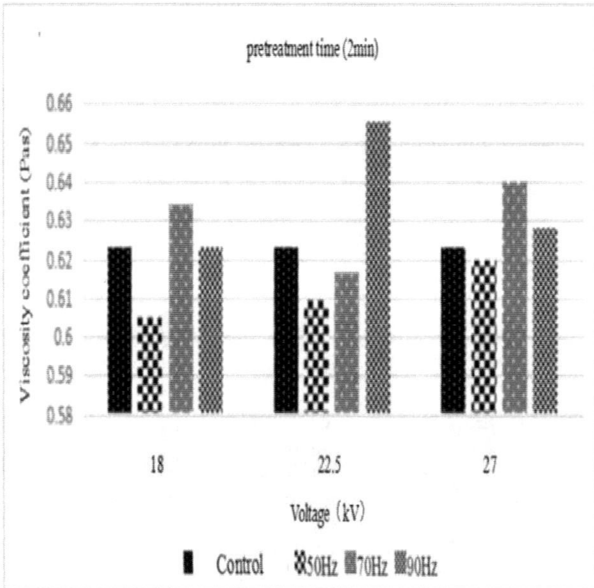

Figure 6. Effect of high voltage pulsed electric field on the viscosity coefficient of tap water.

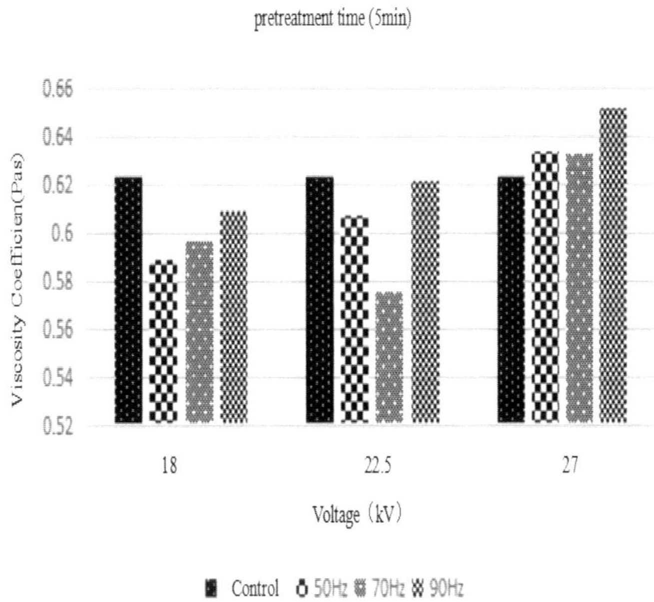

Figure 7. Effect of high voltage pulsed electric field on the viscosity coefficient of tap water.

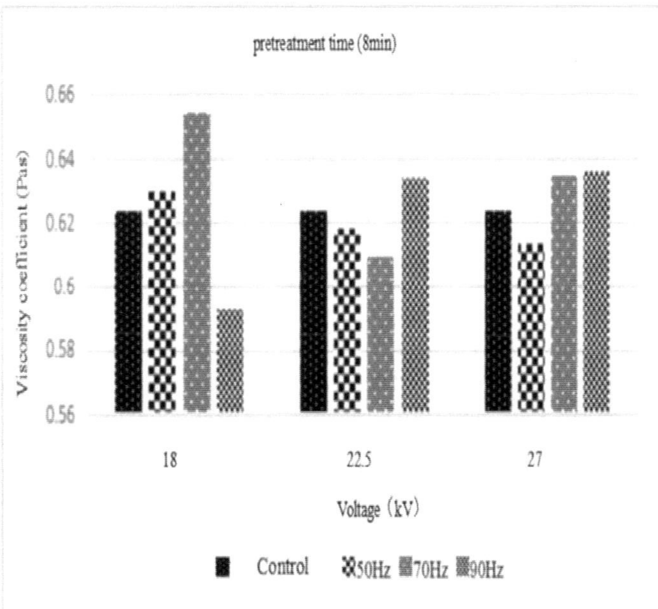

Figure 8. Effect of high voltage pulsed electric field on the viscosity coefficient of tap water.

2.6. Effect of high voltage pulsed electric field on the viscosity coefficient of tap water

Figure 6-8 show the results of the viscosity coefficient of tap water. According to the results, HPEF pretreatment can decrease the viscosity coefficient of tap water significantly. Under a 22.5kV voltage at a frequency of 70 Hz for 5min, the viscosity coefficient of pretreatments is 7.59% lower compared with that of the control group.

3. Conclusion

The the viscosity coefficient and the surface tension coefficient of the top water can be decreased by high-voltage pulsed electric fields pretreatment, and while high voltage pulsed electric field pretreatment was applied for 22.5kV at a frequency of 50Hz, the surface tension coefficient of the pretreatment treatment were reduced 13.1% and while high voltage pulsed electric field pretreatment was applied for 22.5kV at a frequency of 70Hz, the viscosity coefficient of pretreatments were reduced 7.5%, respectively.

References

[1] Schossler K, Jaeger H, Knorr D 2012 Innov Food Sci Emerg. 16 113
[2] Fabiano A N, Fernandes L F, Rodrigues S 2008 Ultrasonic Sonoche. 15 1049
[3] Nowacka M, Wiktor A, Sledz M, Witrowa-Rajchert M 2012 J. Food Eng. 113 427
[4] Angersbach A, Heinz V, Knorr D 1997 LVT. 42 195
[5] Ade-Omowaye B I O, Rastogi N K, Angersbach A, Knorr D 2002 J. Food Eng.54 35
[6] Gachovska T K, Adedeji A A, Ngadi M, Raghavan G V S 2008 Drying Technol. 26 244
[7] Shynkaryk M V, Lebovka N I, Vorobiev E I 2008 Drying Technol. 26 695
[8] Soliva-Fortuny R, Balasa A,Knorr D 2009 Trends Food Sci Tech. 20 544
[9] Wu Y L, Guo Y M, Zhang D G, Drying Technol. 29 1714

Experimental and numerical investigation on transient liquid velocity in hilly-terrain wet natural gas pipelines

Pengbo Yin[1], Xuewen Cao[1,*], Yuhao Li[2], Sisi Qin[1], Chenyang Fu[1] and Wenjuan Sun[1]

[1]China University of Petroleum, Qingdao, China
[2] Binzhou University, Binzhou, China

*Corresponding author e-mail: caoxw@upc.edu.cn (X. Cao)

Abstract. In the present work, the transient liquid velocity in hilly-terrain pipelines is experimentally and numerically investigated through ultrasonic Doppler technique and CLSVOF method. An ultrasonic transducer mounted on upward inclined pipeline is employed for transient liquid velocity measurement, and the results agree well with simulations. It is found that the slug flow is only presented in the inclined pipeline. The superficial gas and liquid velocity has a significant influence on the slug frequency while it has a slight influence on the slug translational velocity.

1. Introduction

The gas gathering pipelines between well sites, gas gathering stations and natural gas processing facilities can be divided into wet and gas pipelines. In most wet-gas pipelines, small amount of free water and condensate, resulting from pressure and temperature decrease, leads to the formation of a two-phase flow [1, 2, and 3]. Natural gas resources in China are mostly located in basins surround by mountains. Due to the terrain effect, the two phase flow is prone to liquid accumulation and slug flow with a low flow rate, which cause severe liquid holdup and pressure fluctuation and flow-dependent corrosion [4]. Therefore, it is important to understand and predict the slug flow behavior in a hilly-terrain pipeline [5].

Ultrasonic Doppler velocimetry (UDV), a non-invasion flow field measurement method with the advantage of high precision and flexibility, has been used to measure flow velocity in some studies. Takeda [6] extended the application range of UDV method to general fluids for the first time, which was originally applied in the medical diagnosis. Treenuson [7] used the ultrasonic velocity profile method for flow rate measurement in the downstream of double bent pipe. Fachun Liang [8] measured the velocity profile of the liquid film using the UDV in a gas-liquid swirling flow.

A coupled Level Set and volume of fluid method (CLSVOF) [9] has been employed for two phase flow simulation in recent years. Kharangate [10] simulated the annular flow behavior of a condensation system in a vertical pipeline using the CLSVOF method. Shang [11] used the CLSVOF method in the transition of two-phase flows in a vertical pipeline through validation of single bubble cases.

In this work, ultrasound Doppler measurement and CLSVOF simulation are conducted for the research of transient liquid flow velocity in a hilly-terrain wet gas pipeline. A multiphase flow experimental loop with an inside diameter of 40 mm is employed, and a three-dimensional numerical model is simultaneously carried out, which is verified through comparison with the experimental results. The researches on the transient velocity characteristic of liquid slug and liquid film contribute to the improvement of theoretical model and numerical simulation

2. Experimental apparatus

2.1. Multiphase flow loop

A schematic of the multiphase flow loop is presented in Fig. 1. The inner diameter of the pipe is 40mm. Air and water are employed as the experimental medium. The energy of fluids is supplied from a screw compressor and a centrifugal pump. The gas and liquid phase are separately measured in the measuring section, and are mixed through a three-way mixture tee. At the end of the pipeline, the fluid enters a cyclone separator where air and water are separated. The air is vented to the atmosphere, while the water flows back to storage tank. The superficial gas velocity range is 1.985m/s-10.076m/s, and the superficial liquid velocity range is 0.0204m/s-0.1003m/s.

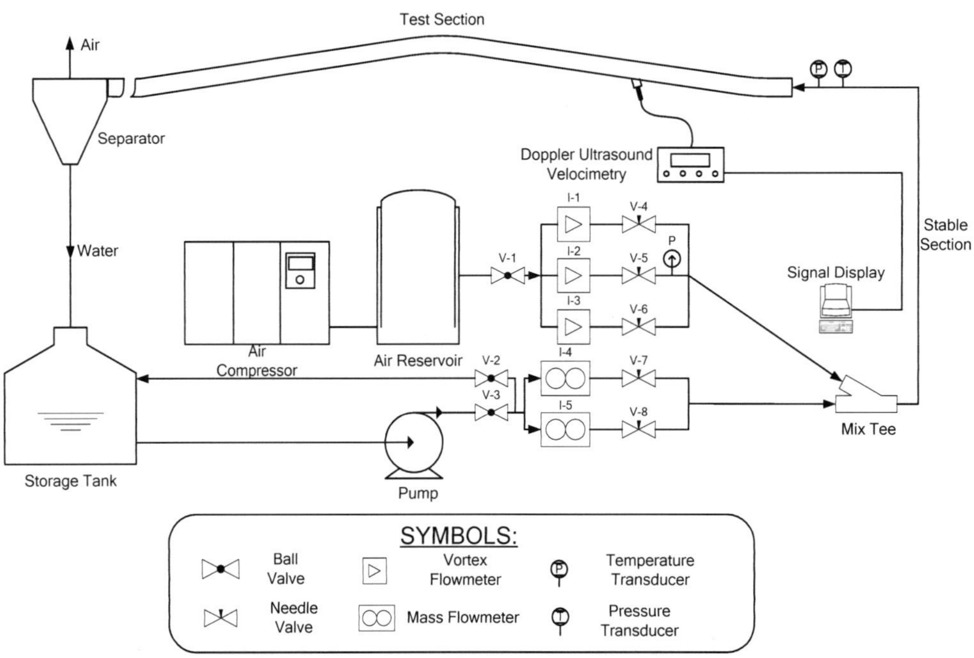

Figure 1. Schematic of the experimental apparatus.

The test section consists of three parts: three sections of horizontal pipe, one ascending pipe and one descending pipe, which are made of polymethyl methacrylate (PMMA). Since slug flow is only presented in the upward inclined pipeline, an ultrasonic transducer is installed on the position of 1.5 m from the elbow, as shown in Fig.2. The trace particles made of copolyamide have a diameter of 75 μm and a concentration from 100 to 1000 ppm.

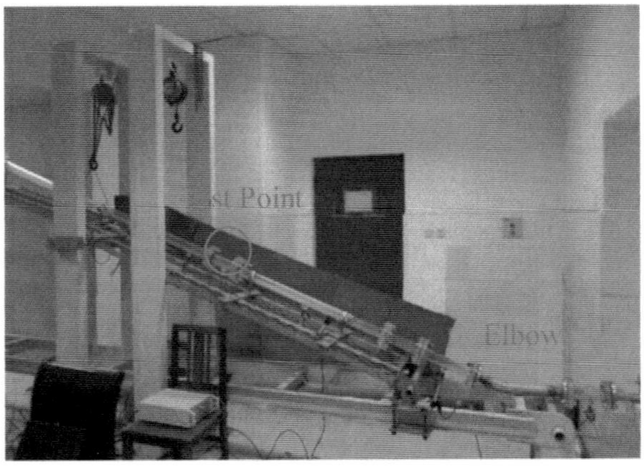

Figure 2. Photograph of testing pipeline.

2.2. Ultrasonic Doppler velocimetry

The working principle of Ultrasound Doppler Velocimetry is to detect and process many ultrasonic echoes issued from pulses reflected by micro particles contained in a flowing liquid. The measurement time lapse between the emission of ultrasonic bursts and the reception of the pulse gives the position of particles. By measuring the Doppler frequency in the echo as a function of time shifts of these particles, a velocity profile after few ultrasonic emissions is obtained.

The velocity of the particles is given as follows:

$$u_{real} = \frac{cf_d}{2f_e \cos\beta} \tag{1}$$

Where, u_{real} is the velocity of the particles. c is the sound velocity of the ultrasonic wave in fluid medium. f_d is the phase shift of the received echo. f_e is the emitting frequency. β is the installation angle of ultrasonic transducer.

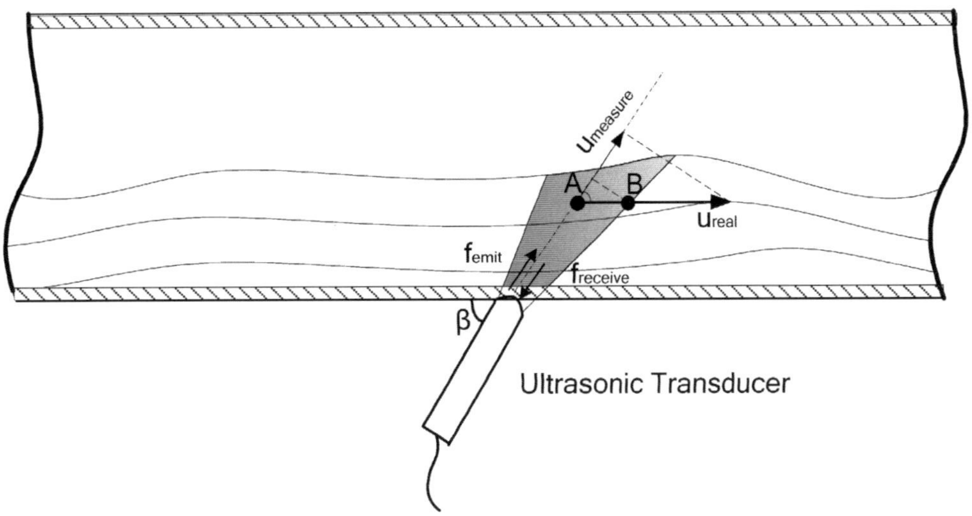

Figure 3. Velocity measurement principle of UDV.

3. CFD simulation

3.1. Geometry model

The geometry model of the simulation is shown in Fig. 4. The slope of the inclined pipeline is 10°, and the inner diameter is 40 mm. The pipeline lengths are noticed in the figure. The radian in the junction of the horizontal and inclined sections is 5D.

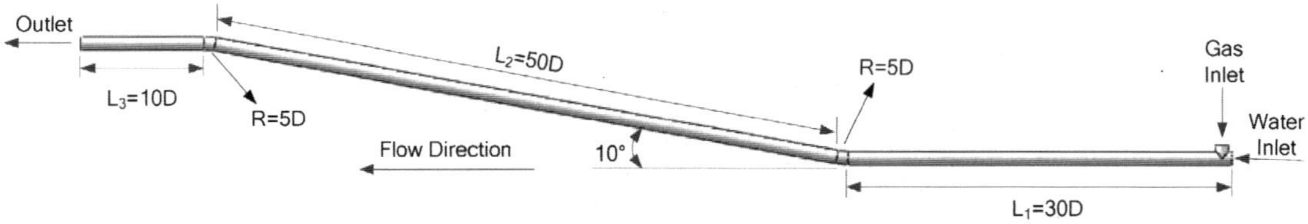

Figure 4. Geometry of simulation domain.

3.2. Multiphase flow model

A coupled Level Set and volume of fluid method (CLSVOF) approach is applied for multiphase flow simulation in this work. The governing equations are as follows:

Continuity:

$$\nabla \cdot \vec{u} = 0 \tag{2}$$

Where, $\vec{u}$ is the underlying velocity field.

The evolution of the level-set function:

$$\frac{\partial \varphi}{\partial t} + \nabla \cdot \left(\vec{u}\varphi\right) = 0 \tag{3}$$

Where, φ is the level-set function, defined as a signed distance to the interface. t Is the dimensionless time.

Momentum:

$$\frac{\partial \vec{u}}{\partial t} + \vec{u} \cdot \nabla \vec{u} = -\frac{\nabla p}{\rho} + \frac{1}{\rho} \nabla \cdot \left(2\mu D\right) + \rho \vec{g} - \frac{1}{\rho} \sigma \kappa \delta\left(\varphi\right) \nabla \varphi \tag{4}$$

Where, Continuity:

$$\delta\left(\varphi\right) = \begin{cases} \dfrac{1 + \cos\left(\pi\varphi / a\right)}{2a} & \left|\varphi\right| < a \\ 0 & \left|\varphi\right| \geq a \end{cases}, \quad a = 1.5\ h \tag{5}$$

Where, p is the total pressure. $\vec{g}$ is the gravitational acceleration. σ is the surface tension coefficient. κ is local mean curvature, $\kappa = \nabla \cdot \left(\nabla\varphi / \left|\nabla\varphi\right|\right)$. D is the rate of deformation tensor, $D = \nabla\vec{v} + \left(\nabla\vec{v}\right)^T$. h Is the grid spacing. ρ And μ is the density and viscosity, which are written as:

$$\rho = \delta(\varphi)\rho_l + (1 - \delta(\varphi))\rho_g \text{ And } \mu = \delta(\varphi)\mu_l + (1 - \delta(\varphi))\mu_g \tag{6}$$

Where, the subscript l and g represent the liquid phase and the gas phase respectively.

3.3. *Realized k-ε model*

The realized k-ε model has been used to deal with the turbulent effect. The transport equations of turbulent kinetic energy k and dissipation rate ε are as follows:

$$\frac{\partial(\rho k)}{\partial t} + \frac{\partial(\rho k v_i)}{\partial x_i} = \frac{\partial}{\partial x_j}\left[\left(\mu + \frac{\mu_t}{\sigma_k}\right)\frac{\partial k}{\partial x_j}\right] + G_k + G_b - \rho\varepsilon - Y_M \tag{7}$$

$$\frac{\partial(\rho\varepsilon)}{\partial t} + \frac{\partial(\rho\varepsilon v_i)}{\partial x_i} = \frac{\partial}{\partial x_j}\left[\left(\mu + \frac{\mu_t}{\sigma_\varepsilon}\right)\frac{\partial\varepsilon}{\partial x_j}\right] + \rho C_1 E\varepsilon - \frac{\rho C_2\varepsilon^2}{k + \sqrt{v\varepsilon}} + C_{1\varepsilon}\frac{\varepsilon}{k}C_{3\varepsilon}G_b \tag{8}$$

$$C_1 = max\left(0.43, \frac{\eta}{\eta + 5}\right), \quad \eta = \left(2E_{ij}\cdot E_{ij}\right)^{1/2}\frac{k}{\varepsilon}, \quad E_{ij} = \frac{1}{2}\left(\frac{\partial v_i}{\partial v_j} + \frac{\partial v_j}{\partial v_i}\right) \tag{9}$$

Where, v_i is the velocity in the i direction. μ_t Is the turbulent viscosity. σ_k And σ_ε are the turbulent Prandtl numbers of k and ε. G_k is the generation of turbulence kinetic energy due to the mean velocity gradients. G_b is the generation of turbulence kinetic energy due to buoyancy. Y_M is the contribution of the fluctuating dilatation in compressible turbulence. C_2, $C_{1\varepsilon}$, $C_{3\varepsilon}$ are constants.

3.4. *Mesh generation*

A computational grid with 560,000 hexahedral cells is built through the pre-processor ICEM-CFD, and the meshing quality of the cross and elbow sections is shown in Fig. 5. The cross section includes 1045 elements, and boundary layer and elbow meshes are added to guarantee the capture of complex flow behavior around the boundary.

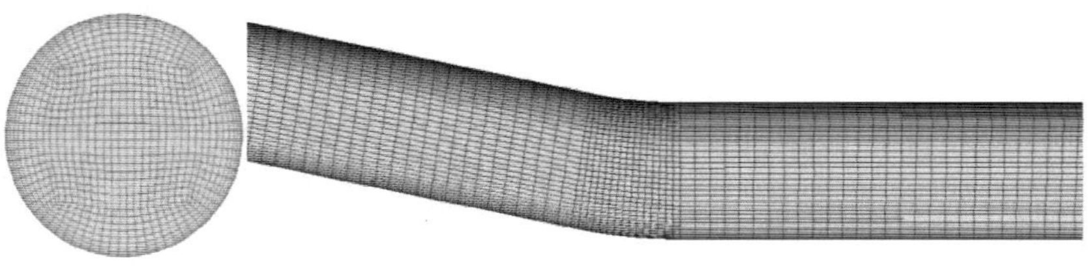

Figure 5. Mesh quality.

4. Results and discussion

The transient liquid velocity can present the character of velocity change for liquid film, liquid slug and their correlation in the ascending pipe. Considering the requirement of liquid phase continuity, the dynamic characteristics of liquid flow velocity at measuring point which is 5 mm from the bottom of

IOP Conf. Series: Journal of Physics: Conf. Series **916** (2017) 012030 doi:10.1088/1742-6596/916/1/012030

the pipe inner wall on the pipe centerline that is perpendicular to the pipe flow is studied. The velocity simulation values at the experimental measuring point were processed with the FLUENT post-processing software. The comparison of the experiment and simulation is shown in Fig. 6. The simulation results agree well with the experiments.

The variation of transient liquid velocity is similar regardless of the different gas and liquid velocity in this study. Due to the entrainment effect of gas and liquid, the liquid velocity suddenly increases to its peak when the fronts of the liquid slugs reach the measuring point, and then decrease gradually owing to the slug body and liquid film throughout.

At the superficial gas velocity of 1.985 m/s, as the superficial liquid velocity raises, the liquid volume in the elbow and upward pipeline increases, which lead to a higher slug frequency. Owing to the inertia effect of the liquid phase, the peak velocity has a little change. The liquid slugs with a low peak velocity increase, and it is prone to dissipate and cause serious backflow in the inclined pipeline. At the superficial liquid velocity of 0.0204m/s, as the superficial gas velocity raises, the enhancement of liquid-carrying cause the decrease of slug frequency. The peak value of liquid velocity has small variations since a pseudo slug is formed in the pipeline.

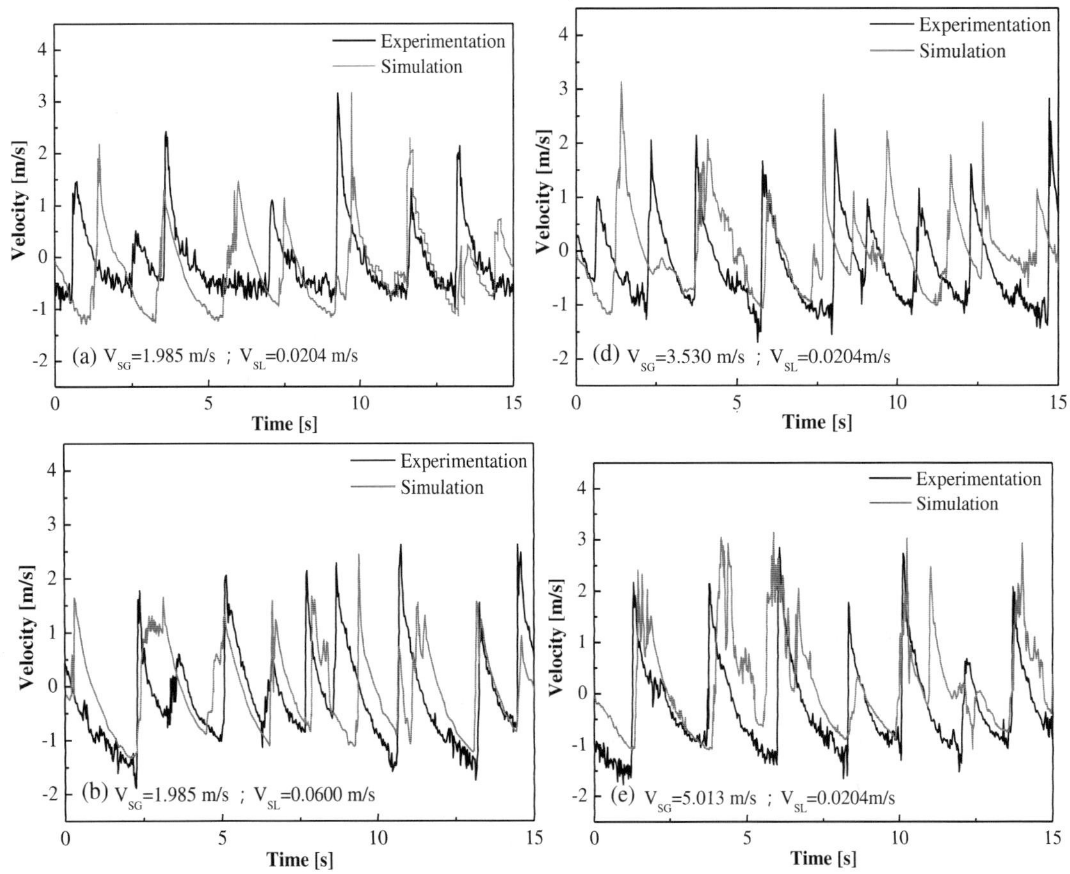

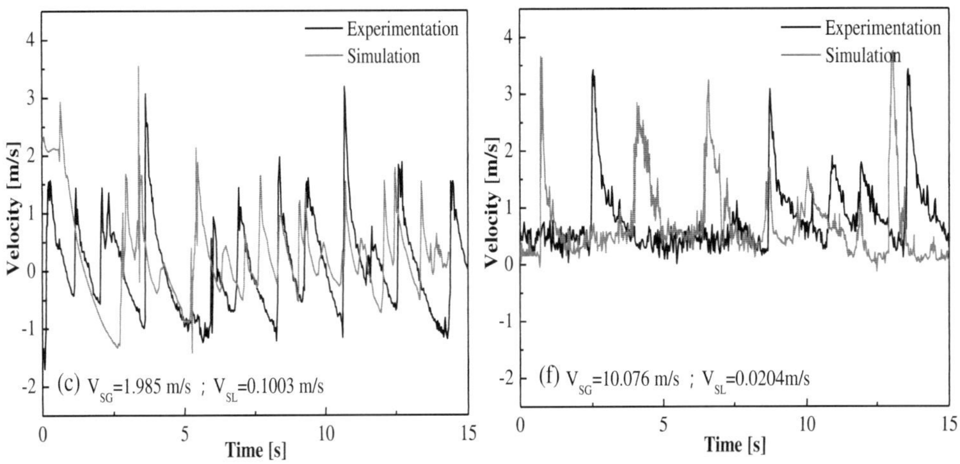

Figure 6. Experimental and simulated transient liquid velocity in the inclined pipe.

5. Conclusion

The simulation model shows a reliable prediction of transient liquid velocity behavior in hilly-terrain wet gas pipelines

The slug frequency increases with the increase of superficial liquid velocity, and decreases with the increase of superficial gas velocity. The peak value of liquid velocity, namely the translational velocity of slug flow, changes little in the research range.

Acknowledgments

This work was financially supported by national key research and development program of China (No.2016YFC0802302) and scientific research innovation projects of China University of Petroleum (16CX06004A).

References

[1] H.Q. Zhang, C. Sairca, Low liquid loading gas/liquid pipe flow, J. Nat. Gas Sci. Eng. 3 (2011) 413-422.

[2] L.F. Ayala, M.A. Adewumi, Low-liquid loading multiphase flow in natural gas pipelines, ASME J. Energy Res. Technol. 125 (2003) 284-293.

[3] S. Badie, C.P. Hale, C.J. Lawrence, G.F. Hewitt, Pressure gradient and holdup in horizontal two-phase gas-liquid flows with low liquid loading, Int. J. Multiphase Flow 26 (2000) 1525-1543.

[4] T. Sydberger, Flow dependent corrosion: mechanisms, damage characteristics, and control, Br. Corros. J. 22 (1987) 83-89.

[5] Y. Yang, J. Li, S. Wang, C. Wen, Understanding the formation process of the liquid slug in a hilly-terrain wet natural gas pipeline, J. Environ. Chem. Eng. 5 (2017) 4220-4228.

[6] Y. Takeda, Velocity profile measurement by ultrasonic Doppler method, Exp. Therm. Fluid Sci. 10 (1995) 444-453.

[7] W. Treenuson, N. Tsuzuki, H. Kikura, M. Aritomi, S. Wada, K. Tezuka, Accurate flowrate measurement on the double bent pipe using ultrasonic velocity profile method, The Japanese Society for Experimental Mechanics 13 (2013) 200-211.

[8] F. Liang, Z. Fang, J. Chen, S. Sun, Investigating the liquid film characteristics of gas–liquid swirling flow using ultrasound Doppler velocimetry, Aiche Journal 63 (2017) 2348-2357.

[9] M. Sussman, E. Puckett, A coupled level set and volume-of-fluid method for computing 3D and axisymmetric incompressible two-phase flows, J. Comput. Phys. 162 (2000) 301-337.

[10] C.R. Kharangate, H. Lee, I. Park, I. Mudawar, Experimental and computational investigation of vertical upflow condensation in a circular tube, Int. J. Heat Mass Transfer 95 (2016) 249-263.
[11] Z. Shang, J. Lou, H. Li, Simulations of flow transitions in a vertical pipe using coupled level set and VOF method, Int. J. Comput. Methods 14 (2016) 1-8.

IOP Conf. Series: Journal of Physics: Conf. Series **916** (2017) 012031 doi:10.1088/1742-6596/916/1/012031

Flow Field Analysis of Wind Power Generation System

Chao Zhou[1], Wei Huang[1], Daohua Zhu[1], Yan Li[1], Liwei Wu[2]

[1]State Grid Jiangsu Electric Power Research Institute, Nanjing, 211100, China
[2]Nanjing University, Nanjing, 210000, China

Abstract. The purpose is to solve the problem of wind power curtailment in the wind power generation system. A new type of wind power generation system made of flexible blades is proposed. Combined with the experimental method and numerical simulation method, the impeller blades are analyzed by flow field. The values of the lift coefficient of the leaves under different working conditions were obtained and analyzed. The optimum curvature and the range of the best angle of attack are determined. The influence of the parameters such as the curvature of the blade and the angle of attack on the hydrodynamic performance of the blade and impeller in the power generation system is obtained. The results showed that the best curvature K was 0.019 and the best angle of attack was 20°. Therefore, it can be concluded that it reveals the influence of blade curvature and angle of attack on the lift coefficient.

1. Introduction

As the world's energy consumption increases, carbon dioxide emissions are also increasing. It causes global warming. The resource depletion and environmental degradation have brought severe challenges to countries all over the world [3]. Since the 10th Five-Year Plan, China's total energy consumption has grown too fast. The pollutant emissions ranked the forefront of the world [7]. The greenhouse gas emissions account for 25% of the world's total. China's economy has reached a critical stage of "transition and development" [5]. Since the implementation of China's sustainable development strategy, its main line can be expressed as "science, green, low carbon." Energy conservation, efficiency improvement and reasonable control of energy demand are important tasks of energy strategy [2].

With the continuous increase of energy consumption, the wind power industry has developed rapidly. Wind energy resources are abundant and widely distributed in china. It is mainly distributed in the northern and southeastern coastal areas. According to the evaluation of Chinese Academy of meteorological sciences, China's wind energy reserves is about 253GW [1]. Offshore wind energy is far more abundant than land-based wind energy, which is about three times more than it. Therefore, as a clean and renewable energy, wind energy has important value in the sustainable development.

2. State of the art

Since 2006, the installed capacity of wind power generation around the world has shown an upward trend [4]. From 2007 to 2009, the growth rate of new installed capacity of the whole world is larger, and the maximum can reach 45%. The maximum installed capacity can reach 32%. Since 2010, due to the impact of the global economy, the growth of new installed capacity has slowed down. The growth rate of the total installed capacity has dropped to 20%. By the end of 2011, the global cumulative installed capacity of wind power generation reached 237669MW. China's new installed capacity growth has gradually slowed down. The number of new installed capacity is 17630.9MW. The cumulative installed capacity of China is 62364.3MW. It indicates that China's wind power industry from the rapid development stage to the adjustment period in 2011 [6].

During the working of electric system, the hydropower station generating unit is in the functions of peak load regulation of power grid and emergency standby. Hydro turbine generating unit has two kinds of braking methods: electric braking and mechanical braking. In general, the mechanical braking method creates the braking effect through the fraction drag generated by the touching of brake valve and brake ring, whereas the electric braking works through consuming the main power in a form of electricity and generating reverse electromagnetic force in the motor. The wind power generation system has a wide range of applications. In the 80s of the last century, a great deal of research was done abroad. However, the early development of wind power generation system has a high risk, which led to private companies do not want to take risks alone. Therefore, the research funding of MEMS gyroscope mostly comes from the government abroad. The United States relies mainly on universities [5]. The main MEM research university is Stanford University, Georgia Institute of Technology, University of California, Los Angeles, and Dreiberi Laboratory (CSDL). Europe has a national research in France LETI (Information and Electronic Technology Laboratory). Germany has a non-profit Fraunhofer Institute and the German Institute for wind power generation system. In addition, there are some universities and research institutes in Switzerland. In 2002, ADI published a gyro system program. The gyro adopts the capacitive micromechanical gyroscope of the body processing method. The micro machined structure and the readout circuit are integrated on the same substrate

The generating unit employs traditional braking method: mechanical braking method, which is reliable, convenient and widely applicable, with low energy consumption. Its braking effect will not be influenced by faults like power interruption or short-circuit of line owing to its high reliability and safety. However, this mechanical braking method also has some shortcomings, for example, the powder generated during the braking process may enter the ventilation ducts, and powder accumulation during long periods of time will downsize the cross-section area of ventilation ducts and result in loss in cooling effect. During the process of braking, the surface temperature of the braking ring will increase rapidly, leading to thermal distortion and even crack. The airlock sometimes cannot fall itself, so our personnel will inspect the wind tunnel at every stop of it, which also lowers its level of automation.

The increase of installed capacity of wind power will inevitably lead to the increase of power generation in wind power generation system. However, due to the seasonal and environmental factors, wind power has its own volatility and randomness, which has a certain impact on grid connection. At the same time, the wind resources are unevenly distributed, and the large wind field is far away from the load center. Due to the limitation of peaking capacity, the wind power consumptive capacity is weakened. The phenomenon of wind power rationing has become a prominent problem. In order to solve this problem, in the "12th Five-Year" period, the government put forward the requirements of wind power consumption. The relevant policies have been introduced to support decentralized development of wind power generation. The development of small-scale distributed wind power generation is actively promoted. Therefore, a new type of wind turbine generator has been developed, whose blade is made of flexible material [10]. When the force is different, the curvature of the blade will change in different degrees, and the lift effect of the fluid can be taken into full use to extract the wind energy. At the same time, a wind collecting hood with wind collecting effect is arranged outside the impeller of the wind power generation system to increase the wind speed at the entrance of the fan. It provides more power for the wind turbine to do its work. The wind turbine has the advantages of simple structure, low starting wind speed, and strong mobility and flexibility. The utility model is suitable for providing power generating power for residents in the covered blind area of China's power grid, and establishing a separate power supply system, which solves the regional non-electric problems caused by the remote area and the difficult construction of the power grid [9]. At the same time, it can also take full advantage of the local wind resources, to avoid the wind power curtailment condition and other problems.

3. Methodology

In all wind power system equipment, the most important component is the impeller part. The impeller is the most important part that affects the efficiency of wind turbine. The shape of the blade on the impeller directly affects the aerodynamic performance of the blade. As the core part of the wind turbine, the pneumatic performance of the blade directly dominates the conversion efficiency of the wind turbine.

3.1. Experimental study method

The experimental object consists of two cascades. Each leaflet contains 3 identical blades. Taking the middle of the leaves as a test blade, the test blade is made of rigid material. The span length is 200mm, the section arc length is 60mm, the thickness is 3mm, and the curvature of each group is different. At the face of the blade pressure and suction, the ten holes are drilled evenly and inserted into the piezo metric tube and numbered [8].

The ambient temperature is 25 ℃, the atmospheric pressure is 88.92kpa, the air density is 1.04Kg / m^3, and the wind speed v_ of the closed experimental section is set to 5m / s. There were 5 kinds of leaves, and each group of leaves were tested under 6 angles of attack. The experimental conditions are shown in Table 1.

Tab.1 Experimental condition

K	*α*/°	*V* /m/s	d/mm	h/mm	Total number of test groups
0	5				
0.009	10				
0.013	15				
0.017	20	5	200	70	30
0.026	25				
	30				

3.2. Numerical simulation method

CFD is an abbreviation for Computational Fluid Dynamics. Through the numerical calculation of the computer, the numerical simulation method is used to replace the fluid experiment, and the fluid flow state and the heat conduction are solved and analyzed.

The cascade geometry is established, and the cascade structure is the same as that of the cascade in the test experiment. They are composed of two cascades. Each leaflet contains 3 identical blades. The blade length is 200mm, the section arc length is 60mm, and the thickness is 3mm. The size of the calculated domain is the same as that of the experimental wind tunnel. The distance between the inlet end and the outlet end is 3000mm. Width and height are 900mm.

The geometric model is meshed. The unstructured grid is more practical for boundary complex problems. Therefore, the tetrahedral unstructured mesh is used as the basic element, and the mesh of the mesh surface is refined. The total number of total domains is 3.1 million. The quality of the grid is checked, and the distortion of Skewness is 0.25. The maximum value is 0.42. Therefore, the quality of the grid is considered to meet the calculation requirements. The fluid is set to air and the reference back pressure is set to 88.92 kPa. The inlet boundary is set to the velocity inlet boundary and the inlet speed is 5 m / s. The exit boundary is set to free outflow. The other surfaces of the blade and the computational domain are non-slip wall boundary conditions, and they are adiabatic. For turbulence computation, the RNG model is adopted, and the second-order upwind scheme is chosen in the difference scheme to ensure a high computational accuracy.

The straight blade and the blade with 0.017 curvature are chosen as the object to carry out numerical simulation, so as to obtain the pressure difference on the blade. Then, it is substituted into the lift coefficient calculation formula to solve the lift coefficient value. The lift coefficient is compared with the experimental results. The comparison results show that the simulation results are basically consistent with the experimental results, and the relative average error is about 5%. Thus, it is considered that the numerical simulation method is correct. Through the experimental study of the lift characteristics, the range of the optimum curvature and the optimum angle of attack are found. In order to accurately determine the optimum working conditions of blade, 14 kinds of blades with curvature between 0.013 and 0.026 are chosen as the object of study, and three kinds of angle of attack are chosen at 15°, 20° and 25° respectively. The specific calculation conditions are shown in Table 2, and the number of calculation groups is 42. By numerical simulation, the optimum blade shape and optimum angle of attack are obtained by comparing the magnitude of each lift system.

Tab.2 working condition of calculation

K	*α*/°	*V* /m/s	d/mm	h/mm	Total number of test groups
0.013~0.026	15/20/25	5	200	70	42

4. Result analysis and discussion

4.1. Experimental results

Through the blade surface pressure test, the pressure difference F in the normal direction of the blade surface was obtained. From the previous analysis, the blade F deformed by the force F can be decomposed into the horizontal direction of the component and vertical force. Assuming that each test point on the blade is a micro-area, the component F_{Li} of the pressure difference F_i in each vertical area is obtained experimentally. The F_{Li} is brought into the lift coefficient calculation formula (1), and the lift coefficient on the region is approximated. Finally, the average value C_L of the lift coefficient of the 10 regions is obtained, and the advantages and disadvantages of the lift characteristics are represented.

$$C_L = \frac{F_L}{\frac{1}{2}\rho v_{\infty}^{2} c} \tag{1}$$

Leaf lifting coefficient C_L is influenced by curvature K and angle of attack α. The law of the lift coefficient with the angle of attack is shown in Figure 1 and Figure 2.

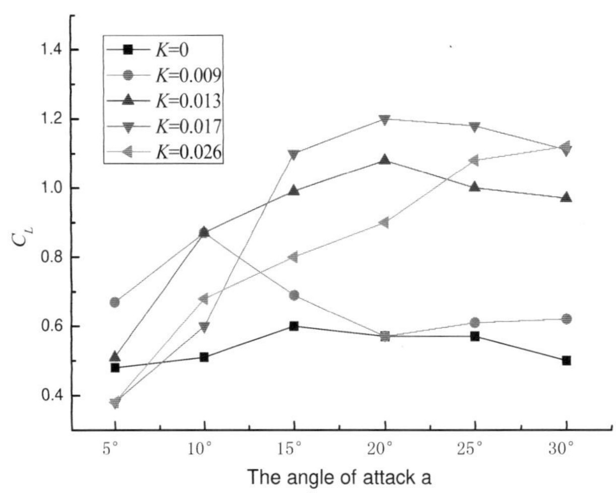

Fig.1 The relationship between lift coefficient and attack angle on layer blade

As can be seen from Figure 1, the lift coefficient of all leaf types increases first and then decreases with the increase of the angle of attack α. The angle of attack of the maximum blade and the maximum lift coefficient of curvature of 0.009 is less than 15°. The other three kinds of leaf lift coefficient corresponding to the maximum angle of attack are greater than 20°. The lift coefficient is much larger than the first two leaves. The leaf with a curvature of 0.017 has the largest lift coefficient at an angle of attack of 20°.

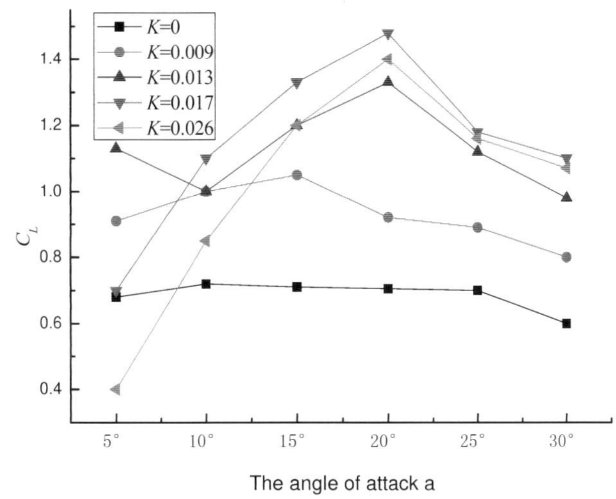

Fig.2 The relationship between lift coefficient and attack angle on back blade

As can be seen from Figure 2, the value of the lift coefficient of the back leaves is similar to that of the front layer, and they all increase first and then decrease. When the angle of attack is the same, the lift coefficient of the later blade is larger than the lift coefficient of the front blade. Combined with the above two figures, the best angle of attack is between 15° and 25°.

Figure 3 and Figure 4 is the lift coefficient with the curvature of the law. When the angle of attack is fixed, with the increase of the curvature of the blade, the lift coefficient increases first and then decreases. From the figure, the curvature of the blade is 0.017, and the lift coefficient is the largest, so the best curvature appears between 0.013 and 0.026.

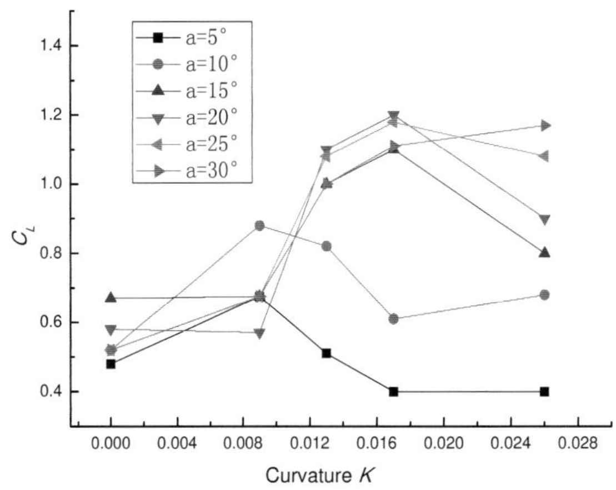

Fig.3 The relationship between lift coefficient and curvature on layer blade

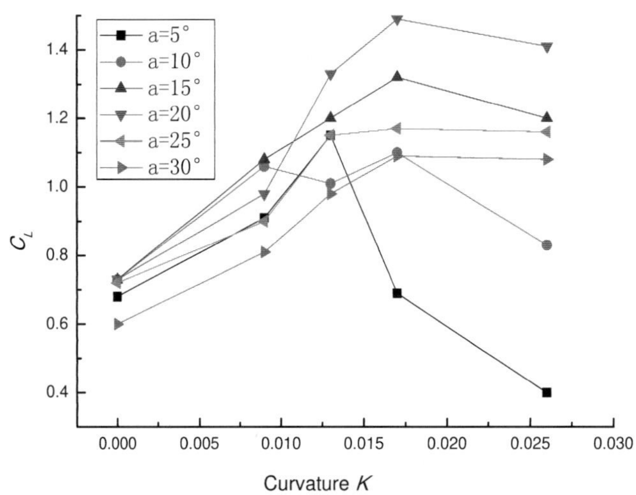

Fig.4 The relationship between lift coefficient and curvature on back blade

Through the above analysis, the lift coefficient of the back layer leaves is larger than the lift coefficient of the front layer leaves under the corresponding working conditions. This indicates that the lift characteristics of the late leaves are better than those of the front leaves. When the curvature of 0.017 leaves at an angle of attack of 20°, the lift coefficient of the front and back leaves is the largest in all working conditions. Thus, the curvature of 0.017 and the angle of attack of 20° were studied.

4.2. Numerical simulation results

The numerical values of the lift coefficient of each group were calculated by numerical simulation. Figure 5 and Figure 6 show the influence of the curvature and the angle of attack on the lift coefficient.

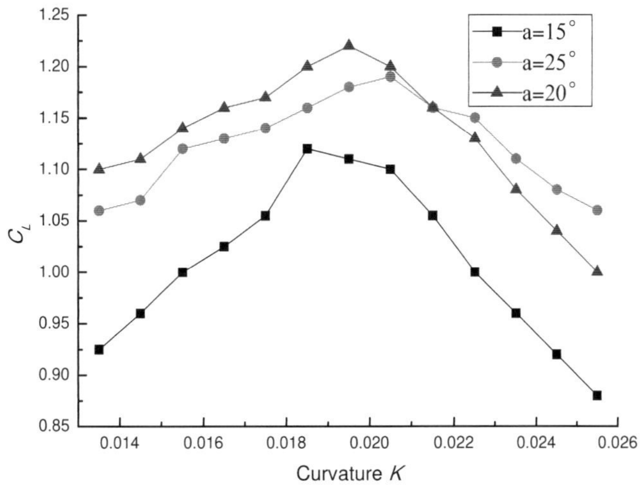

Fig.5 Relationship among lift coefficient, curvature and attack angle on layer blade

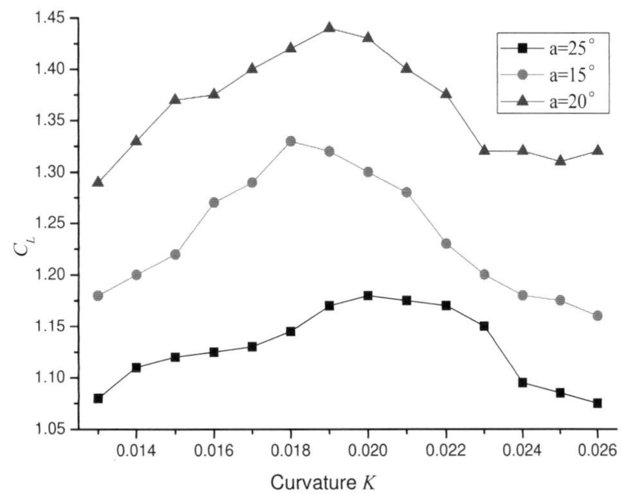

Fig.6 Relationship among lift coefficient, curvature and attack angle on back blade

From the figure, when the angle of attack is 15°, the best curvature is 0.018. When the angle of attack is 20°, the best curvature is 0.019. When the angle of attack is 25°, the best curvature is 0.020. Thus, with the angle of attack increased from 15° to 25°, the best curvature from 0.018 to 0.020 gradually transition. In all the calculated cases, when the curvature of 0.019 leaves is at an angle of attack of 20°, the lift coefficient reaches the maximum. As can be seen from Figure 5, when the angle of attack is 15°, the lift characteristics of the front blade are the worst. When the angle of attack is 20°, the lift characteristics are in the best condition. When the angle of attack is 25°, the lift characteristic is at the normal level. As can be seen from Figure 6, when the angle of attack is 25°, the lift characteristics of the back leaves are the worst. When the angle of attack is 20°, the lift characteristics are in the best condition. When the angle of attack is 15°, the lift characteristic is between 20° and 25°. In summary, the best curvature of the blade K is 0.019, and the best angle of attack is 20°.

5. Conclusion

Through the combination of experiment and numerical simulation, the flow field analysis of flexible blade wind turbine impeller in new wind power generation system is carried out. The pressure values obtained by experimental tests. The lift coefficients of the blades under different working conditions are obtained and compared, and the optimum curvature and the optimum angle of attack are determined. The results show that the optimum curvature appears between 0.013 and 0.026, and the optimum angle of attack is between 15° and 25°. In the experimental conditions, when the curvature of 0.017 leaves at an angle of attack is 20 °, the lift coefficient reaches its maximum. In order to determine the optimum working conditions of the blade more accurately, 14 kinds of blades are selected in the optimum range of curvature, and three angles of attack are set at 15°, 20° and 25°. A numerical simulation model was established. The numerical simulation is carried out and the correctness of the model is verified. By analyzing the results, the best curvature K of the leaves is 0.019, and the best angle of attack is 20°. It reveals the influence of blade curvature and angle of attack on the lift coefficient.

References

[1] Yao, Jun, et al. "Enhanced control of a DFIG-based wind-power generation system with series grid-side converter under unbalanced grid voltage conditions." IEEE Transactions on power electronics 28.7 (2013): 3167-3181. https://doi.org/10.1109/tpel.2012.2219884

[2] Li, Rui, and Dianguo Xu. "Parallel operation of full power converters in permanent-magnet direct-drive wind power generation system." IEEE Transactions on Industrial Electronics 60.4 (2013): 1619-1629. https://doi.org/10.1109/tie.2011.2148684

[3] Syahputra, Ramadoni, Imam Robandi, and Mochamad Ashari. "Performance analysis of wind turbine as a distributed generation unit in distribution system." International Journal of Computer Science & Information Technology 6.3 (2014): 39. https://doi.org/10.5121/ijcsit.2014.6303

[4] Lodermeyer, Alexander, et al. "Phase-locked flow field analysis in a synthetic human larynx model." Experiments in Fluids 56.4 (2015): 77. https://doi.org/10.1007/s00348-015-1942-6

[5] Balduzzi, Francesco, et al. "Critical issues in the CFD simulation of Darrieus wind turbines." Renewable Energy 85 (2016): 419-435. https://doi.org/10.1016/j.renene.2015.06.048

[6] Borg, Michael, Andrew Shires, and Maurizio Collu. "Offshore floating vertical axis wind turbines, dynamics modelling state of the art. Part I: aerodynamics." Renewable and Sustainable Energy Reviews 39 (2014): 1214-1225. https://doi.org/10.1016/j.rser.2014.07.096

[7] Jeon, Minu, Seungmin Lee, and Soogab Lee. "Unsteady aerodynamics of offshore floating wind turbines in platform pitching motion using vortex lattice method." Renewable Energy 65 (2014): 207-212. https://doi.org/10.1016/j.renene.2013.09.009

[8] Tummala, Abhishiktha, et al. "A review on small scale wind turbines."Renewable and Sustainable Energy Reviews 56 (2016): 1351-1371. https://doi.org/10.1016/j.rser.2015.12.027

[9] Li, Qing'an, et al. "Study on power performance for straight-bladed vertical axis wind turbine by field and wind tunnel test." Renewable Energy 90 (2016): 291-300. https://doi.org/10.1016/j.renene.2016.01.002

[10] Mohamed, M. H., A. M. Ali, and A. A. Hafiz. "CFD analysis for H-rotor Darrieus turbine as a low speed wind energy converter." Engineering Science and Technology, an International Journal 18.1 (2015): 1-13. https://doi.org/10.1016/j.jestch.2014.08.002

IOP Conf. Series: Journal of Physics: Conf. Series **916** (2017) 012032 doi:10.1088/1742-6596/916/1/012032

Fault Study on Valve Based on Test Analysis and Comparision

Li Cheng, Wukui Yang*, **Tao Liang, Yu Xu, Chao Chen**

Airforce Engineering University, Xi'an710038, China

*Corresponding author e-mail:yangwukui1975@163.com

Abstract. The valve of a certain type of small engine often has the fault phenomenon of abnormal vibration noise and can't close under the specified pressure, which may cause the engine automatic stop because of valve incomplete close leading to fuel leakage during test and startup on the bench. By test study compared to imported valve with the same use function and test condition valve, and put forward the thinking of improving valve structure, compared no-improved valve to improved valve by adopting Fluent field simulation software. As a result, improved valve can restore close pressure of valve, restrain abnormal vibration noise phenomenon, and effectively compensate compression value of spring because of steel ball contacting position downward with valve casing.

1. Introduction

As a common mechanical equipment in the flow conveying and control system, the valve mainly regulates and controls the flow rate of the fluid mass, the pressure and the flow direction to meet the requirements of the working system[1]. As the components of the hydraulic system, the variety and quantity of the valve are quite large proportion, so it is necessary to study the valve fault or failure.

At present, in order to suppress the noise is generated by the valve, the research on the valve mainly shows the theory and design of the noise[2,7], The research object is mainly for the valve shell, but few for the force of spool named functional parts, the influence analysis of flow field and fault pieces repair. So it is of great significance to carry out this research.

This type of valve mentioned in the paper often test failed and cause automatic stop failure because of being not closing fully during test on the bench, so the repair scientific research has been carried out, on the basis of structure and test comparison with the valve of introduction, put forward the technical program of the valve repair structure improvement. The valve is easy to cause the valve closed pressure deviation from the design state after use of a TBO life.

2. Structure improvement of the fault valve

2.1. Prototype valve

The prototype valve structure mentioned in this paper is shown in Fig. 1, an engine is fitted with a valve. The working principle of the valve as shown in Fig. 2, the valve assembly as shown in Fig.3, when the small engine startup to work, with the system fuel pressure increasing, the fuel begin to push the ball (spool) to overcome the spring force work, the valve will close in the fuel pressure 15kPa to

25kPa, then the system fuel enter internal of the small engine for combustion by the fuel screw. When the engine stop, the valve will gradually open, the system fuel will change the direction of flow, discharge residual fuel from the small engine casing through the valve. If the valve is vibration screaming during the work, the valve cannot be completely closed, and cannot supply the amount of fuel required for the engine to work, finally causing engine unsuccessful startup or automatically stop.

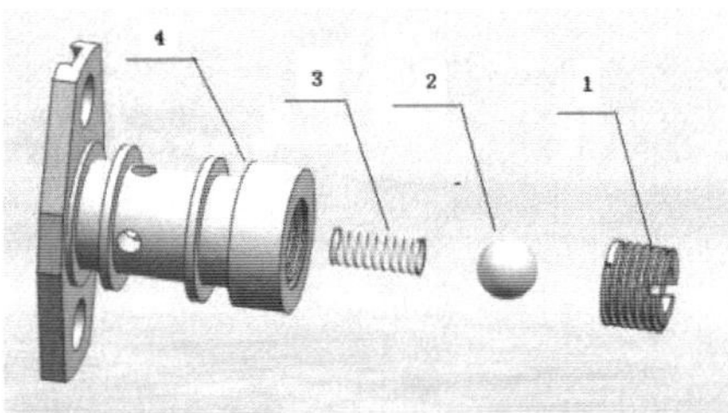

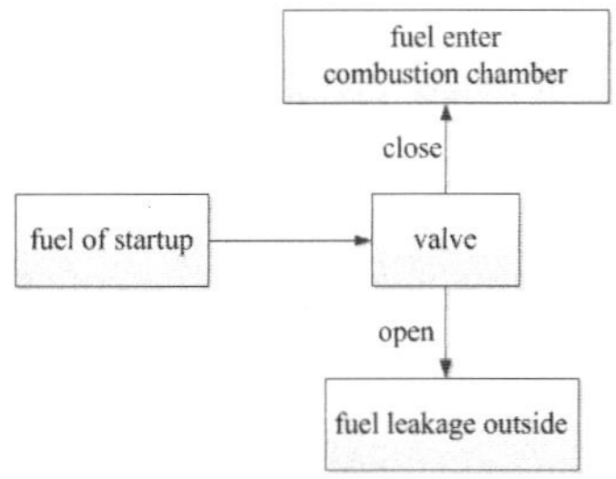

1 Screw, 2 Steel ball, 3 spring, 4 Shell

Fig.1 Schematic diagram of valve structure **Fig.2** Schematic diagram of valve work principle

Fig.3 Cross-section schematic diagram of valve assembly on the engine

The brief diagram of the prototype valve structure at work as shown in Fig. 4, the fuel flow from inlet P_1 to outlet P_2, when the hydraulic pressure before the steel ball (spool) is greater than the resultant force of hydraulic pressure in the casing P_2 and spring together after the steel ball (spool), the ball (spool) will move back to gradually close under the fuel pressure difference, and finally inlet P_1 and outlet P_2 is not connected. If omit the movement friction resistance of the ball (spool), it must meet the following relationship.

$$p_1A_1 > p_2A_2 + F_H \tag{1}$$

Where, p_1 —fuel inlet pressure or control pressure, A_1—the projection area of the steel ball in the horizontal direction under P_1, p_2—fuel outlet pressure or controlled pressure, A_2—the projection area of the steel ball in the horizontal direction P_2, F_H —spring force.

When the ball (spool) is just closed, the value P_2 is reduced to 0kPa, and the support force of the valve cone surface face steel ball (spool) is 0N, the amount of spring compression reaches the maximum, and the spring force reaches the maximum, then the Eq.1 becomes.

$$p_1A_1 > F_{H\max} = \mathrm{K}\Delta L_{\max} \tag{2}$$

Where, $F_{H\max}$ —maximum force of spring, K —spring elasticity coefficient, $\Delta L_{\max}$ —Maximum compression of spring.

It can be seen from the Eq.2, the minimum position of the spring compression is further reduced with the valve closing because of the tangential position of the steel ball (spool) and the cone surface moving downward in the valve, that is value $\Delta L_{\max}$ further increases, the spring force increases, the right side of the Eq.2 becomes larger, when reaches a certain amount, the Eq.2 may not exist to establish.

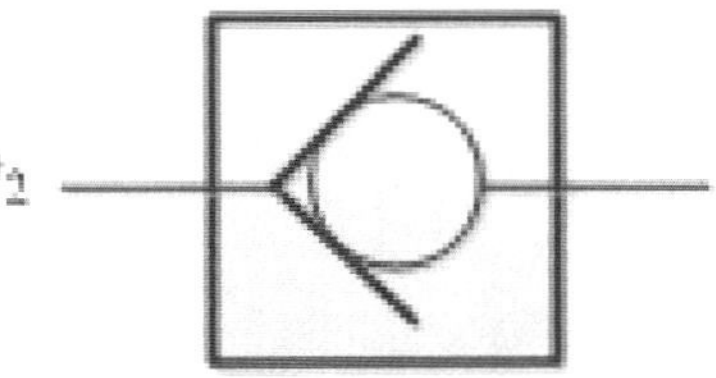

Fig.4 Schematic diagram of valve simplification symbol

2.2. Sprototype failure valve

During the repair process, the valve parts are further disassembled and checked. Under the premise of spring and ball qualified, it is found that there was a slight tendency to increase the cone angle of the inner surface of the valve, as shown in Fig. 5.After analysis, the reason is that the material for the stainless steel ball repeatedly collision and contact with the cone surface inside the aluminum shell during valve use a TBO time, because the material of aluminum shell is soft, the collision process can cause micro-deformation, resulting in tangential position of the steel ball and cone surface downwards, the amount of spring compression increased, causing the spring force much more when the valve closing, the corresponding fuel closure pressure also increased.

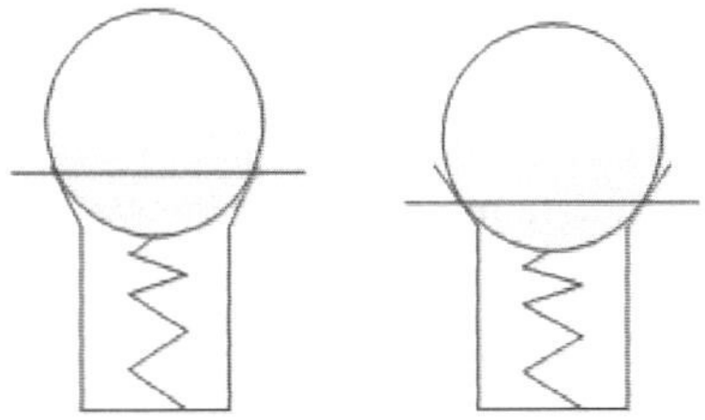

Fig.5 Comparison schematic diagram of steel ball tangent to valve casing

2.3. Improved valve

Compared with the valve of introduction and the prototype valve with both the same function of use and the same parameters of test, the steel ball (spool) technical status of two kinds of valve are the same, the difference is maily as following, First, the spring of the introduction valve is in three groups, but the spring of prototype valve only one, Second, the shell depth of introduction valve is about 1.0mm deeper than the prototype valve. To sum up the reasons above, the pressure adjustable range of introduction valve is wider, and the shell is deeper, spring compression deformation amount is relatively small, which explains test pass rate of the introduction valve is much more than the prototype valve.

According to the above analysis, in order to restore the function of the failure valve, the spring or the shell must be improved, but because the processing of spring improvement is complex, so we can refer to the depth size of the introduction valve shell, processing 1.0mm deep for the prototype valve shell. At the same time, repair and surface treatment the cone surface of the failure valve, the processing flow chart of improved valve is shown in Fig. 6.

In order to avoiding depth processing of the shell is too large, and the valve closing pressure is less than the lower limit of the technical documents, in the specific engineering applications, only the upper limit of the valve closing pressure as a reference object to calculate the amount of processing, the depth processing amount of the shell as following.

$$\Delta L = \frac{(P_1 - 25)A_1}{K} \quad (0 < \Delta L < 1.0\text{mm}) \tag{3}$$

In the Eq.3, P_1 is the unqualified pressure measurement value when the valve is closed.

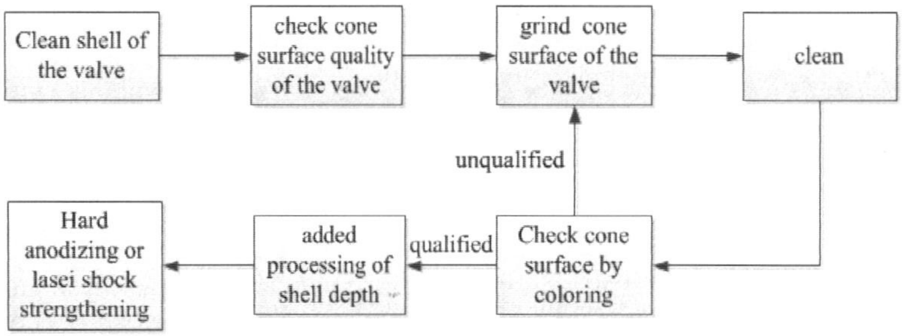

Fig.6 Flow diagram of improved valve process

3. Simulation analysis of internal flow field before and after Improvement of Valve

3.1. Physical model establishment

According to the product structure, the establishment of steel ball (spool) valve model shown in Fig. 7, and grid shown in Fig. 8. Set the steel ball (spool) opening value is 0.3mm at actual work, when the fuel inlet pressure P_1 is 20kPa,under 20℃ fuel temperature condition, the dynamic viscosity $\mu = 2.500\text{N} \cdot \text{s} / \text{m}^2$, density $\rho = 0.785 \times 10^3\text{kg} / \text{m}^3$. Fluent software is used to respectively simulate anslysis and compare flow field before and after the valve improvement, the simulation is steady-state simulation, and total iteration 20 thousands steps.

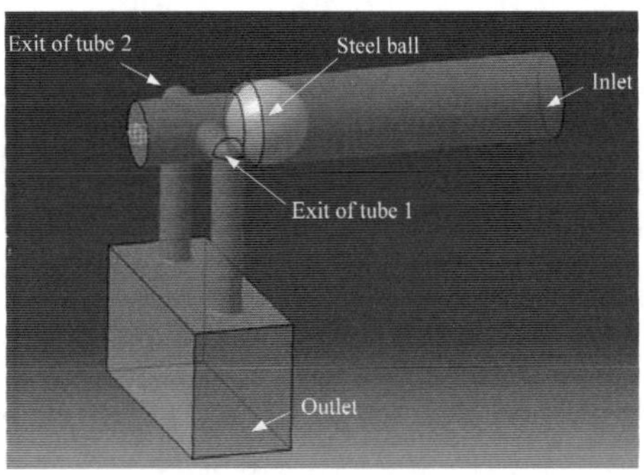

Fig.7 Model of simply valve structure

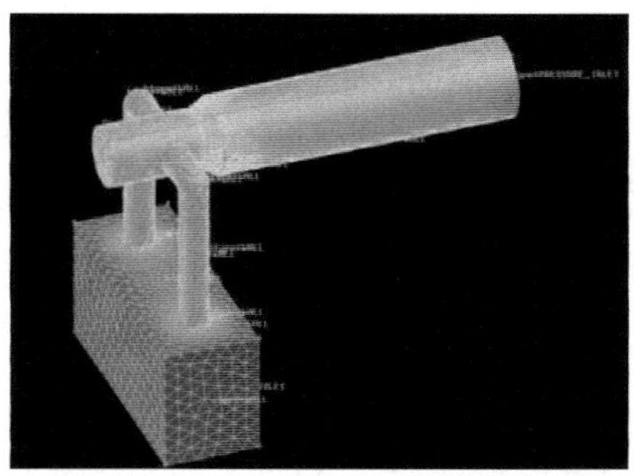

Fig.8 Schematic diagram of grid

3.2. Set boundary condition

Inlet boundary: opening x = 0.3mm, pressure inlet P_1 = 20kPa;

Exit boundary: pressure outlet P = 0kPa(relative pressure).

The process of the ball (spool) opening from the maximum rapid drop to 0.3mm and the valve on the fuel blocking are abstracted same as transient flow in the opening 0.3mm of the valve, transient simulation is based on stabilized simulation and initial condition for iterative calculation.

3.3. Analysis results of flow field

It can be seen from Fig. 9 and Fig. 10,when the opening of the steel ball (spool) is 0.3mm, the inlet fuel is encountered by the steel ball , making the hydraulic pressure contact with the steel ball (spool) is the largest, while the speed is the smallest, because of continuous inflow of inlet fuel into the valve casing, making the fuel into the channel between the steel ball (spool) and the cone surface, finaliy expansion to discharge, so the pressure is continuous reduced to the minimum, speed continues to increase, and finally to the maximum.

It can be seen from Fig. 9 and Fig. 11 ,when the opening of the steel ball (spool) is 0.3mm, the fuel is contracted and expanded inside the valve because of the obstruction of the steel ball (spool), causing the fuel to be compressed before the steel ball and forming partial high pressure zone,while the fuel to be expanded after the steel ball and forming partial low pressure zone, because of the influence of the cone taper and the spherical geometry of the steel ball, the fuel is separated at the edge of the rear surface of the valve channel, resulting in eddy current, and the turbulence intensity increases first, then decreases, finally reaches the strongest in the valve cone channel because of throttling.

At the same time, it can be seen that the pressure cloud diagram, velocity cloud diagram and turbulence intensity are almost unchanged before and after improvement of the valve, that is compared with before improvement of valve, when the opening degree is the same as before the improvement, improved valve does not affect and change the original internal flow field distribution essentially. Therefore, in the valve closed fuel pressure recovery and improvement, the key consideration to eliminate the extra spring force is mentioned, indirectly proved that the Eq.3 is similar to the actual rationality, that is, the results of calculation is consistent with the actual compensation trend.

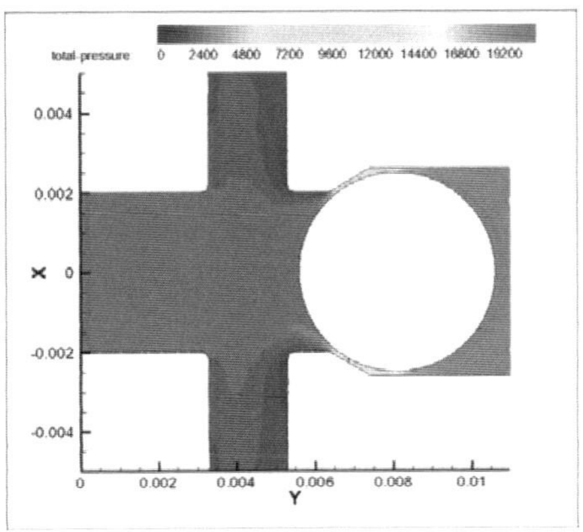

(a) Before improvement

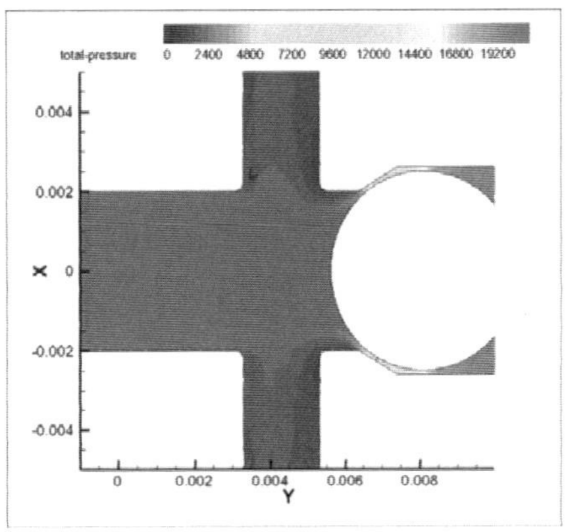

(b) After improvement

Fig.9 Comparision cloud schematic diagram of total pressure

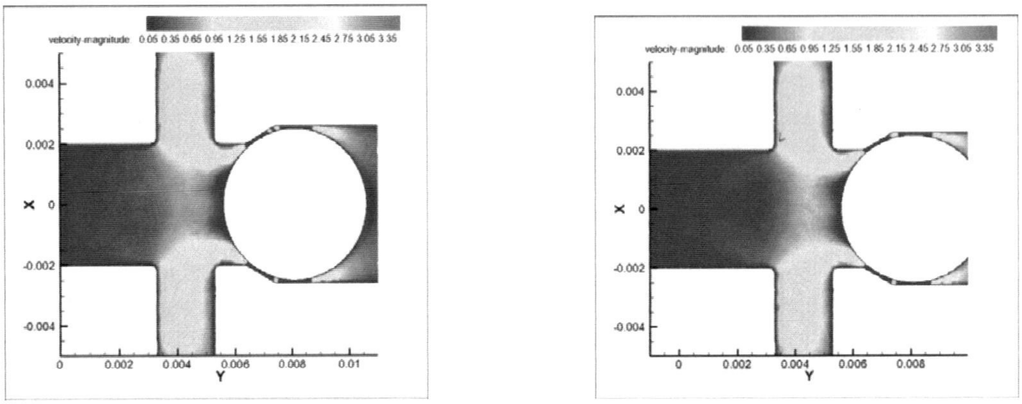

(a) Before improvement (b) After improvement

Fig.10 Comparision cloud schematic diagram of speed

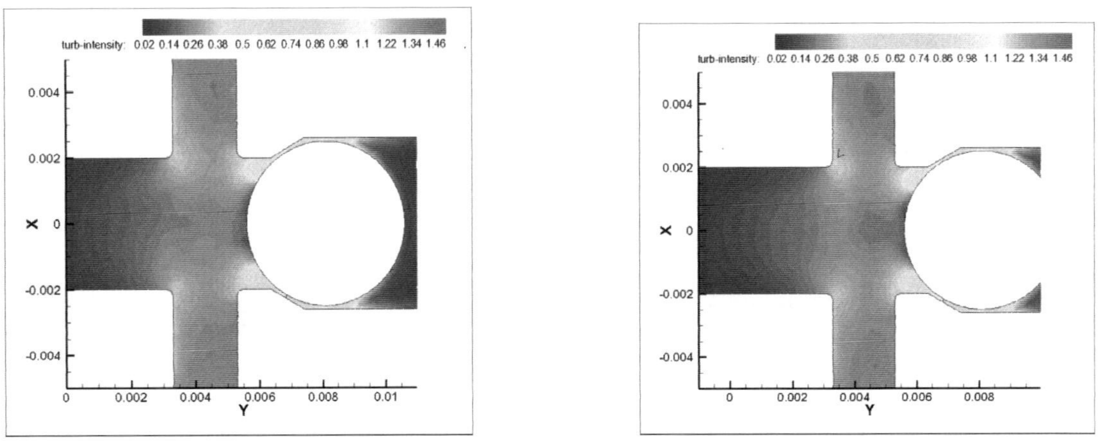

(a) Before improvement (b) After improvement

Fig.11 Comparison cloud schematic diagram of turbulence intensity

4. Test content and comparison analysis

4.1. Close pressure test comparison

Respectively select a new shell of the introduction valve, a new shell of domestic prototype valve, a failure shell of domestic prototype valve and a improved shell of domestic prototype failure valve that processing depth 1.0 mm, have valve closed tests for 30 times respectively, the test results as shown in Fig. 12.

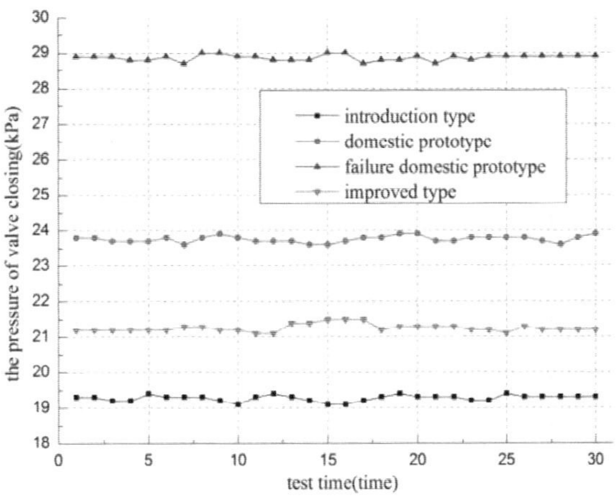

Fig.12 Schematic diagram of different valve close test

Test result of Fig.12 shows that, closing pressure test of the new part of domestic prototype valve is qualified, after use a TBO time, under the premise of the spring, steel ball technical state meeting the original design requirements, the valve closing pressure becomes larger and beyond the technical documents value 15kPa to 25kPa, proved that cone surface of the shell changed in the course of use ,at last affects the valve pressure. Closing pressure test of the improved valve is qualified, It is proved that the improvement of the depth of the shell can improve the closing pressure of the valve, and the improvement measures are effective. At the same time found that as long as the valve closing pressure is qualified, vibration & screaming phenomenon will be disappeared in the test process.

4.2. *Test verification*

Select the three pieces of failure valve, under the premise of the spring mechanical properties, the appearance quality of the steel ball qualified, calculate the amount of processing of the valve shell by the Eq.3, have the test for 50 times respectively, select the average closing pressure to test data comparison, as shown in Table1 and Fig.13.

Table.1 Statistical list of fault removal and test checking

valve No.	unqualified closing pressure(kPa)	processing amount(mm)	average closing pressure(kPa)
1	27.5	0.25	23.1
2	29.0	0.30	23.4
3	30.0	0.33	23.5

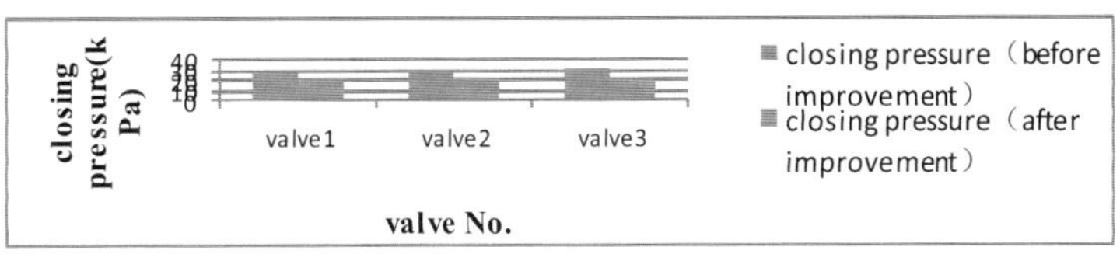

Fig.13 Schematic diagram of no-improved valve close test comprision to improved

The results of the test results in Table1 and Fig. 13 show that the processing amount of the valve carried out according to the Eq.3 can achieve the normal closing of the valve, and reflect the reason that the valve closing pressure is less than 25kPa compensation because not considering the friction and hydrodynamic forces of the steel ball (spool) in the Eq.3 [8], but the hydrodynamic force in the actual working process is the same direction as the steel ball (spool), and is greater than the friction force, that is to say, the hydrodynamic force urges the spool to close , So the actual closing pressure value is actually less than the theoretical compensation target value 25kPa. At the same time found that vibration & screaming phenomenon of the valve disappeared during the test process. Installation the three valves for the test,three types of small engine all can successfully startup, performances are stable and the valve can normal open or close. After the test run, inspeced and found no damage on the cone surface of three valve shells.

According to the test results, the root cause of the failure is that the steel ball in the valve repeatly contact with the shell cone surface in the work process, the shell comes up micro-deformation because of collision, the tangent position of the steel ball and the shell moves downwards, opening range of the valve becomes widened, and resulting in instability of the valve work.

5. Conclusion

Through the test comparison, the following conclusions are drawn:

1) The failure analysis of the valve is reasonable, the formula deducted of the valve processing is effective, more accurately reflect the work of the valve characteristics.

2) Improved valve can effectively solve and compensate for the problem of tangent position moving downwards of the steel ball (spool) and the cone surface.

3) Through the test, the test parameters of the improved valve are verified. Through the comparison of the test, the closing pressure of the improved valve is less than the prototype valve and the failure valve, but still higher than the introduction valve.

4) The improved valve can meet the functional test requirements, stable work and reliable performance in the process of use, so the effect of improvement is obvious.

References

[1] LIU cuiwei, LI yuxing, WANG wuchang, XIE ZHUZHU.Analysis on the mechanism of aero-acoustic noise generated by gas flow through valves of natural gas pipelines[J].Journal of Vibration and Shock,33(2):152-157.

[2] MAO weiping. Hydraulic Valve.Beijibng, Chemical Industry press, 2009.

[3] Powell A.Theory of vortex sound [J]. Journal of the Acoustical Society of America.1964,36(1):177-195.

[4] Francesco P.,Marco B.,Antonio Z.,Matteo S.Experiments on shells under base excitation [J]. Journal of Sound and Vibration,2016, 369 :209-227.

[5] V.D. kubenko,P.S. Koval'chuk.Influence of initial geometric imperfections on the vibrations and dynamic stability of elastic shells[J].international Applied mechanics.2004,40(8):847-877.

[6] L.R.Koval,Effect of longitudinal resonance on the paramemetric stability of an axially excited cylindrical shell [J]. The Journal of the Acoustical Society of America.1974,55(1) :91-97.

[7] M.strozzi,L.I. Manevitch,F.Pellicano V.V.Smirnov,D.S.Shepelev.Low-frequency linear vibrations of single-walled carbon nanotubes[J]. Journal of Sound and Vibration.2014, 333(13) :2936-2957.

[8] LI yulin. Hydraulic Component and System Design [M]. Beijing, Beijing University of Aeronautics and Astronautics Press, 1991.

Effect of Cryogenic Distillation and Chemical Absorption on the Argon Concentration in Krypton Purification

Zhiliang Wang, Gaoping Feng, Mingdong Wang and Yanji Hong *

Space Engineering University, Beijing, China

*Corresponding author e-mail: hongyanji@vip.sina.com

Abstract. Cryogenic distillation and chemical absorption are two important means of inert gas purification. Experiments show that the order of the two methods has an effect on the final purification result. In order to verify the relationship between the retention of argon and the order of the two methods, experimental verification was designed. At the same time, the numerical simulation was carried out by using the principle of low temperature distillation, and the influence results of the two methods on the argon content after gas purification was obtained.

1. Introduction

Gas cryogenic distillation refers to the enrichment and separation of one or more components of the mixed gas in the conditions of different low temperature by using different saturated vapor pressures of different gases. During the distillation process under low temperatures, the following to meet the relationship is met all the time[1]:

$$n_i = P_{i,0} X_i \frac{V_g}{RT} + LX_i$$
$$V_{tot} = \frac{(\Sigma n_i - L)RT}{\Sigma P_{i,0} X_i} + \Sigma LX_i \frac{m_i}{d_i} \tag{1}$$

The relevant parameters are shown as the table below.

Table 1. Physical properties of air samples under 77K low temperature compression systems[2]

Gas	Volume fraction (%)	Moore quality (g/mol)	Gas liquid density (g/mL)	Saturated vapor pressure(77K, Pa)
N_2	78.1	28.01	0.808	97056
O_2	21	32.00	1.141	19632
Ar	0.93	39.94	1.400	25750
Kr	1.14×10^{-4}	83.80	2.740	220

In order to understand the principle of low temperature distillation and master the relevant parameters of the final degree of low temperature distillation, the following simulation is carried out:

Assuming a closed container volumed of 100 mL, 10 L of the sample was compressed into the vessel and the vessel was immersed in liquid nitrogen. Under such condition the sealed container had both liquid and gas phases. The continuous low-temperature distillation process is carried out. Assuming that the low-temperature distillation process is an infinite vapor phase extraction processe, it is assumed that 10% volume of the gas phase in the vessel is extracted and the gas phase is immediately added in after each extraction, after a large number of iterations has been carried out ,the relationship between the ratio of the retention value (the ratio of the remaining amount to the total amount of the component in the original sample) and the number of iterations of the various components in the container is shown in the **Figure 1**.

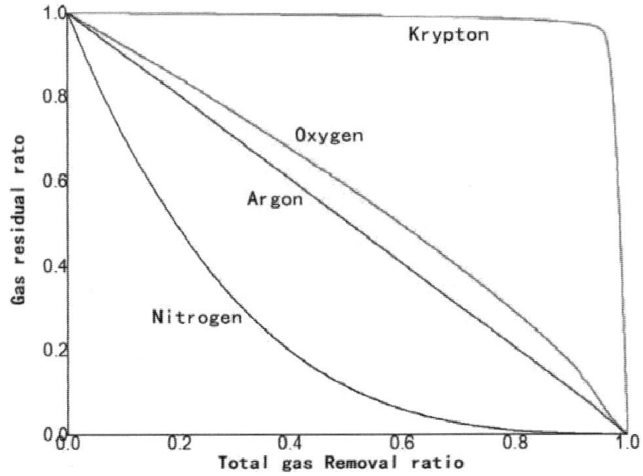

Figure 1. Numerical Simulation of Air Low Temperature Distillation

The use of sponge titanium to remove nitrogen and oxygen in gas samples is one of the most commonly used methods in the inert gases purification[3]. Sponge titanium lump is composed of numerous small sponge titanium stacking accumulated macro body[4]. Loose porous structure gives it a large specific surface area[5]. It's conducive to its chemical reaction with the gas. Titanium sponge is covered with a layer of oxide, mainly titanium dioxide, under room temperature, it hinders the chemical reaction between different gases and titanium. So the performance of the titanium is stable and room temperature[6]. However, the protective layer will rupture when the temperature gradually increased. With the temperature rising, the reaction rate of the gases with the titanium sponge was accelerated. So the reaction of titanium with gases and temperature are directly related. Under different temperatures, titanium can produce stable, irreversible titanium compounded with gases like oxygen and nitrogen, other than inert gases. Under atmospheric pressure of about 500 ℃, sponge titanium begin to produce titanium dioxide and titanium oxide with oxygen. Titanium dioxide and carbon are produced with carbon dioxide. React with water to produce titanium dioxide with hydrogen released. Under 800 to 900 ° C sponge titanium react to form titanium nitride with nitrogen[7].

2. Experimental Methods

Two experiments were designed as a comparison. Each controls 1L gas into the system. The total volume of the system is measured as 815mL. In the first group of experiments, the first step is to freeze the gas in the liquid nitrogen temperature, and extract air with a mechanical pump, next perform the operation of low-temperature distillation, then put the remaining gas into the sponge heating furnace to heat, and finally the total pressure inside the system is measured to reflect the effectiveness of argon removal.

In the second group of experiments, gas will be put into the titanium sponge firstly to make, gas and the hot sponge titanium completely react, and then pass the remaining gas through the cold trap of activated carbon under liquid nitrogen temperature, next, perform the operation of low temperature distillation. In the meantime, record the final gas pressure in the system. Compared with the first set of experiments.

3. Results and Discussion

3.1. Results

After two sets of experiments, the results were as follows:

Table 2. Experimental results on argon residual ratio

	Start pressure value (Pa)	Final pressure (Pa)	Pressure difference (Pa)	Argon residual rate
The first group of experiments	53Pa	128Pa	75Pa	8.06%
The second group of experiments	50Pa	301Pa	261Pa	27.9%

3.2. Simulation

The effect on the argon content of the order of adsorptive distillation and sponge titanium determines the concentration of argon the process of adsorptive distillation. When the adsorptive distillation is carried out firstly, the concentration of argon is the same as in the atmosphere, the value is 0.934%. If the sponge titanium reaction is performed in the first place, argon concentration will greatly increase.

In the first group of experiments, the concentration of the gas component was consistent with the concentration of air components. In order to know the concentration of argon in the second group of experiments, the composition of the gas subjected to the sponge titanium absorption operation was first analyzed and the following conclusions were obtained as **Figure 2.**

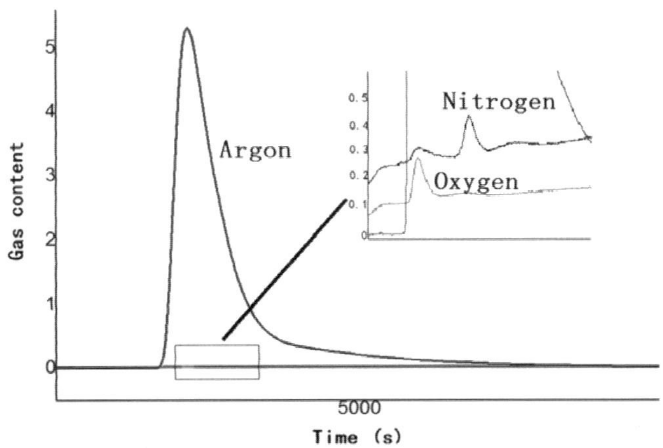

Figure 2. Air component after sponge titanium reaction

It can be said that after the sponge titanium reaction of the gas, nitrogen, oxygen content is minimal, the main component turns intointo argon, and the volume is also reduced by onea hundred timeses. Therefore, assume that the simulation conditions are as follows: After the sponge titanium reaction, the sample gas is 10mL and the volume of the activated carbon cold trap is 5mL. The volume fraction of each component ofin the sample gas is shown as the Table 3:

Table 3. Sample component volume fraction of the adsorption Distillation Simulation assumption

Ingredient	Ar	N_2	O_2	Kr
Volume fraction	0.95	0.04	0.01	10^{-4}

Using the related data in formula 1 and Table 1, for the result of numerical simulation calculation shows, the relationship between constituent retention rate and the total distillation time between the relationships as below.

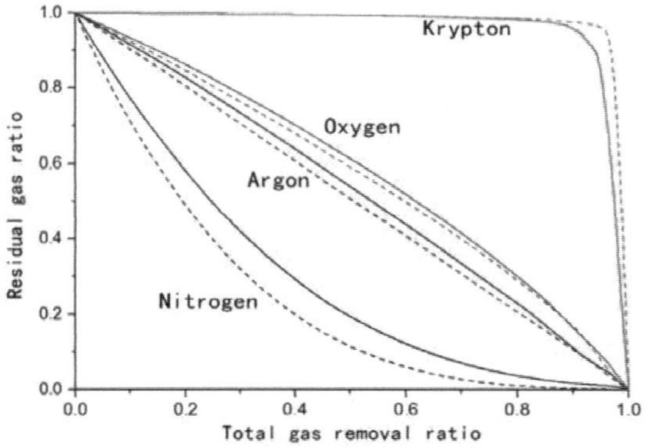

Figure 3. Comparison of low temperature distillation processes of different air content.

As shown, in order to ensure the recovery rate of krypton unchanged, argon needs to be surplus. By analyzing the results of the simulation data, the conclusion is reached that the concentration of krypton began to decrease rapidly at 92%, in the first group of experiments, the residual rate of argon was 1.7% when the total amount of krypton was 92%. In the he second group of experiments, the residual rate of argon was 9.1% when the total amount of krypton was 92%. The experimental results are in good agreement with the simulation results.

4. Conclusion

1. Low temperature distillation has a strong effect on the separation and purification of krypton gas. After this process, the concentration of krypton in the gas can be improved by one hundred times.

2. Sponge Titanium has a strong gas absorption capacity, and through the sponge titanium reaction nitrogen and oxygen gas is almost absorbed completely.

3. The sequence of low temperature distillation and sponge titanium adsorption has an effect on the removal of argon in the gas. The first low-temperature distillation operation can lower the proportion of argon in the gas.

References

[1] Sun Hongxiang. Low-temperature gas purification process and equipment [J]. Low temperature and special gas, 1990 (02): 44.
[2] Si Wenying. Talk about rare gas [J] .Chinese juvenile (study of youth education), 2013 (16): 7.
[3] Quan Jing Zeng. Sponge Titanium high temperature degassing purification of argon [J]. Nonferrous Smelting, 1987 (03): 30-33.
[4] Ozima M, Podosek F A. Noble gas geochemistry [M].[S.l.]: Cambridge University Press, 1983.
[5] Xu Weichun, Sun Huimin. Titanium structure and microstructure analysis [J]. Rare Metals, 2009 (06): 161-166.
[6] Yang Xiaochan (translation) Titanium Sponge [J] Modern Materials Dynamic, 2008 (12): 21-22.
[7] Zhang Yiyan. Rare earth alloy and sponge titanium mixture purification argon, nitrogen, hydrogen [J]. Low temperature and special gas, 2015 (04): 25-28.

IOP Conf. Series: Journal of Physics: Conf. Series **916** (2017) 012034 doi:10.1088/1742-6596/916/1/012034

Numerical simulation for different densities multi-phase fluids with an energy law preserving method

L Li[1] and Y Y Jiang [2, 3, *]

[1] School of Mathematics and Physics, University of South China, Hengyang 421001, China
[2] Beijing Time Technologies Co., Ltd., Kaituo Road 17, Beijing 100085, China
[3] School of Civil & Environmental Engineering, University of Science and Technology Beijing, Beijing 100083, China

* Corresponding author e-mail: jiangyongyue1987@sina.com

Abstract. In this paper, we present a numerical investigation for simulating fluid-fluid interface of isothermal multi-phase fluids with a phase field model. Unlike the previous work, we use continuous finite element method to compute the governing equations and get a discrete energy law equation with a special designed scheme which is almost the same as the continuous energy law. We deal with the stability in pressure by using a penalty formulation to rewrite the continuity condition. The accuracy of calculation is examined by numerical simulation of coalescence of two kissing bubbles. And the accuracy of the discrete energy law is validated as the error in the energy law of numerical examples falls below 10^{-10} and this energy law at the discrete level is preserved at each time step.

1. Introduction

Multi-phase fluids have been widely applied in biology, chemistry, physics and engineering, such as the application of nan fluids in cooling equipment's, boiling tube in nuclear power station and liquid-vapor separator and so on [2]. Multiphase fluids have attracted more and more attention to investigate the mechanism of flows for satisfying the high-speed developing technology.

There are two kinds of theoretical models named sharp interface model and phase field model widely used for multi-phase fluids with clear interface. Sharp interface model [3] uses the single phase fluid model to simulate each phase and treats the interface as the boundary which makes the jump of boundary conditions exists at the interface due to the different property parameters of different phase. As boundary conditions at interface not continuous any more, it adds the difficulty in the computing of the system. Many people investigate multi-phase fluids numerically with sharp interface model as this model is closer to the reality [4], even though they have to deal with the jump condition.

Compared with sharp interface, phase field model which is also called diffuse interface model not considering the interface as the boundary introduces the hypothesis in the continuity of interface. It has thickness between one phase to the other at the interface and the physical property parameters of different phases such as density, viscosity change continuously during the interface. It also avoids the appearance of jump of boundary conditions and treats the multi-phase fluids as one fluid with different material properties. An order parameter is applied to identify different phase which is mostly uniform in the bulk energy and this also makes the numerical calculation more easily [5]. In recent years, numerical results with phase field model are better than sharp interface model like the droplet coalescence, pinch off rising and deformations in shear flows.

In addition, there are more and more available numerical methods used to solve the phase field model including spectral method [6], moving mesh method [7], finite element method [8] and finite difference method. A physical energy law usually exists in the phase field model and the energy stable scheme can make a better result. H. Yu and X. Yang developed two efficient and unconditionally energy stable numerical schemes for Navier Stokes Cahn Hilliard and got an energy law at discrete level but inequality [9]. Actually, the preservation of energy law turns out to be important in designing a numerical method as the preserving energy law method can produce the correct solution even though a rapid change or singularity occurs in the computation. An energy law equality at discrete level for the variable density two-phase flow model is obtained by Guo et al. [10] with continuous finite element method due to a special temporal scheme. In addition, the difference between the numerical solution obtained by preserving energy law method with a relatively coarse mesh and numerical results obtained by other methods with finer grid is almost zero.

In this paper, we develop an energy law preserving continuous finite element method for the variable density two-phase flow model proposed by Abels et al. [1]. A special temporal scheme is designed to discretize the weak form of Navier-Stokes-Cahn-Hilliard equations to obtain a discrete energy law equality based on the work in [10].

2. Multi-phase model, weak form and energy law

The model for the vary density multi-phase fluids proposed by Abels et al. [1] is composed of Navier Stokes equations and Cahn-Hilliard models as follows:

$$\frac{\partial \phi}{\partial t}+\mathbf{v}\cdot\nabla\phi=\nabla\cdot(M\nabla\mu) \tag{1}$$

$$\nabla\cdot\mathbf{v}=0 \tag{2}$$

$$\frac{\partial(\rho\mathbf{v})}{\partial t}+\nabla\cdot(\rho\mathbf{v}\otimes\mathbf{v})=-\nabla p+\eta\Delta\mathbf{v}+\mu\nabla\phi-\nabla\cdot(\mathbf{J}\otimes\mathbf{v}) \tag{3}$$

$$\mu=\sigma\varepsilon^{-1}\psi'(\phi)-\sigma\varepsilon\Delta\phi \tag{4}$$

With boundary conditions of $\mathbf{v}=0, \partial_{\mathbf{n}}\phi=0, \partial_{\mathbf{n}}\mu=0$. Here $\mathbf{v}=(v_1, v_2)$ represents the velocity of the mixture, p is the hydrostatic pressure and ϕ is the order parameter used to mark different phases ($\phi=1$: fluid 1; $\phi=-1$: fluid 2). $\eta, M, \sigma, \varepsilon, \mu$ Denote fluid viscosity, phenomenological mobility, physical surface tensor, interface thickness and chemical potential, respectively. The density ρ and mass flux $\mathbf{J}$ due to the variable density has the expression of $\rho=((\rho_2+\rho_1)+(\rho_2-\rho_1)\phi)/2$ and $\mathbf{J}=(\rho_1-\rho_2)M\nabla\mu/2$ where ρ_1, ρ_2 represent the fluid density of each phase before mixed, respectively.

Please to see more details in [10] for the finite element space, the weak form reads: Find $\mathbf{v}\in\mathbf{W}^{1,3}$, $p\in L_0^2, \phi, \mu\in W^{1,3}$ such that

$$\int_\Omega\left(\frac{\partial\phi}{\partial t}\varphi+(\mathbf{v}\cdot\nabla\phi)\varphi+M\nabla\mu\cdot\nabla\varphi\right)d\mathbf{x}=0 \qquad \forall\psi\in W^{1,3} \tag{5}$$

$$\int_\Omega\left(\mu\chi-\sigma\varepsilon^{-1}\psi'\chi-\sigma\varepsilon\nabla\phi\cdot\nabla\chi\right)d\mathbf{x}=0 \qquad \forall\chi\in W^{1,3} \tag{6}$$

$$\int_\Omega (\nabla\cdot\mathbf{v}) q d\mathbf{x} = 0 \qquad \forall q \in L_0^2 \tag{7}$$

$$\int_\Omega \left(\frac{\partial(\rho\mathbf{v})}{\partial t}\cdot\mathbf{u} + \nabla\cdot(\rho\mathbf{v}\otimes\mathbf{v})\cdot\mathbf{u} - p(\nabla\cdot\mathbf{u}) + \eta\nabla\mathbf{v}:\nabla\mathbf{u} - \mu\nabla\phi\cdot\mathbf{u} + \nabla\cdot(\mathbf{J}\otimes\mathbf{v})\cdot\mathbf{u} \right) d\mathbf{x} = 0 \quad \forall \mathbf{u}\in\mathbf{W}^{1,3} \tag{8}$$

We take $\varphi = \mu$, $\mathbf{u} = \mathbf{v}$, $q = p$, $\chi = \partial\phi/\partial t$ into equation (5)-(8) and the weak form becomes

$$\int_\Omega \left(\frac{\partial\phi}{\partial t}\mu + (\mathbf{v}\cdot\nabla\phi)\mu + M|\nabla\mu|^2 \right) d\mathbf{x} = 0 \tag{9}$$

$$\int_\Omega \left(\mu\frac{\partial\phi}{\partial t} - \sigma\varepsilon^{-1}\frac{\partial\psi}{\partial t} - \frac{\sigma\varepsilon}{2}\frac{\partial}{\partial t}(|\nabla\phi|)^2 \right) d\mathbf{x} = 0 \tag{10}$$

$$\int_\Omega (p\nabla\cdot\mathbf{v}) d\mathbf{x} = 0 \tag{11}$$

$$\int_\Omega \left(\frac{\partial}{\partial t}\left(\rho\frac{|\mathbf{v}|^2}{2}\right) - p(\nabla\cdot\mathbf{v}) + \eta|\nabla\mathbf{v}|^2 - \mu\nabla\phi\cdot\mathbf{v} \right) d\mathbf{x} = 0 \text{ due to } \frac{\partial\rho}{\partial t} + \nabla\cdot(\rho\mathbf{v}+\mathbf{J}) = 0 \tag{12}$$

And the continuous energy law derived from the equation (9) - (10) + (11) + (12):

$$\frac{\partial}{\partial t}\left(\frac{\rho}{2}\|\mathbf{v}\|_{L^2}^2 + \frac{\sigma\varepsilon}{2}\|\nabla\phi\|_{L^2}^2 + \sigma\varepsilon^{-1}\int_\Omega \psi d\mathbf{x} \right) = -\left(\eta\|\nabla\mathbf{v}\|_{L^2}^2 + M\|\nabla\mu\|_{L^2}^2 \right) \tag{13}$$

3. Special finite element scheme and discrete energy law

In the computation of the system, we rewrite equation (2) with $\nabla\cdot\mathbf{v} + \delta p = 0$ with $\delta = 10^{-6}$ as the penalty treatment to increase the stability of pressure. The special finite element scheme called preserving energy law method and the fully discretized weak form are given as follows:

$$\int_\Omega \left(\nabla\cdot\left(\sqrt{\rho}\mathbf{v}\right)_h^{n+1} + \delta p_h^{n+1/2} \right) q \, d\mathbf{x} = 0 \tag{14}$$

$$\begin{aligned} \int_\Omega \Bigg(& \sqrt{\rho}_h^{n+1/2}\left(\sqrt{\rho}\mathbf{v}\right)_{\bar{t}}^{n+1}\cdot\mathbf{u} + \left(\rho_h^{n+1/2}\left(\sqrt{\rho}\mathbf{v}\right)_h^{n+1} + \mathbf{J}_h^{n+1/2}\right)\cdot\nabla\left(\sqrt{\rho}\mathbf{v}\right)_h^{n+1}\cdot\mathbf{u} + \frac{1}{2}\nabla\cdot\left(\rho_h^{n+1/2}\left(\sqrt{\rho}\mathbf{v}\right)_h^{n+1} + \mathbf{J}_h^{n+1/2}\right)\left(\sqrt{\rho}\mathbf{v}\right)_h^{n+1}\cdot\mathbf{u} \\ & - p_h^{n+1/2}(\nabla\cdot\mathbf{u}) + \eta\nabla\left(\sqrt{\rho}\mathbf{v}\right)_h^{n+1} : \nabla\mathbf{u} - \mu_h^{n+1/2}\nabla\phi_h^{n+1/2}\cdot\mathbf{u} \Bigg) d\mathbf{x} = 0 \end{aligned} \tag{15}$$

$$\int_\Omega \left(\phi_{\bar{t}}^{n+1}\varphi + \left(\sqrt{\rho}\mathbf{v}\right)_h^{n+1}\cdot\nabla\phi_h^{n+1/2}\varphi + M\nabla\left(\omega_h^{n+1/2} + c\phi_h^{n+1/2}\right)\cdot\nabla\varphi \right) d\mathbf{x} = 0 \tag{16}$$

$$\int_\Omega \left(\mu^{n+1/2}\chi - \frac{\sigma}{\varepsilon}\psi\left(\phi_h^{n+1}, \phi_h^n\right)\chi - \sigma\varepsilon\nabla\phi_h^{n+1/2}\cdot\nabla\chi \right) d\mathbf{x} = 0 \tag{17}$$

Where

$\sqrt{\rho}_h^{n+1/2} = \left(\sqrt{\rho_h^{n+1}} + \sqrt{\rho_h^n}\right)/2$, $\rho_h^{n+1/2} = \left(\rho_h^{n+1} + \rho_h^n\right)/2$, $\left(\sqrt{\rho}\mathbf{v}\right)_h^{n+1} = \left(\sqrt{\rho_h^{n+1}}\mathbf{v}_h^{n+1} + \sqrt{\rho_h^n}\mathbf{v}_h^n\right)/\left(\sqrt{\rho_h^{n+1}} + \sqrt{\rho_h^n}\right)$

$\mu_h^{n+1/2} = \left(\mu_h^{n+1} + \mu_h^n\right)/2$, $\phi_h^{n+1/2} = \left(\phi_h^{n+1} + \phi_h^n\right)/2$, $p_h^{n+1/2} = \left(p_h^{n+1} + p_h^n\right)/2$, $\left(\sqrt{\rho}\mathbf{v}\right)_{\bar{t}}^{n+1} = \left(\sqrt{\rho_h^{n+1}}\mathbf{v}_h^{n+1} - \sqrt{\rho_h^n}\mathbf{v}_h^n\right)/\Delta t$,

$\rho_h^{n+1} = (\rho_1 + \rho_2)/2 + (\rho_2 - \rho_1)\phi_h^{n+1}/2$, $\psi\left(\phi_h^{n+1}, \phi_h^n\right) = \left(\left|\phi_h^{n+1}\right|^2 + \left|\phi_h^n\right|^2 - 2\right)\left(\phi_h^{n+1} + \phi_h^n\right)/4$, $\phi_{\bar{t}}^{n+1} = \phi_h^{n+1} - \phi_h^n/\Delta t$,

$\mathbf{J}_h^{n+1/2} = \left(\mathbf{J}_h^{n+1} + \mathbf{J}_h^n\right)/2$. Still we take $\varphi = \mu^{n+1/2}$, $\mathbf{u} = \left(\sqrt{\rho}\mathbf{v}\right)_h^{n+1}$, $\chi = \left(\phi_h^{n+1} - \phi_h^n\right)/\Delta t$, $q = p_h^{n+1/2}$ into equation (14)-(17) and we get the energy law at discrete level as

$$\frac{E^{n+1} - E^n}{\Delta t} = -\left(\eta\left\|\nabla\left(\sqrt{\rho}\mathbf{v}\right)_h^{n+1}\right\|_{L^2}^2 + \delta\left\|p_h^{n+1/2}\right\|_{L^2}^2 + M\left\|\nabla\mu^{n+1/2}\right\|_{L^2}^2\right) \tag{18}$$

Where $E^i = \left\|\sqrt{\rho_h^i}\mathbf{v}_h^i\right\|_{L^2}^2/2 + \sigma\varepsilon\left\|\nabla\phi_h^i\right\|_{L^2}^2/2 + B\int_\Omega\left(\sigma\left(\left(\phi_h^i\right)^2 - 1\right)^2/4\varepsilon\right)d\mathbf{x}$ the total free energy at time i. Compared with equation (13), equation (18) is almost the same as it obviously.

4. Numerical example and discussion

In this section, we compute coalescence of kissing bubbles as the numerical example. We use 64×32 relative coarse grid to discretize the computing domain and the value of some parameters are $\rho_1 = 1$, $\rho_2 = 10$, $\eta_1 = 0.01$, $\eta_2 = 0.06$, $M = 0.0005$, $\sigma = 1.96$, $\varepsilon = 0.01$, $c = 0.5$, $\Delta t = 10^{-5}$. Figure 1 shows the interface structure of kissing bubble at different time. As the value of ϕ varies from -1 to 1 at the vicinity of the interface, the result would show the thickness apparently according to the different color and here we choose $\phi = 0$ to show in Figure 1. The dentate interface due to the relative coarse grid comes out in the initial time, but the interface becomes smooth with the preserving energy method. As the interface of kissing bubbles overlapped, they coalesce into one large bubble and become an elliptical bubble duo to the surface tension. The result for the velocity field is shown in Figure 2. Compared the result in [10], they agreed well and also validate the accuracy of the numerical solution.

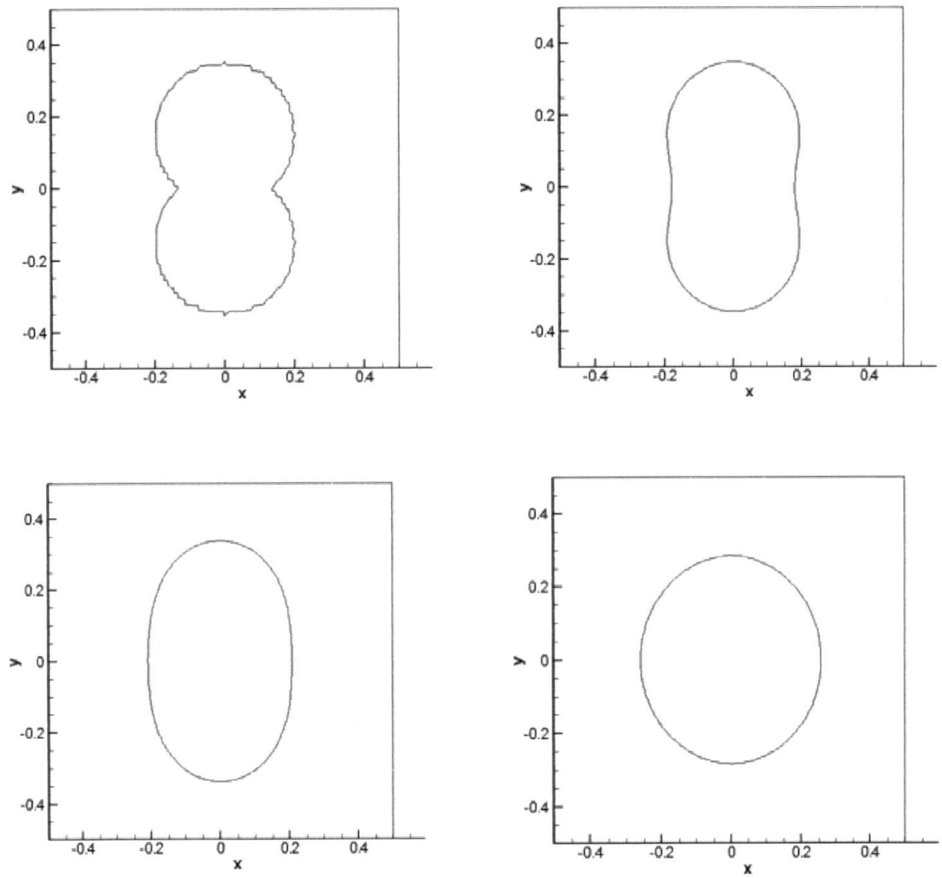

Figure 1. The evolution of interface ($\phi = 0$) for two kissing bubbles at time $t = 0, 0.03, 0.1, 0.15$.

IOP Conf. Series: Journal of Physics: Conf. Series **916** (2017) 012034 doi:10.1088/1742-6596/916/1/012034

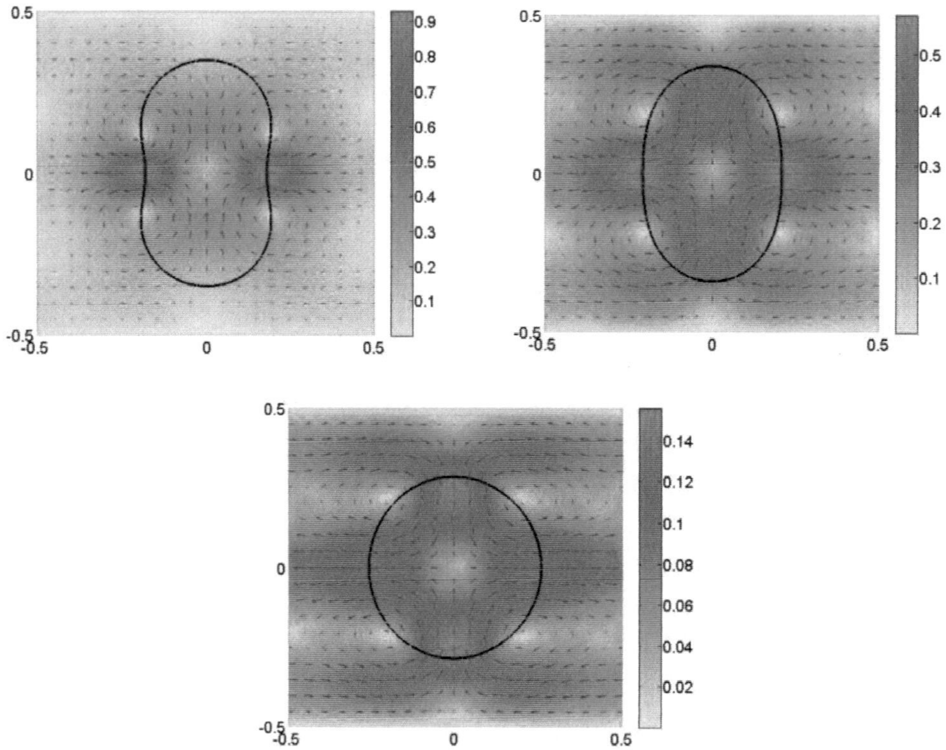

Figure 2. The velocity field for two kissing bubbles at time $t = 0.03, 0.1, 0.15$.

We also examine the time evolution of the total free energy and the energy law at discrete level defined in equation (18). The total free energy is shown in Figure 3 (a) and the decreasing result match with the equation (18). The result in Figure 3 (b) shows the error in the discrete energy law (equation (18)). The discrete energy law can be preserved at each time and the accuracy of equation (18) can be validated as the error in the discrete energy law falls down to 10^{-10}.

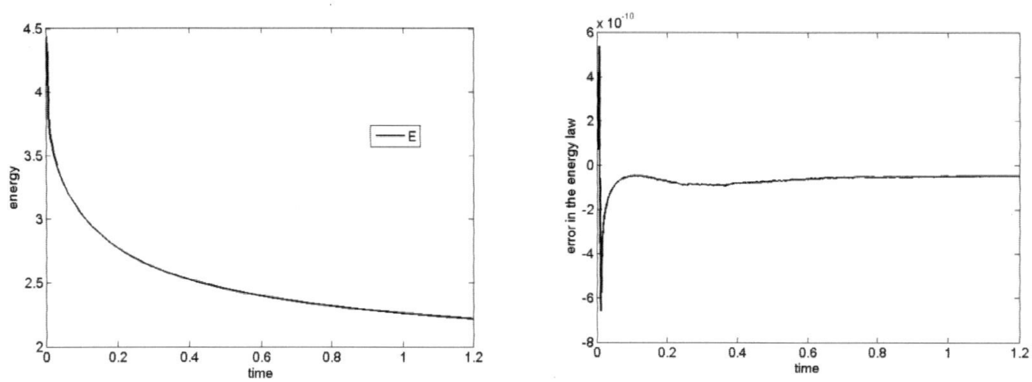

(a) (b)

Figure 3. a: The evolution of total free energy; b: The error in the discrete energy law.

5. Conclusion

In this paper, we numerical investigate fluid-fluid interface of isothermal multi-phase fluids with a preserving energy law finite element method. The Navier-Stokes-Cahn-Hilliard equations are discretized by this method and a discrete energy law which is almost the same as the continuous energy law is derived. This discrete energy law equality of this phase model is never established before with finite element method. With the examination of numerical simulation of two kissing bubble, velocity field, the evolution of total free energy and the error in the energy law, the accuracy of the preserving energy law method and the discrete energy law equality are validated apparently.

References

[1] H. Abels, H. Garcke and G. Grun, Thermodynamically consistent, frame invariant, diffuse interface models for incompressible two-phase flows with different densities, Math. Models Meth. Appl. Sci. 22 (2012) 11-25.
[2] Y. Jiang, P. Lin, Z. Guo and Dong, Numerical simulation for moving contact line with continuous finite element schemes, Commun. Comput. Phys. 18 (2015) 180-202.
[3] L. Li, Y. Jiang and A. Chen, Numerical simulation of nanofluids based on power-law fluids with flow and heat transfer, Earth Environ. Sci. 61 (2017) 012106-012114.
[4] L. Han, X. Hu and N. Adams, Adaptive multi-resolution method for compressible multi-phase flows with sharp interface model and pyramid data structure, J. Comput. Phys. 262 (2014) 131-152.
[5] J. Hua, P. Lin, C. Liu and Q. Wang, Energy law preserving C0 finite element schemes for phase field models in two-phase flow computations, J. Comput. Phys. 230 (2011) 7115-7131.
[6] J. Luo, X. Hu and N. Adams, Curvature boundary condition for a moving contact line, J. Comput. Phys. 310 (2016) 329-341.
[7] H. Lee, J. Lowengrub and J. Goodman, Modeling pinchoff and reconnection in a hele-shaw cell, Phys Fluids. 14 (2002) 514-545.
[8] Z. Guo, P. Lin, S. Wise and J. Lowengrub, Mass conservative and energy stable finite difference methods for the quasi-incompressible Navier-Stokes-Cahn-Hilliard system: primitive variable and projection type schemes, Comput. Phys. Commun. 326 (2017) 144-174.
[9] P. Spelt, Shear flow past two-dimensional droplets pinned or moving on an adhering channel wall ar moderate Reynolds numbers: A numerical study, J. Fluid Mech. 56 (2006) 439-463.
[10] Z. Guo and P. Lin, A thermodynamically consistent phase-field model for two-phase flows with thermocapillary effects, J. Fluid Mech. 766 (2015) 226-271.

IOP Conf. Series: Journal of Physics: Conf. Series **916** (2017) 012035 doi:10.1088/1742-6596/916/1/012035

3D CFD simulation of Multi-phase flow separators

Zhiying Zhu*

School of petroleum engineering, China University of petroleum, Beijing, China

*Corresponding author e-mail:zhiying_2008@hotmail.com

Abstract. During the exploitation of natural gas, some water and sands are contained. It will be better to separate water and sands from natural gas to insure favourable transportation and storage. In this study, we use CFD to analyse the effect of multi-phase flow separator, whose detailed geometrical parameters are designed in advanced. VOF model and DPM are used here. From the results of CFD, we can draw a conclusion that separated effect of multi-phase flow achieves better results. No solid and water is carried out from gas outlet. CFD simulation provides an economical and efficient approach to shed more light on details of the flow behaviour.

1. Introduction

During the exploitation of natural gas, some water and sands are contained. And the existence of water and sands caused kinds of damages in the transportation and storage. For instance, water can reduce the delivery capacity of the pipeline [1]; sands can result in wearing down the facility of transportation, and sometimes blocking the outlet of the facility. Therefore, it is very important to remove the water and sands from natural gas.

The work process of separators is, firstly, natural gas with impurities flows into separate chamber. And then, because of gravity and/or centrifugal force, liquid and solid particles settle down to the bottom of chamber. Natural gas without impurities flows out from the outlet [2]. However, the complex flow behavior in the separator such as gas carry-under and efficiency should be considered also. Computational fluid dynamic (CFD) simulation provides an economical and efficient approach to shed more light on details of the flow behavior. Commonly, there are two types of separator, vertical type[3] and horizontal type[4]. Because we study natural gas with less water, and desand is a key need, we consider vertical type here. In this study, we design a separator by detailed geometrical parameters; utilize CFD software-Fluent to simulate three-phase flow in the separator. And analyze whether sand or water are carried out from the outlet with natural gas or not.

2. Modelling

The separator is shown schematically in Figure 1 and detailed geometrical parameters and explanations are illustrated in Table 1.

The part grid configuration is shown in Figure 2. Kinds of grid scheme were tested and we chose one of them here. Each circumference in z direction is set by interval count equals 60. And then apply Tet/hybrid scheme to the whole volume. Grid refinement study was conducted to obtain grid independent solution. The study indicated that a large number of grid points were required to obtain accurate solution for this 3D case. Other grids results are ignored in this paper.

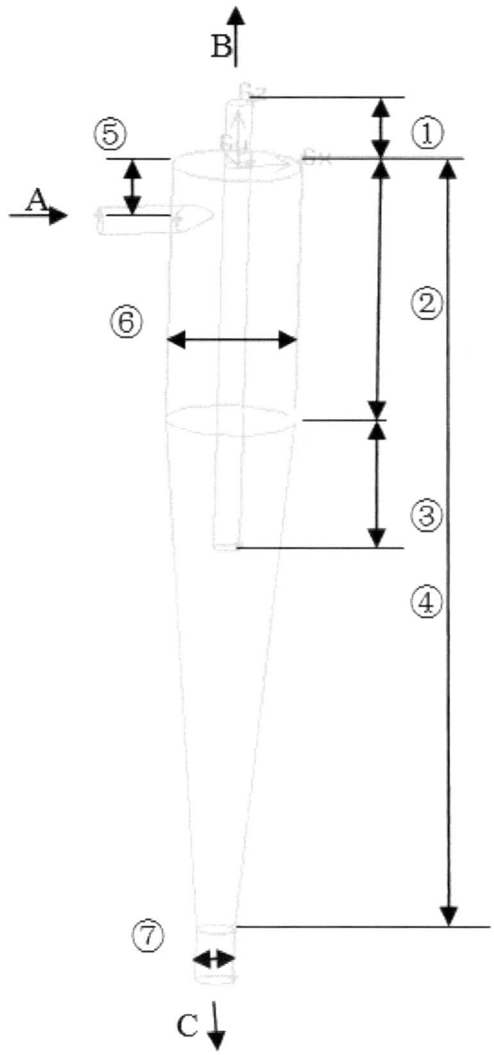

Figure 1 Schematic of the separator configuration.

Table 1. Detailed geometrically parameters (units: mm) and explanations.

H①	H②	H③	H④	H⑤	D⑥	D⑦	A	B	C
50	200	100	600	50	101.6	30	inlet	Outlet	Outlet

(H means height, D means inner diameter. The inner diameter of A and B is 20.)

IOP Conf. Series: Journal of Physics: Conf. Series **916** (2017) 012035 doi:10.1088/1742-6596/916/1/012035

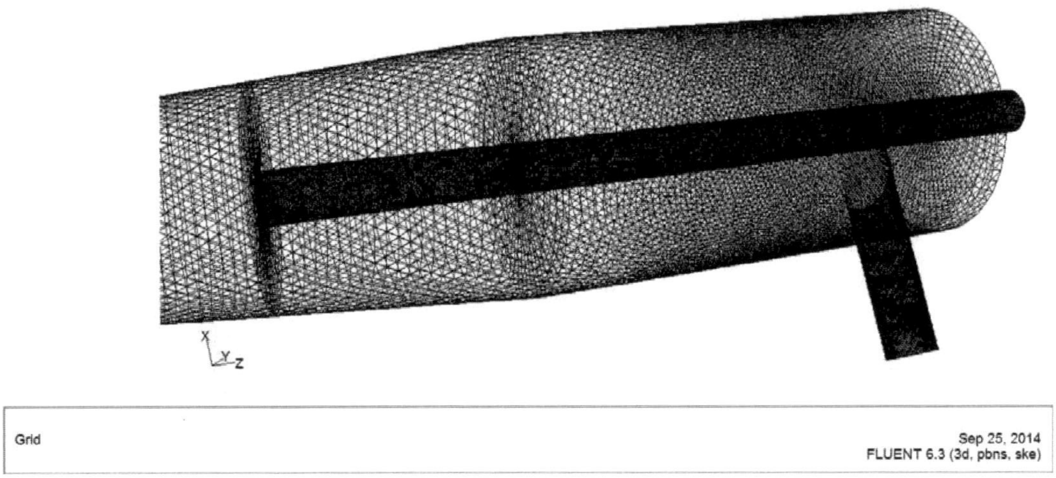

Figure 2 The grid configuration of the separator

3. CFD simulation

Parameters information is as follows: Multi-phase flow in inlet is consist of natural gas, water and sands. Natural gas flow rate is 300,000m3/d, density is 1.293kg/m3 and viscosity is 0.0187mPa • s(standard condition, an atmospheric pressure, 20℃). Water flow rate is 60 m3/d and density is 1,000 kg/m3. Sand flow rate is 5-10 m3/d, density is 2700 kg/m3, thermal conductivity is 0.6W/m • k and specific heat is 900J/Kg • K. There are three kinds of sand with unique particle diameter, they are 0.15mm, 0.42mm and 0.21mm, and the proportion is 7%, 80% and 13% respectively. Inlet pressure is 32MPa, temperature is 90. B outlet pressure is 31.5-31.8MPa. Multi-phase flow contains not only fluid but also solid in this study, therefore, we solve the problem of gas and water firstly and then add solid to the gas-liquid condition. Gas and water are also two-phase, we use VOF model to solve it. And then we utility DPM model to analyze motion trail of discrete phase-sand.

4. Result and discussion

Figure 3 and Figure 4 show velocity contours on part separator and velocity contours through the cross section of the inlet center in z direction. We can see that the fastest velocity appears in the inlet. And in Figure 4 we know that it is faster in near-wall region than far-wall region. This can help solid and liquid drop towards the bottom of the separator.

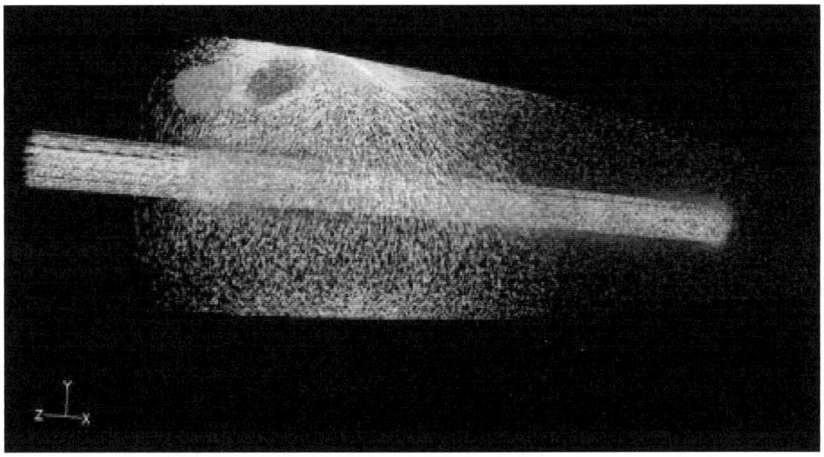

Figure 3 The velocity contours (part separator)

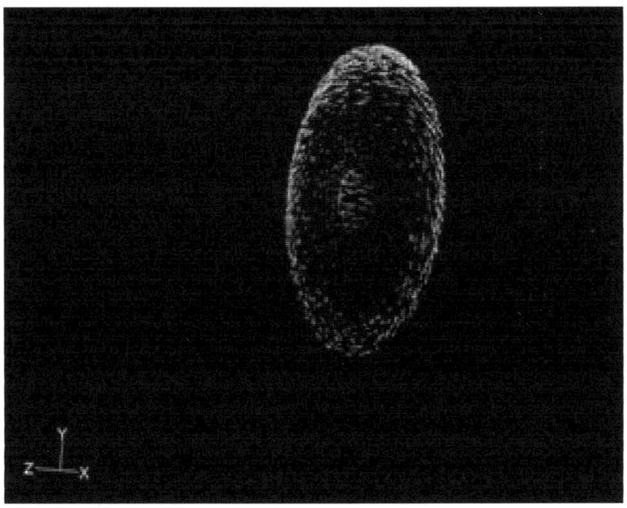

Figure 4 Velocity contours through the cross section of the inlet center in z direction

Figure 5 is the DPM result, it shows the trail of three kinds of snad particles. We can find that the bigger sands moves toward the bottom of the separator, the smallest particles move toward the top of the separator first and then get to the bottom, too. None of them flows out from gas outlet. Separated effect of multi-phase flow can achieve better results.

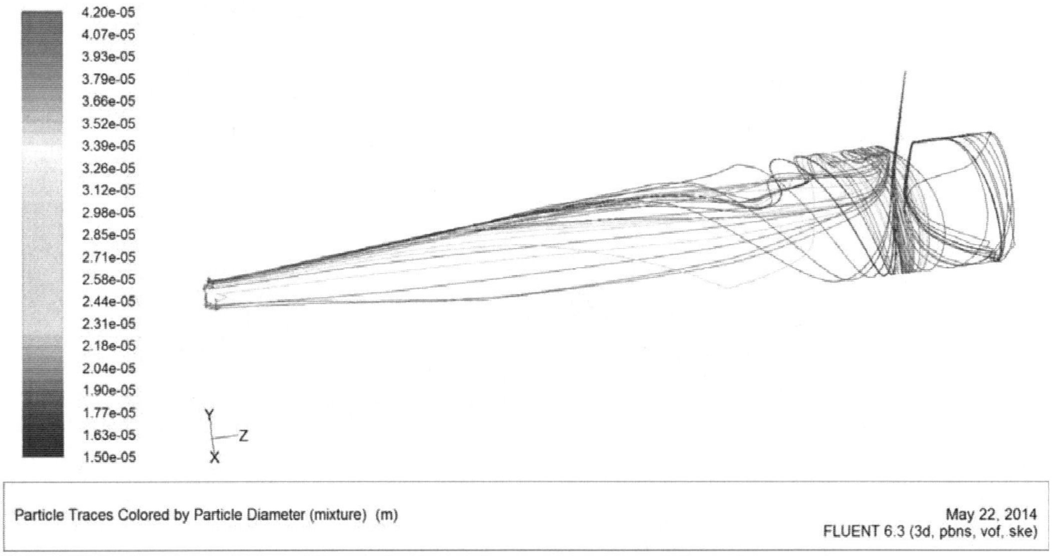

Figure 5 The trail of three kinds of sand particles

5. Conclusion

We can draw a conclusion that designed separator perform well in CFD simulation. No solid and water is carried out from gas outlet. CFD simulation provides an economical and efficient approach to shed more light on details of the flow behavior. We don't need to make a real object and check again and again and make some changes. We can firstly do CFD simulation, and if the model run well, and then make a real object. CFD cannot work without real object experiment although.

Here, we use designed parameters, we can optimize the parameter in the future.

Acknowledgments

The research reported herein is partly supported by China University of Petroleum (Beijing) (grants No. ZX20150195). These financial supports are gratefully acknowledged.

References

[1] Chuang Wen, Xuewen Cao, Yan Yang, Effects of operating parameters on flow characteristics of natural gas in supersonic separators, Offshore Technology Conference (2013), OTC 24086.

[2] Liu Chao, Development of natural gas separator, Agricultural science and technology and equipment, No.227,2013, pp 29-30. (in chinese)

[3] Ferhat M.Erdal, Siamach A, Shirazi, Ovadia Shoham, Gene E.Kouba, CFD simulation of single-phase and two-phase flow in gas-liquid cylindrical cyclone separators, SPE-36645-PA, 1996.

[4] Seyed Ehsan Rafiee, M.M.Sadeghiazad, Experimental and 3D CFD analysis on optimization of geometrical parameters of parallel vortex tube cyclone separator, Aerospace science and technology, 63 (2017), pp 110-122.

IOP Conf. Series: Journal of Physics: Conf. Series **916** (2017) 012036 doi:10.1088/1742 6596/916/1/012036

Research on Heat Dissipation of Electric Vehicle Based on Safety Architecture Optimization

Chao Zhou[1], Yajuan Guo[1], Wei Huang[1], Haitao Jiang[1], Liwei Wu[2]

[1]State Grid Jiangsu Electric Power Research Institute, Nanjing, 211100, China
[2]Nanjing University, Nanjing, 210000, China

Abstract. In order to solve the problem of excessive temperature in the discharge process of lithium-ion battery and the temperature difference between batteries, a heat dissipation of electric vehicle based on safety architecture optimization is designed. The simulation is used to optimize the temperature field of the heat dissipation of the battery. A reasonable heat dissipation control scheme is formulated to achieve heat dissipation requirements. The results show that the ideal working temperature range of the lithium ion battery is 20℃~45℃, and the temperature difference between the batteries should be controlled within 5℃. A cooling fan is arranged at the original air outlet of the battery model, and the two cooling fans work in turn to realize the reciprocating flow. The temperature difference is controlled within 5℃ to ensure the good temperature uniformity between the batteries of the electric vehicle. Based on the above finding, it is concluded that the heat dissipation design for electric vehicle batteries is safe and effective, which is the most effective methods to ensure battery life and vehicle safety.

1. Introduction

The electric vehicle powered by lithium-ion battery has become a hot spot in the international competition. The research and development of the heat dissipation scheme of the battery is very important for the safe and reliable operation of the electric vehicle. Compared with traditional fuel vehicles, lithium-ion battery electric vehicles have the advantages of no vibration, no noise and no pollution. Its development prospect is broad, and has become one of the important development directions of automobile industry in the future. However, lithium ion batteries generate a large amount of heat when they are used, resulting in an increase in the temperature of the battery and an uneven temperature inside the battery. This will affect not only the performance of the battery, but also the life of the battery and the safety of the battery [1]. Therefore, the thermal performance analysis and scientific heat dissipation management of lithium-ion batteries are the most effective methods to ensure battery life and vehicle safety. It can not only ensure the whole temperature of the battery in the ideal temperature range, but also ensure the temperature difference between each battery monomer is within the acceptable range. At the same time, the new technology can keep the performance of the power battery in the optimum state [2].

IOP Conf. Series: Journal of Physics: Conf. Series **916** (2017) 012036 doi:10.1088/1742-6596/916/1/012036

2. Literature review

In order to make the battery work within the optimum temperature range, scholars at home and abroad have carried out related research work. Yang Naixing and others designed the cylindrical structure lithium battery to improve the heat radiation performance. YAMADA uses paraffin as a phase change material and develops a hybrid cooling system for pipe thermal anomalies and thermal runaway. Firstly, ANSYS thermal analysis software is used to simulate the heat generating temperature field of the cell under natural convection. The accuracy of the model is verified by single charge discharge experiments, and the whole model of the battery pack is built [3]. The CFD simulation software is used to simulate the temperature field of the battery under forced air cooling condition, and the temperature distribution of the battery is observed and measured. The heat dissipation effect is analyzed, thus changing the heat dissipation mode and optimizing the temperature field of the battery [4]. Realizing the effective heat dissipation of the battery can ensure the good performance and sufficient service life of the power battery, and has a milestone significance for the safe driving of the electric vehicle.

3. Methods

3.1. Structure and working principle of lithium-ion battery

The internal structure of the lithium ion battery is shown in figure 1. When the battery is charged, the lithium in the cathode material falls off and passes through the membrane into the negative graphite. When the battery is discharged, the lithium ion is removed from the negative graphite and passed through the diaphragm back into the positive material. The charged lithium ions are constantly inserted and removed from the anode and cathode. This phenomenon is called "rocking chair battery" [5].

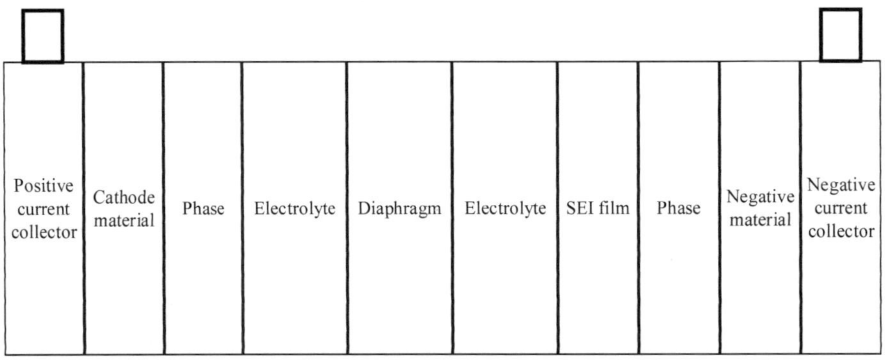

Figure 1. Internal structure of battery

3.2. Analysis of batteries temperature characteristics

According to the principle of battery reaction, from the electrochemical point of view, it can be seen that the heat of the battery comes from four parts, respectively: electrochemical reaction heat Qr, Joule heat Qj, polarization reaction heat Qp and deputy reaction heat Qs. According to the basic principle of internal structure and heat transfer of lithium ion battery, it is known that the heat produced by the reaction of the battery in the battery is heat transfer. And the heat eventually reaches the surface of the battery. There is a temperature difference between the surface temperature of the battery and the ambient temperature. A high temperature object transfers heat to a lower temperature object, so that the surface of the cell is convection heated with the outside. Objects with high temperature can generate electromagnetic waves around them, so the radiation heat transfer is also carried out on the surface of the battery [6].

The temperature has a great influence on the performance of the battery. Analysis of the temperature characteristics of the battery is mainly to study the effect of temperature on the battery capacity, internal resistance, life, as well as charging and discharging performance [7]. The thermal effect of lithium ion monomer is experimentally investigated, and the thermal characteristics of the battery are further understood, which provides a specific experimental verification process for the subsequent battery simulation.

3.3. Three-dimensional thermal model of lithium-ion battery

IFR26650 cylindrical lithium iron phosphate battery is selected. Its size is: 26 mm diameter, 65 mm high, 0.25 mm battery shell thickness. The thermal physical parameters such as specific heat, thermal conductivity and density of the battery are calculated according to the parameters of each material in the battery. A three-dimensional thermal model of lithium-ion battery was established by using ANSYS thermal analysis software. When the model is heated, the loads, the natural convection boundary condition and the initial temperature 25℃ are generated. The natural convection heat transfer coefficient is 5W/m2K. The solution method is set to transient analysis, and the simulation time and time step are defined. Under the natural convection environment, the temperature rise of the battery is obtained by the discharge of 1C, 2C, 3C and 4C, and the simulation is carried out at last.

3.4. Simulation and analysis of temperature field of battery pack CFD

The basic definition of computational fluid dynamics (CFD) is: Based on fluid dynamics, it uses the three conservation laws of mass, energy and momentum as the basic governing equations to simulate the flow state of fluid, and finally obtains the physical quantities of complex fluid in each position [8]. Fluent is the CFD software package which is widely used in the world, highly commercialized and highly specialized. It can simulate complex flow phenomena, heat conduction problems and fluid structure coupling problems. At the same time, it can calculate the flow of fluid using various structured and unstructured grids. According to the physical characteristics of each flow phenomenon, the calculation method suitable for it is chosen. Fluent can also control the speed of calculation, and make the precision and stability of the result reach the optimum value [9]. The whole calculation process is shown in figure 2.

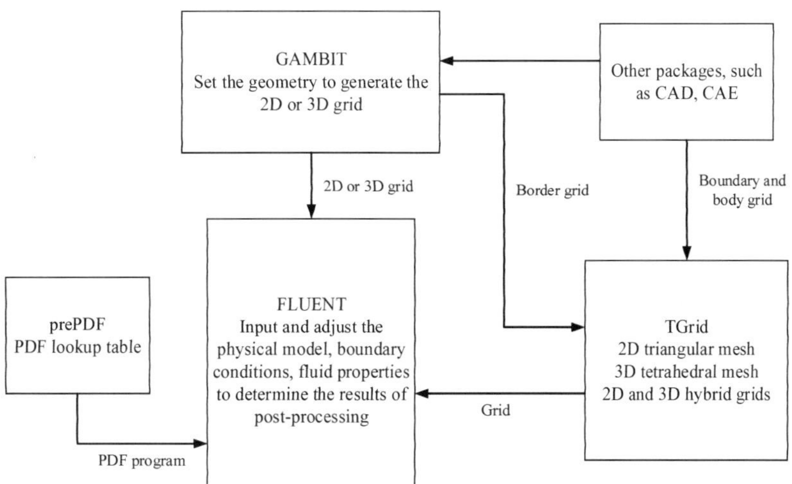

Figure 2. Basic program diagram of Fluent simulation

A power cell grouping module with upper and lower layers is designed, and a parallel unidirectional flow ventilation scheme is selected. In the pre-processing software gambit, the grid is divided, and the material, physical attributes and boundary conditions of the model are set up. After the solution model is established, the flow field is initialized and the residual in the calculation is set up. The Fluent software is used to simulate the flow field.

3.5. Heat dissipation control strategy for lithium-ion batteries

The temperature field of the reciprocating flow cooling of the battery model is analyzed when the ambient temperature is 32 and 40℃. The reciprocating cycle is 300 s. Through the simulation analysis of the battery temperature field under different environmental temperature, a reasonable heat dissipation control scheme is set up and verified by experiments.

4. Results and discussion

4.1. Analysis of temperature characteristics of lithium-ion batteries

At different temperatures, the relationship between the battery capacity and temperature is shown in figure 3. When the temperature is low, the battery capacity is also low. In addition, as the temperature drops, the battery capacity also drops. When the battery temperature is below 5℃, the capacity of the battery is rapidly attenuated. On the contrary, when the temperature is over 20℃, the discharge capacity of the battery changes slowly and basically reaches the rated capacity of the battery. When the temperature is between RT~45℃, the capacity of the battery is always maintained at 100%.

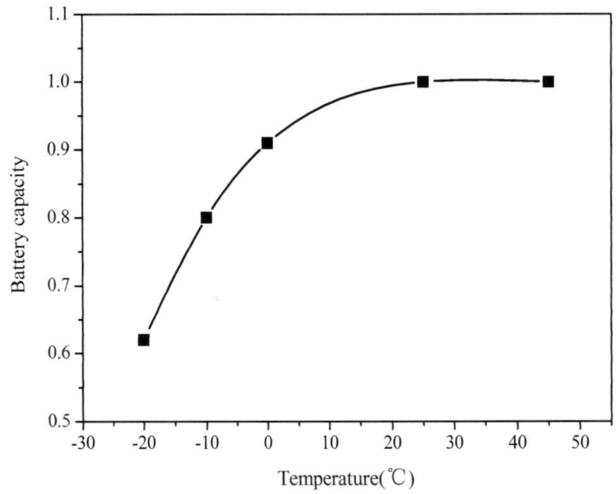

Figure 3. Change curve of battery capacity and temperature

The temperature difference causes the internal resistance of the battery to change (shown in figure 4), and the test result shows that the temperature is constant. When the SOC is between 0.2~1.0, the battery internal resistance is almost constant, and its value is smaller. When the temperature is between 25 and ~ 40℃, the temperature has little influence on the internal resistance of the battery. The experiment shows that the lithium-ion battery is within its ideal operating temperature range, and there is little difference in the internal resistance of the battery under the condition of no excessive discharge.

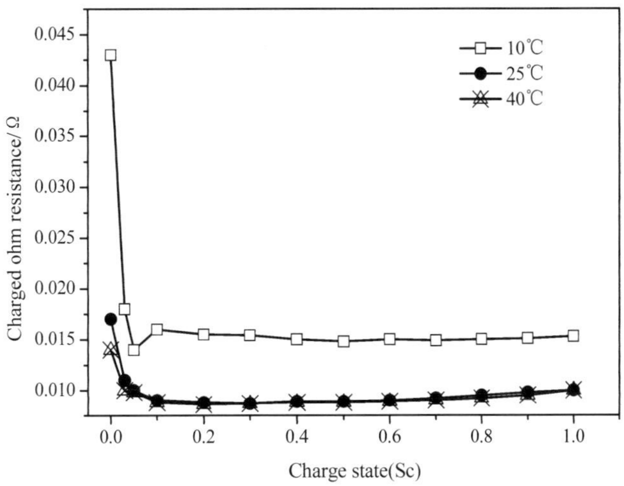

Figure 4. Relationship between total internal resistance of batteries and SOC at different temperatures

4.2. Establishment and verification of thermal effect model

Adjusting the experimental environment at room temperature, the temperature rise test of lithium-ion battery under different rate discharge was carried out. The discharge rates are 1C, 2C, 3C, 4C. The average temperature of four channels of each cell is averaged, and the experimental data are processed. As shown in figure 5, the temperature rise curve of the battery discharging at different rates is obtained.

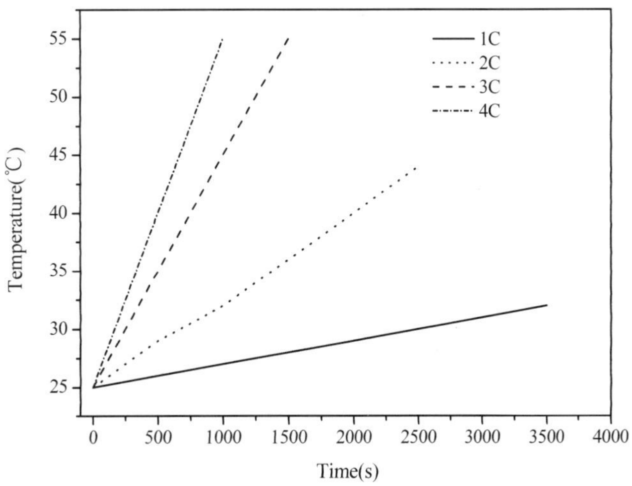

Figure 5. Surface temperature rise curve of battery at different rate discharge

As the battery is discharged, the temperature of the battery rises gradually. The greater the discharge rate of the battery is, the greater its heat output is. The more heat gathered inside the battery is, the greater its temperature rise rate is. After the battery was discharged at 1C, 2C, 3C and 4C, the surface temperature of the battery reached 31.4℃, 39.30℃, 48.05℃ and 56.2℃, respectively. The higher the battery temperature is, the greater the impact on its performance is. Therefore, the effective heat dissipation of the battery will greatly improve the performance of the battery.

Table 1 is the thermal physical parameters of the battery obtained by calculation. The experimental results of the same discharge rate and the same ambient temperature are compared with the simulation results. The error analysis is shown in figure 6. When the batteries are discharged at different discharge rates 1C, 2C, 3C and 4C, the average temperature of the surface of the battery is compared. The difference between the experimental results and the simulation results is 0.12℃, 0.72℃, 0.85℃, 1.41℃, respectively. Although there are some errors in the results, the errors are small and within the range of 1.5℃. This shows that the thermal model is accurate, and the calculation of thermal physical parameters and the setting of parameters in simulation are correct. It is feasible to study the thermal characteristics of the battery with this thermal model.

Table 1. Thermophysical properties of lithium-ion batteries

Battery component	Materials	Density Kg/m^3	Specific heat KJ/(Kg·K)	Thermal conductivity W/(m·K)
Battery core	Mixing of various materials	2066	1167.24	$\lambda_{r,\theta}$=0.95
Battery case	Stainless steel material	8030	502.48	λ_z=25.86 16.27

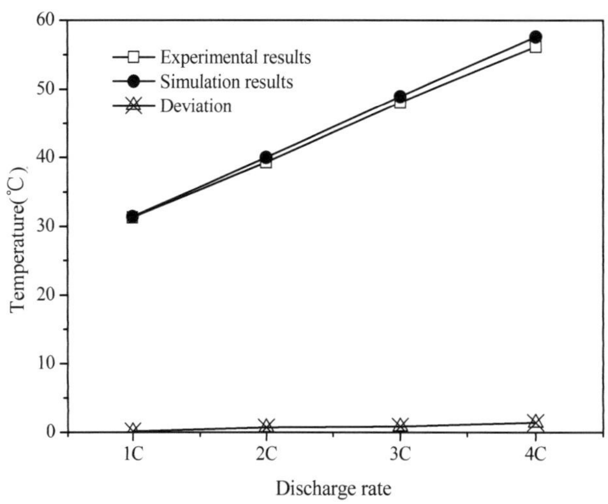

Figure 6. Error comparison between experimental and simulated values

4.3. Simulation analysis of temperature field CFD of lithium-ion battery

The temperature difference between the battery model and the discharge rate at different discharge magnifications is shown in figure 7. When the battery discharge at a small rate of 1C, because the heating rate is small and the surface temperature of the battery is not high, its heat is only several degrees higher than the ambient temperature and the temperature difference is not large. The heat dissipation air with a speed of 0.3m/s can give a good heat dissipation effect on the battery model. The temperature difference of the 2C rate discharge cell increased much, and the inlet wind speed of 0.5m/s can reduce the maximum temperature of the battery to an ideal value. However, the temperature difference between the batteries reaches 4.31℃, and the temperature uniformity between the batteries is not ideal. The temperature difference between the same batteries is within 2℃, and it is within the acceptable range. The temperature difference between the upper and lower batteries of 3C rate discharge is great, and the maximum temperature of the 1m/s can be controlled at the ideal temperature range by the heat radiation wind speed of the battery. But the temperature difference between the batteries exceeds the ideal value, reaching 7.10℃, which will seriously affect the performance of the battery.

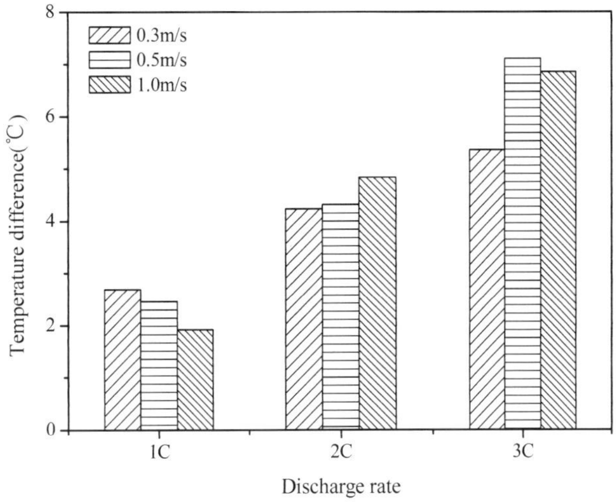

Figure 7. Surface temperature difference at different rate discharge cells

4.4. Verification of heat dissipation control scheme

Figure 8 shows the cooling effect of reciprocating flow cooling at 32℃ and 40℃. When the ambient temperature is 32℃, the battery will discharge at 1C, 2C and 3C rate, and the reciprocating flow with different wind speed can control the temperature of the battery model within 45℃. The temperature difference is less than 4℃, which achieves the temperature control target of the battery, and ensures the temperature uniformity of the battery. When the ambient temperature reaches 40 ℃, the temperature of the battery is difficult to decrease. When the battery is discharged at 1C and 2C rate, the maximum temperature of the battery can be reduced to the ideal temperature by using the reciprocating environment wind with larger speed. When the discharge rate of the battery is 3C, it is difficult to keep the temperature of the battery down by ambient air cooling.

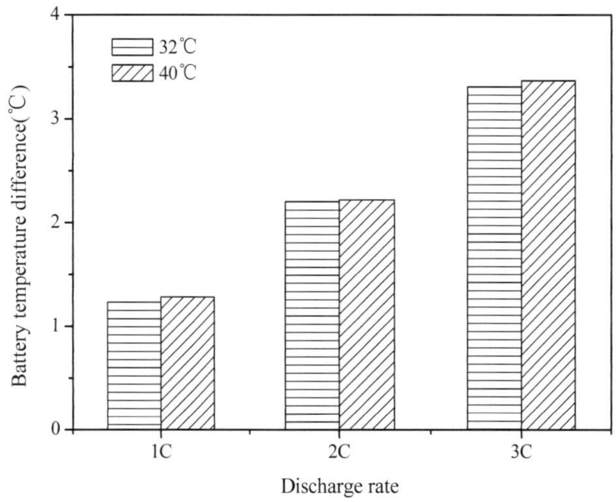

Figure 8. Comparison of heat dissipation effects of reciprocating flow under different temperature conditions

In order to verify the above scheme, when the external temperature is 28℃ and 32℃, respectively, the maximum temperature of the battery reaches 38.30 and 41.91℃ after the reciprocating flow heat is over. This is 10 and 13.07℃ lower than that without heat dissipation, respectively. The temperature difference of the battery is controlled within 3℃ in the whole discharge process, and the maximum temperature of the battery and the temperature difference between the batteries are all in the ideal temperature range, which shows that the heat dissipation control scheme is feasible.

5. Conclusion

IFR26650 lithium-ion battery is selected as the object of study. In order to solve the battery heat problem, the influence of temperature on the battery capacity and internal resistance is analyzed. In order to get good battery performance, the ideal operating temperature range of the lithium ion battery is 20~45℃, and the temperature difference between the batteries should be controlled within 5℃. The three-dimensional thermal model of lithium-ion battery is established by theoretical calculation of thermal physical parameters needed for the simulation of lithium-ion battery. The discharge experiment of single cell is carried out, and the simulation results show that the model has practical value. A power cell grouping module with an upper and lower layered structure is designed, and the optimized reciprocating flow heat dissipation scheme is selected. The maximum temperature and the temperature difference of the battery can be well controlled so as to meet the requirement of heat radiation management.

This research has only analyzed the heat dissipation of the battery in the high temperature environment, and the low temperature has great influence on the performance of the battery. Throughout the study, the inside of the battery is treated as a homogeneous heater, and each component is an independent heat source. Therefore, the next step should establish a fine thermal model to analyze the impact of low temperature on the performance of the battery. Based on this, the thermal management of the power battery in the full temperature range can be achieved.

References

[1] Perner, A., and J. Vetter. "Lithium-ion batteries for hybrid electric vehicles and battery electric vehicles." Advances in Battery Technologies for Electric Vehicles (2015): 173.

https://doi.org/10.1016/b978-1-78242-377-5.00008-x.
[2] Xie, Jinhong, et al. "Structural Optimization of Lithium-ion Battery Pack with Forced Air Cooling System." Applied Thermal Engineering (2017).
[3] https://doi.org/10.1016/j.applthermaleng.2017.07.143.
[4] Yang, Naixing, et al. "Assessment of the forced air-cooling performance for cylindrical lithium-ion battery packs: A comparative analysis between aligned and staggered cell arrangements." Applied Thermal Engineering 80 (2015): 55-65.
[5] https://doi.org/10.1016/j.applthermaleng.2015.01.049.
[6] YAMADA, Tatsuya, et al. "Analysis of a lithium-ion battery cooling system for electric vehicles using a phase-change material and heat pipes." Journal of Thermal Science and Technology 12.1 (2017): JTST0011-JTST0011. https://doi.org/10.1299/jtst.2017jtst0011.
[7] Cai, Wayne. "LITHIUM-ION BATTERY MANUFACTURING FOR ELECTRIC VEHICLES: A CONTEMPORARY OVERVIEW." Advances in Battery Manufacturing, Service, and Management Systems (2016): 1. https://doi.org/10.1002/9781119060741.ch1.
[8] Kurzweil, P. "Post-lithium-ion battery chemistries for hybrid electric vehicles and battery electric vehicles." Advances in Battery Technologies for Electric Vehicles (2015): 127-172. https://doi.org/10.1016/b978-1-78242-377-5.00007-8.
[9] Waag, Wladislaw, Christian Fleischer, and Dirk Uwe Sauer. "Critical review of the methods for monitoring of lithium-ion batteries in electric and hybrid vehicles." Journal of Power Sources 258 (2014): 321-339. https://doi.org/10.1016/j.jpowsour.2014.02.064.
[10] Saw, Lip Huat, et al. "Computational fluid dynamics simulation on open cell aluminium foams for Li-ion battery cooling system." Applied Energy (2017). https://doi.org/10.1016/j.apenergy.2017.04.022.
[11] Chen, Kai, et al. "Configuration optimization of battery pack in parallel air-cooled battery thermal management system using an optimization strategy."Applied Thermal Engineering 123 (2017): 177-186. https://doi.org/10.1016/j.applthermaleng.2017.05.060.

Analysis of the autonomous problem about coupled active non-Newtonian multi-seepage in sparse medium

Shuxian Deng, Hongen Li *

Department of Basic Courses, Zhengzhou University of Industrial Technology, Xinzheng, 451150, China

*Corresponding author e-mail: 857931663@qq.com

Abstract. The flow field of non-Newtonian fluid in sparse medium was analyzed by computational fluid dynamics (CFD) method. The results show that the axial velocity and radial velocity of the non-Newtonian fluid are larger than those of the Newtonian fluid due to the coupling of the viscosity of the non-Newtonian fluid and the shear rate, and the tangential velocity is less than that of the Newtonian fluid. These differences lead to the difference in the sparse medium Non-Newtonian fluids are of a special nature. The influence of the weight function on the global existence and blasting of the problem is discussed by analyzing the non-Newtonian percolation equation with nonlocal and weighted non-local Dirichlet boundary conditions. According to the non-Newtonian percolation equation, we define the weak solution of the problem and expound the local existence of the weak solution. Then we construct the test function and prove the weak comparison principle by using the Grown well inequality. The overall existence and blasting are analyzed by constructing the upper and lower solutions.

1. Introduction

The famous Darcy law marks the birth of porous media fluid mechanics, and the French mathematical mechanic Joseph Boussinesq gives the differential equation of the diving seepage movement, laying the foundation for the theory of unsteady flow. According to the similarity between the seepage problem and the heat conduction equation problem in the mathematical model, the Russian mathematical mechanics N.E. Rukovsky strongly promoted the development of porous media fluid mechanics by means of the heat conduction problem.

Since Kalashinikow, Oleinik and Czou put forward the porous media equation equation $u_t = \Delta\phi(u)$, many scholars at home and abroad have devoted themselves to the research of this problem. They use various theoretical models and research methods such as fractal theory, heat flux solid coupling theory, mixed theory and micro Pole theory and so on, and made a lot of research results.

In recent years, Castro, Gancedo and Orive [1] obtained the existence and uniqueness of subcritical solutions by studying the incompressible porous dielectric equations for finite energy cases. In supercritical cases, the solutions with small initial values and the asymptotic behavior of the solution in $L^p(p>2)$ space is given. A and B [2-4] through the study in the critical Besov space $B_{p,1}^{3/p}(R^n)$, the critical existence and uniqueness of the solution to the critical case.

By means of fractional Laplace operator and Riesz transform, we can see that the above equation is similar to the rotation form of the three - dimensional Euler equation, that is, its level set is similar to the eddy line of the three - dimensional Euler equation. Therefore, the above problems can be sub-critical, critical and supercritical three cases were studied in recent years has made a number of related results. For example, for the subcritical case, Constantin and Wu [5] prove the global existence of the smooth solution of the equation, Carrillo, Lucas and Ferreira [6] studied the asymptotic behavior of the solution. For the critical case, Kiselev, Nazarov and Volberg [7] demonstrate the overall well-being of the solution. In addition, many scholars [2-7] are committed to the study of this equation and make valuable conclusions. However, there are still many problems to be solved in the overall well-being of the supercritical situation. Through the study of the above problems, we can get a deeper understanding of the Eulerian equation and the Navier-Stokes equation from another perspective.

2. Preliminaries

We focus on a non-Newtonian fluid flows in a porous medium, its trajectory function is as follows:

$$\frac{\partial X}{\partial t}(\alpha,t)=u\left(X(\alpha,t),t\right)$$

Where $X(\cdot,t):\alpha\in R^N\to X(\alpha,t)\in R^N$ is the position of a particle of the fluid at time t and its initial position is $\alpha=(\alpha_1,\alpha_2,\cdots,\alpha_N)$, $X(\Omega,t)=\{X(\alpha,t):\ \alpha\in\Omega\}$ is the region of the initial region $\Omega\in R^N$ arrives after time t. Since we assume that the fluid considered is a non-Newtonian incompressible fluid, $V_t\left(X(\Omega,t)\right)=V_0(\Omega)$, where the volume mapping $\left(X(\cdot,t)\right)$ of the fluid is an incompressible map, which is equivalent to $\nabla\cdot u(x,t)=0$. It has the following three properties:

1. $\frac{\partial}{\partial t}\int_{X(\Omega,t)} f(x,t)\,dx=\int_{\Omega}[f_t+div_x(uf)]\,dx$, $\Omega\in R^N$ is bounded smooth open set.

2. $\frac{\partial J}{\partial t}=(div_x u)J(\alpha,t)\nabla_\alpha X(\alpha,t)$

In which $J(\alpha,t)=\det(\nabla_\alpha X(\alpha,t)),\nabla_\alpha=\left(\frac{\partial}{\partial x_1},\frac{\partial}{\partial x_2},\cdots,\frac{\partial}{\partial x_n}\right)$.

3. $\forall\Omega\in R^n, V_t\left(X(\Omega,t)\right)=V_0(\Omega)\Leftrightarrow\nabla\cdot u(x,\ t)=0\Leftrightarrow J(\alpha,t)=1$

For the sake of discussion, we give the following definition:

Definition 1 suppose $d\geq0$, $u\in B\left(R^N\right)$, $\overline{u}$ is the Fourier transform of u, $\left\|u,B\left(R^n\right)\right\|<\infty,\overline{u}\in L^2\left(R^n\right),\left\|u,B\left(R^n\right)\right\|\overset{\Delta}{=}\underset{\gamma\in\mathbb{R}^n}{ess\sup}|\gamma|^d|\overline{u}(\gamma)|$.

Definition 2 for any $u(x,t),\delta(x)\in B\left(R^n\right)$,and

(1) $u(x,t)$ is weakly continuous in $B\left(R^N\right)$,that is $\langle u(x,t)-u(x,0),\delta(x)\rangle\to0,(t\to0)$.

(2) $\lim_{t\to t_0}\left\|u(x,t)-u(x_0,t)\right\|_{B(R^n)}=0,(\forall t,t_0>0)$

Then the space composed of $u(x,t)$ is $C\left([0,+\infty);B\left(R^n\right)\right)$.

3. Main Conclusion

We consider the following mathematical model for a class of incompressible non-Newtonian porous media problems:

$$\begin{aligned}&\frac{\partial T}{\partial t}+u\cdot\nabla T+\sigma(-\Delta)^{\tau}T=f\\&\frac{1}{r}\frac{\partial}{\partial r}\left(r\frac{\partial\omega}{\partial r}\right)+\lambda\omega=-\Phi(x,r,t)\\&\nabla\cdot u=0\\&u_0(x)=u(x,t)\big|_{t=0},T_0(x)=T(x,t)\big|_{t=0}\end{aligned}\tag{1}$$

σ is the viscosity coefficient. $T(x,t)$ and $f(x,t)$ are the real functions about t and x, $\tau\in[0,p)$.

We study the suitability of the equation and the asymptotic behavior of the solution. Note that the dissipation mechanism is given by the fractional Laplacian operator and that the velocity is related to the pressure, and we will use the new tools and methods to get the following conclusions:

Theorem $\forall\varepsilon>0$, The kernel function $H(x)$ has the following point estimation

$$|H(x)|\le M(1+|x|)^{-(n+\varepsilon)},x\in\Omega$$

And $\forall q\in(1,+\infty)$, we have

$$H^{(k)}(x)=(-\Delta)^{\frac{k}{3}}H(x),H_t^{(k)}(x)=(-\Delta)^{\frac{k}{3}}H_t(x),t\in(0,+\infty)$$

Furthermore, suppose that $Q(x)$ is the unique positive solution of equation $div\left(|\nabla Q|^q\nabla Q\right)=a$, $x\in\Omega$, $Q(x)_{x\in\partial\Omega}=0$, then exists a $\kappa>0$, such that $\kappa=\sup\limits_{\Omega}Q(x)<+\infty$

$$\|u^k-H(x)u_h^k\|_{w.2}\leqslant C(h^{l+1}+\tau^{2-\alpha}+\varepsilon)\tag{2}$$

4. Proof of Conclusion

We give the main conclusion of the proof.

Proof of theorem: $\forall\vartheta(x)\in B(\Omega)$ and $t=0$,

$$\|H_\alpha(t)T_0-T_0,\vartheta(x)\|-\left|\int_\Omega(e^{-|\alpha|^pt}-1)T_0(\alpha)\vartheta(\alpha)d\alpha\right|$$

$$\leqslant C_1(t)\|T_0;B(\Omega)\|\,\||\alpha|^{2q-1}\vartheta(x),L^\infty(\Omega)\|\to0,\ t\to0\tag{3}$$

this means $\forall t\in[0,+\infty)$, $H(\cdot)T_0$ is continuously.

On the other hand

$$\xi(x,t)=\tau\left(\frac{\omega^m}{1-\omega^m}\right)^{\frac{1}{m}}=\tau\left(\frac{\mu-1}{1-\omega^m}\right)^{\frac{1}{m}}\omega\leqslant\tau\left(\frac{\omega^m}{1-\omega^m}+1\right)^{\frac{1}{m}}\int_\Omega u(x,t)dt$$

$$\leqslant\int_\Omega u(x,t)\tau\left(\frac{\omega^m}{1-\omega^m}+\frac{\Gamma(t)}{p+\varepsilon}\right)^{\frac{1}{m}}dt\leqslant\int_\Omega\tau u(x,t)\xi(x,t)dt\tag{4}$$

If $P(x,t)\in C^2([0,+\infty);B(\Omega))$, it is obviously that

$$L^2(\Omega)\in C^2([0,+\infty);B(\Omega)),\quad\mathcal{L}(P)\in L^\infty([0,+\infty);\Omega)$$

In fact, $\| L^2(\Omega) \| \leqslant \sup_{t>0} \operatorname{ess\,sup}_{\xi\in\Omega} \int_0^1 e^{-t(1-\sigma)|\xi|^{2\sigma}} t \| P(\sigma t,x);B(\Omega) \| d\sigma$

$$\leq t \sup_{1<\sigma<p} \left\| P(x,t);B(\Omega) \right\| \tag{5}$$

From Green formula and the definition of interpolation

$$\| \partial_t L(P);\Omega \| \leqslant \sup_{t>0} \operatorname{ess\,sup}_{\xi\in\Omega} |\alpha|^{\lambda} e^{-t|\alpha|^{\frac{\tau}{2}}} \int_0^1 e^{-\sigma t|\xi|^{2\sigma}} t\, d\sigma + \| P(x,t) \|$$

$$\leqslant ess \sup_{1<\kappa<p+1} |\alpha|^{\lambda} e^{-t|\alpha|^{\frac{\tau}{2}}} \| P(\kappa x,t) \| + \| P(x,t) \| \tag{6}$$

(1) if $\kappa < 1 + \dfrac{q+m}{p-1}$, let $\lambda = \max\left\{ u(x,t)\,|_{t=0}, |\Omega| M^{2\alpha-1}, \left(\dfrac{\gamma^q}{1-\gamma^\tau}\right)^{\frac{\alpha}{q+n}} \right\}$, by the comparision principle, $\xi(x,t)$ satifies the conditions, there holds the conclusions.

(2) If $\kappa = 1 + \dfrac{q+m}{p-1}$, let $|\Omega| \leqslant \left(\dfrac{1+v^p}{C(1+\rho^q)} \right)$, $\lambda = \max\{u(x,t)\,|_{t=0}, |\Omega| M^{2\alpha-1}\}$

$$\xi_t - div(|\nabla\xi|^{q+1}\nabla\xi) - \beta\xi^p \int_\Omega \xi(x,t)dx$$

$$= \alpha \left(\frac{\rho^{\frac{q-1}{p}}}{M} \right)^m - \tau^{q+1-\mu} \left(\frac{\gamma^q}{1-\gamma^p} + \frac{H(x)}{\rho} \right)^{\frac{m+1}{p}} \cdot \int_\Omega \kappa^{\alpha+2-\lambda} \left(\frac{\gamma^m - 1}{\gamma^{q-2}} + \frac{|\Omega|}{|H(x)|} \right)^{\frac{p+1-\alpha}{q+1}} dx$$

$$\leqslant \kappa \left(\frac{\varepsilon}{|\Omega|} \right)^{\frac{p-1}{p}} \left(\frac{1}{\lambda^{q-1}} - \left(\frac{1}{1-\gamma^{q-1}} \right)^{\frac{p-1}{p}} |\Omega| \right) \tag{7}$$

By the comparision principle, $\xi(x,t)$ satifies the conditions, there holds the conclusions.

(3) If $\kappa > 1 + \dfrac{q+m}{p-1}$, Let $\min_{x\in\Omega} |\xi(x,t)| \leqslant \lambda \leqslant \left(\kappa^{-(q+2)} (1-\gamma^m)^{\frac{q+n}{p+m}} |\Omega|^{\alpha} \right)^{\frac{m(p+1)}{q+m}}$

by the comparision principle, $\xi(x,t)$ satifies the conditions, there holds the conclusions.

Ulteriorly,

$$|\sigma T| \leqslant M \int_\Omega |\xi|^{-p} |\xi - \eta|^{-\frac{1}{p}} d\eta \| \sigma; C([0,+\infty);B(\Omega)) \| \, \| T; C([0,+\infty);B(\Omega)) \|$$

$$\leqslant M |\xi - \eta|^{q+2\alpha} \| \sigma; C([0,+\infty);B(\Omega)) \| \, \| T; C([0,+\infty);B(\Omega)) \| \tag{8}$$

From direct calculation, one can get

$$\| A(\sigma,T); C([0,+\infty);\Omega) \| = \sup_{t>0} \operatorname{ess\,sup}_{\xi\in\Omega} \left| \int_0^t |\alpha|^{\lambda} e^{-(t+v)|\xi|^{\delta}} (\sigma\cdot\nabla T)(\delta,\xi) dt \right|$$

$$\leqslant \varepsilon \sup_{t>0} \operatorname{ess\,sup}_{\xi\in\Omega} |\xi|^{\frac{p-1}{p}} \left| \int_0^t e^{-(t+v)|\xi|^{\delta}} dt \right| \|\sigma\|_{C[0,+\infty)} \|T\|_{C[0,+\infty)}$$

$$\leq \delta|\xi|^{\frac{p-1}{p}}\left|\int_0^t e^{-(t+v)|\xi|^\delta}dt\right|\|\sigma\|_{C[0,+\infty)}\|T\|_{C[0,+\infty)} \tag{9}$$

From the capillary pressure law

$$P_c(S_g) = P_g - P_w \tag{10}$$

We introduce the following coefficients

$$\rho_j(S_g,P)=\rho_j(P_j(S_g,P)) \tag{11}$$

$$\lambda(S_g,P)=\rho_w(S_g,P)\lambda_w(S_w)+\rho_g(S_g,P)\lambda_g(S_g) \tag{12}$$

$$\alpha(S_g,P)=\rho_w(Sg,P)\rho_g(S_g,P)\lambda_w(S_w)\lambda_g(S_g)/\lambda(Sg,P) \tag{13}$$

Note that the choice of $S = S_g$ as primary saturation variable is motivated by the fact that $S_g \to P_c$ (S_g) is an increasing function.

$$\begin{aligned}\Delta_{m,p}u &= |\nabla u^m|^{\gamma+1}\triangle u^m + \frac{p-2}{m\gamma}|\Delta u^m|^{m+1}(\nabla u^m)^\lambda(G_x(u^m))\nabla u^m \\ &= |\nabla u^m|^{\gamma+1}\triangle u^m + \frac{p-2}{m\gamma}|\Delta u^m|^{m+1}(\nabla u^m)^\lambda\frac{\sqrt{\eta_w(S_w)\rho_g(P_g)}}{\mu(P_g,S)}\end{aligned} \tag{14}$$

$\forall\varphi(x)\in B(\Omega)$,From Gagliardo-Nirenberg inequality, we can get $\langle P(u,t),\varphi(x)\rangle \to 0$.

Boundary conditions: let $\Omega\subset R^n$ be a bounded, Lipschitz domain with its boundary divided in two parts, $\partial\Omega = \Gamma_{in}\cup\Gamma_{im}$, where Γ denotes the injection boundary, and impervious one. Let [0, T] be the time interval of interest and $Q_T= \Omega\times [0, T]$. Such correction does not influence the system behaviour in physically admissible range of pressures, and therefore our discussion does not limit the applicability of our result.

$$|\Delta_{m,p}u| \leq \frac{p-2}{B(S-t)^{\lambda+2/3}} + \left(\frac{|\Omega|}{(S-t)^{q-1}}\right)^{n+1-2\alpha} \leq \frac{(q-1)\|\nabla S_g\|^{1/2}|\xi|^{p+q-2}}{B(P-t)^{(q+n)(\lambda-2)}} \tag{15}$$

From(8) (13), and assumption discribed above, by Cauchy-Shwartz inequality and Young inequality with ε, one can get

$$\begin{aligned}\int_\Omega u^n(x,t)e^{-\frac{\lambda}{2}}dx &\leq \frac{|\xi|}{(S-t)^{\gamma/2}}\int_\Omega P^n\frac{\|\nabla u\|^{1/2}}{(S-t)^\tau}dx \\ &\leq \frac{|\xi|}{(S-t)^{q+n}}\int_\Omega S^{n(p+1)}\left(\frac{|\xi|}{T-t}\right)^{(p-1)/2}dt \leq C_1(\varepsilon)\end{aligned} \tag{16}$$

Let $g(x,t) = \int_0^\infty G_\alpha(\tau)\nabla\cdot f(x,t-\tau)V^n(\xi)d\tau$,Without losing the generality, we take a range(a,b) for t_0,that is confine t,in which

$$f(x,t-\tau)=uT(t-\tau)\phi^{\frac{\alpha}{2}}(t),\qquad g(x,t)=g_1(x,t)+g_2(x,t)$$

$$\begin{aligned}\|\partial_t g_1(x,t);\ B(\Omega)\| &\leqslant \left\|\frac{d}{dt}\int_\Omega e^{-(T-t)/2}f_1(x,t)\,|g_1(x,t)|dt;C([0,+\infty);\Omega)\right\|\\ &\leqslant \operatorname*{ess\,sup}_{\xi\in\Omega}\left|\frac{d}{dt}\int_\Omega e^{m\gamma+2\sigma}g_1(x,t)dt+\rho_w(S_g,P)S_w+\|\nabla P\|\right|\\ &\leqslant M\frac{m(p-1)}{t(q+n)}\|u;C([0,+\infty);\Omega)\|+\sigma div(a(S_g,P)\nabla S_g)\end{aligned}\tag{17}$$

We can also talk about the continuity of $g_2(x,t)$, We may wish to assume $a<t<b$, one can get

$$\begin{aligned}g_2(x,t)-g_1(x,t)&=\int_0^t S(t-\sigma)\left[f_2[x,t-\sigma]-f_1[x,t-\sigma]\right]d\sigma-div(\Gamma_w(S_g,P)\kappa\nabla P)\\ &\quad+\int_0^t P(t-\eta)f_2(x,t-\eta)d\eta-\frac{\sqrt{\lambda_w(S_w)\lambda_g(S_g)}}{\lambda(S_g,P)}\\ &=A+Q\end{aligned}\tag{18}$$

For A, if $1\leqslant P(\xi)\leqslant(\alpha+\sqrt{A})^{\frac{p+1}{m}}$, $P'(\xi)\leqslant 0$

$$\begin{aligned}\|A;C([0,\infty);\Omega)\|&\leqslant M\operatorname*{ess\,sup}_{\xi\in\Omega}\|\xi\|^{-2\tau}\left|\int_0^t e^{r(t-\tau)\frac{q-1}{p}}d\tau\right|\|u;0,t;L^q(\Omega)\|\,\|T\|_{L^q(\Omega)}\\ &\leqslant M_1\|\xi\|^{-2\tau}\left|\int_0^t e^{r(t-\tau)}d\tau\right|\|u;0,t;L^q(\Omega)\|\,\|T\|_{L^q(\Omega)}\end{aligned}\tag{19}$$

$$\begin{aligned}\|A(0,P)\|&\leqslant\iint_\Omega(u\phi_t-|\nabla u^m|^{p-2}\nabla u^m\nabla\phi-u^q\phi)dxdt\\ &\leqslant\int_0^t\left(\int_\Omega V^n|u|^q|\xi|d\xi\right)dt\leqslant\int_\Omega P^n|u|^q|\xi|d\xi=M\end{aligned}\tag{20}$$

In a similar way, we can deal with Q and combine with (16)- (20), hence

$$\begin{aligned}&\iint_\Omega\eta(t)\mathrm{sgn}_+(u-v)\frac{\partial(u-v)}{\partial t}dxdt+\int_0^{S_g}f_w(s,P)P_c'(s)ds\\ &\leqslant\int_\Omega(u(x,s_2)-v(x,s_2))dx+\int_\Omega(u(x,s_1)-v(x,s_1))dx\\ &\leqslant u-\Delta_{m,p}u-\sigma u^q\int_\Omega e^{\|u(x,t)\|}u^n(x,t)dx+\int_{s_1}^{s_2}s^{2-\gamma}P_c'(s)s^{\gamma-1}f_w(s,p)|\xi|^{\frac{p-q+1}{2}}ds\\ &\leqslant C\int_{s_1}^{s_2}e^{\|u(x,t)\|}s^{\gamma-1}ds+S_1(P_c(S_2)-P_c(S_1))-\int_{s_1}^{s_2}sP_c'(s)ds\\ &\leqslant\frac{C}{\gamma}|S_1-S_2|^\gamma e^{-\tau t}\leqslant Me^{-\tau t}\end{aligned}\tag{21}$$

IOP Conf. Series: Journal of Physics: Conf. Series **916** (2017) 012037 doi:10.1088/1742-6596/916/1/012037

Let $t \to \infty$, and notice $s_1 \to 0$, $\tau = const$, we come to the conclusion.

5. Simulation and Numerical Experiments

We use COMSOL software to simulate the parameter setting of the above equation and the analysis of the results. As can be seen that its large time behavior is stable. To achieve the optimal order and convergence effect, which is consistent with the previous theoretical analysis of the results

6. Conclusion

In this paper, we study the asymptotic behaviorof the fluid equation in the porous medium with nonlinear boundary source nonlinear diffusion of incompressible fluid. Based on the non-Newton fluid equation of the incompressible porous media, the priori estimates in the measure space are discussed by using the properties of the operator semigroup and the measure space and the principle of the compression image. Through the diffusion regularization method and the ω-limit compact method, we study the overall asymptotic behavior of the solution of the equation in Ω space.

References

[1] CHEN Yarzhe. On finite diffusing speed for uniformly degenerate quasilinear parabolic equations[J]. ChinAnn Math, 1986, 7B: 318-329.

[2] Pelipenko S, Frigaard I A. Computational method forsimulation of eccentric annular displacements of non-Newtonian fluids. submitted to Elsevier. Preprint,March, 2003: 1-17.

[3] Szabo P, Hassager O. Displacement of one Newtonian Fluid another: density effects in axial annular flow. International. Journal Multiphase Flow, 2017 (1):113-129.

[4] Bakker A, Laroche R D, Wang M H, et al. Sliding mesh simulation of laminar flow in stirred reactors[J]. Tians. IChemE.: PartA, 2017, 75: 42-44.

[5] Reyna LG, Ward MJ. On exponential ill-conditioning and internal layer behavior. Numerical Functional Analysis and Optimization. 2015, 16:475-500.

[6] A.V. Ivanov. Holder sestimats for quasilinear doubly degenerate parabolic equations[J], J. Sov. Math., 2011(56): 2330-2347.

[7] V. A. Galaktionov and H. A. Levine, On critical Fujita exponents for heat equations with nonlinear flux conditions on the boundary, Israel J. Math., 94 (2016), 125-146.

The principle and applications of Bernoulli equation

Ruqiong Qin[a], Chunyi Duan[b]

Guangxi Polytechnic of Construction, Nanning 530003, China

[a]qinruqiong@163. com , [b]dcyjoyce@126. com

Abstract. Bernoulli equation is one of the most important theories of fluid mechanics, it involves a lot of knowledge of fluid mechanics, and is used widely in our life. This paper comprehensives the research present situation of Bernoulli equation at home and abroad, introduces the principle of Bernoulli equation and some applications in our life, and provides direction of the application for the future.

Keywords: Bernoulli equation; Principle; application.

1. Introduction

Bernoulli's principle is an important theory in fluid mechanics, involves a lot of knowledge of fluid mechanics, and is Daniel Bernoulli (d. Bernoulli, Swiss physicists, mathematicians, 1700 ~ 1782) in 1726, is three basic equation of hydrodynamics another, it is the embodiment of objects mechanical energy conversion of hydraulics.

Bernoulli equation solves the problem of force and energy which is often involved in engineering practice, which lays the theoretical foundation for solving hydraulic calculation of actual engineering.

The application of this equation runs through the course of hydraulic mechanics.

Based on the principle of Bernoulli's equation, this paper summarizes the research status of its application at home and abroad, and provides direction for the future application of Bernoulli equation.

1.1. The Bernoulli equation

As shown in figure l, the ideal fluid for steady flow in the gravitational field is to take a thin flow tube, and the area of the two cross sections, and the height of their relative horizontal reference plane.

Since the ideal fluid is incompressible, the quality of the fluid between the two is equal to the mass between.

The increment of mechanical energy is the increment of mechanical energy in the process of position/flow to position /.

$$\begin{aligned}\Delta E &= (E_k + E_p)_2 - (E_k + E_p)_t \\ &= \left[\frac{1}{2}(\Delta m)v_2^2 + (\Delta m)gh_2\right] - \left[\frac{1}{2}(\Delta m)v_1^2 + (\Delta m)gh_1\right] \\ &= \left(\frac{1}{2}v_2^2 + gh_2 - \frac{1}{2}v_1^2 - gh_1\right)\Delta m\end{aligned} \quad (1)$$

The pressure of the surrounding fluid is the total work of the fluid block:

$$\Delta A = \Delta A_1 + \Delta A_2 = p_1 S_1 v_1 \Delta t - p_2 S_2 v_2 \Delta t \quad (2)$$

According to the continuity principle, and

$$\Delta m = pS_2 v_1 \Delta t = pS_2 v_2 \Delta t \quad (3)$$

$$\text{So: } \Delta A = (p_1 - p_2)\frac{\Delta m}{\rho} \quad (4)$$

According to the principle of function, $\Delta E = \Delta A$ so,

$$p_1 + \frac{1}{2}pv_1^2 + \rho gh_1 = p_2 + \frac{1}{2}pv_2^2 + \rho gh_2 \quad (5)$$

The following relationship is always true for any section of the same flow tube:

$$p + \frac{1}{2}pv^2 + \rho gh = cons\tan t \quad (7)$$

Both of these are known as Bernoulli's equations, and they describe the fundamental laws of the ideal fluid for stationary flow. If the ideal fluid flows along the horizontal flow tube, then

$$p + \frac{1}{2}\rho v^2 = cons\tan t \quad (8)$$

When the ideal fluid flows along the horizontal pipe, the area of the pipe is small and the velocity is small and the area of the pipeline is small and the pressure is strong.

$$\text{So: } p_1 + \frac{1}{2}\rho v_1^2 + \rho gh_1 = p_2 + \frac{1}{2}\rho v_2^2 + \rho gh_2 \quad (9)$$

IOP Conf. Series: Journal of Physics: Conf. Series **916** (2017) 012038 doi:10.1088/1742-6596/916/1/012038

Take two points in the density of the static fluid, and make a list of Bernoulli's equation:

$$p_A + \rho g h_1 = p_B + \rho g h_B ORp_A - \mathrm{p}_B = \rho g h_a - \rho g h_B \quad (10)$$

If the height of two points is equal, then the upper equation shows that the pressure in the rest liquid is equal to the pressure in the two points [1-2].

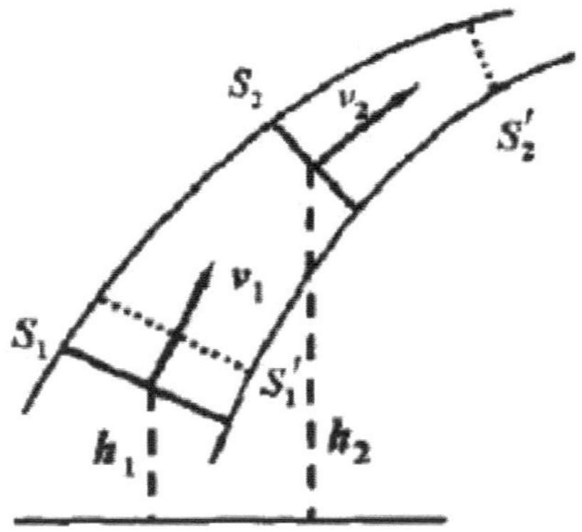

Fig. 1 schematic diagram of Bernoulli equation

If there is no heating or cooling fluid flow in the system, you can ignore the fluid viscosity, fluid density is constant, so, this is an ideal incompressible fluid system of isentropic flow system, the internal energy remains constant. The ability of this system is calculated, only the mechanical energy balance is required. The energy equation for the mechanical energy balance in this case is the Bernoulli equation.

In the case of incompressible heavy fluid, Bernoulli equation is the natural result of the first integral of the motion equation, which is the most important expression of energy relations in fluid mechanics [3].

The principle of Bernoulli's equation is in the same fluid, the velocity is large and the pressure is small. Small flow rate, strong pressure. Flow is automatically flowing from high pressure to low pressure. Its simple narrative is: when the ideal fluid as a stationary flow, if the level of fluid flow or the influence of the height difference was not significant (such as gas flow), Bernoulli equation can be expressed as $P + \rho V^2 / 2 = \mathrm{constant}$, namely in the fluid flow, velocity V of the local pressure P Small, velocity V of small pressure P big [4].

2. The application of the Bernoulli equation

Bernoulli equation, the principle of using a lot of, play football or play table tennis in the stagnation pressure, inside the chimney flue gas flow rate, water pump, water power, spray flow rate and pressure of siphon and radial flow between circular plate, etc. , some of the example below.

2.1. Agricultural spray gas

Agricultural sprayer is a kind of machine, the liquid dispersed currently use sprayer knapsack, stretcher and matched with the tractor traction model, rural pesticides commonly used simple and knapsack.

How can a sprayer spray water or liquid into a fog? As shown in figure 2, hard to push the horizontal tube, air from through a narrow place, by the Bernoulli principle. The narrow A larger flow rate and pressure is small, and in A vertical tube above the CB, CB tube, the pressure is less than the bottom pressure above the solution will be along the tube, powder under the action of wind pressure through the door, the water plug connector, iv reaches the nozzle. As the cross section shrinks suddenly and the flow rate increases, the air from the fan outlet passes through here to generate negative pressure. Under positive pressure and negative pressure, the liquid is ejected from the nozzle, and it coincides with the high speed air flow from the nozzle.

Because the velocity of the two is very different, and the direction is perpendicular, it is cut into small droplets of 100-150 diameter by the high-speed airflow and blown into mist at the outlet of the thin tube.

This paper USES the characteristics of large liquid velocity and small pressure in the narrow part of the pipe, and the phenomenon of external liquid suction is called air suction.

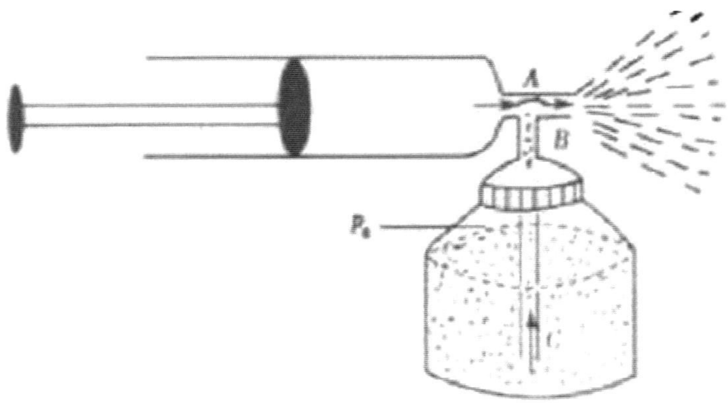

Fig. 2 A schematic diagram of the sprayer for agriculture

2.2. Cooling effect

Bernoulli's equation: $p + \frac{1}{2} pv^2 + \rho gh = cons\tan t$

For horizontal flow pipes, there are:

$$p + \frac{1}{2} \rho v^2 = cons\tan t \tag{11}$$

The ideal gas is used as the ideal fluid for horizontal flow pipes.

For an ideal gas, the pressure formula is the following [8]

$$p = \frac{1}{3} mn\bar{v}^2 = \frac{2}{3} n\bar{\varepsilon} = \frac{2}{3} n \bullet \frac{3}{2} kT = nkT \tag{12}$$

The Bernoulli equation in the horizontal flow tube

$$nkT + \frac{1}{2} \rho v^2 = cons\tan t \tag{13}$$

(In this formula is the molecular density of the gas, the density of the gas is $\rho = mn$, the Boltzmann constant is k.)

Soya can see from the upper formula that the lower the flow velocity, the lower the temperature of the gas. When the flow velocity is large enough to a certain value, the temperature of the gas in the tube will be lowered to the freezing point and crystallized in the wall. The combustible ice (gas hydrate) produced in natural gas pipelines is an example. When the tube is not uniform, it is known from the $Sv = cons\tan t$, continuity principle that the velocity of the detail is greater than the velocity of the thick part. According to the above formula, the temperature in the detail is lower than the temperature in the thick.

In order to understand the cooling effect of flow gas more clearly, the formula derived from above is changed to:

$$nkT_1 + \frac{1}{2}\rho v_1^2 = nkT_2 + \frac{1}{2}\rho v_2^2 \tag{14}$$

It is assumed that the density of the gas in the flow process is constant, so n is also invariable. For convenience, make the $v_1=0$,

So:

$$T_1 = T_2 + \frac{1}{2}\frac{\rho}{nk}v_2^2 = T_2 + Kv_2^2 \tag{15}$$

In the equation, $K = \frac{1}{2}\frac{\rho}{nk} \approx 2\times10^{-3}\,Km^{-2}s^2$ (The values are for ordinary air, different for different gases)

Substitute the different air velocity into the top formwork can get the Different numerical of $\Delta T = T_1 - T_2 = Kv_2^2$,

The list below:

Table 1. The Numerical table of ΔT

$v_2(m/s)$	1	5	10	20	30	50	70	100	200	300	400
$\Delta T(^0C)$	0. 002	0. 05	0. 2	0. 8	1. 8	5	9. 8	20	80	180	(320)

It can be seen from the table that the temperature drop is not obvious when the wind speed is small. When the wind speed is high, the temperature will decrease obviously. When the gas flow is very high, the temperature of the gas will drop very low. When the gas is very high, the temperature of the gas will be very low, and the gas cannot be considered as an ideal gas. Because for the general gas, only the pressure is not too high and the temperature is not too low to be considered an ideal gas.

Therefore, the last part of the list is written in brackets, where the temperature drop value cannot be calculated using the derived formula. Second, is derived from the formula may draw a conclusion: when the gas forming orderly movement, his disorderly thermal motion will be reined in, the higher the speed of the orderly movement, the lower the degree of disorder of thermal motion.

2.3. The collision of two ships near the road is easy to collide

As is known to all, in the fall of 1912, the "Olympic" in parallel with the cruiser HMS hack, although both are about 100 m, "hack" also is very small, but as if by ship took to the boat, don't obey the helmsman manipulation. Race 1 vigorously to the "Olympic". Finally, the ship's bow hit the side of the Olympic and hit the Olympic with a big hole.

The reason for this accident is as shown in figure 6. According to the Bernoulli principle of hydrodynamics, the pressure of the fluid is related to its velocity. The higher the velocity, the smaller the pressure; and vice versa. It turns out that when two ships parallel forward two ships in the middle of the water is faster than the outside of the water flow rate, according to the principle of Bernoulli equation, in the middle of the water on the pressure of the two internal au 5 is smaller than the lateral pressure of two ship lateral.

Then, under the pressure of the outside water, the two boats approached and collided. Again due to the small "g", in the same size pressure, under the action of it to two ship speed is much faster, therefore, the "hack" crash "Olympic" number of accidents. This phenomenon is now referred to as "ship suction".

3. Summary

Bernoulli's equation is in a fundamental position in fluid mechanics. Although it is based on the incompressible ideal fluid, on the basis of stationary flow, but when a lot of fluid flow can be approximately regarded as ideal fluid, using Bernoulli equation can get many useful conclusions.

In practical engineering, the use of Bernoulli equation can solve many problems, such as the lift of airplane wings and the air pump. Bernoulli equation in water conservancy, shipbuilding, chemical industry, aviation and other departments have a wide range of applications, as long as a careful study of these application instance, everyone has the ability to further for the application of Bernoulli equation to find other innovations, for example, reservoir gate drain flow calculation; The calculation of the force of the pipe bend in the horizontal plane; Put a Ping-Pong ball in the inverted funnel, and blow with the mouth into the mouth of the funnel. The Ping-Pong ball will not only not fall but will be pressed more tightly with the funnel.

It can be seen that learning a little Bernoulli's knowledge is of great significance for developing intelligence and applying physics knowledge.

Acknowledgements

Project source: This paper is one of the results of Guangxi scientific research projects in 2017. Project name: Sponge campus and low impact development of rainwater use typical assessment measures and application research. The approval number: GUI teach [2017] 16, scientific research.

References

[1] Dawei Song. Physical application basis [M] . beijing: People's Medical Publishing House. 2008, 2—4.
[2] Li xin Gu, Yuqing Liu. physics [M] . Henan: Elephant press, 2007, 10—11.
[3] Zhiquan Liang. Fluid mechanics [M] . Sichuan: Chongqing university press, 2002, 3—5.
[4] Xinjun Zhao. The basis of medical physics [M] . Henan: Henan science and technology press. 2007, 30—32.
[5] Hong Wei. Working principle of motor sprayer [0L] . http: //www. chinabaik, 2008.
[6] Gaihua Zhao, Weiyin Luo. New concept physics tutorial: 1st edition of mechanics [M]. Beijing: education press, 1995, 235—244.
[7] Jiatai Guo, Yanfeng Jing. The application of Bernoulli equation-cold and cool [J] . Wuhan: mathematics and medicine journal, 2008, 43—44.
[8] Xinmin Hu. Medical physics: version 6 [M]. Beijing: people's health press, 2004, 107—109.
[9] Lixie Chen. The indomitable captain - Bernoulli principle [J]. Encyclopedia of knowledge, 2001.
[10] Yanli Chen. The principle and application of Bernoulli equation [J]. Henan: journal of luohe vocational and technical college, 2012, 30—31.

A New technique of Initial Boundary Value Problems Using Homotopy Analysis Method

D. M. Wang [1,*], W. Zhang [2], M. H. Yao [2], Y.L.Liu [1]

[1]College of science, Tianjin University of Science and Technology, Tianjin, China
[2]College of Mechanical Engineering, Beijing University of Technology, Beijing, China.

*Corresponding author e-mail: mutouwang@yeah.net

Abstract. In this paper, a new homotopy analysis technique which is applying to solve initial boundary value problems of partial differential equations by admitted both the initial and boundary conditions in the recursive relation to obtain a good approximate solution for the problem is proposed. The proposed iterative scheme finds the solution without any discretization, linearization, or restrictive assumptions. Furthermore, we can easily control and adjust the convergence domain and rate of series solutions by the convergence control parameter. The effectiveness of the approach is verified by several examples.

1. Introduction

Many important physical phenomena on the engineering and science fields are frequently modeled by partial differential equations. However, it is difficult to obtain closed-form solutions for them, especially for nonlinear ones. In most cases, only approximate solutions (either analytical ones or numerical ones) can be expected. Numerical solutions got by numerical techniques such as Finite Difference Method, Finite Element Method and so on are only the discrete points approximation. It is difficult to have a comprehensive understanding of the nature through the numerical solutions of the problem. Nevertheless, analytical expressions can display different physical quantities are investigated and the variation tendency, so it can reflect the essence of the problem. Thus, analytic methods to obtain approximate solutions have gained importance in recent years. There are Various techniques for seeking analytical solutions to the partial differential equations such as perturbation techniques [1], Darboux transformation [2], the tanh method [3], homogenous balance method [4], Adomian Decomposition Method (ADM) [5], Variation iteration method (VIM) [6], homotopy perturbation method (HPM) [7] and so on.

Comparing with those methods, the HAM initially proposed by Liao [8] who employed the basic idea of the homotopy in topology in his PH.D. thesis is a general analytic approach to get solutions of partial differential equations and has many advantages [9] as follows, first, it is independent of any small or large quantities such as perturbation techniques. So the HAM can be applied no matter if governing equations and boundary/initial conditions contain small or large quantities or not. Second the HAM does not require specific algorithms and complex calculation such as ADM which has difficulties to calculate the so-called Adomian polynomials and VIM which has difficulties of optimal identification of Lagrange multipliers via the variational theory. Furthermore, the HAM always provides us with a family of solution expressions in the auxiliary parameter c_0, which provides us with a simple way to adjust and control the convergence region and rate of convergence of the series solution. We can find a proper value of c_0 by the so called c_0-curves, as suggested by Liao [10]. Liao [10-11] pointed out that ADM and HPM are special case of the homotopy analysis method. When the convergence control parameter $c_0 = -1$, HAM becomes the so-called HPM. Literature [11] showed that sometimes the approximation series given by HPM are not convergent. So the HAM is a general analytic approach to solve differential equations.

In recent yeas, this method has been successfully employed to solve many types of problems in science and engineering such as the viscous plows of non-Newtonian fluids[12], the KdV-type equations [13], a new technique of using homotopy analysis method for solving high order nonlinear differential equations [14], solving two point boundary value problems by homotopy analysis method [15] and Fourth-order parabolic partial differential equation with variable coefficients [16], The HAM approach is applied obtaining the solution of the one-phase inverse Stefan problem by using the homotopy analysis method [17], Solving the convection–diffusion equation by means of the optimal q-homotopy analysis method [18]. All of these successful applications verified the validity, effectiveness and flexibility of the HAM.

Yet to solve partial differential equations with initial and boundary conditions, the classical HAM among some other series solution methods, build the recurrence scheme of solution using only one type of the problem conditions: either the initial conditions or the boundary conditions. Our aim here is to propose a new HAM technique that incorporates both types of conditions in the scheme to solve initial-boundary value problems (IBVP) over finite domains.

2. The Classical HAM

Consider the following equation

$$N(u(x,t)) = 0, \quad (x,t) \in \Omega, \tag{1}$$

along with boundary conditions:

$$B(u, \frac{\partial u}{\partial n}) = 0, \ (x,t) \in \Gamma \ . \tag{2}$$

Where N is a nonlinear operator, $u(x,t)$ is an unknown function, B is a boundary operator, and Γ is the boundary of the domain Ω, x and t denote the spatial and the temporal independent variables, respectively. The operator N can be generally divided into linear and nonlinear parts, say L and F. Therefore (1) can be written as

$$L(u) + F(u) = 0 \tag{3}$$

By means of generalizing the traditional homotopy method, Liao constructs a homotopy $\varphi(x,t;p) : \Omega \times [0,1] \to R$ which satisfies the so-called zero-order deformation equation

$$(1-p)L[\varphi(x,t;p) - u_0(x,t)] = pc_0 N[\varphi(x,t;p)], \qquad (x,t) \in \Omega \tag{4}$$

Where $p \in [0,1]$ is an embedding parameter (called homotopy-parameter), c_0 is a nonzero auxiliary parameter which is called convergence-control parameter, L is an auxiliary linear operator, $u_0(x,t)$ which satisfy initial conditions or boundary conditions is an initial guess of

$u(x,t)$, $\varphi(x,t;p)$ is an unknown function, respectively. It is important, that one has great freedom to choose auxiliary things in HAM. Obviously, from (2.4), when $p=0$ and $p=1$ one has

$$\varphi(x,t;0)=u_0(x,t) \quad ; \quad \varphi(x,t;1)=u(x,t); \tag{5}$$

Respectively. Thus, as p increases from 0 to 1, the solution $\varphi(x,t;p)$ varies from the initial guess $u_0(x,t)$ to the solution $u(x,t)$.Expanding $\varphi(x,t;p)$ in Maclaurin series with respect to p , one has

$$\varphi(x,t;p)=u_0(x,t)+\sum_{m=1}^{\infty}u_m(x,t)p^m \tag{6}$$

Where $$u_m(x,t)=\frac{1}{m!}\frac{\partial^m\varphi(x,t;p)}{\partial p^m}\Big|_{m=0} \tag{7}$$

Here the series (6) is called homotopy-series and Eq.(7) is called the mth-order homotopy-derivative of φ .If the auxiliary linear operator, the initial guess and the convergence-control parameter c_0 are so properly chosen, the homotopy-series (6) converges at $p=1$, then using the relationship $\varphi(x,t;1)=u(x,t)$, one has the so-called homotopy-series solution

$$u(x,t)=u_0(x,t)+\sum_{m=1}^{\infty}u_m(x,t) \tag{8}$$

Which must be one of solutions of original equation, as proved by Liao[10]. As $c_0=-1$, Eq.(4) become

$$(1-p)L[\varphi(x,t;p)-u_0(x,t)]+pN[\varphi(x,t;p)]=0, \qquad (x,t)\in\Omega \tag{9}$$

Which is used in the homotopy perturbation method.
Let us define the vector

$$\vec{u}_m=\{u_0(x,t),u_1(x,t),\cdots u_m(x,t)\} \tag{10}$$

Substituting (6) into (4) and equating the terms with identical powers of p , we have the so-called mth-order deformation equation

$$L[u_m(x,t)-\chi_m u_{m-1}(x,t)]=c_0R[\vec{u}_{m-1}(x,t)] \tag{11}$$

Where

$$R(\vec{u}_{m-1})=\frac{1}{(m-1)!}\frac{\partial^{m-1}N[\varphi(x,t;p)]}{\partial p^{m-1}}\Big|_{p=0} \tag{12}$$

And

$$\chi_m = \begin{cases} 0, & m \le 1, \\ 1, & m \ge 2. \end{cases} \tag{13}$$

It should be emphasized that $u_m(x,t)$ for $m \ge 1$ is governed by the linear equation (11) with the linear boundary conditions that come from the original problem, which can be easily solved by symbolic computation software such as Maple and Mathematica . The solutions $u(x,t) = u_0(x,t) + \sum_{m=1}^{\infty} u_m(x,t)$ is approximated by the truncated series

$$U_n(x,t) = \sum_{i=0}^{n} u_i(x,t) \tag{14}$$

3. The HAM technique for IBVP over finite domains

In this section, we propose the HAM technique for solving IBVP over finite domains. Suppose (1) can be written

$$Qu + Mu + F = 0 \tag{15}$$

Associated with initial and boundary conditions where Q denotes the highest-order partial derivative with respect to t, M denotes the highest-order partial derivative with respect to x and F is a function of x, t, u , and its temporal and spatial partial derivatives of order less than the order of Q and M , respectively. Then, to include both the initial and boundary conditions in the solution, we construct the two homotopies:

$$(1-p)Q[\varphi(x,t;p) - u_{0_1}(x,t)] = c_0 pN[\varphi(x,t;p)]; \tag{16}$$

$$(1-p)M[\varphi(x,t;p) - u_{0_2}(x,t)] = c_0 pN[\varphi(x,t;p)]; \tag{17}$$

$u_{0_1}(x,t)$ satisfies initial conditions and $u_{0_2}(x,t)$ satisfies boundary conditions respectively. for each homotopy the corresponding powers of p are compared to obtain two systems of partial differential equations with the prescribed conditions. We assume the solutions of problem (15) in the form

$$u = \sum_{i=0}^{\infty} u_i \tag{18}$$

Where u_i is given by

$$u_i = \frac{u_{i_1} + u_{i_2}}{2} \quad i = 0,1,\cdots, \tag{19}$$

Where u_{i_1} and u_{i_2} are solutions of the ith equations in the PDE systerms obtained from the homotopies (16) and (17), respectively.

4. Numerical Implementation

In this section, two numerical examples are presented to validate the proposed solution scheme. The results are calculated using the symbolic software maple.

Example 4.1 Consider the wave equation:

$$\begin{cases} \dfrac{\partial^2 u}{\partial t^2} = \dfrac{\partial^2 u}{\partial x^2} + \sin 3\pi x, \qquad 0 < x < 4, t > 0 \\ u(0,t) = u(4,t) = 0 \\ u(x,0) = \sin(\pi x), \ \dfrac{\partial u}{\partial t}(x,0) = \sin 2\pi x \end{cases} \tag{20}$$

According to the homotopies (16) and (17), the two following systems of PDEs are obtained

$$\frac{\partial^2 u_0}{\partial t^2} = 0, \qquad u_0(x,0) = \sin \pi x, \ \frac{\partial u_0}{\partial t}(x,0) = \sin 2\pi x,$$
$$\frac{\partial^2 u_m}{\partial t^2} = \begin{cases} (1+c_0)\dfrac{\partial^2 u_{m-1}}{\partial t^2} - c_0\left(\dfrac{\partial^2 u_{m-1}}{\partial x^2} + \sin 3\pi x\right), & m = 1 \\ (1+c_0)\dfrac{\partial^2 u_{m-1}}{\partial t^2} - c_0\dfrac{\partial^2 u_{m-1}}{\partial x^2}, & m > 1 \end{cases} \tag{21}$$
$$u_m(x,0) = 0, \ \frac{\partial u_m}{\partial t}(x,0) = 0, \qquad (m \geq 1)$$

$$\frac{\partial^2 u_0}{\partial x^2} = 0, \qquad u_0(0,t) = 0, \ u_0(4,t) = 0,$$
$$\frac{\partial^2 u_m}{\partial x^2} = \begin{cases} (1+c_0)\dfrac{\partial^2 u_{m-1}}{\partial x^2} - c_0\left(\dfrac{\partial^2 u_{m-1}}{\partial t^2} - \sin 3\pi x\right), & m = 1 \\ (1+c_0)\dfrac{\partial^2 u_{m-1}}{\partial x^2} - c_0\dfrac{\partial^2 u_{m-1}}{\partial t^2}, & m > 1 \end{cases} \tag{22}$$
$$u_m(0,t) = 0, \ u_m(4,t) = 0, \qquad (m \geq 1)$$

Solving (21) and (22), we can get any order homotopy solutions for problem (20). Fig.1. shows the c_0-curve for 3th-order HAM approximation of $u(0.2,0.6)$ of the problem(20). Fig.2. gives the comparison at t=0.6 between the NHAM 3rd-order approximate solutions of the problem(20) and the exact solution $u(x,t) = \cos \pi t \sin \pi x + \frac{1}{2\pi}\sin 2\pi t \sin 2\pi x + \left(-\frac{1}{9\pi^2}\cos 3\pi t + \frac{1}{9\pi^2}\right)\sin 3\pi x$. From Fig.1. and Fig.2., we can find that when $c_0 \in [-0.25, -0.3]$, the NHAM3[rd]-order approximate solutions meet the exact solution closely.

Example 4.2 Consider the heat-like equation:

$$\begin{cases} \dfrac{\partial u}{\partial t} - \dfrac{1}{2}x^2 \dfrac{\partial^2 u}{\partial x^2} = 0 \qquad 0 < x < 1, t > 0 \\ u(0,t) = 0, \qquad u(1,t) = e^t \\ u(x,0) = x^2 \end{cases} \tag{23}$$

According to the homotopies (16) and (17), the two following systems of PDEs are obtained

$$\begin{aligned} &\frac{\partial u_0}{\partial t} = 0, \qquad u_0(x,0) = x^2, \\ &\frac{\partial u_m}{\partial t} = (1+c_0)\frac{\partial u_{m-1}}{\partial t} - c_0 \cdot \frac{x^2}{2}\frac{\partial^2 u_{m-1}}{\partial x^2}, \\ &u_m(x,0) = 0, \quad (m \geq 1) \end{aligned} \tag{24}$$

$$\begin{aligned} &\frac{\partial^2 u_0}{\partial x^2} = 0, \qquad u_0(0,t) = 0,\ u_0(1,t) = e^t, \\ &\frac{\partial^2 u_m}{\partial x^2} = (1+c_0)\frac{\partial^2 u_{m-1}}{\partial x^2} - c_0 \cdot \frac{2}{x^2}\frac{\partial u_{m-1}}{\partial t}, \\ &u_m(0,t) = 0, u_m(1,t) = 0, \qquad (m \geq 1) \end{aligned} \tag{25}$$

Solving (24) and (25), we can get any order homotopy solutions for problem (23). Fig.3. depicts the c_0-curve for 5th-order HAM approximation of u(0.2,0.5) of the problem (23). Fig.4. gives the comparison at t=0.5 between the NHAM 5rd-order approximate solution and the exact solution $u(x,t) = x^2 e^t$ of the problem (23) . From Fig.3. and Fig.4., we can find that when $c_0 \in [-0.5,-0.2]$, the NHAM5[rd]-order approximate solutions meet the exact solution closely. Especially, when $c_0 = -0.3$, the NHAM5[rd]-order approximate solution curve and the exact solution curve are coincident completely.

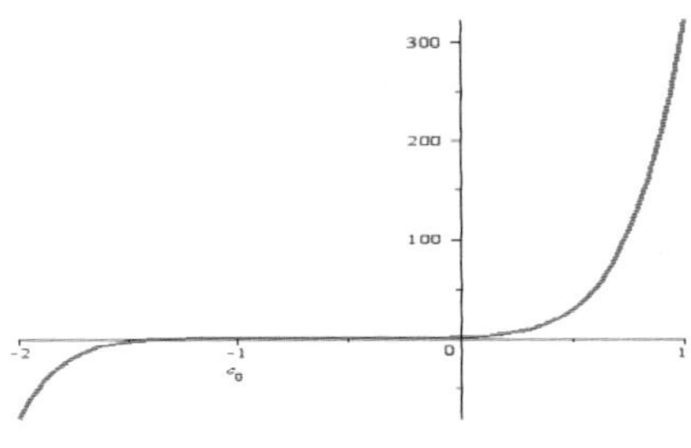

Figure 1. The c_0-curve for 3th-order NHAM approximation of $u(0.2,0.6)$

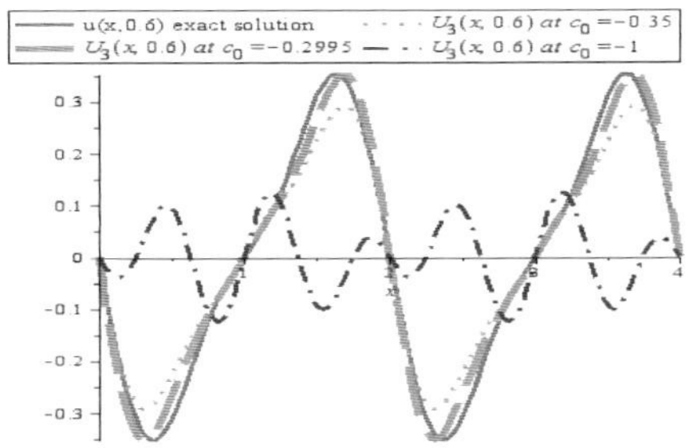

Figure 2. Comparison at t=0.6 between the NHAM3[rd]-order approximate solution (at $c_0 = -0.35, -0.2995, -1$) and the exact solution of problem(23) .

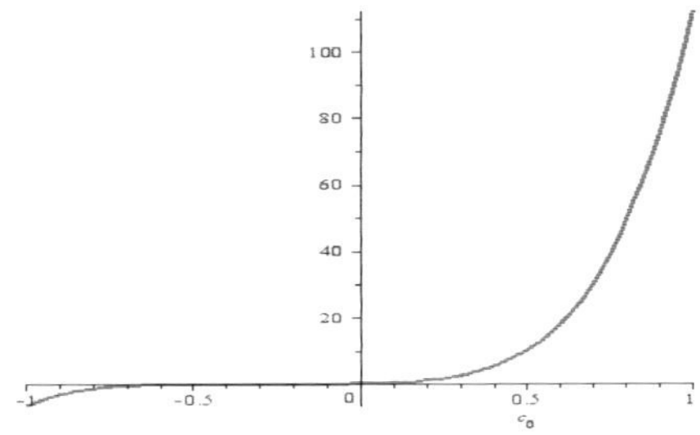

Figure 3. The c_0-curve for 5th-order NHAM approximation of $u(0.2, 0.5)$

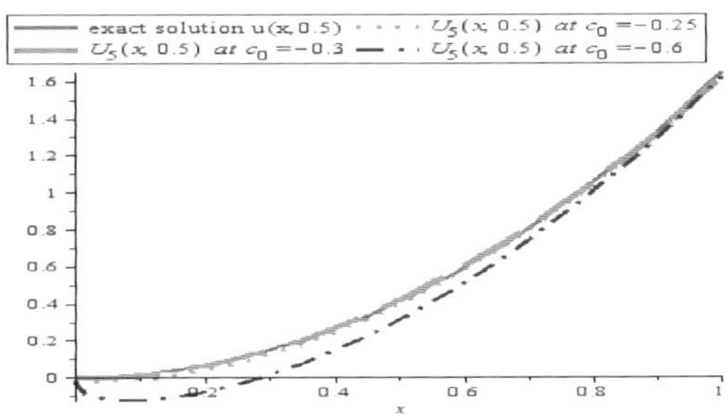

Figure 4. Comparison at t=0.6 between the NHAM5[rd]-order approximate solution (at $c_0 = -0.3, -0.6, -0.25$) and the exact solution of problem(26) .

5. Conclusion

In this paper, a new analytical technique based on the HAM to solve IBVP over finite domains has been proposed. The advantage of this technique is to include both the initial and boundary conditions in the recursive relation, so that we can obtain a good approximate solution for the problems. Otherwise, the results obtained in the numerical examples show that the auxiliary parameter c_0 play an important role on the solution series. According to the c_0-curve, we can find a so proper value of c_0 that we can get exact results of the problem only by a few truncated series. Furthermore, the results obtained in the numerical examples show that, as a special case of the HAM when $c_0 = -1$, the HPM cannot ensure the convergence of solution series and might give useless results.

Acknowledgments
This work was financially supported by the National Natural Science Foundation of China (NSFC) through grant Nos. 11502165, 11290152 and 11427801. Fund.

IOP Conf. Series: Journal of Physics: Conf. Series **916** (2017) 012039 doi:10.1088/1742-6596/916/1/012039

References

[1] Nayfeh AH, Perturbation methods. New York: John Wiley & Sons; 2000.

[2] Engui Fan, Darboux transformation and soliton-like solutions for the Gerdjikov-Ivanov equation,J.Phys.A:Math.Gen.33, p6925-6933, 2000.

[3] Abdul-Majid Wazwaz, The tanh method for traveling wave solutions to the Zhiber-Shabat and other related equations, Communications in Nonlinear Science and Numerical Simulation 13, p584-592, 2008.

[4] M.Senthilvelan, On the extended applications of Homogenous Balance Method, Applied Mathematics and computation 123, p381-388, 2001.

[5] Elaf Jaafar Ali, A New Technique of Initial Boundary Value Problems Using Adomian Decomposition Method, International Mathematical Forum 7, p799-814, 2012.

[6] B.Batiha, M.S.M.Noorani, I.Hashim, K.Batiha, Numerical simulations of PDEs by variational iteration method, Physics Letters A 372, p822-829, 2008.

[7] Yasir Khan, Qingbiao Wu, Homotopy perturbation transformation method for nonlinear equations using He's polynomials, Computers &Mathematics with Applications 61, p1963-1967, 2011.

[8] S.J.Liao, The proposed homotopy analysis techniques for the solution of nonlinear problems. Ph.D. dissertation, Shanghai Jiao Tong University, Shanghai,1992.

[9] S.J.Liao, Beyond perturbation: Introduction to the homotopy analysis method. Boca raton: CRC Press, Chapman and Hall, 2003.

[10] Shijun Liao, Comparison between the homotopy analysis method and homotopy perturbation method, Applied Mathematics and Computation 169, p1186-1194, 2005.

[11] Songxin Liang and David J. Jeffrey, Comparison of homotopy analysis method and homotopy perturbation method through an evolution equation, Communications in Nonlinear Science and Numerical Simulation 14, p4057-4064, 2009.

[12] Hayat T, Javed T and Sajid M, Analytic solution for rotating flow and heat transfer analysis of a third-grade fluid, Acta Mechanica 191, p219-229, 2007.

[13] Abbasbandy S, The application of the homotopy analysis method to solve a generalized Hirota-Satsuma coupled KDV equation, Physics Letters A 361, p478-473, 2007.

[14] Hassan HN and El-Tawil MA, A new technique of using homotopy analysis method for solving high-order nonlinear differential equations. Mathematical Methods in the Applied Science 34, p728-742, 2011a.

[15] Hassan HN and El-Tawil MA, An efficient analytic approach for solving two-point nonlinear boundary value problems by homotopy analysis method, Mathematical Methods in the Applied Science 34, p977-989, 2011b.

[16] Najeeb Alam, Asmat Ara, Muhammad Afzal and Azam Khan, Analytical Aspect of Fourth-order Parabolic Partial Differential Equations with Variable Coefficients, Mathematical and Computational Applications 15, p481-489, 2010.

[17] Edyta Hetmaniok, Damian Słota, Roman Wituła, Adam Zielonka, Solution of the one-phase inverse Stefan problem by using the homotopy analysis method, Applied Mathematical Modelling 39, p6793-6805, 2015.

[18] Mohamed S. Mohamed, Yasser S. Hamed, Solving the convection–diffusion equation by means of the optimal q-homotopy analysis method (Oq-HAM), Results in Physics 6, p20-25, 2016.

IOP Conf. Series: Journal of Physics: Conf. Series **916** (2017) 012040 doi:10.1088/1742-6596/916/1/012040

Control for well-posedness about a class of non-Newtonian incompressible porous medium fluid equations

Shuxian Deng [1], **Xinxin Ge** [2,*]

[1]Department of Basic Courses, Zhengzhou Shengda University of Economics, Business & Management, Zhengzhou, China
[2]School of management engineering, Henan Institute of Engineering, Zhengzhou, China

*Corresponding author e-mail: 857931663@qq.com

Abstract. Considering the non-Newtonian fluid equation of incompressible porous media, using the properties of operator semigroup and measure space and the principle of squeezed image, Fourier analysis and a priori estimate in the measurement space are used to discuss the non-compressible porous media, the properness of the solution of the equation, its gradual behavior and its topological properties. Through the diffusion regularization method and the compressed limit compact method, we study the overall decay rate of the solution of the equation in a certain space when the initial value is sufficient. The decay estimation of the solution of the incompressible seepage equation is obtained, and the asymptotic behavior of the solution is obtained by using the double regularization model and the Duhamel principle.

1. Introduction

Porous media the fluid mechanics branch is a discipline with rock mechanics, porous media physics, surface physics, physical chemistry, and thermodynamics and so on. Therefore, the situation is more complex, mainly in: porous media pore area larger, viscous effect is obvious, because the flow of fluid in porous media by the pressure, so its compression cannot be ignored. Due to the role of capillary phenomenon, the role of molecular force is also more complex in our study. Thus, porous media fluids are often accompanied by complex physical and chemical processes [1-4]. Porous media Fluid mechanics is widely used in water conservancy, marine, petroleum, geothermal, soil, environment, aviation, aerospace and other fields.

Using the Darcy rule, we can establish the fluid equation model of the porous medium as $u=-E(\nabla p+grT)$, $\frac{\partial T}{\partial t}+u\bullet\nabla T^{\alpha}+\kappa(-\Delta)^{\beta}=F$, where E is the matrix based on the different viscous media, T C is the fluid temperature, $r\in R^n$ is the unit vector, $0<\alpha\le 1$, $\kappa>0$ is the viscosity coefficient, the temperature $T(t,x)$ and the external force $F(t,x)$ are the true value function of about t and x, which are the incompressible porous fluid equations[5-8]. We call it subcritical as $\alpha>1/2$, critical as $\alpha=1/2$, and supercritical as $0<\alpha<1/2$.

2. Preliminaries

We focus on the flow of a class of incompressible non-Newtonian fluids in porous media, and find out the suitability and gradual behavior of a class of non-Newtonian fluids in porous media [9-10].

When a non-Newtonian fluid flows in a porous medium, its trajectory function is as follows:

$$\frac{\partial X}{\partial t}(\alpha,t)=u\left(X(\alpha,t),t\right)$$

Where $X(\cdot,t):\alpha\in R^N\to X(\alpha,t)\in R^N$ is the position of a particle of the fluid at time t and its initial position is $\alpha=(\alpha_1,\alpha_2,\cdots,\alpha_N)$, $X(\Omega,t)=\{X(\alpha,t):\ \alpha\in\Omega\}$ is the region of the initial region $\Omega\in R^N$ arrives after time t.

3. Main Conclusion

We consider the following mathematical model for a class of incompressible non-Newtonian porous media problems:

$$\begin{aligned}
&\frac{\partial T}{\partial t}+u\cdot\nabla T+\sigma(-\Delta)^{\tau}T=f\\
&\frac{1}{r}\frac{\partial}{\partial r}\left(r\frac{\partial\omega}{\partial r}\right)+\lambda\omega=-\Phi(x,r,t)\\
&\nabla\cdot u=0\\
&u_0(x)=u(x,t)\big|_{t=0},T_0(x)=T(x,t)\big|_{t=0}
\end{aligned}\tag{1}$$

σ is the viscosity coefficient. $T(x,t)$ and $f(x,t)$ are the real functions about t and x, $\tau\in[0,p)$

We study the suitability of the equation and the asymptotic behavior of the solution. Note that the dissipation mechanism is given by the fractional Laplacian operator and that the velocity is related to the pressure, and we will use the new tools and methods to get the following conclusions:

Theorem suppose $P_0(x,0)\in B(R^3)$, $Q(x,t)\in C^3\left([0,+\infty),B(R^3)\right)$, $\forall\eta>0$, $\exists\delta>0$, such that $\left\|P_0(x)B,C^3\left([0,+\infty),B(R^3)\right)\right\|_{L^2(\Omega)}+\left\|Q(x,t);C^3\left([0,+\infty),B(R^3)\right)\right\|_{L^2(\Omega)}=\varepsilon$, and $3\delta\varepsilon<2$. Further, if the solution of the equation $P(x,t)$ satisfies $\left\|P(x,t);C^3\left([0,+\infty),B(R^3)\right)\right\|_{L^2(\Omega)}<\varepsilon$, then $P(x,t)$ are the solutions of the equations(1), and they are unique. That is the solutions have well posedness in $C^3\left([0,+\infty),B(R^3)\right)$.

4. Proof of Conclusion

In order to give the main conclusion of the proof, we first introduce the following lemma[11]

Lemma 1 suppose $A:\Omega\times\Omega\times\cdots\cdots\times\Omega\to\Omega$ is a bounded linear operator for Banach space Ω, there exists a constant μ, for any $\xi_1,\xi_2,\cdots,\xi_n\in\Omega$,there holds[7]

$$\left\|A(\xi_1,\xi_2,\cdots,\xi_n)\right\|_\Omega\le\mu\left\|\xi_1\right\|_\Omega\left\|\xi_1\right\|_\Omega\cdots\cdots\left\|\xi_1\right\|_\Omega\tag{2}$$

$\forall\delta>0$, and $n\delta^{2n-1}\mu<1/3$. For any $z,z_1\in X$, $\|z\|_\Omega,\|z_1\|_\Omega\le\delta$, the following equations

$$\xi = z + A(\xi,\xi,\cdots,\xi) \tag{3}$$

$$\eta = z + A(\eta,\eta,\cdots,\eta) \tag{4}$$

Have unique solutions ξ,η, $\|\xi\|_\Omega, \|\eta\|_\Omega \le \delta$, and

$$\|\xi - \eta\| \le \frac{2}{3 - n\delta^{2n-1}\mu + \sigma}\|z - z_1\| \tag{5}$$

Proof: use compressed image principle, for small positive number δ, construct an inspection functions $S = H(\xi - \eta)$, $S|_{\partial(\Omega)} = 0$, $\forall \tau \in [0,T]$, $E = \{\xi : \xi \in \Omega, \|\xi\| \le \delta\}$, let $H = \max\left(1, \min(\frac{z}{\delta}, 1)\right)$, difine $d(\xi,\eta) = \|\xi - \eta\|_\Omega$, and let

$$\beta\xi = z + A(\xi,\xi,\cdots,\xi) \tag{6}$$

For any $\xi \in E$,

$$\iint_\Omega H(\xi^m - \eta^m)dxdt + \iint_\Omega (|\nabla\xi^m|^{p-2}\nabla\xi^m - |\nabla\eta^m|^{p-2}\nabla\eta^m)\ \nabla H(\xi^m - \eta^m)dxdt$$
$$\leqslant \iint_\Omega \lambda\left(\xi^q \int_\Omega \xi^n dy - \eta^q \int_\Omega \eta^n dy\right) H(\xi^m - \eta^n)dxdt \tag{7}$$

Notice that $\|\xi\|_\Omega \le \delta$, $\forall z \in \Omega$, we have

$$\begin{aligned}\|\rho z\|_\Omega &\le \|z\|_\Omega + \|A(\xi,\xi,\cdots,\xi)\|_\Omega \le \|z\|_\Omega + \mu\|\xi\|_\Omega^\lambda \\ &< \delta + n\delta^{2n-1}\mu < 3\delta\end{aligned} \tag{8}$$

Hence $\rho z \in \Omega$, $\forall \xi,\eta \in \Omega$, $m,n \in N$

$$\left\langle |\xi|^{p-2}\xi - |\eta|^{p-2}\eta, \xi - \eta\right\rangle \ge \upsilon_0|\xi - \eta|^p \tag{9}$$

On the other hand, by the nature of the film, the direct calculation is available:

$$\|\rho\xi - \rho\eta\| \le \|A(\xi,\xi - \rho\eta,\cdots,\xi)\| + \|A(\xi,\eta,\xi - \rho\eta,\cdots,\xi)\| + \cdots + \|A(\xi - \eta,\rho\eta,\cdots,\eta)\|$$

$$\le \frac{3 - n\delta^{2n-1}\mu + \sigma}{2}\|\xi - \eta\|_\Omega \tag{10}$$

Because of $\frac{3-n\delta^{2n-1}\mu+\sigma}{2}\le 1$, So ρ is the self-compression map on metric space (Ω,d). On the other hand, let $\delta\to 0$, we have

$$\iint_{\Omega}(\xi-\eta)_t\alpha[\xi>\eta]dxdt\leqslant\iint_{\Omega}\eta\left(\xi^q\int_{\Omega}\xi^n dy-\eta^q\int_{\Omega}\eta^n dy\right)\alpha(\xi>\eta)dxdt$$
$$\leqslant\iint_{\Omega}\alpha\left(\int_{\Omega}\xi^q dy\right)(\xi^n-\eta^n)\kappa(\xi>\eta)dxdt$$
$$\leqslant\iint_{\Omega}\alpha\xi^q\left(\int_{\Omega}(\xi^q-\eta^q)dy\right)\kappa(\xi>\eta)dxdt \tag{11}$$

By the principle of compressed image, ρ has a unique fixed point $\xi\in\Omega$, so the equation

$$\beta\xi=z+A(\xi,\xi,\cdots,\xi)$$

Has unique solution $\xi\in\Omega$.

From (11), we can get:

$$\int_{\Omega}\xi(x,t)-\eta(x,t)dx$$
$$\leqslant\int_{\Omega}(\xi(x,0)-\eta(x,0))dx+\iint_{\Omega}\alpha\left(\int_{\Omega}\xi^n dy\right)f(x,t)(\xi-\eta)dxdt$$
$$+\iint_{\Omega}\alpha\xi^p\left(\int_{\Omega}g(x,t)(\xi-\eta)dy\right)dxdt \tag{12}$$

From (10) (11) (12), one can get:

$$\|\xi-\eta\|=\|z-z_1\|_{\Omega}+\|A(\xi,\xi-\eta,\cdots,\xi)\|_{\Omega}+\|A(\xi,\eta,\xi-\eta,\cdots,\xi)\|_{\Omega}+\cdots+\|A(\xi-\eta,\rho\eta,\cdots,\eta)\|_{\Omega} \tag{13}$$

Since $\xi,\eta,f(x,t),g(x,t)$ is bounded on Ω and positive, and $\xi-\eta\le 0$, from Grownwall inequality,

$$\|\xi-\eta\|\le\frac{2}{3-n\delta^{2n-1}\mu+\sigma}\|z-z_1\| \tag{14}$$

$$\int_{\Omega}(\xi(x,t)-\eta(x,t))dx\leqslant M_1\iint_{\Omega}f(x,t)(\xi-\eta)dxdt+M_2\iint_{\Omega}f(x,t)(\xi-\eta)dxdt \tag{15}$$

Lemma 1 is proved.

Proof of theorem: In fact, if $T_0\in B(\Omega)$, We are easy to judge that $H_\alpha(\cdot)T_0\in C([0,\infty);B(\Omega))$, in which $\alpha=n+\lambda-1$.

$$\|H_\alpha(\cdot)T_0;B(\Omega)\|\le ess\sup_{\alpha\in\Omega}|\alpha|^{\varepsilon}|T(\beta)|^{\delta}=\|T_0;B(\Omega)\| \tag{16}$$

Which shows that $H_\alpha(\cdot)T_0\in L^\infty([0,\infty);B(\Omega))$

$$\left\|\partial_t H_\alpha(\cdot)T_0;B(\Omega)\right\| \le ess\sup_{\alpha\in\Omega}|\alpha|^{\varepsilon}|\alpha|^{\tau}e^{-t|\alpha|^{\delta}}|T_0(\alpha)|^{\varepsilon} \le C(t)\left\|T_0;B(\Omega)\right\| \tag{17}$$

Let $\xi(x,t)=\lambda\left(\frac{\gamma^m}{1-\gamma^m}+\frac{Q(x)}{K}\right)^{\frac{1}{p}}$, hence

$$\begin{aligned}
&\xi_t - div\left(|\nabla\xi|^{q-2}\nabla\xi\right)-\lambda\xi\int_\Omega \xi(x,t)dx \\
&=\lambda\left(\frac{K^n}{M}\right)^{p-1}-\kappa^p\left(\frac{\gamma^m}{1-\gamma^m}+\frac{H(x)}{M}\right)^{\frac{p}{m}}\int_\Omega \kappa^q\left(\frac{\kappa^m}{1-\kappa^m}+\frac{H(x)}{M}\right)^{\frac{q}{m}}dx \\
&\leqslant \lambda\left(\frac{\kappa^m}{M}\right)^{\frac{1}{p}}-\lambda\kappa^{p+n}\left(\frac{1}{1-\tau^m}\right)^{\frac{n+2}{q}}V(\Omega)
\end{aligned} \tag{18}$$

On the other hand, From direct calculation, one can get

$$\begin{aligned}
\langle P(u,t),\varphi(x)\rangle &\leqslant C\int_\Omega\int_0^t e^{-(\eta-t)|\xi|^{p+1}}|\xi|^{\frac{p-1}{2}}d\xi d\eta \\
&\leqslant C\int_\Omega (1-e^{2\alpha-n+1})|\xi|^{\frac{p+q-1}{2\alpha}}d\xi
\end{aligned} \tag{19}$$

From(14), (15), (18) and (19), one can get $\langle P(u,t),\varphi(x)\rangle \to 0$.
In a similar way, we can deal with Q and get

$$\begin{aligned}
\|Q;C([0,\infty);\Omega)\| &\leqslant ess\sup_{\xi\in\Omega}\left|\int_0^t e^{r^{\frac{q-1}{p}}}|\xi|^r(u,\nabla T)(\tau,\xi)d\tau\right| + \|u;L^q(\Omega)\|^{\frac{1}{m}} \\
&\leqslant \iint_\Omega (|\nabla u^m|^{p-2})\nabla u^m - |\nabla v^m|^{p-2}\nabla v^m(\nabla u^m-\nabla v^m)w'\phi dxdt \\
&\leqslant \iint_\Omega (|\nabla u^m|^{p-2})\nabla u^m - |\nabla v^m|^{p-2}\nabla v^m w_n\nabla\phi dxdt \\
&\leqslant \iint_\Omega (u^m-v^m)\frac{\partial(u-v)}{\partial t}w_n\phi dxdt + div(|\nabla u^m|^{p-2}\nabla u^m) \\
&\leqslant \iint_\Omega w_n\phi(u^m-v^m)dxdt + div(|\nabla u^m|^{p-2}\nabla u^m) \\
&\leqslant M_2\ln\frac{\alpha}{t(2t-\alpha)}\|u;L^q(\Omega)\|\,\|T;L^q(\Omega)\|
\end{aligned} \tag{20}$$

Therefor $\|Q;C([0,\infty);\Omega)\| \leqslant M_2\|u;L^q(\Omega)\|^{\frac{1}{m}}\|T;L^q(\Omega)\|^{\frac{1}{q}}$ (21)

On the basis of (18)-(21) and Grownwall inequality

$$u_t - \Delta_{m,p}u + A(S_g,P)\kappa\nabla\theta - au^q\int_\Omega u^n(x,t)\nabla u^m dx$$

$$\leqslant \left|\iint_{\Omega} (|\nabla u|^{p-2}\nabla u^m - |\nabla v|^{p-2}\nabla v^m)\right| \eta(t)\sigma\left(\frac{x}{k}\right)dxdt$$
$$\leqslant \frac{p-1}{2k}\iint_{\Omega} (|\nabla u^m|^{p-1} + |\nabla v^m|^{p-1})|\nabla_y \eta(y)|\,\|\xi\|_{L^\infty(\Omega)}(u^m - v^m)dxdt$$
$$\leqslant \iint_{\Omega} k_m|\xi|^2 |\nabla_y\eta(y)|\,\|\xi\|(1-S_g)(u^m - v^m)dxdt$$
$$\leqslant \iint_{\Omega} k_m\|\xi\|(1-S_g)(u^q - v^q)\eta\theta\omega' dxdt \leqslant C\left|\int_{s_1}^{s_2}\sqrt{\lambda_g(s)\lambda_w(s)}ds\right|^{\tau} \quad (22)$$

Combine with (20), (22), hence

$$\int_{\Omega}(u(x,s_2) - v(x,s_2))dx \leqslant \int_{\Omega}(u(x,s_1) - v(x,s_1))dx$$
$$\leqslant S_1(P_c(S_2) - P_c(S_1)) - \int_{s_1}^{s_2} sP_c'(s)ds$$
$$\leqslant \int_{s_1}^{s_2} s^{2-\gamma}P_c'(s)s^{\gamma-1}f_w(s,p)\,|\xi|^{\frac{p-q+1}{2}}ds$$
$$\leqslant C\int_{s_1}^{s_2} s^{\gamma-1}ds \leqslant \frac{C}{\gamma}|S_1 - S_2|^{\gamma} \quad (23)$$

Let $t \to \infty, s_1 \to 0$, we come to the conclusion.

5. Simulation and Numerical Experiments

In order to verify the correctness of the theoretical analysis of the above equation, we simulate and simulate the experiment. We use COMSOL software to simulate the parameter setting of the above equation and the analysis of the results. It can be shown that grid 1 divides region Ω along the X-axis direction and Y-axis direction (n is a positive integer) to form a parallel linear triangular mesh. Mesh 2 is a parallel linear triangular mesh formed by dividing the region by an N-division in the X-axis direction and dividing the M-divisions in the Y-axis direction with a mesh ratio of $m/n = 8$

As can be seen from the data in the table, the error order of grid 1 and grid 2 is the same, and its large time behavior is stable. To achieve the optimal order and convergence effect, which is consistent with the previous theoretical analysis of the results

Table 1. Result of the calculation at t = 1.0 at grid 1

$n\times m$	8×8	16×16	32×32	64×64	ε
$\|u-u_h\|_0$	1.3736006	0.4308764	0.2569566	0.9012883	1.6588750
$\|u-u_h\|_h$	0.1871856	0.6474759	0.2407519	1.9218672	0.6128201

Table 2. Result of the calculation at t = 1.0 at grid 2

$n\times m$	2×16	4×32	8×64	16×128	ε
$\|u-u_h\|_0$	0.2146650	0.6729420	0.1448862	0.3229337	0.9802149
$\|u-u_h\|_h$	1.7257903	1.4105861	1.4076397	1.8032814	0.7397161

IOP Conf. Series: Journal of Physics: Conf. Series **916** (2017) 012040 doi:10.1088/1742-6596/916/1/012040

6. Conclusion

In this paper, we study the global fitness of the solution of the fluid equation in the porous medium with nonlinear boundary source nonlinear diffusion of incompressible fluid. Based on the non-Newton fluid equation of the incompressible porous media, the Fourier analysis method and the priori estimates in the measure space are discussed by using the properties of the operator semigroup and the measure space and the principle of the compression image. The suitability of the solution. The existence and uniqueness of solutions for a class of non - Newtonian seepage equations with nonlinear boundary conditions are proved. Through the diffusion regularization method and the ω -limit compact method, we study the overall well - posedness of the solution of the equation in Ω space when the initial value is sufficient small.

References

[1] V. A. Galaktionov and H. A. Levine, On critical Fujita exponents for heat equations with nonlinear flux conditions on the boundary, Israel J. Math., 94 (2016), 125-146.
[2] GUO Bo-ling, ZHU Pei-cheng. Algebraic L^2 decay for the solution to a class system of non-Newtonian fluid in R^n [J]. J Math Phys, 2016, 41: 349-356.
[3] O. A. Oleinik, E. V. Radkevic, Second order diferential equations with nonnega-tine characteristic form, American Mathematical Society, Rhode Island and PlenumPress, New York, 2013.
[4] SOLANOV K N. On Some modification Navier-Stokes equations[J]. Nonlinear Analysis, TMA, 2013, 52:769-793.
[5] WIEONER M. Decay results for weak solutions of the Navier-Stokes equations in R^n[J]. J London Math Soc, 2007, 35: 303-313.
[6] J. R. Anderson, N. Su and H. F. Zhang, Existence and uniqueness of solutions of degenerate parabolic equations in exterior domains. Nonlinear Anal. 44(4)(2011)， 453-468.
[7] J. R. Anderson, Local existence and uniqueness of solutions of degenerate parabolic equations. Comzn. Partial Defrential Equations 16 (1)(2011), 105-143.
[8] F. luiros and J. D. Rossi, Blow-up sets and Fujita type curves for a degenerate parabolic system with nonlinear boundary conditions. Indiana Univ. Math. J. 50(1)(2014)， 629-654.
[9] A. de Pablo, F. Cluiros and J. D. Rossi, Asymptotic simplification for a reaction diffusion problem with a nonlinear boundary condition. IMA J. APpI. Math. 67(1)(2012)， 69-98.
[10] AQUIRRE J, ESCOBEDO M. On the blow-up of solutions of a connective reaction diffusion equation[J]. Proc Roval Soc Edinburgh, 2013. 123A:433-460.
[11] CHEN Yarzhe. On finite diffusing speed for uniformly degenerate quasilinear parabolic equations[J]. ChinAnn Math, 2016, 7B: 318-329.

Research of Non-linear 2.5D Fluid-structure Interaction Model of Gravity Dam

Yiran Li [1, 2], Menghua Xu [2,*], Hao Sun [2]

[1]Hunan Province Key Laboratory of Key Technologies for Water Power Resources Development, Changsha, China.
[2]School of Civil engineering and Transportation, South China University of Technology, Guangzhou, China

*Corresponding author e-mail: mhxu@scut.edu.cn

Abstract. Aseismic safety of dam involves people's lives and properties, even has a great impact on national economic development. The key to this project is to have a fairly complete fluid-structure interaction (FSI) theory and a calculation method with a good accuracy. Therefore, focused on the FSI dynamic problem of gravity dam under seismic action, this paper presented the non-linear 2.5D dynamic theoretical model of concrete gravity dam established by segmenting fluid model and numerical solution method, compared the numerical simulation result under preliminary gravity action with 3D model. The study showed that the 2.5D dynamic calculation model could well reflect the seismic dynamic FSI behavior and effect of concrete gravity dam in the respects of displacement and stress and benefited the large-scale non-linear 3D aseismic calculation of gravity dam because of the effective reduction of 3D FSI numerical calculation time, which improved the calculation efficiency.

1. Introduction

Violent earthquakes often happen to China. When large engineering structures suffer an intense earthquake, it will cause not only direct engineering loss, but also secondary disasters what are hard to expect and threat to people's lives and properties, even a great impact on national economic development. So higher requirement for critical engineering and lifeline engineering is necessary.The key to this project is to have a fairly complete FSI theory and a calculation method with a good accuracy. Since H.M.Westergaard published his famous paper *Water pressure on dams during earthquakes*, it has caught many researches' attention for a long time, because of the significance of this project in engineering practice.

FEM applied to solving FSI problem is presented first in *The finite element method*, a noted book by iZeknieMcz[1]. After developing from low level to high level and simplicity to complexity, FEM is extensively and profoundly practical in engineering projects, as the most efficient analytic method. According to FSI research project, large finite element software is usually used to build theoretical models and do simulation calculation and analysis, where the adopted models are most 2D and 3D models. Ghaedi[2] builds 2D model to analyze the concrete dam under seismic action. And Puigdellivol[3] also builds a 2D model to study a topology optimization method for laminated busbars in power converters and he points out that Comparison with 3D FEM proves the 2-D approach to be faster while remaining accurate and a perfect method. However, it is obvious that considering the complexity of large engineering projects like dam, there are too many factors what should be taken into account that 2D theoretical model cannot meet the accuracy requirement. In Crosta' [4] study about collapse and rockslide–reservoir interaction, a fully 3D rockslide–water reservoir simulation has been considered to provide a more realistic simulation. But 3D simulation calculation of large engineering has a high requirement for computer and that will takes a long time, which greatly decreases the calculation efficiency. However, 2.5D simulation calculation can increase the accuracy and efficiency of calculation at the same time. This 2.5D theory has extensive application and research in many fields. But at present, the research of 2.5D theory in hydraulic engineering is mainly about large aqueduct [4], little attention is paid to gravity dam. However, finding a rational and effective FSI dynamic calculation methods is of great significance to the large-scale seismic dynamic analysis of gravity dam.

Therefore, the main objectives of this study are (1) to present a 2.5D fluid-structure interaction dynamic model of gravity dam, (2) to compare the calculation results in non-linear numerical test method with 3D model to verify the feasibility, rationality and accuracy of 2.5D simulation, and(3) to discuss how to build a 2.5D model with the best simulation.

2. Case study

2.1. Concept of 2.5D Numerical Models of Gravity Dam

The FSI 2.5D theory means that the model is consist of 3D solid structure and 2D fluid structure, where 2D fluid structure simulates the FSI effect between structure and fluid. Structure part is the system of 3D dam body-2D dam shells, fluid part is 2D fluid coupling with 2D dam shells, FSI effect is transferred by the interaction between 2D fluid and 2D dam body completely consolidated with 3D dam body. The model concept is shown in Figure 1.

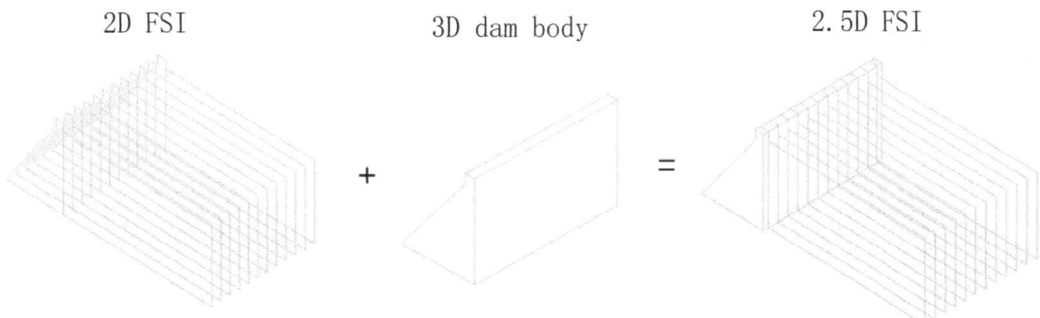

Figure.1 The concept of 2.5D model

2.2. Introduction of studied dam

Because the practical application of the 2.5D theory in gravity dam has not been studied at all and this paper is mainly focused on the feasibility, rationality and accuracy when 2.5D theory is used to solve problems of gravity dam, a geometrically simple gravity dam under a comparatively ideal circumstance was simulated and analyzed.

The concrete dam is 120m high with upstream slope vertical, downstream slope gradient 0.75, crest dam 1.5 times the length of dam height and 0.1 times the breadth of dam height. Dam body of gravity dam uses homogeneous concrete material whose elastic modulus of concrete E equals 2.85E10 Pa, Poisson's ratio v equals 0.15, density d equals 2600 Kg/m3, tensile strength f_t equals 1.96E6 Pa and compressive strength f_c equals 22E6 Pa.

Fluid uses water adopting laminar flow model whose parameters are shown as follows: density equals 1000 Kg/m3, viscous coefficient equals 0.002(Pa·s), bulk elastic modulus equals 0.002(Pa·s),and the fluid is set as incompressible liquid with the height of 108 m and the length of 240 m.

2.3. The finite element model

This paper adopted the large finite element software ADINA to accomplish numerical solution. With regard to concrete gravity dam, the adopted numerical solution characteristics of FSI 2.5D theoretical model was as follows:

the 3D solid elements were applied to dam body structure which was set gravity and Rayleigh damping;

the 2D solid elements were applied to fluid whose amount and location of 2D shells corresponded with 2D dam shells, and the model was built with top surface as free surface, surface contacting dam as FSI surface, surface on the inverse side as uniform surface and bottom surface as wall surface;

the material of 2D dam shells is linear elastic material with no rigidity to avoid impacting the rigidity of dam body and at the same time contribute to transfer the action of fluid for dam body, 2D dam shells are completely consolidated with 3D dam body, plane-strain model is adopted, calculated length of each dam shell equals to the width of the 3D solid dam body.

This paper respectively established three 2.5D model with 6, 12, 24 shells. The calculation results of 2.5D models were all compared with 3D model. The 2.5D model with 6 shells is shown in **Figure.2** (2.5D models with 12 and 24 shells only had more equally spaced shells; 3D model had the same parameter set as 2.5D models):

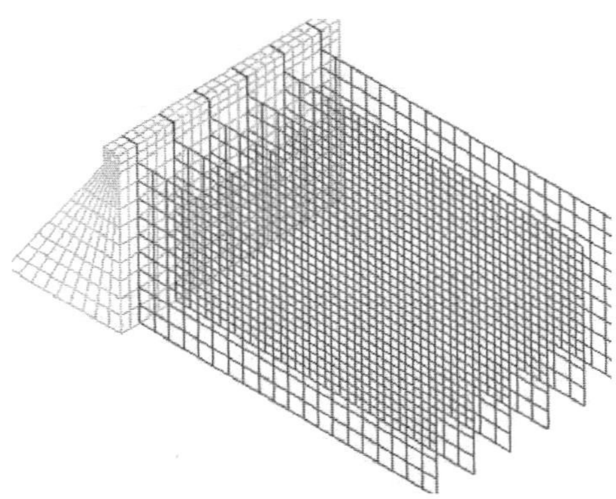

Figure.2 The 2.5D model with 6 shells

The test picked up the displacement data of points A, B, C, D, E on upstream sideline of dam crest and stress data of points F、G、H、I、J on upstream sideline of dam bottom and points K, L, M, N, O on downstream sideline of dam bottom to compare. The positions of these points are shown in **Figure.3** .

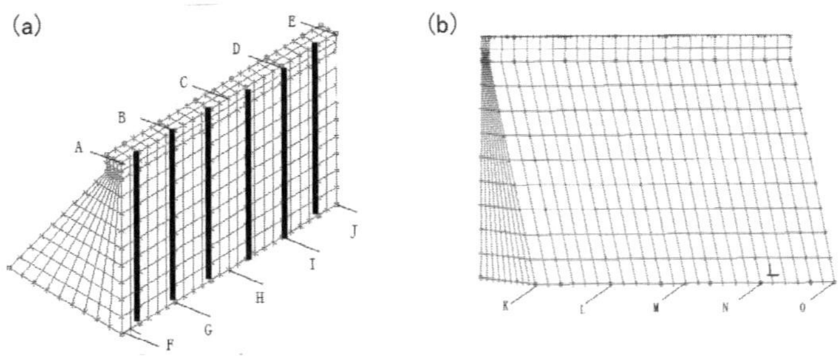

Figure.3 The position of picked pointed

3. Results analysis and discussion

3.1. Displacement data comparison of points on upstream sideline of dam crest

As is shown in below charts, **Figure 4** shows the displacement data comparison of point B and C in static response analysis, and **Table 1** shows the comparison of maximum error values of stable data.

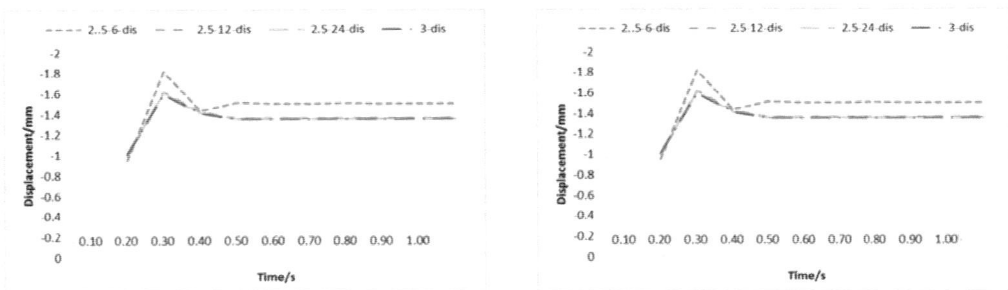

(a) (b)

Figure.4 The comparison of static displacement process lines of point B

Tab.1 The Comparison of maximum error values of displacement data

Point	A	B	C	D	E
6 shells	11.09%	11.16%	11.13%	11.16%	11.09%
12 shells	0.47%	0.46%	0.46%	0.46%	0.47%
24 piece	0.08%	0.08%	0.07%	0.08%	0.08%

Under the only action of gravity, calculation models did not reach a stable with calculation time less than 0.5s, and the data of the 0.5s do not have research value. So **Table 1.** Shows the comparison of maximum error values of displacement data without the first 0.5s, and all the below comparison only picks the stable calculated data like here.

As in depicted in **Figure 4**, displacement data of 2.5D models with 12 and 24 shells are similar to 3D model in trend and numerical value, but that of model with 6 shells have similar trend with obvious difference in numerical value. **Table 1** shows that maximum error value is decreasing gradually with 2D shells in 2.5D model increasing, and the errors in the models with 12 and 24 shells are far less than the model with 6 shells.

IOP Conf. Series: Journal of Physics: Conf. Series **916** (2017) 012041 doi:10.1088/1742-6596/916/1/012041

It can be concluded that under gravity, the 2.5D model with more 2D shells will simulate better 3D model in displacement data on upstream sideline of dam crest.

3.2. Stress data comparison of points on downstream sideline of dam bottom

As is shown in below charts, **Figure 5** shows the displacement data comparison of point M (a) and L (b) in static response analysis, and **Table 2** shows the comparison of maximum error values of stable data.

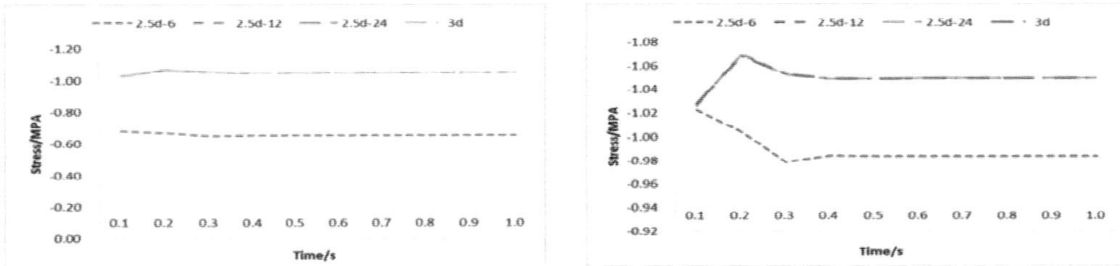

(a) (b)

Figure.5 The comparison of static stress process lines of point M

Tab.2 The Comparison of maximum error values of stress data

Point	K	L	M	N	O
6 shells	6.27%	37.51%	6.27%	37.51%	6.27%
12 shells	0.02%	0.02%	0.02%	0.02%	0.02%
24 shells	0.02%	0.02%	0.02%	0.02%	0.02%

As is depicted in **Figure 5**, the models with 12 and 24 shells simulate well with 3D model in stress data, but there is a great difference between the model with 6 shells and 3D model; **Table 2** shows that errors of the models with 12 and 24 are far less than the model with 6 shell where errors of models with 12 and 24 shells are very close to each other; as for the model with 6 shells, data errors of points L and N are bigger than the others.

It can be concluded that under gravity, the 2.5D model with more 2D shells will simulate better 3D model in stress data on downstream sideline of dam bottom. And that Points L and N have bigger errors than the other points in the model with 6 shells is because points L and N are located on the 2D dam shells and then directly participate in the transfer of FSI effect.

3.3. Stress data comparison of points on upstream sideline of dam bottom

As is shown in below charts, Figure 6 shows the displacement data comparison of point H and G in static response analysis, and Table 3 shows the comparison of maximum error values of stable data.

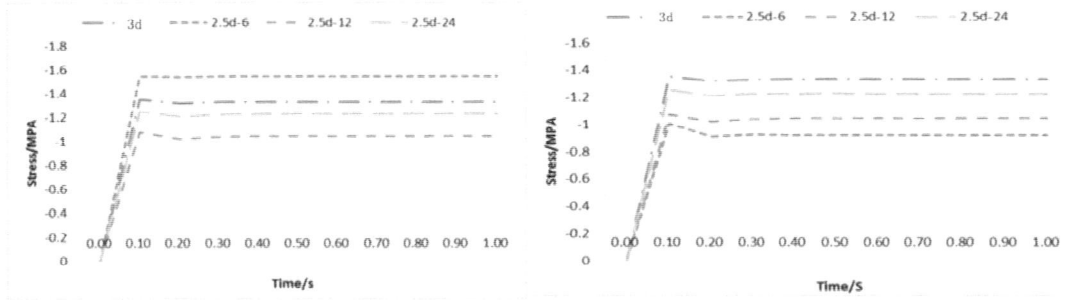

(a) (b)

Figure.6 The comparison of static stress process lines of point H.

Tab.3 The Comparison of maximum error values of stress data

Point	F	G	H	I	J
6 shells	30.86%	16.24%	30.86%	16.24%	30.86%
12 shells	21.74%	21.74%	21.74%	21.74%	21.74%
24 shells	7.79%	7.79%	7.79%	7.79%	7.79%

It can be seen from **Figure 6 (a)** that under gravity, the 2.5D model with more 2D shells simulates better 3D model in stress data on upstream sideline of dam bottom and the data process lines have the same trend; **Table 3** also shows that error values of the models with 6 to 24 shells are decreasing gradually. But as is depicted in **Figure 6 (b)** and **Table 3**, point G has a different performance from point H that process line of the model with 6 shells is above the others and maximum error value obviously decreases. The reason why this happened is the same to what causes the performance of **Table 2**, which is that point G is located on the 2D dam shell. However, this time it makes the error decrease.

It can be concluded that under gravity, the 2.5D model with more 2D shells will simulate better 3D model in stress data on upstream sideline of dam bottom. As for the reason why point G and I in the model with 6 shells have less errors, this paper suggests that when the stress condition of points on fluid of 2.5D model is similar to 3D model, the corresponding points on dam body directly respond to the force similar to that from 3D fluid model, then the stress data is closer to that of 3D model; for the points on downstream sideline of dam bottom, they are not in direct contact with fluid, which is why their errors are bigger.

4. Conclusion

This paper presented a 2.5D fluid-structure interaction dynamic model of gravity dam and compared the calculation results in non-linear numerical test method with 3D model to verify the feasibility, rationality and accuracy of 2.5D simulation. After the comparison and analysis, conclusions are shown as follows:

1. 2.5D FSI model of concrete dam simulates well 3D model in respects of displacement on upstream sideline of dam crest and stress on upstream and downstream sidelines of dam bottom;

2. 2.5D model with 24 shells has the best simulation effect in displacement and stress, which shows that under static condition, more 2D shells to transfer FSI will make numerical test results closer to that by 3D model.

3. As for the special points on 2D shells, the data error of displacement on the upstream sideline of dam crest are similar to the others, that of stress on the downstream sideline of dam bottom are bigger and on the upstream sideline of dam bottom, the data error of stress are smaller than the others, which is because of the local stress concentration.

Acknowledgments
This work was financially supported by Hunan Province Key Laboratory of Key Technologies for Water Power Resources Development (PKLHD201401).

References

[1] Zienkiewjcz O.C., The finite element method, McGRAW-HILl, 1977.
[2] Ghaedi K, Jameel M, Ibrahim Z, et al. (2016) Seismic analysis of Roller Compacted Concrete (RCC) dams considering effect of sizes and shapes of galleries. KSCE JOURNAL OF CIVIL ENGINEERING, 20(1), 261-272.
[3] Puigdellivol O, Meresse D, Le Menach Y (2017) Thermal Topology Optimization of a Three-Layer Laminated Busbar for Power Converters, IEEE TRANSACTIONS ON POWER ELECTRONICS, 32(6), 4691-4699
[4] Crosta GB, Imposimato S, Roddeman D. (2016) Landslide Spreading, Impulse Water Waves and Modelling of the Vajont Rockslide. ROCK MECHANICS AND ROCK ENGINEERING ,46(6), 2413-2436.

Parallel simulation of dam-break flow by OpenMP-based SPH method

Zhao Luo[1, 2,*], Qihe Wu[1, 2] and Lei Zhang[1, 2]

[1] CCCC Second Harbor Engineering Co., Ltd. National Enterprise Technology Center, Hubei Wuhan, China
[2] CCCC Highway Bridges National Engineering Research Centre Co., Ltd., Beijing, China

*Corresponding author e-mail: 490458835@qq.com

Abstract. Smoothed particle hydrodynamics (SPH), a Lagrangian mesh-free particle numerical method, is suitable for simulating strong impact and large deformation problems. In the method, quantities of particles can ensure high precision. However, with the increase of the particle numbers, the calculation efficiency becomes a challenge for applying the method to the engineering practice. OpenMP, a Portable Shared Memory Parallel Programming, is a great solution to improve the efficiency of SPH algorithm. In the paper, dam-break flow is simulated by SPH method. Vortex centre is discovered, the role of numerical technique is compared. At the same time, two parallel schemes for SPH algorithm is introduced, the speedup ratio with respect to number of particles or threads are revealed.

1. Introduction

Smoothed particle hydrodynamics (SPH) is a lagrangian mesh-less method originally developed to deal with astrophysics problems [1, 2]. Without requirement for generating mesh and capturing free surface, SPH is successfully extended to a variety of fluid-dynamic systems [3] and multiphase flow system [4, 5]. As an interpolation method, the accuracy of the SPH method often depends on the number of particles in the support domain. In theory, when the number of particles approaches infinity, the computational accuracy will be very close to reality. However, as an adaptive method, SPH has to run a large amount of computation at each time step, which is quite time-consuming and limit the development of the method.

Parallel computing, also known as high performance computing, can coordinate multiple processor to solve one problem together. For example, if your computer has 8 processors, the speedup would be 8X after parallel computing in ideal, but actually it can't be so high because of data exchange. OpenMP, a standard for implementing the thread model by adding directives to the code, has been applied in a number of fields [6, 7]. Compared with MPI or GPU, OpenMP is more cheap and easy to parallelize an existing serial code [8]. SPH algorithm can be paralleled by means of several OpenMP directives. These directives serve as instructions for a compiler which blocks and loops in the source code are to be executed in parallel by available processors. OpenMP directives can divide the loop into several blocks, which are computed by different processors. In the paper, the processor is Intel (R) Xeon (R) CPU E3-1230 V3, four core and eight thread processor, dominant frequency 3.30GHZ, computer memory 8G. The software platform for parallelizing program execution is Microsoft Visual Studio 2012, and the compiler system is the Intel Fortran compiler.

In SPH algorithm, the loop statements can be divided into two categories, i.e. particles number loop and particle pairs number loop. The parallel schemes for the loop statements are introduced in the paper, and computational efficiency is studied. At the same time, dam-break flow is simulated by OpenMP-based SPH method, the pressure evolution on the wall is studied.

2. Basic model

2.1. SPH formulation

In the model, the water satisfies balance laws for conservation of mass

$$\frac{\mathrm{D}\rho}{\mathrm{Dt}} = -\rho\nabla \cdot \boldsymbol{u}, \tag{1}$$

And momentum

$$\rho\frac{D\boldsymbol{u}_\eta}{Dt} = \nabla \cdot \boldsymbol{\sigma}_\eta. \tag{2}$$

In the SPH method, the problem domain is dispersed into many particles. Every particle holds physical characteristics such as volume, mass, and field function information. A field function of particle can be approximated by summation over the neighboring particles in the support domain, as

$$f_a = \sum\nolimits_{b\in P} V_b f_b w_{ab}, \tag{3}$$

so as the derivative of field function

$$\nabla f_a = \sum\nolimits_{b\in P} V_b(f_a + f_b) \cdot \nabla w_{ab}, \tag{4}$$

Where f_a and f_b is the field function of particle a and b, V_b volumn of particle b, and P assemblage of all particles. The equation (4) is the classic symmetric form of SPH formulation. w_{ab} Is called smoothed kernel function with symmetry and normalization. With the increasing of distance between two particles, the value of smoothed kernel function decreases, given by

$$w(r, h) = \begin{cases} \alpha_D\left(1 - \frac{q}{2}\right)^4(2q + 1), & 0 \le q \le 2 \\ 0, & q > 2 \end{cases} \tag{5}$$

where $q = |r_{ab}| / h$; $r_{ab} = r_a - r_b$. r_a And r_b is position vectors of particles a and b, and h is smoothing length. α_D Is a parameter to ensure the normalization of the smoothed kernel function, equal to $7/(4\pi h^2)$ in two-dimension model.

In SPH method, the water is regraded as weak compressible fluid, the water pressure is related to density by equation of state, given by

$$p = \frac{\rho_0 c_0^2}{\xi}\left[\left(\frac{\rho}{\rho_0}\right)^{\xi} - 1\right], \tag{6}$$

where ρ_0 is initial density of water, a constant 1000kg/m^3; $\xi = \text{constant}$ normally set to seven; c_0 artificial sound speed, depends on the case. The continuity equation and momentum equation can be dispersed. Numerical techniques, such as artificial viscosity, MLSPH and XSPH, are also added in the model to make sure a great precision. Besides, ghost particles boundary treatment method is applied to the boundary. A brief introduction to SPH method is presented in this section and detail can be found in [4].

2.2. *OpenMP parallel scheme*

In the SPH method, the problem domain is dispersed into many particles. Every particle has a special number, like the number as shown in the fig.1. Every particle has a support domain, like the circle as shown in the fig.1. In the support domain, particle pair will be formed by the particle and another one, like the p1, p2, p3 in the figure. In SPH algorithm, the loop statements can be divided into two categories, i.e. particles numbers loop and particle pair's numbers loop.

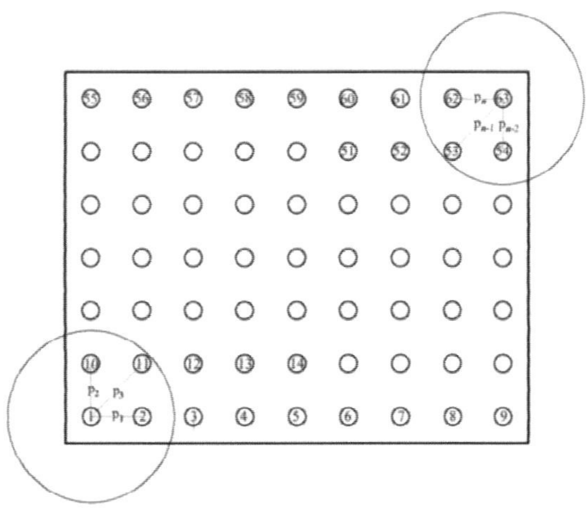

Fig.1 particle number and particle pair number

2.2.1. Particles numbers loop. Link-list search algorithm, a popular nearest neighboring particle searching (NNPS), has high efficiency, but still takes about half computing time. So, parallelization of link-list algorithm is imperative. In the particles numbers loop, we divided the particles according to the number of threads at first. Then variables, which are called and assigned at the same time, are defined as private variables. Those private variables can be calculated separately in different processors, thus avoiding data conflict. The private data will be accumulated before ending parallel. And we will organize memory of data for the subsequent calculations. Flow chart of parallel code design as shown in fig.2(a).

2.2.2. Particles pair's numbers loop. Continuity equation and momentum equation are calculated by the particles pair's numbers loop, which is complicated. Every particles pair has two particles, and two different particles pairs could have same particle. It is no problem in the serial computing. But in parallel calculation, the particle information may be assigned and called simultaneously by multiple processors, which is forbidden. For example, when the first thread is going to take the function value of a particle, the value may have been modified by another thread. To solve the problem, we define the function as a private variable and accumulate the field function from different threads before ending parallel. Flow chart of parallel code design as shown in fig.2 (b).

3. Simulation and results analysis

In this section, the SPH method are validated in the simulation of dam-break and impact problem. The effect of numerical techniques, like artificial viscosity, MLSPH and XSPH, is discussed. The pressure evolution on the wall is studied.

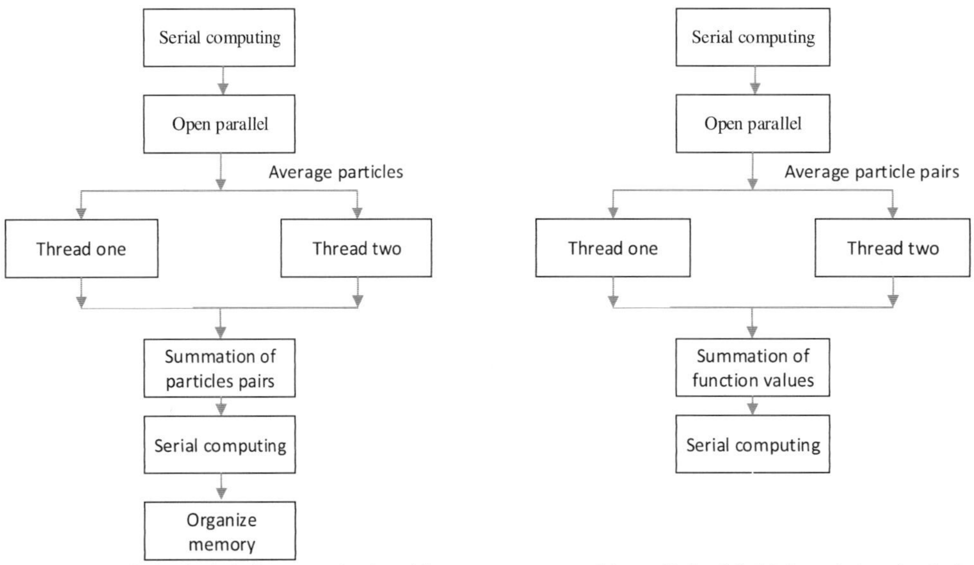

(a) Parallel of Link-list search algorithm (b)parallel of field function calculation

Fig.2 Flow chart of parallel code

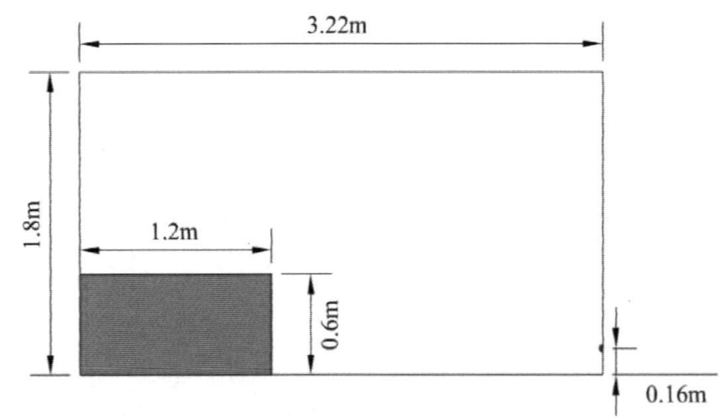

Fig.3 Dam-break model

IOP Conf. Series: Journal of Physics: Conf. Series **916** (2017) 012042 doi:10.1088/1742-6596/916/1/012042

3.1. Dam-break flow against a vertical wall

As sketched in Fig.3, in the 3.22m×1.8m tank, 1.2m×0.6m water column exists in the lower left corner. After computing, the water column begins to move under gravity. The point on the right wall is pressure probe.

Fig.4 displays snapshots of the flow evolution up to the plunging wave closure and the subsequent generation of a cavity. At t=0.63s, water begins to impact the wall and move upward. Under the force of the wall, the water has left velocity. Then, jet flow is formed as shown in the fig.4(c). Finally, impacting the surface of the water, the jet flow splashes water around and forms holes.

Fig.5 displays the evolution of velocity vector when the dam-break flow impacts the wall. The water particle move anticlockwise as a whole, but the water particle near the bottom right corner moves clockwise and forms vortex. As time goes on, vortex center increase gradually, which we can see from the circle in the figure.

Fig.6 is the simulation profile at 1.4s, which shows the role of numerical techniques. Artificial viscosity, XSPH technology and MLSPH technology, have reduce numerical oscillations from acceleration, velocity and density three respects. The combined application of the three numerical techniques can simulate dam break flow more accurately.

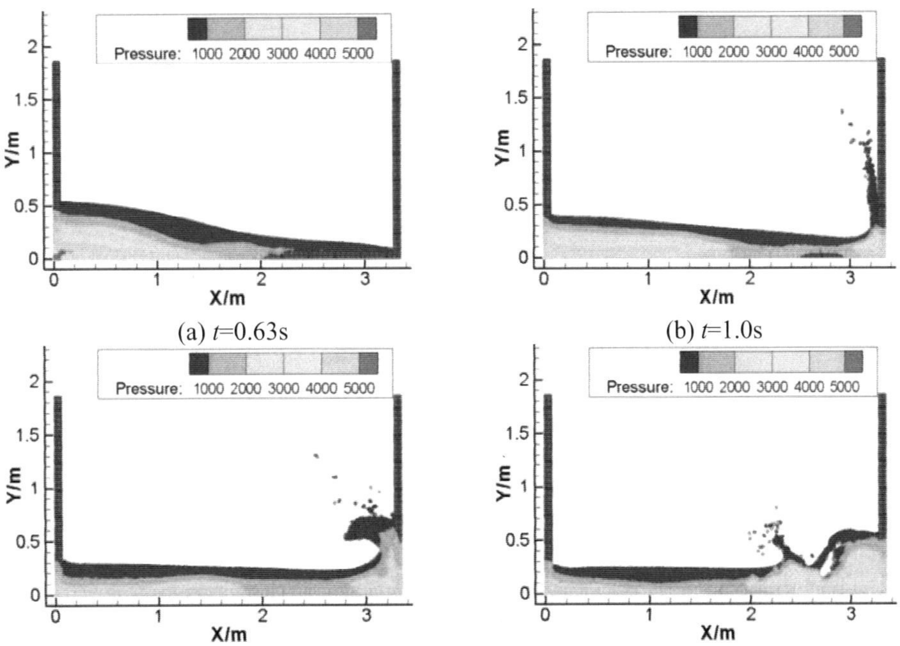

(a) t=0.63s (b) t=1.0s

(c) t=1.4s (d) t=1.7s

Fig.4 Snapshots of the dam-break flow against a vertical wall.

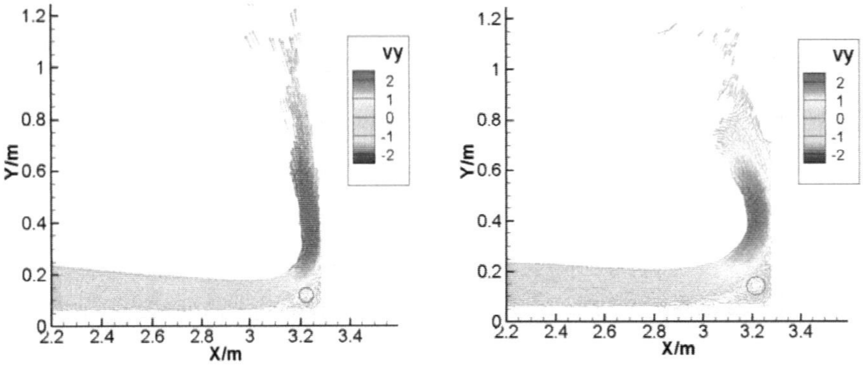

(a) t=1.0s (b) t=1.2s

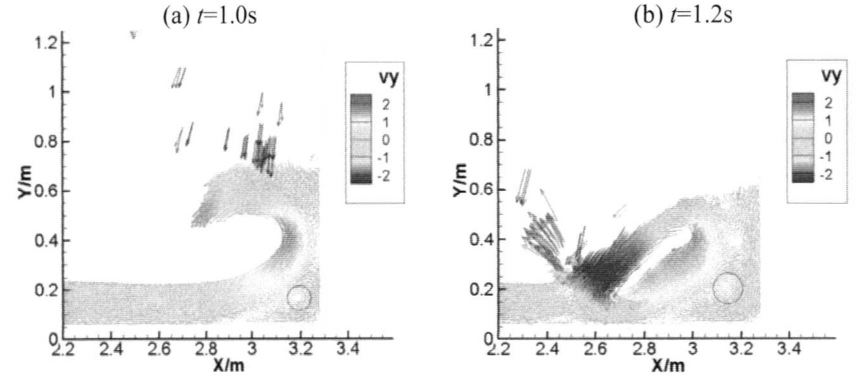

(a) t=1.4s (b) t=1.6s

Fig.5 Velocity vector and vortex evolution

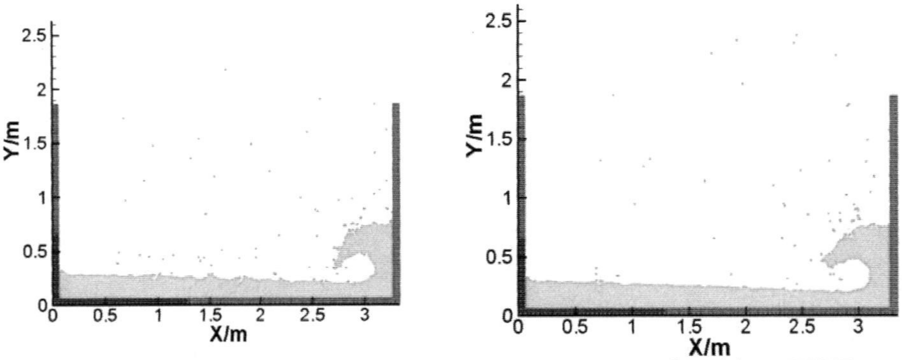

(a) Without artificial viscosity (b) without XSPH

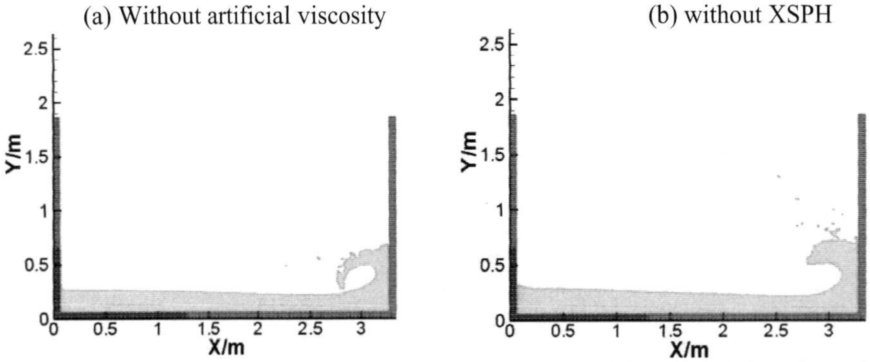

(c) Without MLSPH (d) with artificial viscosity, XSPH, MLSPH

Fig.6 The role of numerical techniques

Fig.7 displays the Comparison between the pressure loads measured experimentally and predicted by the numerical model. Simulation result is basically in line with experiment value. However, numerical oscillation problem still occurs in the strong impact, especially in the early stage of contact. The numerical techniques can help to decrease oscillation at the cost of increasing energy dissipation, which the pressure is lower than the experimental value. We think the oscillation problem can be eliminated by optimized boundary treatment.

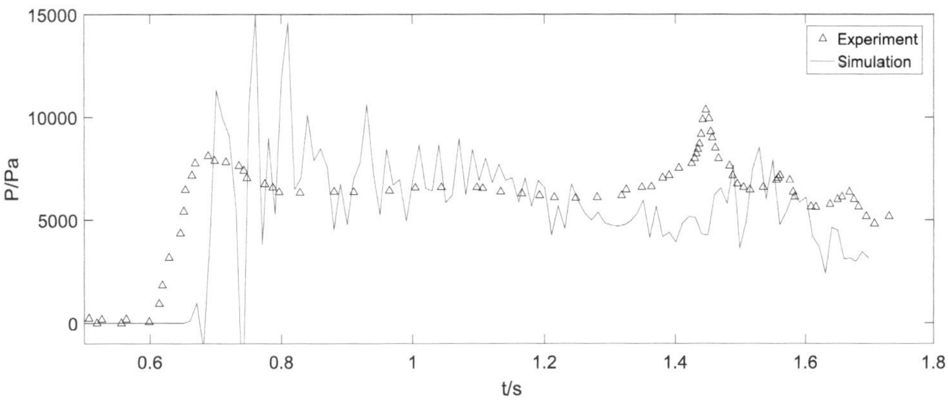

Fig.7 Comparison between the pressure loads measured experimentally and predicted by the numerical model

3.2. Parallel result analysis

For an OpenMp parallel program, computation time is generally consumed in the following three sections: serial computing, parallel computing and data transfer between different processors. Even in parallel program, there must be many serial statements, such as input and output statements. In addition to the serial part, the parallel efficiency is decided by the time consumed by parallel computation and the data transfer between different processors. In general, the more threads in parallel, the less time consumed by parallel computing, but the more time it takes to transfer data between different threads. This means that more processors may have lower computational efficiency. The parallel efficiency is often determined by the speedup ratio S_p and the efficiency of the individual processor E_p, as defined below:

$$S_p = \frac{T_1}{T_p}; \qquad E_p = \frac{S_p}{p},$$

Where subscript refers to the number of processors. T_1 Is the time cost by serial computing? T_p Is the time cost by parallel computing under p threads? Changing the number of particles or threads, we can get different speedup radio.

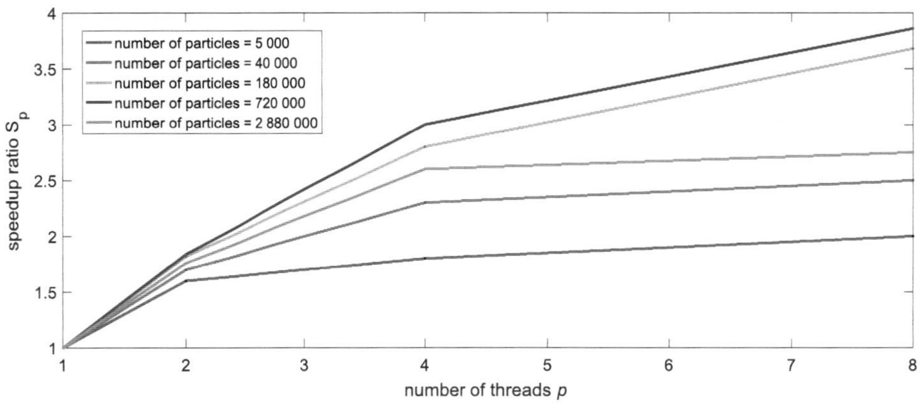

Fig.8 S_p evolution with respect to the number of threads

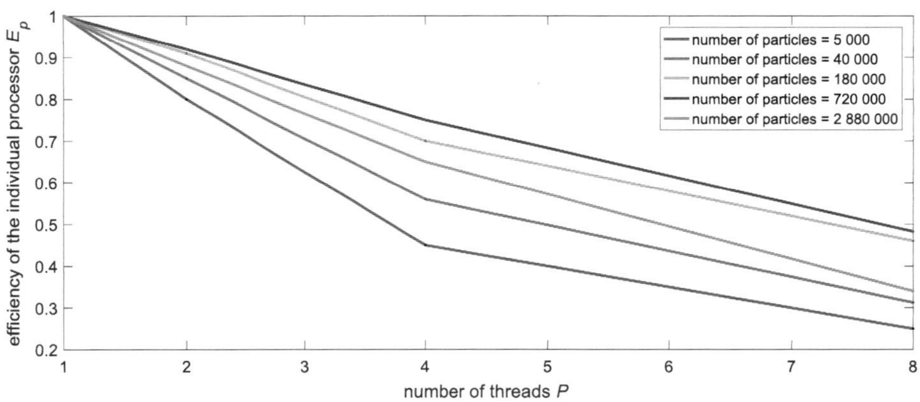

Fig.9 E_p evolution with respect to the number of threads

Fig.8 displays the speedup ratio evolution with respect to the number of threads. When the number of particles is constant, the speedup ratio is directly proportional to number of threads. When the number of threads is constant, the speedup ratio is directly proportional to number of threads within certain range. When the number of particles is too large, such as 2 880 000, computer may consume too much time in data transferring between different threads, resulting in a low parallel efficiency.

Fig.9 displays the efficiency of the individual processor evolution with respect to the number of threads. The efficiency of a single processor always decreases as the number of threads increases, because of the time consuming of data transferring.

4. Conclusion

In the paper, the dam-break flow is simulated successfully by SPH method. Jet water and cavity have been captured accurately. By observing velocity vector of particles, vortex center near the wall has been discovered. The role of numerical techniques such as artificial viscosity, MLSPH and XSPH have been studied, and combined application of the three numerical techniques can achieve better results. Compared with the pressure loads measured experimentally, the pressure on the wall predicted by the numerical model is basically consistent.

Two parallel schemes for SPH algorithm is introduced, and parallel efficiency and the efficiency of single thread have been studied. The speedup ratio is basically proportional to number of threads within certain range. However, when the number of particles is too large, the parallel efficiency will be poor.

References

[1] L.B. Lucy, A numerical approach to the testing of the fission hypothesis, J. The astronomical journal. 82 (1977): 1013-1024.

[2] R.A. Gingold, J.J. Monaghan, Smoothed particle hydrodynamics: theory and application to non-spherical stars, J. Monthly notices of the royal astronomical society, 181.3 (1977): 375-389.

[3] Z. LUO, C. WANG, Unified semi-analytical wall boundary treatment in SPH and irregular particle distribution, J. Chinese Journal of Hydrodynamics, 32.2 (2017): 189-197.

[4] C. Wang, Y. Wang, C. Peng, et al, SPH simulation of water-soil mixture flows, J. ASCE Journal of Hydraulic Engineering, 142.10 (2016): 04016032.

[5] C. Wang, Y. Wang, C. Peng, et al, Two-fluid smoothed particle hydrodynamics simulation of submerged granular column collapse, J. Mechanics Research Communications, 79 (2017): 15-23.

[6] D. Cheng, W.H. Wang, Application of OpenMP in SAR Image Processing, J. Computer science. 44 (2017): 161-163,187.

[7] J. Guan, S. Yan, J.M. Jin, An OpenMP-CUDA Implementation of Multilevel Fast Multipole Algorithm for Electromagnetic Simulation on Multi-GPU Computing Systems, J. IEEE Transactions on Antennas & Propagation, 61.7 (2013):3607-3616.

[8] S. Yeylaghi, B. Moa, P. Oshkai, et al, ISPH modelling of an oscillating wave surge converter using an OpenMP-based parallel approach, J. Journal of Ocean Engineering & Marine Energy. 2.3 (2016):1-12.

IOP Conf. Series: Journal of Physics: Conf. Series **916** (2017) 012043 doi:10.1088/1742-6596/916/1/012043

Flow analysis of engine intake manifold based on computational fluid dynamics

Jianmin Xu*

School of mechanical and automotive engineering, Xiamen University of Technology, Xiamen, China

*Corresponding author e-mail: xujianmin1020@163.com

Abstract. Intake manifold is a part of intake system, which has an important impact on intake uniformity of the multi-cylinder engine. Three kinds of intake manifold are designed and a three-dimensional flow field numerical simulation for them is conducted using software FLUENT. The flow characteristics of intake manifold are analyzed and the intake unevenness of intake manifold is calculated. According to the flow field contours, the reason of generating non-uniformity of intake system is analyzed. The results show that the inlet end arranged in the middle of the intake manifold can reduce the non-uniformity of intake system. And it can improve the engine intake and combustion quality. The calculation results provide a theoretical basis for designing the complex intake manifold

1. Introduction

Intake manifold should provide a uniform fresh charge for each cylinder of the engine, which is an important factor affecting the quality of engine combustion. Intake manifold is one of the main parts of the engine. Its main role is to provide uniform and stable intake air for the engine so as to ensure the intake uniformity of each cylinder. Uneven intake will produce unstable torque output, engine vibration, emissions increase and other problems. The rational design of the intake manifold not only can reduce the inlet pressure loss and increase the intake air quantity, but also can ensure the intake uniformity of each cylinder. Therefore, designing the intake manifold is one of the key technologies of ensuring the engine power, economy, and reliability and emission quality. The flow characteristics of air-fuel mixture flowing in various designs of manifold of IVECO 6 cylinder heavy-duty engine were studied [1]. Internal flow field analysis of engine intake manifold was conducted based on RE and CFD [2]. The internal flow characteristic in the intake manifold of a diesel engine was investigated computationally for the different intake manifold (Helical, Spiral and Helical-Spiral) configurations [3]. The transient behavior of air flow through the intake manifold of a heavy duty diesel engine was researched according to the step of valve opening [4]. A methodology for optimization of gas injector orientation for better thermal efficiency was emerged [5]. The flow distribution in an intake manifold under steady state turbulence conditions was studied [6]. A novel Dual Intake Manifold system was designed and developed which improves the torque at POT condition by making the charge to flow through the longer path and at WOT condition the charge flows through the shorter path[7]. The flow field of intake system of CW6200 gas engine was simulated [8-9]. And the intake uniformity of each cylinder was studied. The gas liquidity and the gas flow quality in branched pipe can be improved by changing the air inlet direction and regulator chamber volume and other geometric parameters [10-13].

In this paper, the three-dimensional models of three kinds of new intake manifold for engine were established. The model was divided with tetrahedral mesh using software GAMBIT. Three-dimensional flow simulation was conducted using software FLUENT and the velocity contours of intake manifold were obtained. The influence of intake manifold structures on the intake non-uniformity was researched. The study provides the basis for the optimization design of the intake manifold.

2. Mathematical Model

2.1. Equations

For the stable dimensional compressible flow, there are mass and momentum conservation equations which are Reynolds-averaged.

$$\frac{\partial}{\partial x_j}(\rho u_j) = 0 \tag{1}$$

$$\frac{\partial}{\partial x_j}(\rho u_j u_i - \tau_{ij}) = -\frac{\partial p}{\partial x_j} + s_i \tag{2}$$

Where S_i is the source term, it represents resistance; τ_{ij} is the stress tensor, and there are the following equations for Newtonian flow.

$$\tau_{ij} = 2\mu(s_{ij} - \frac{1}{3}\frac{\partial u_k}{\partial x_k}\delta_{ij}) - \overline{\rho u'_i u'_j} \tag{3}$$

Where μ is the molecular dynamic viscosity coefficient; δ_{ij} is Kroneker number; $\overline{\rho u'_i u'_j}$ is Reynolds stress tensor; s_{ij} is fluid strain rate tensor, which is determined by the following equation.

$$s_{ij} = \frac{1}{2}(\frac{\partial u_i}{\partial x_j} + \frac{\partial u_j}{\partial x_i}) \tag{4}$$

2.2. Turbulence model of muffler flow field

The Reynolds stress is calculated using a standard κ-ε model to close the flow control equations:

$$\overline{\rho u'_i u'_j} = -2\mu_t s_{ij} + \frac{2}{3}(\mu_t \frac{\partial u_k}{\partial x_k} + \rho k)\delta_{ij} \tag{5}$$

Where μ_t is the turbulent viscosity coefficient, which is determined by the following equation.

$$\mu_t = \frac{c_\mu \rho \kappa^2}{\varepsilon} \tag{6}$$

Where κ, ε is turbulent kinetic energy and turbulent energy dissipation rate respectively. The transport equation is as follows.

$$\begin{aligned} &\frac{\partial}{\partial x_j}\left(\rho u_j \kappa - \frac{\mu_{eff}}{\sigma_\kappa}\frac{\partial \kappa}{\partial x_j}\right) = \\ &\mu_t s_{ij}\frac{\partial u_i}{\partial x_j} - \rho\varepsilon - \frac{2}{3}\left(\mu_t \frac{\partial u_i}{\partial x_i} + \rho\kappa\right)\frac{\partial u_i}{\partial x_i} \end{aligned} \tag{7}$$

$$\begin{aligned} &\frac{\partial}{\partial x_j}\left(\rho u_j \varepsilon - \frac{\mu_{eff}}{\sigma_\varepsilon}\frac{\partial \varepsilon}{\partial x_j}\right) \\ &= c_{\varepsilon 1}\frac{\varepsilon}{\kappa}\left\{\mu_t s_{ij}\frac{\partial u_i}{\partial x_j} - \frac{2}{3}\left(\mu_t \frac{\partial u_i}{\partial x_i} + \rho\kappa\right)\frac{\partial u_i}{\partial x_i}\right\} \\ &-c_{\varepsilon 2}\rho\frac{\varepsilon^2}{\kappa} + c_{\varepsilon 4}\rho\varepsilon\frac{\partial u_i}{\partial x_i} \end{aligned} \tag{8}$$

Where $\mu_{eff} = \mu + \mu_t$; The empirical coefficients of C_μ , σ_k , σ_ε , $C_{\varepsilon 1}$, $C_{\varepsilon 2}$ and $C_{\varepsilon 4}$ are determined according to Table 1.

Table 1. The coefficient of experience

c_μ	σ_k	σ_ε	$c_{\varepsilon 1}$	$c_{\varepsilon 2}$	$c_{\varepsilon 4}$
0.09	1.0	1.22	1.44	1.92	-0.33

3. Calculation model and boundary conditions

3.1. Geometric model

The three-dimensional models of 3 kinds of the intake system are shown in figure 1 to 3. The intake system is composed of an inlet end, regulator chamber and intake manifold. The role of intake manifold is to distribute the gas to each cylinder evenly. The structural parameters of the engine intake manifold are shown in Table 2. Three kinds of intake manifold structure are proposed. The inlet ends of design scheme 1 and 2 are located at the side of the regulator. The inlet end of design scheme 3 is located in the middle of regulator.

Table 2. The parameters of intake manifold

The length of regulator cavity(mm)	275
The width and height of regulator cavity(mm)	80
Intake manifold diameter(mm)	50

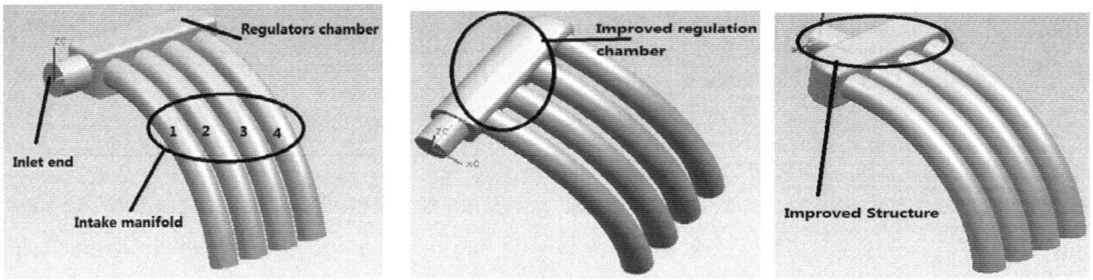

(a) The design scheme 1 (b) The design scheme 2 (c) The design scheme 3

Figure1. The three-dimensional model of intake pipe system

3.2. Meshing and boundary conditions

Meshing quality will directly affect the level of convergence speed, accuracy and computation time. It has a significant impact on the entire simulation. In order to make the final calculations close to the real situation, the quality of meshing should be improved. The tetrahedral mesh is used to divide the model of four intake manifolds, and the hexahedral mesh is applied in rest parts. Computing grid model of intake manifold systems is shown in figure 2. The SIMPLE algorithm is used to couple pressure and velocity field. The tube air flow is set to incompressible viscous turbulent flow. The standard turbulence model $k-\varepsilon$ is applied. Wall function method is used to describe the fluid velocity and pressure distribution of boundary layer. Turbulence intensity is set to 0.03, the turbulence length is set to 0.05. The front surface of the inlet end is set to mass flow inlet boundary, outlet face of each manifold is set to pressure outlet boundary and others are set to wall boundary.

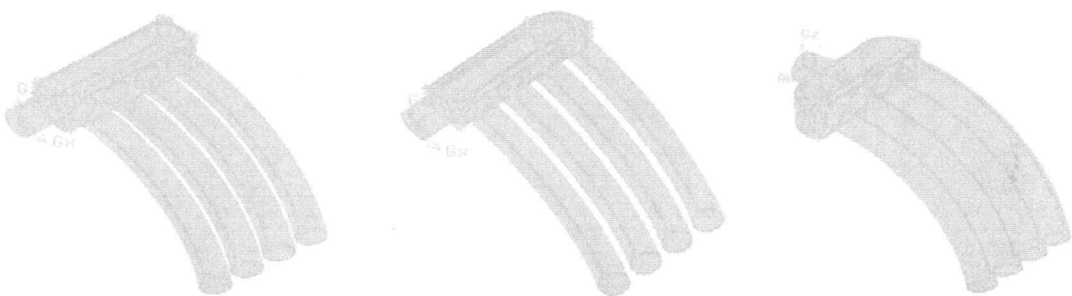

(a) The design scheme 1 (b) The design scheme 2 (c) The design scheme 3

Figure 2. Computing grid model of intake manifold systems

4. Analysis of calculation results

4.1. Evaluation standard

Intake manifold flow coefficient of each outlet is usually the common method of evaluating intake uniformity. The maximum unevenness E is adopted to represent engine intake unevenness. The formula is as follows.

$$E=(Q_{\max}-Q_{\min})/Q_{me} \tag{9}$$

Where Q_{max} the maximum flow is rate of export quality; Q_{min} is the minimum flow rate of export quality; Q_{me} is the average mass flow.

4.2. The results and discussion

4.2.1. Velocity distribution. Figure 3 shows the velocity contours of intake manifold for design scheme 1. The figure shows that the airflow velocity of the manifold which is far away from the inlet end is large and the airflow velocity near the entrance end is small. When gas flows into the intake manifold from the regulator chamber, the vortex is generated in the intersection of the regulator chamber and manifold because a sudden change in flow cross-sectional area. The air distribution of each intake manifold is uneven. The intake rate of the third and fourth manifold is big and the intake air is sufficient. Since the shaft of regulator manifold is perpendicular to each branch pipe axis, intake swirl is generated so as to result in feed gas shortage of the first and second manifold. According to the simulation result, the unevenness of the calculated intake is 9.31 percent. This shows that the intake uniformity of scheme 1 is poor. So the intake manifold is designed unreasonable.

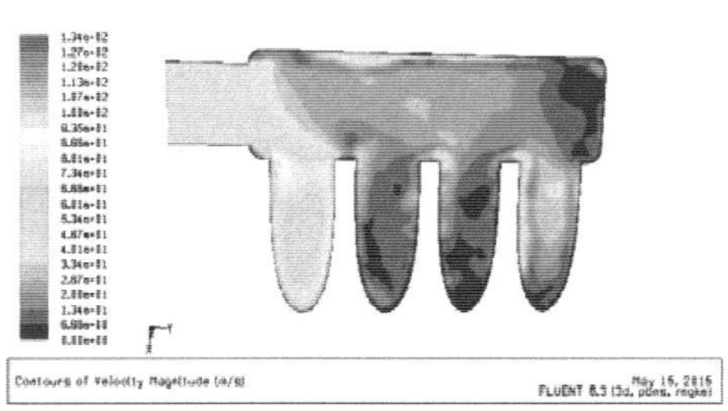

(a) The intake manifold 1 is open

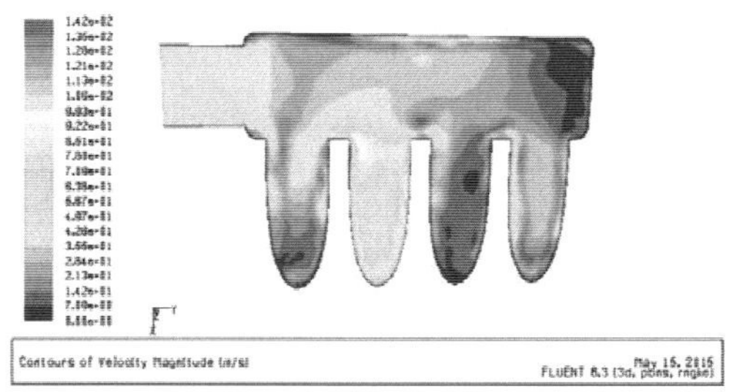

(b) The intake manifold 2 is open

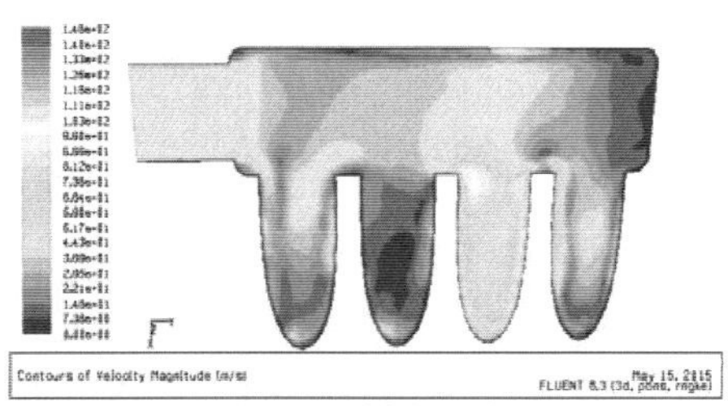

(c) The intake manifold 3 is open

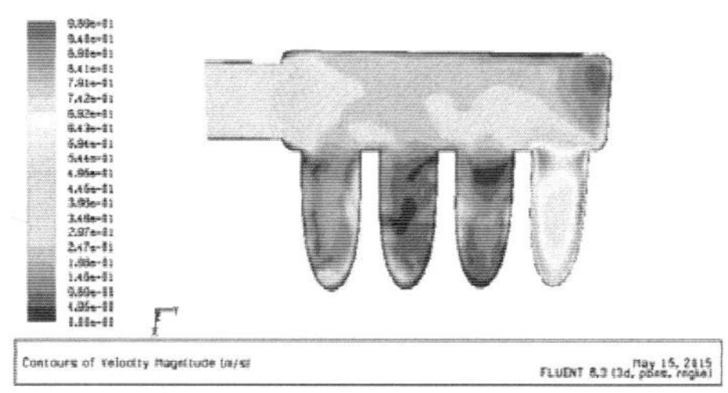

(d) The intake manifold 4 is open

Figure 3. The velocity contours of intake manifold for design scheme 1

Figure 4 shows the velocity contours of intake manifold for design scheme 2. The intake quality of intake manifold 3 and 4 is very good and the gas reflux phenomenon appeared in manifold 1 and 2. Reflux phenomenon causes an increase of flow resistance so as to result in reduction of flow coefficient. So the intake amount of entering the cylinder is reduced and the combustion process is affected. According to the simulation result, intake unevenness is 7.78 percent, so the intake uniformity of scheme 2 is poor.

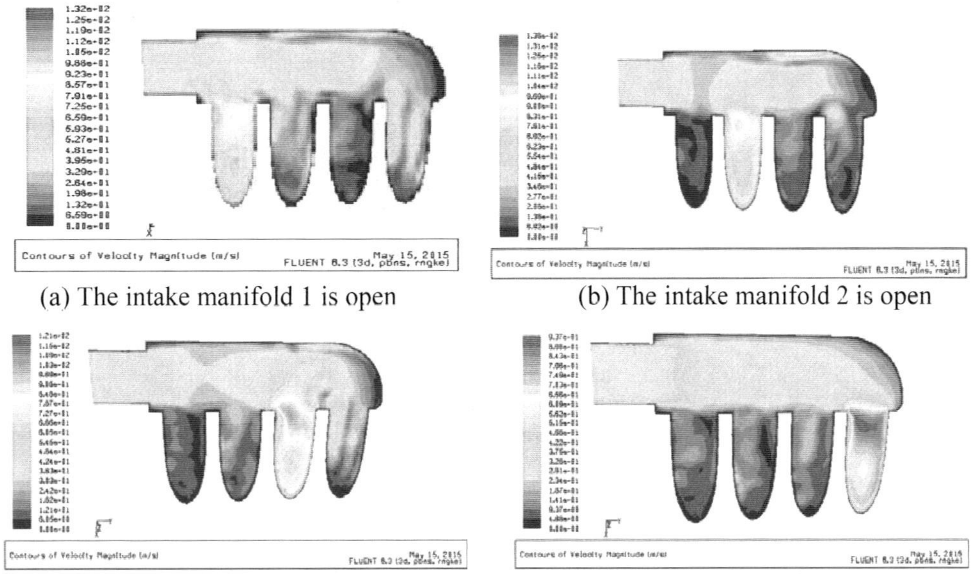

(a) The intake manifold 1 is open (b) The intake manifold 2 is open

(c) The intake manifold 3 is open (d) The intake manifold 4 is open

Figure 4. The velocity contours of intake manifold for design scheme 2

The design scheme 3 is that the inlet end is located in the middle of the regulator. The velocity contours of each manifold are shown in figure 5. Intake air flow rate of each manifold is relatively uniform and no vortex phenomenon appears. According to the simulation results, the intake unevenness E is 4.716 percent, which shows that the intake uniformity of scheme 3 is very good.

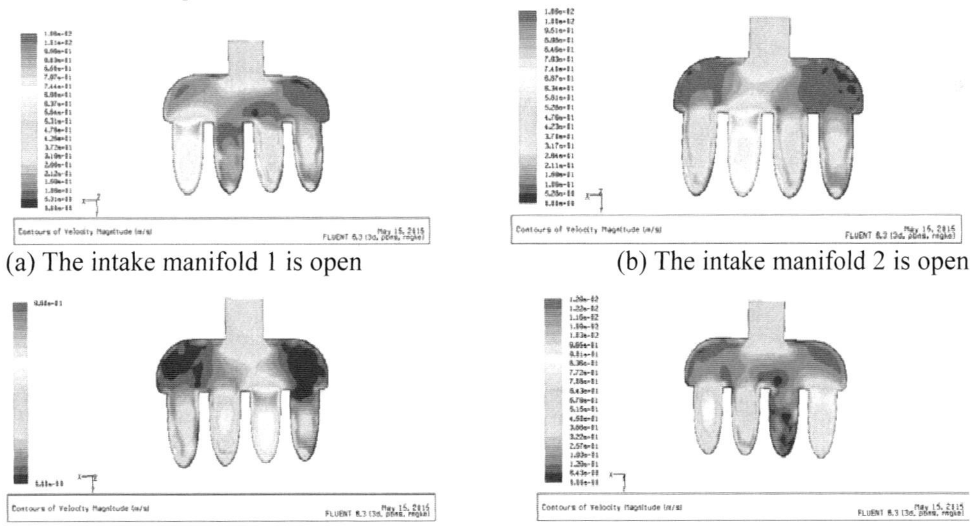

(a) The intake manifold 1 is open (b) The intake manifold 2 is open

(c) The intake manifold 3 is open (d) The intake manifold 4 is open

Figure 5. The velocity contours of intake manifold for design scheme 3

4.2.2. Pressure loss analysis. Figure 6 shows the pressure loss comparison of 3 kinds of intake manifold. As can be seen from the figure 6 that the improving effect of the program 3 is the best. The pressure loss of each manifold is reduced and the pressure loss uniformity of the respective manifold is greatly increased. The intake air of each manifold is relatively uniform. The improving effect of the intake manifold in program 3 is feasible.

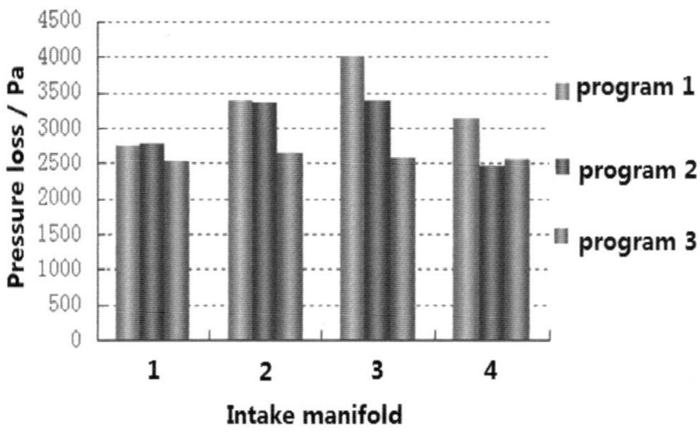

Figure 6. The pressure loss comparisons of 3 kinds of intake manifold

5. Conclusion

(1) Numerical Simulation of three kinds of intake system is conducted and flow field of intake manifold is analyzed. The results show that the inlet end arranged in the middle of the regulator can greatly improve intake uniformity so as to improve the intake efficiency of the engine and combustion quality.

(2) Based on computational fluid dynamics theory, three-dimensional numerical simulation of the engine's intake manifold system is carried out, which is an effective and feasible method of analysis.

(3) The results show that the pressure loss and flow uniformity of intake manifold are two major evaluation indicators, and the scheme 3 meets these two design specifications.

References

[1] Jemni M A, Driss Z, Kantchev G, Design and Modeling of Mechanical Systems,CA: Springer Berlin Heidelberg,pp.331-338,2013.

[2] Wang C Y, He Z X, Wang Q. "Internal Flow Field Analysis of Engine Intake Manifold Based on RE and CFD", Advanced Materials Research, vol. 1079, pp. 930-934, 2015.

[3] Rahiman A, AD M S, Ramis M K. "CFD Analysis of Flow Field Development in a Direct Injection Diesel Engine with Different Manifolds", American Journal of Fluid Dynamics,Vol.4, No.3, pp.102-113,2014.

[4] Srisattha T, Issarakul S, Wechsatol W. "Transient Behavior of Air Flow through the Intake Manifold of a Heavy Duty Diesel Engine", Applied Mechanics and Materials, Vol. 87, pp.146-150,2011.

[5] Chintala V, Subramanian K A. "A CFD (computational fluid dynamics) study for optimization of gas injector orientation for performance improvement of a dual-fuel diesel engine", Energy,Vol.57, No2.pp.709-721,2013.

[6] Thamaraikanan R, Anish M, Kanimozhi B. "Design and Analysis of an Intake Manifold in an IC Engine", Applied Mechanics and Materials. Vol.766, pp.1021-1027, 2015.

[7] Saravanan D, Gokhale A, Karthikeyan N. "Design and Development of a Flow Based Dual Intake Manifold System", SAE International Journal of Engines,Vol.8, No1. , pp. 68-74,2014.

[8] Zhang Q, Li N. "Numerical analysis of intake uniformity of biogas engine", Transactions of the Chinese Society of Agricultural Engineering, Vol.26, No.8, pp.145-149,2010.

[9] Wang Han, Cai Yixi, Mao Xiaoping. "3D numerical simulation of intake unevenness for engines", Small Internal Combustion Engine and Motorcycle, Vol.36,No.3, pp.41—44,2007.

[10] Fenzhu J, Keshuai G. "Numerical simulation of liquidity and uniformity of gasoline engine intake manifold" Journal of Beijing University of Aeronautics and Astronautics, Vol.40,

No.2,pp.154-159,2014.
[11] Ceviz M, Akim M. "Design of a new SI engine intake manifold with variable length plenum",Energy Conversion and Management, Vol. 51, No11. ,pp. 2239 – 2244,2010.
[12] David C, Pascal C. "Fluid dynamic modeling of junctions internal combustion engine inlet and exhaust systems", Journal of Thermal Science, Vol. 19, No.5 , pp.410 –418,2010.
[13] Tong-qun H A N. "Design and Exploitation of Intake Pipe of Engine Equipped with EFI", Chinese Internal Combustion Engine Engineering, Vol.26, No.3, pp.22-26,2005.

IOP Conf. Series: Journal of Physics: Conf. Series **916** (2017) 012044 doi:10.1088/1742-6596/916/1/012044

Three dimensional numerical simulation for air flow field in melt blowing

Sheng Xie[1,*], Wanli Han[1] and Guojun Jiang[2]

[1]College of Material Textile Engineering, Jiaxing University, Jiaxing 314000, China
[2]Zhijiang College of Zhejiang University of Technology, Hangzhou 310024, China

[*]Corresponding author e-mail: shengxie16@163.com

Abstract. Melt blowing is one-step approach for producing microfibrous nonwovens. In this study, the characteristics of air flow field in different die melt blowing were studied. CFD simulation analysis was adopted to complete the air flow field simulation. The characteristics of the air flow fields in different die melt blowing were revealed and compared. Meanwhile, the fiber paths in these two melt-blowing processes are recorded with a high-speed camera. The fiber path profiles, especially the whipping, are obtained and presented.

1. Introduction

Melt blowing is an important one-step approach for producing nonwovens with fiber diameter of several micrometers. There are different kinds of dies are used in melt blowing process to produce microfibrous nonwovens. The most common used such as slot diea, annular dies and swirl dies.

Different die geometries create different air flow fields. The air flow fields in melt blowing have been studied with experimental method and computational fluid dynamics (CFD) simulation [1-3]. In recent years, CFD analysis has been developed to be a useful and essential method on searching the air flow field. Researchers [4-8] simulated the air flow field below different kinds of dies. Unfortunately, most of their simulations were based on two-dimensions. In this paper, the air flow fields in slot die and swirl die melt blowing were numerical simulated by approach of CFD, Three-dimensional model was built and the simulated air flow field such as stream line was correspondingly illustrated in three-dimensions. Meanwhile, the fiber paths in different die melt blowing were demonstrated. This paper showed the fundamental relationship between characteristics of air flow field and the corresponding fiber motion.

2. Air flow field

2.1. Numerical Simulation

We used a calculation domain of a quarter of the total air flow field due to the symmetrical characteristic of the field. The computational domain and the corresponding grids were created by Gambit, and 862710 cells were yield. The air flow field was obtained by solving the N-S equations through the commercial software FLUENT 6.3. The Reynolds Stress Model (RSM) was applied to run the simulations. And turbulence parameters $C_{1\varepsilon}$ and $C_{2\varepsilon}$ were set as 1.24 and 2.05 based on Krutka's study [4]. Figure 1 shows the three-dimensional model of the slot die. Figure 3(a) shows the corresponding mesh dividing, quadrilateral mesh is applied in this model.

For the swirl die, the calculation domain is 1/6th of the total air flow field, which is symmetric about the centerline (i.e., the z-axis). The computational domain and the corresponding grids were created by Gambit, and 545903 cells were yield. Simulations were run using the SST k-ω model. The SST k-ω model is a popular two-equation eddy-viscosity model that combines k-ω model in the near-wall region and reverts to k-ε in the free-stream via a blending function. Figures 2 and 3(b) illustrate the numerical model and the mesh dividing for the air flow field in swirl die melt blowing, different with the method of mesh dividing, hybrid meshes are applied in the simulation of swirl die.

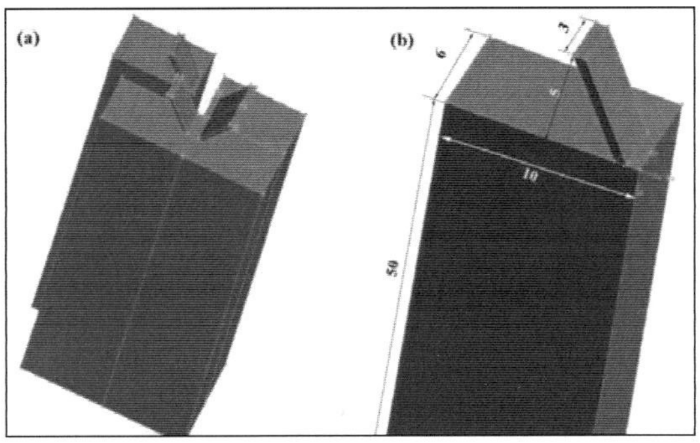

Figure 1. Numerical simulated model of air flow field in slot die. (a) is the whole air flow field, (b) is the 1/4 model used in numerical simulation.

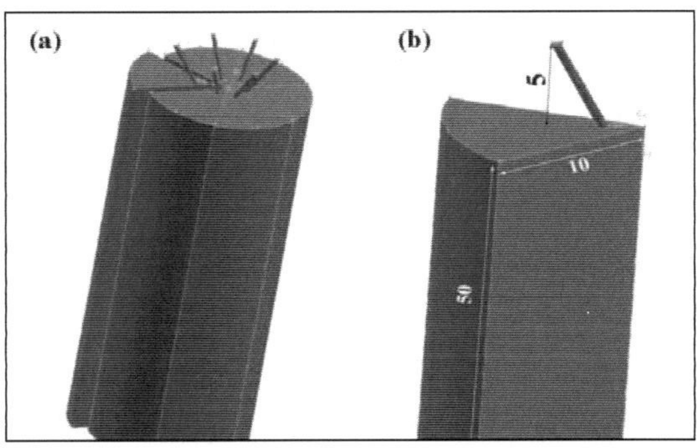

Figure 2. Numerical simulated model of air flow field in swirl die. (a) is the whole air flow field, (b) is the 1/6 model used in numerical simulation.

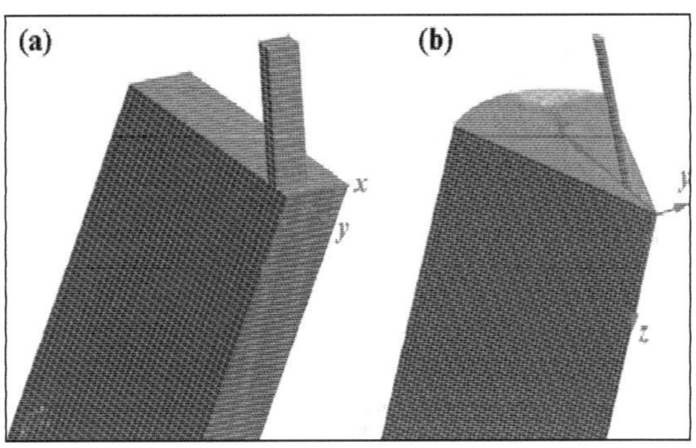

Figure 3. The mesh dividing of the model for (a) slot die and (b) swirl die melt blowing respectively.

IOP Conf. Series: Journal of Physics: Conf. Series **916** (2017) 012044 doi:10.1088/1742-6596/916/1/012044

2.2. Profiles of Simulated Air Flow Field

Figure 4 shows the path lines emanating from the inlets in the slot die and the swirl die melt blowing air flow fields, respectively. The path line profiles for the two dies are apparently different. For the slot die, Figure 4a shows that the air issued from the inlets moves almost parallel to the z axis.

Figure 4b shows that in the swirl die air flow field, at a short z-position of several millimeters below the die, the six streams of swirling air merge together and become a combined swirling air stream. Unlike the slot die, each air orifice in a swirl die has a twist angle θ which can create swirling air.

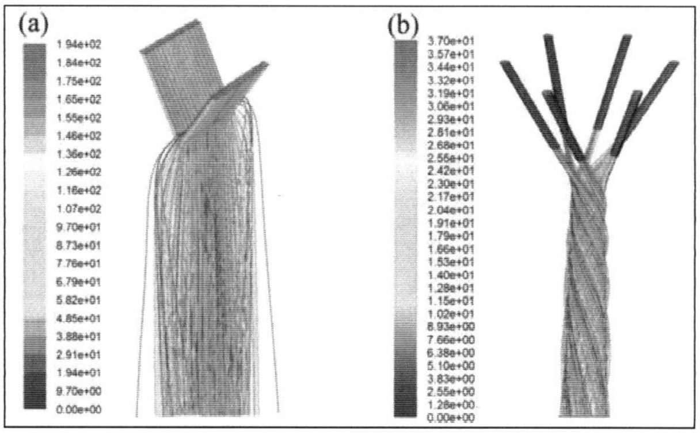

Figure 4. Path lines emanating from the inlets in (a) the slot die and (b) the swirl die melt-blowing air flow fields.

3. Fiber paths

Figure 5(a) shows the time evolution of fiber paths in the slot-die air flow field. Whipping in the slot-die melt-blowing process appears to be a fiber path with two groups of loops moving downward [9]. And the whipping path in the region near the die is almost in the y-z plane.

Figure 5(b) shows the time evolution of the fiber path in three dimensions in swirl die melt blowing. Whipping in the swirl-die process appears as a spiral path in three dimensions [10], which is different from the two-dimensional bending motion in the slot-die melt-blowing process. The aerodynamic characteristics of a melt-blowing swirl die are crucial in determining the fiber motion behavior. Unlike the slot die, each air orifice in a swirl die has a twist angle which creates swirling air.

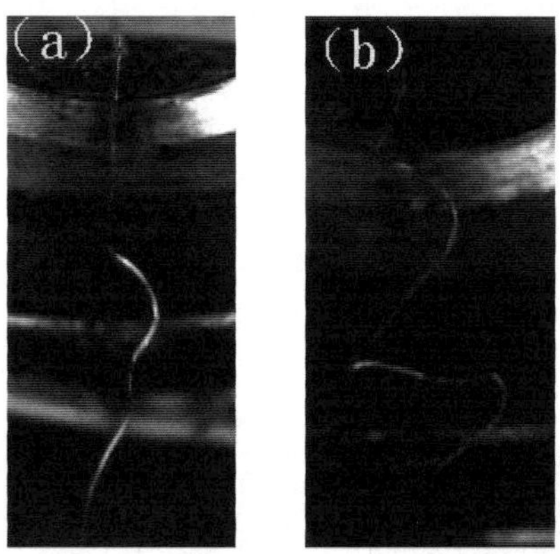

Figure 5. Path lines emanating from the inlets in (a) the slot die and (b) the swirl die melt blowing air flow fields.

4. Conclusion

Three dimensional numerical simulation models were built for simulating the air flow field in melt blowing. Meanwhile, the fiber pathes were cited for illustrating the relationship between the air flow field and the fiber path. The result shows that the evolution of fiber path in melt blowing was dominated by the given air flow field, in addition, the different air flow field was achievable via different die geometries.

This study illustrated that the nonwoven fibers can be controlled by the die design.

Acknowledgments
This research was supported by the National Natural Science Foundation of China (Grant 11702113 and Grant 51506075), Zhejiang Provincial Natural Science Foundation (Grant LQ18E030013).

References

[1] M. A. J. Uyttendaele, R. L.Shambaugh, The flow field of annular jets at moderate Reynolds numbers, Ind. Eng. Chem. Res. 28 (1989) 1735-1740.
[2] A. S. Harpham, R. L. Shambaugh, Flow Field of Practical Dual Rectangular Jets, Ind. Eng. Chem. Res. 35 (1996) 3776-3781.
[3] A. S. Harpham, R. L. Shambaugh, Velocity and Temperature Fields of Dual Rectangular Jets, Ind. Eng. Chem. Res. 36 (1997) 3937-3943.
[4] H. M. Krutka, R. L. Shambaugh, D. V. Papavassiliou, Effects of a Melt-Blowing Die: Comparison of CFD and Experiments, Ind. Eng. Chem. Res. 41 (2002) 5125-5138.
[5] H. M. Krutka, R. L. Shambaugh, D. V. Papavassiliou, Effects of Temperature and Geometry on the Flow Field of the Melt Blowing Process, Ind. Eng. Chem. Res. 43 (2004) 4199-4210.
[6] H. M. Krutka, R. L. Shambaugh, D. V. Papavassiliou, Analysis of Multiple Jets in the Schwarz Melt-Blowing Die Using Computational Fluid Dynamics, Ind. Eng. Chem. Res. 44 (2005) 8922-8932.
[7] H. M. Krutka, R. L. Shambaugh, D. V. Papavassiliou, Analysis of the Temperature Field from Multiple Jets in the Schwarz Melt Blowing Die Using Computational Fluid Dynamics, Ind. Eng. Chem. Res. 45 (2006) 5098-5109.
[8] E. M. Moore, R. L. Shambaugh, D. V. Papavassiliou, Analysis of Isothermal Annular Jets: Comparison of Computational Fluid Dynamics and Experimental Data, J. Appl. Polym. Sci. 94 (2004) 909-922.
[9] S. Xie, Y. C. Zeng, Online Measurement of Fiber Whipping in the Melt-Blowing Process, Ind. Eng. Chem. Res. 52 (2013) 2116-2112.
[10] S. Xie, Y. C. Zeng, Fiber Spiral Motion in a Swirl Die Melt-Blowing Process, Fiber. Polym. 15 (2014) 553-559.

Study on casing treatment and stator matching on multistage fan

Chuangliang Wu[1, a], Wei Yuan[2], Zhe Deng[1]

[1] Master Degree Candidate, School of Energy and Power Engineering, Beihang University, Beijing 100191, China
[2] Associate Professor, School of Energy and Power Engineering, Beihang University, Beijing 100191, China

[a]Email:946375937@qq.com

Abstract. Casing treatments are required for expanding the stall margin of multi-stage high-load turbofans designed with high blade-tip Mach numbers and high leakage flow. In the case of a low mass flow, the casing treatment effectively reduces the blockages caused by the leakage flow and enlarges the stall margin. However, in the case of a high mass flow, the casing treatment affects the overall flow capacity of the fan, the thrust when operating at the high speeds usually required by design-point specifications. Herein, we study a two-stage high-load fan with three-dimensional numerical simulations. We use the simulation results to propose a scheme that enlarges the stall margin of multistage high-load fans without sacrificing the flow capacity when operating with a large mass flow. Furthermore, a circumferential groove casing treatment is used and adjustments are made to the upstream stator angle to match the casing treatment. The stall margin is thus increased to 16.3%, with no reduction in the maximum mass flow rate or the design thrust performance.

1. Introduction

An important benchmark for developing fourth-generation advanced turbofan engines is to improve their thrust-to-weight ratio. Recent advancements in high-load fan technologies are particularly important for achieving this goal. Currently, the three-stage fan technology is widely used in developed countries. For example, the total pressure ratio of GE's F110 three-stage turbofan reaches 3.2, whereas that of Pratt and Whitney's three-stage F119 turbofan reaches 4.0[1]. Higher-load fans used in next-generation engines remain in the laboratory phase of research. In 1987, the United States military initiated the Integrated High Performance Turbine Engine Technology [2] program, which achieved great conceptual strides towards the design of high-load compressor fans. In 2002, NASA's Green Centre achieved a stage pressure ratio exceeding 3[3]. In the early 1990s, several Chinese labs performed numerical simulations and experimental studies to investigate the performance of high-load inlet-stage fans. Two experiments on single-stage fans have been successfully completed, achieving pressure ratios of 2.2–2.3[4-5].

For achieving a simple structure, good stability and high reliability, casing treatments have been widely used in aerospace engineering applications. Casing treatments follow two typical structures, with either axial grooves or circumferential grooves. Axial groove casing treatments have good stability-enhancing effects [6-8]; however, they require some axial space in front of the rotor, restricting their use to single-stage fans. To ensure compactness of the fan, circumferential groove casing treatments are usually used on the rear stage of multistage fans. Researchers have devoted considerable attention towards optimizing the casing parameters and understanding the mechanisms behind the stability extension caused by circumferential groove treatments. There still remains a controversy associated with the optimal axial position of circumferential grooves. Bailey [9] and Moore [10] found that the optimum position of circumferential grooves is at the middle of the blade chord. Furthermore, Muller [11] and Raba and Hah [12] suggested that the leading edge of the blade is the best position. Wu Anhui [13] and Bennington [14] reported that the stability extension achieved through circumferential groove treatments decreases as the groove width increases. In addition, Bailey [9] and Bennington [14] reported that the stability extension effect increases with increasing groove depth. Several other groups have attempted to explain the effectiveness of circumferential grooves. From the stable flow point of view, Muller [11] and Wu et al. [13] suggested that the circumferential grooves delay the overflow at the leading edge of blade tips, thereby enlarging the stall margin. Zhao et al. [15] studied the unstable flow effects of the circumferential grooves on the leakage flow at blade tips. Hah [16] found that the unstable fluctuations of the circumferential suction and ejection have an important influence on the stability extension effect.

It has been difficult to overcome the lack of aerodynamic stability for high-load turbofans. Many researchers have extended the stable operating range of fans by installing circumferential slot casings with some success. Dang [17] used circumferential groove slots to reduce the stall point of a high mass-flow rate fan by 11.5%. Duane [18] installed various circumferential groove slots on a high-load fan and increased the stability margin by 7.3%. Li [19], Liu [20] and other researchers have separately studied the optimum application and the mechanism of action for circumferential groove slots installed around high-load compressor fans. However, most studies focused on improving the stall margin of the fan while ignoring the side effects of the casing treatment on the right-hand branch of fan's performance curves. Most studies are further limited by focusing only on single-stage fan designs. In fact, casing treatments in real-world applications will reduce the mass-flow rate through the point of maximum flow while enlarging the stability margin. For high-load compressor fans, the performance design points are often near the right-hand branch of the performance curve. Casing treatments may therefore reduce the design thrust of a turbofan engine. Considering these complications, we study a two-stage high-load transonic fan design. Using numerical simulation tools, we investigate a circumferential groove casing treatment and propose an upstream stator-matching scheme that can eliminate the drawbacks of casing treatments.

2. Research object and methods

2.1. Two-stage fan and casing treatment

This research focuses on a high-load two-stage fan that includes guide vanes. Its pressure ratio is above 4.5 at the design speed, and the tangential speed of the rotor blade tips exceeds 500m/s (Mach 1.46). The load coefficient of the first-stage rotor reaches 0.4, whereas that of the second-stage rotor reaches 0.35.

Figure1. Two-stage high-load fan

Figure 1 shows a three-dimensional (3D) model of the fan. The design parameters of the circumferential casing treatment are listed in Table 1, and a detailed schematic of the circumferential groove casing treatment is shown in Figure 2.

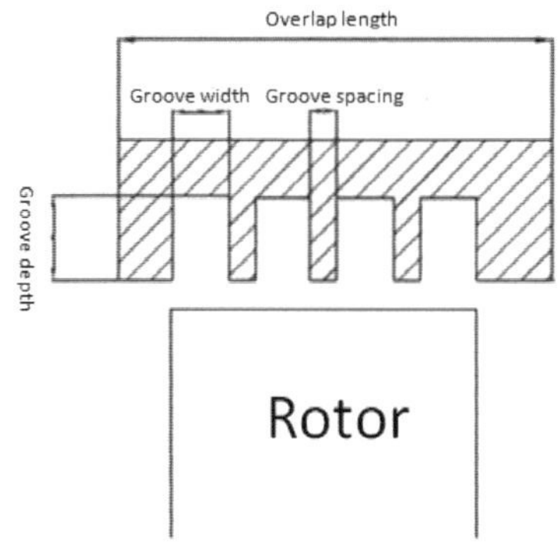

Figure2. Circumferential groove casing treatment

Table 1. Geometric parameters of the circumferential groove casing treatment

Groove number	Groove depth/mm	Groove width/mm
7	10	4
Overlap length/mm	Groove spacing/mm	Distance to leading edge/mm
40	2	2

2.2. Numerical simulation method

We used ANSYS CFX, a numerical simulation software, with a finite-volume method to solve 3D Reynolds-averaged Navies–Stokes equations (N–S equations). In addition, the SST turbulence model was used. Considering the accuracy of our results and the computational load involved, unstable simulations were performed only at the near-stall and design points using the transient blade row (TBR) method in the CFX solver and stable simulations were used for all other conditions. Because the entire computational domain (including the casing treatment) is axisymmetric and the intake conditions are uniform, all calculations were performed with a single channel.

The computational grid (Figure 3) was generated using the IGG module of the NUMECA software. The blade surface adopted an O-type grid, the upstream and downstream sections of the blade adopted an H-type grid, the top clearance of the blade adopted a butterfly grid and the casing treatment adopted an H-type grid. When verifying mesh independence, the mesh structure remains unchanged and the flow field structure remains sufficiently equivalent when the grid number is greater than 2 million. In our final model, the grid number of the solid-wall fan was approximately 2.8 million, that of the circumferential groove casing treatment was approximately 500,000 and the y+ value was less than 3.

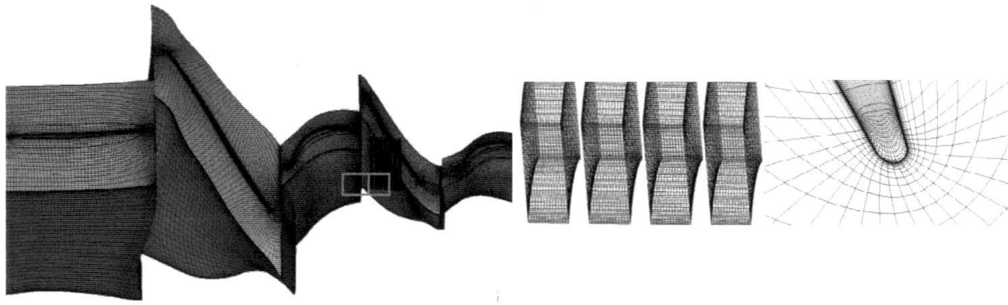

Figure 3. Structure of computational grid

3. Numerical results and analysis

3.1. Flow field analysis of the solid-wall fan

The fan performance curve is shown in Figure 4. The parameters in the plot are dimensionless, and the mass-flow rate, pressure ratio and efficiency are the actual values divided by the values obtained at the design points of the solid-wall fan. The stall margin of the fan is 13.6%. Most of that stall margin is the pressure margin, and there is basically no flow margin with a flow margin of only 0.55%. These preliminary results show that enlarging the stall margin of the fan is quite essential.

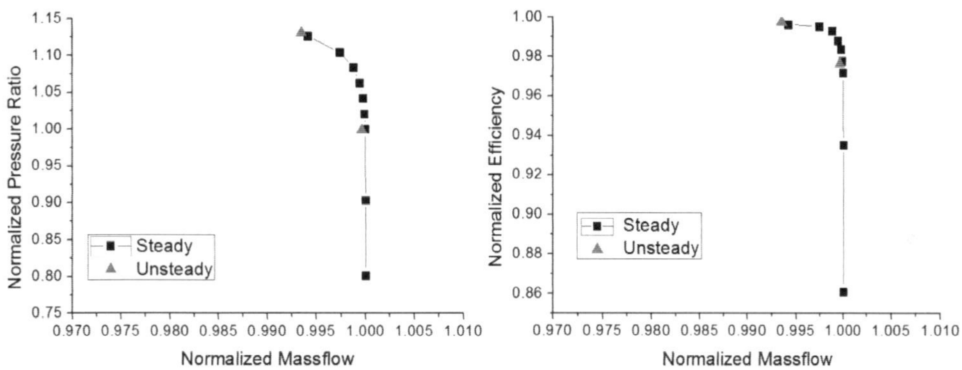

Figure 4. Performance curves of the solid-wall fan

We define the stall margin in Eq. (1):

$$SM = \left[\frac{\pi_{Ks}^{*} / m_{as}}{\pi^{*}_{K0} / m_{a0}} - 1 \right] * 100\% \tag{1}$$

The mass-flow margin is expressed in Eq. (2):

$$MFM = \left[\frac{m_{a0}}{m_{as}} - 1 \right] * 100\% \tag{2}$$

m_{as} And π^*_{Ks} are the mass-flow rate and pressure ratio, respectively, of the fan near the stall point; m_{a0} and π^*_{K0} are the mass-flow rate and pressure ratio, respectively, of the rotor under the designed operating conditions.

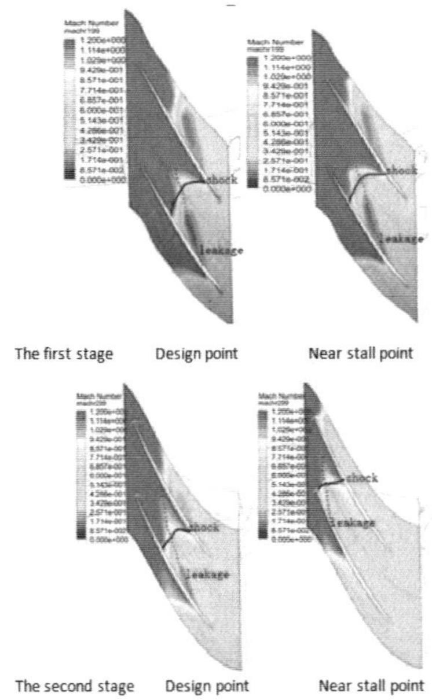

Figure 5. Mach number contour at the blade tip of the solid-wall fan

First, we analysed the blade-tip flow field of the solid-wall fan. In Figure 5, the solid line indicates the shockwave position and the dotted line represents the trajectory of the leakage flow. In transonic rotors, leakage flow will pass through the shockwave at the leading edge of the blade tip. For the first stage rotor, moving from the design point to the near-stall-point plots, the position of the channel shock wave at the blade tip moves forward; however, the shockwave remains in the channel throughout the process. The leakage flow passes through the blade-channel shockwave near the pressure side of the blade and forms a low-velocity zone, and this low-velocity zone also exists at the design point. From the design point to the near-stall point, the low-velocity zone has no tendency to increase and no extensive blockage occurs at the channel; thus, the low-velocity zone cannot be the key factor causing the fan to stall. On the other hand, the leakage-flow trajectory does not significantly move forward from the design point to the near-stall point and the overall leakage-flow structure does not change. For the second-stage rotor, the position of the channel shockwave at the blade tips significantly moves forward from the design point to the near-stall point. In addition, at the near-stall point, the shockwave moves to the leading edge of the blade. This movement of the shockwave forewarns a stall. At the same time, the trajectory of the leakage flow certainly moves forward at the near-stall point so that the axial momentum of the leakage flow drops to such an extent that the leakage flow strikes the pressure side of the adjacent blades at their leading edges, blocking the flow through the channel of the blade tips. Based on the Mach number contour plot at the blade tips, we conclude that stalls begin at the second-stage fan.

3.2. Stability enhancement analysis of circumferential groove casing treatment

Because stalling begins at the second stage of the solid-wall fan, we tried many different schemes of circumferential groove casing treatments. By combining our preliminary simulations with the previous research on circumferential groove casing treatments [5-10], we determined the circumferential groove parameters, as listed in Table 1. Figure 6 shows that the circumferential groove casing treatment clearly enhances the stability. The stall margin (SM) of the fan increases from 13.6% to 16.3%, whereas the mass-flow margin (MFM) of the fan increases from 0.55% to 3.28%. The effect of stability enhancement therefore meets the demands of real-world applications.

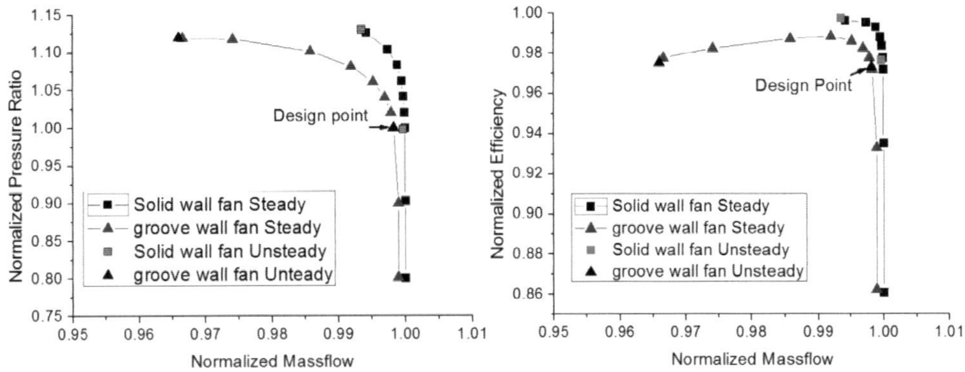

Figure 6. Performance curves for fans with solid walls and groove casing treatments

Comparing the two casings with the same mass-flow rate, Mach contours at the blade tips of the second-stage rotor operating near the stall point are shown in Figure 7. The solid line represents the shock position, whereas the dotted line represents the trajectory of the leakage flow. In general, when the fan functions stably, the shockwave at the blade tip stays within the blade channel. As backpressure gradually increases, the shockwave gradually moves forward; however, when the shockwave moves to the leading edge of the blade, the stability of the fan considerably decreases. Figure 7 shows that the blade-channel shockwave for the solid-wall fan has been pushed to the leading edge of the rotor under near-stall conditions. After the casing treatment, the channel shock wave is clearly located rearwards and the shockwave remains within the blade channel. Therefore, the backpressure of the fan can continues to increase and the stall margin of the fan can be widened. At the same time, the low-velocity zone caused by leakage around the solid-wall fan strikes the pressure side of adjacent blades at the 30% chord, blocking the mainstream flow. Despite the leakage smoothly flows through the blade passage, the leakage flow around the grooved-wall fan does not strike the pressure side of adjacent blade.

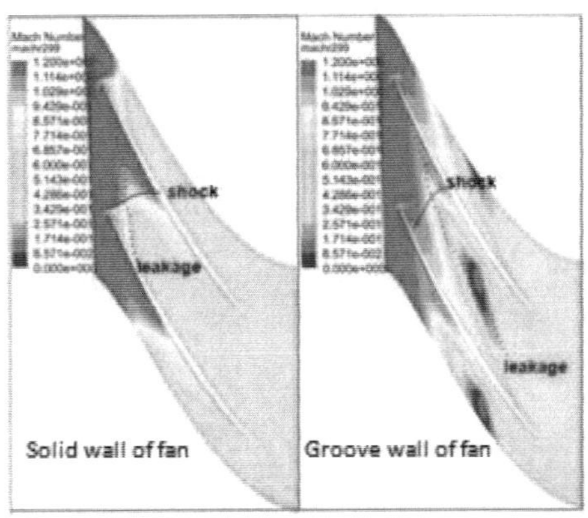

Figure 7. Mach contour at the blade tip of the second rotor

Figure 8 shows the static pressure distribution at the blade tips of the second-stage rotor. First, as the suction-side static pressure rises, the shock stagnation point for the solid-wall fan is clearly located farther forward than that of the grooved-wall fan, which is consistent with the conclusion obtained from the blade-tip Mach contour map. Second, the static pressure for the solid-wall fan is higher than that for the grooved-wall fan, showing that the load at the blade tips of the solid-wall fan is greater than that for the grooved-wall fan. In addition, in the range of 0%–35% blade chord length, the static pressure difference between the pressure and suction sides of the solid-wall fan is significantly greater than that for the grooved-wall fan. The static pressure difference between the pressure and suction sides is the main driving force for the blade-tip leakage flow. Therefore, in the range of 0%–35% blade chord length, the leakage flow around the solid-wall fan is more serious than the leakage flow around the grooved-wall fan.

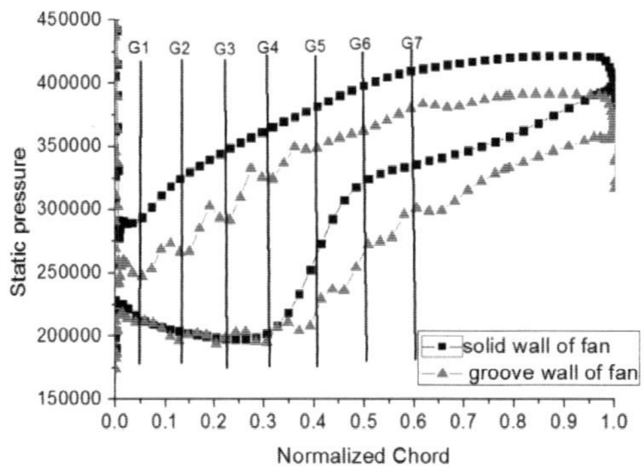

Figure 8. Static pressure distribution at the blade tips of the second-stage rotor

Figure 9 shows the static pressure distribution at the blade tips of the second-stage rotor and the radial velocity distribution in the circumferential grooves at four different moments in a rotational period. The background flow field represents the static pressure contour. The radial velocity distribution in the grooves (the upward direction is positive) clearly shows that the phenomena of 'suction' at the pressure side and 'spray' at the suction side occur. Although the suction and spray effects are different at different moments, they are mainly concentrated in the first five grooves around the casing. Figures 8 and 9 demonstrate that the circumferential grooves form a channel between the pressure and suction sides of the blade tip. The pressure difference between the pressure and suction sides is reduced by the suction and spray effects of the grooves.

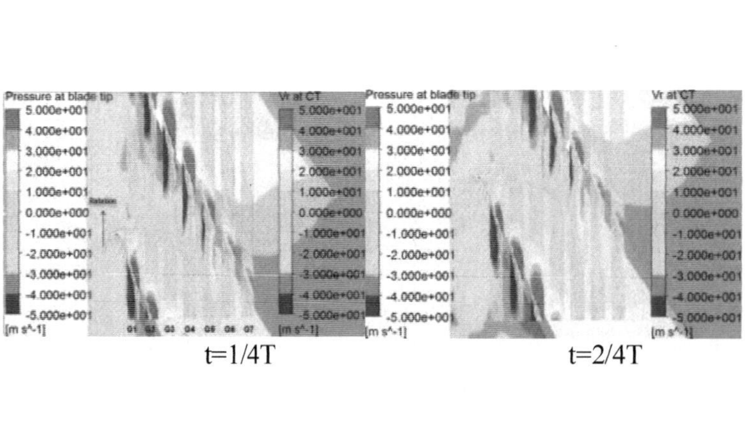

t=1/4T t=2/4T

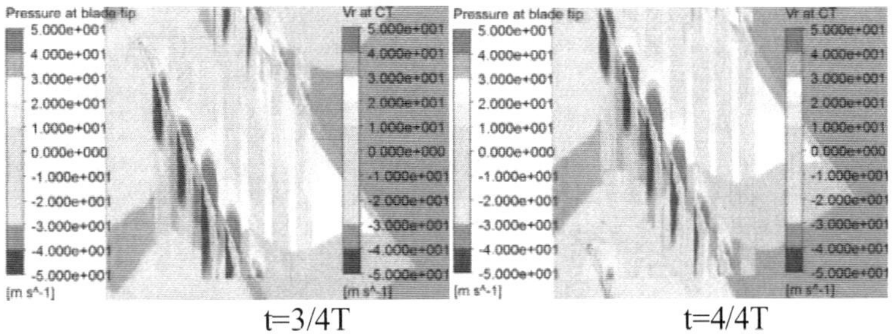

Figure 9. Static pressure contours at the blade tip and radial velocity distribution in grooves

At the same flow point and under near-stall conditions with a solid-wall casing, the blade-tip leakage flow around the second-stage rotor is shown in Figure 10. The leakage-flow structure is divided into three types. Type-A leakage is normal leakage flow due to the existence of tip clearance. The Type-A leakage flow smoothly passes around the blades with the mainstream; this flow commonly occurs throughout the compressor process. Type-B leakage is a kind of leakage flow starting from the leading edge of the blade and flowing to the pressure surface of the adjacent blade before leaving the blade channel with the adjacent blade pressure surface. This leakage flow does not directly form the tip blockage but reduces tip-flow area so that the tip deflects the mainstream flow in the circumferential direction, causing axial momentum to decrease, at which point the tip-flow capacity considerably decreases. Further development of this cascade can lead to a stall. Type-C leakage is a cross-channel leakage flow that occurs at the leading edge of the blade. This leakage flow will pile up at the leading edge of the blade, blocking fluid passage around the leading edge, which causes the compressor to quickly enter the stall process. For a solid-wall casing, Type-B and Type-C leakages may obviously cause stalling. With a casing treatment, the blockage of the blade tips disappears only for the normal Type-A leakage. Therefore, circumferential slot casings improve the tip-leakage flow and enhance blade-tip flow capacity.

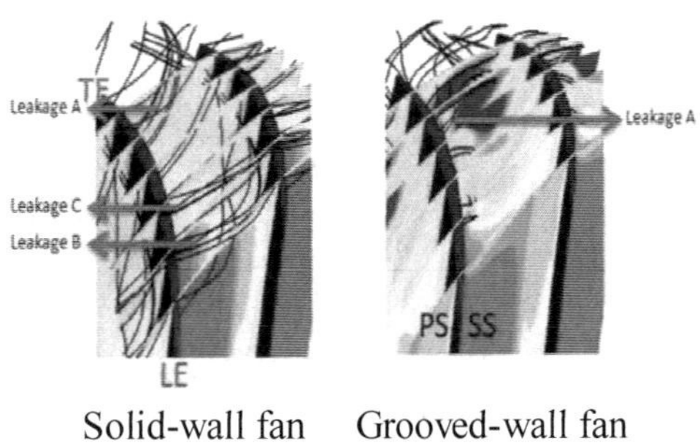

Figure 10. Leakage flow at the blade tips of the second rotor

3.3. Influence of the casing treatment on the performance of the fan at the design point (right branch)

We next compare the performance of the fan with a solid wall and that with the casing treatment (Figure 7). While widening the stability margin, the casing treatment has a negative impact on the right-branch performance of fan. For this fan, the design point is at the large-flow point of the right branch and the casing treatment reduces the design-point flow by 0.2%. The decrease of the flow at the design point reduces the engine cruise thrust, which is considerably detrimental to engine performance. Our analysis of the design-point flow field shows that the key factor reducing the fan flow is the reduction in the blockage at the tip of the second-stage rotor, as shown in the low-speed zone of Figure 11.

The blockage factor B_m is a measure of flow clogging, and our research refers to the definition reported in Khalsa et al. [21]:

$$A_m = \iint (1 - \frac{\rho\omega}{\rho_e\omega_e}) dA \tag{3}$$

$$B_m = \frac{A_b}{A_{exit}} \tag{4}$$

Where ρ and ω are the local density and local axial velocity, respectively, of the blockage area; ρ_e and ω_e are the density and axial velocity, respectively, at the boundary of the blockage area. A_{exit} Is the area of the measuring section, A_b is the blocking area and B_m is the blockage factor. Determining the clogging boundary is the key to analyzing the clogging. Sunder [22] provided a criterion based on the flow velocity gradient:

$$C_s = |\nabla_r(\rho v_m)| + |\nabla_\theta(\rho v_m)| \geq \delta . \tag{5}$$

To ensure that the threshold for judging the boundary is uniform along the direction of the flow, we revise the judgment criterion with reference to the dimensionless treatment LiuBaojie has made to Eq. (5):

$$C_m = \left|\nabla_r(\frac{\rho v_m}{\rho v_{m,avg}})\right| + \left|\nabla_\theta(\frac{\rho v_m}{\rho v_{m,avg}})\right| \geq \delta \tag{6}$$

Where $v_{m,avg}$ is the average flow velocity for each section?

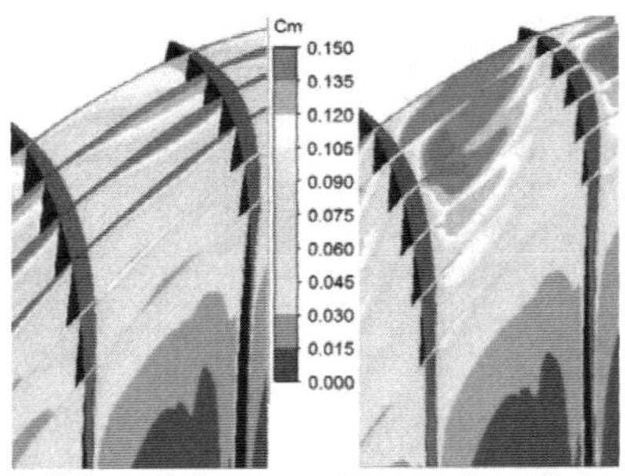

Solid-wall fan Grooved-wall fan

Figure 11. Blockage region with the second rotor at the design point

Figure 11 shows the flow-direction blocking distribution of the second-stage rotor with a solid-wall fan and a groove casing treatment fan at the design point. The red zone in the figure represents a blocking area that shows a clear boundary with the mainstream flow. It is evident from the figure that the clogging of the blade tip is not serious at the design point for the solid-wall fan and that the blockage area is exactly the same as the trajectory of the leakage flow. This result clearly indicates that leakage flow caused the blockage. For the casing treatment fan, the blockage under design-point operation is more serious and the position of the blockage is different from that for the solid-wall fan. In this case, the blockage originates from the leading edge of the blade and expands downstream. In contrast to the radial velocity distribution in the slot shown in Figure 10, the blockage of the blade tip in the casing treatment fan operating at the design point is related to the radial ejection of the circumferential grooves near the suction surface. The ejected fluid blends with the mainstream flow at the blade tip, which hinders the mainstream flow and forms a low-velocity vortex, the area of which dramatically expands at the reverse pressure gradient.

Figure 12 shows a plot of the blockage factor vs. fan flow. For the solid-wall fan, the blockage at the design point and its right branch is relatively small and stable, but increases rapidly near the stall point. The casing treatment reduces the growth rate of blockage in the small-flow area, indicating that the casing treatment's relief effect on the leakage-flow blockage in the small-flow area is much larger than the blocking effect on the mainstream flow due to the slot's ejection. This balance of factors delays the fan stall. However, at the design point and right branch, the result is just the opposite, indicating that the blockage for the grooved-wall fan is larger than that for the solid-wall fan and that the flow at the design point decreases.

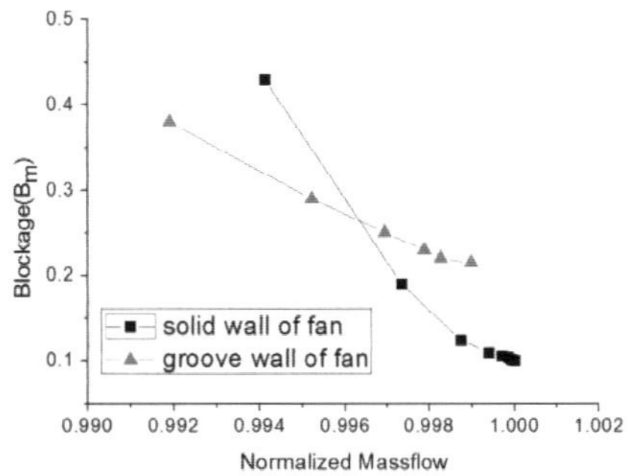

Figure 12. A plot of the blockage at rotor vs. normalised mass flow

Figure 13 shows the axial velocity distribution around the second-stage rotor's entrance at the design point. The axial velocity of the casing treatment fan is obviously less than that of the solid-wall fan above 70% blade height. This result indicates that the blade-tip blockage caused by the casing treatment weakens the circulation of the second rotor above 70% blade height.

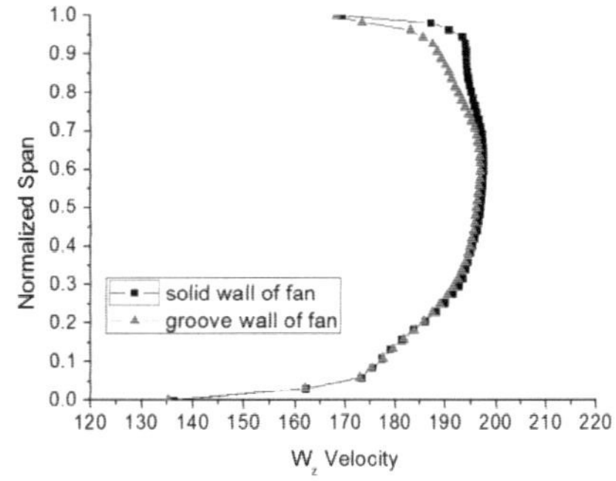

Figure 13. Wz velocity at the entrance of the second rotor

3.4. *Improving the flow at the design point (right branch) by adjusting the stator*

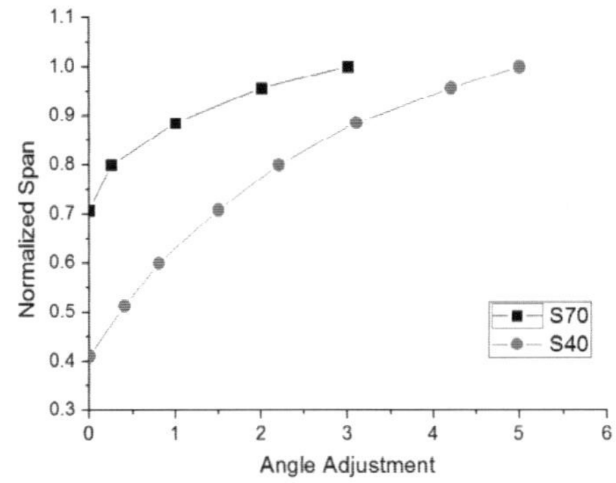

Figure 14. Partial angle adjustment scheme of the first-stage stator

For the two-stage fans, the opening angle of the upstream stator has a considerable impact on the performance of the fans. A local adjustment scheme for the angles of the first-stage stator is shown in Figure 14. The S70 scheme adjusts the first-stage stator above 70% blade height with a maximum adjustment angle of 3°, whereas the S40 scheme adjusts the first-stage stator above 40% blade height with a maximum adjustment angle of 5°

Figure 15 shows the performance curves for the solid-wall fan, along with grooved-wall fans operating with the scheme of not adjusting the stator (S0) and schemes S70 and S40 for stator adjustment. The stability and flow margins for the S0 scheme are 16.3% and 3.31%, respectively. The stability and flow margins for the S70 scheme are 16.5% and 3.04%, respectively. The stability and flow margins for the S40 scheme are 17% and 2.856%, respectively. With increasing stator adjustment range and angle, the fan-flow margin decreases; however, the stability margin increases. Although stator adjustment negatively affects the flow margin, it will increase the pressure ratio for the stall point, which increases the compression ratio margin, thereby increasing the overall stability margin.

From a comparison of the design-point flow with the upstream stator adjustment, as shown in Figure 15, it can be observed that the S70 stator adjustment scheme increases the design-point flow; however, its overall flow is still less than that for the original solid-wall fan. Furthermore, the S40 stator adjustment scheme increases the design-point flow to the level of the original solid-wall fan. Considering the effect of stator adjustment on the stability margin of the fan and the design-point flow, the S40 stator adjustment scheme is an optimal scheme for the stable operation of the fan.

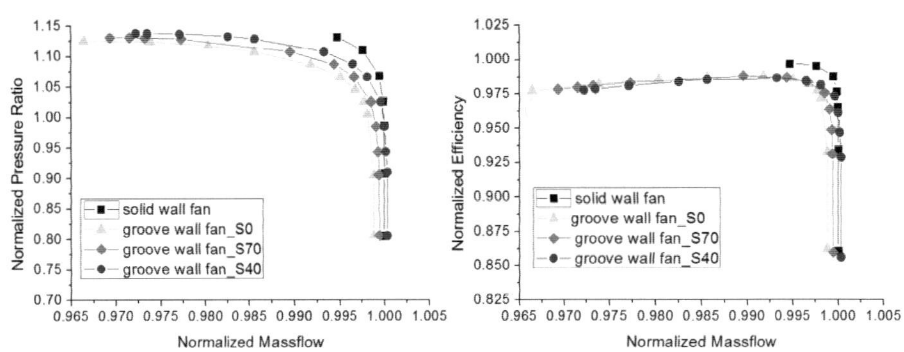

Figure 15. Performance curves for partial angle adjustment of the first-stage stator with casing treatment

For the condition that the flow of the fan is same as the design-point flow, the inlet axial velocity at the second-stage rotor is distributed along the radial direction with a circumferential groove casing, as shown in Figure 16. The figure shows that the local adjustment of the first-stage stator does enhance the flow capacity of the upper part of the second-stage rotor. Comparing this result with Figure 14, we conclude that the local adjustment of the upstream stator can lessen the adverse impact of the casing treatment on the design-point flow.

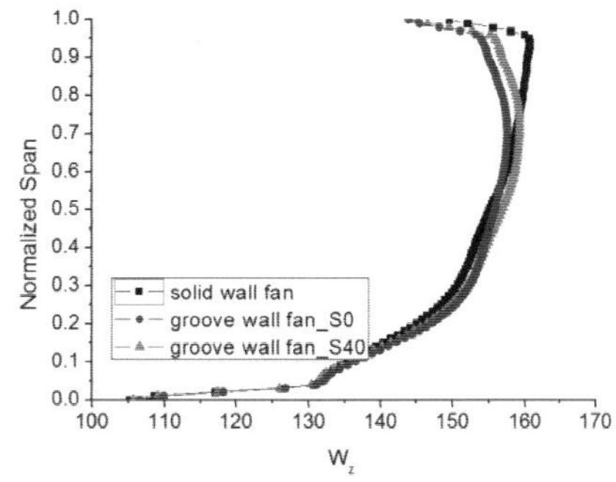

Figure 16. Axial velocity distribution before the second rotor

4. Conclusion

1. To extend the stable operating range of a multistage fan or compressor, the key stage initiating a stall must be determined first and an appropriate casing treatment should be applied around the key stall stage.

2. A channel is formed at both sides of the blade by a circumferential groove casing, and through suction on the pressure surface and spray on the suction surface, the static pressure difference between the pressure and suction surfaces is reduced, which improves the tip leakage flow and effectively inhibits the blade-tip blockage in the small-flow area, thereby increasing the stability margin of the fan.

3. Circumferential groove casings can improve the stability; however, they adversely affect the performance of the fan at the design point. Circumferential grooves cause serious blockage at the blade-tip area when operating at the design point and along the right branch of the performance curve. This blockage weakens the flow capacity of the upper part of the rotor so that total flow at the design point is reduced and the engine thrust is adversely affected.

4. Partial adjustment of the upper part of the upstream stator can enhance the flow capacity of the upper part of the downstream rotor and compensate for the casing treatment's adverse effect on the design-point flow. The combined application of the casing treatment and a locally adjusted upstream stator can extend the stability of an engine while preserving the overall performance.

References

[1] CHEN Zhongguang, ZHANG Zhishu, LI Dewang, SUN Bo. Analysis and evaluation of F119 engine overall performance[J]. Aeronautical Science & Technology,2003,(03):39-42.
[2] Viars P R. The impact of IHPTET on the engine/aircraft system[R].AIAA 89-2137.
[3] Merchant AA. Aerodynamic design and performance of aspirated airfoil[R].ASME Paper 2002-GT-30369.
[4] JIN Hailiang, JIN Donghai, GUI Xingmin. Aerodynamic design and numerical simulation of a highly loaded transonic two-stage fan[J]. Journal of Aerospace Power,2011,02:272-280.
[5] CHEN Baoshi, HU Guorong, WEI Yubing, et al. Design and experimental study of high Mach number and high-loaded single-stage fan[J]. Aeroengine,2000,03:28-35.
[6] WU Yanhui, ZHANG Haoguang, CHU Wuli, SU Zhongliang. Numerical study on structural optimization of axial slot casing treatment for two-stage transonic fan[J]. Journal of Northwestern Polytechnical University,2010,(02): 228-233.
[7] WU Yanhui, ZHANG Haoguang, CHU Wuli, SU Zhongliang. Mechanism of stall margin

improvement of two-stage transonic fan with two-portion axial slot casing treatment[J]. Journal of Propulsion Technology,2010,(02): 181-186+192.

[8] LI Ling, YU Qing, LU Yajun. Experimental research of casing treatment on single stage transonic fan[J]. Journal of Propulsion Technology,2000,(03):49-52.

[9] Bailey E E. Effect of grooved casing treatment on the flow range capability of a single-stage axial-flow compressor[R].NASA TM X-2459,1972.

[10] Moore R D, Kovich G, Blade R J. Effect of casing treatment on overall and blade element performance of a compressor rotor[R].NASA TN-D-6538,1971.

[11] Muller M W, Schiffer H P, Hah C. Effect of circumferential grooves on the aerodynamic performance of an axial single-stage transonic compressor[C]. ASME Paper, 2007, GT2007-27365, 115-124.

[12] Rabe D C, Hah C. Application of casing circumferential grooves for improved stall margin in a transonic axial compressor [C]. ASME paper, 2002, GT2002-30641, 1141-1153.

[13] Wu Y H, Chu W L, Zhang H G,et al. Parametric investigation of circumferential grooves on compressor rotor performance[J]. ASME Journal of Fluids Engineering, 2010, 132(12):121103.

[14] Bennington M A. Tip gap flow and casing treatments in an axial compressor[D]. South Bend: University of Notre Dame,2009.

[15] Zhao S F, Lu X G, Zhu J Q, et al. Investigation for the effects of circumferential grooves on the unsteadiness of tip clearance flow to enhance compressor flow instability[C].ASME Paper, 2010, GT2010-22652, 269- 277.

[16] Hah C, Muller M W, Schiffer H P. Study of convective flow effects in endwall casing treatments in transonic compressor rotors[C]. ASME Paper, 2012, GT2012- 68411, 95-106.

[17] DANG Chunning. Analysis of the effects of circumferential groove casing treatment on a transonic axial flow compressor[J]. Gas Turbine Experiment and Research,2008,(03):18-25.

[18] Duan Z Z, Liu Y W, Lu L P. Influence of circumferential casing grooves on transonic compressor rotor performance[J]. Acta Aeronautica et Astronautica Sinica,2014,35(8):2163-2173.

[19] LI Xiangjun, CHU Wuli, ZHANG Haoguang. Stabilizing mechanism of shallow circumferential grooves on a high loaded transonic axial-flow compressor[J]. Journal of Propulsion Technology,2013,(05):629-637.

[20] LIU Chuanle, CHU Wuli, ZHANG Haoguang. Numerical simulation of influence of groove's depth on stability of transonic compressor[J]. Computer Simulation,2012,(06):394-399.

[21] Khalsa A S, Waitz I A, Tan C S, et al. Endwall blockage in axial compressors[J]. Journal of Turbomachinery, 1999, 121(3):499-509

[22] Suder K L. Blockage development in a transonic, axial compressor rotor. Journal of Turbomachinery, 1998, 120(3):465-476.

[23] LIU B J, ZHANG Z B, YU X J. Experimental investigation on characteristics of tip leakage blockage in an axial compressor[J]. Acta Aeronautica et Astronautica Sinica, 2013, 34(12):2682-2691.

Fluid dynamics simulation for design on sludge drying equipment

Shuiping Li, Wang Liang*, Zhang Kai

School of Mechanical and Electrical Engineering, Wuhan University of Technology, Wuhan, China

*Corresponding author e-mail: wlwhhbch@163.com

Abstract. Sludge drying equipment is a key component in the sludge drying disposal, the structure of drying equipment directly affects the drying disposal of the sludge, so it is necessary to analyse the performance of the drying equipment with different structure .Fluent software can be very convenient to get the distribution of the flow field and temperature field inside the drying equipment which reflects the performance of the structure. In this paper, the outlet position of the sludge and the shape of the sludge inlet are designed. The geometrical model of the drying equipment is established by using pre-processing software Gambit, and the meshing of the model is carried out. The Eulerian model is used to simulate the flow of each phase and the interaction between them, and the realizable turbulence model is used to simulate the turbulence of each phase. Finally, the simulation results of the scheme are compared and the optimal structure scheme is obtained, the operational requirement is proposed. The CFD theory provides a reliable basis for the drying equipment research and reduces the time and costs of the research.

1. Introduction

At present, the method using cement kilns to co-processing sludge is relatively rare, many devices in common use for sludge drying are general waste treatment equipments [1-4].The synergetic treatment of sludge in cement kiln will provide a new method for disposal of sludge due to its advantages of harmlessness, resource utilization and stabilization. The extensive sludge disposal methods will be gradually eliminated. And that will lead to more and more intense competition in the equipment research and technology of sludge disposal. FLUENT software can provide considerable help for the research of drying equipment [5-7]. This paper develops a highly efficient mechanical and automated processing device to realize harmless sludge treatment under the guide of fluid dynamics simulation.

2. Structure and mesh

The drying equipment composed of feed inlet, air inlet, outlet, rotating shaft, blade and other components is the core component of the whole sludge disposal process, as shown in Fig1-a.The rotating shaft and blade is used to smash the sludge. High water content sludge enters through the feed inlet, high temperature gas enters through the air inlet, and the outlet discharges the sludge after the drying treatment. The position of the inlet and outlet need to be determined through simulation.

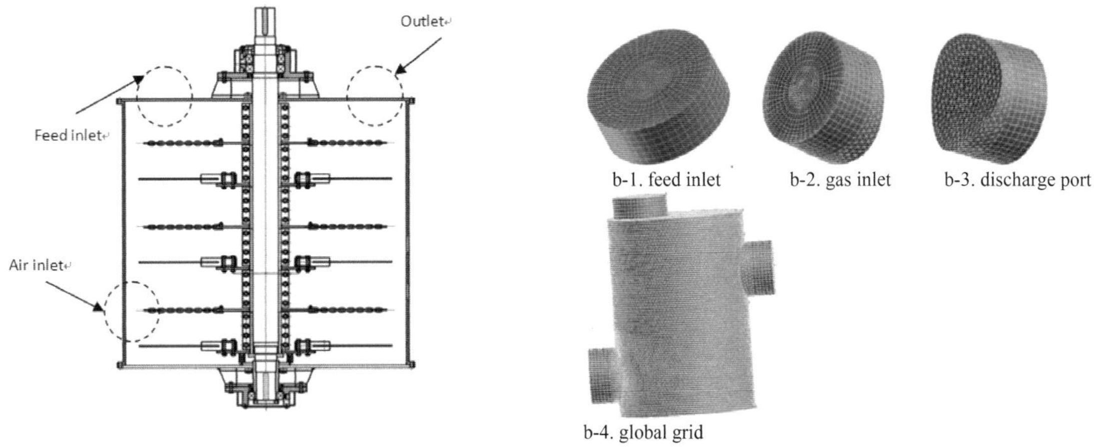

a b

Figure 1. Schematic diagram and grid of the drying equipment

The main size of the drying equipment is as follows: the diameter of the drying equipment is 2000mm, the height is 2850mm; the inlet diameter is 800mm; the outlet diameter is 800mm; the feed inlet is a cylindrical shape with a diameter of 800mm or a square with a width of 200mm; the position of the air inlet and the outlet are not sure. The different geometric models of drying equipment are constructed by FLUENT pre-processing software GAMBIT. In order to make the simulation results more consistent with the actual flow conditions, the drying equipment should be divided into sub-regions, and the grid of each area is shown in Fig1-b. A total of 7200,000 tetrahedral and hexahedral meshes are used to partition the model. After meshing is completed, it's necessary to check the skew of the mesh. In general, the maximum skew of the grid should be less than 0.95, the average should be below 0.33.When the skew of the grid is higher than 0.95, the calculation result is not easy to converge, and it will affect the accuracy of the simulation results. The grid model of the drying equipment is checked, and the maximum skew of the grid is 0.74, which is lower than 0.95.Therefore, the mesh quality meets the requirements of the calculation, and it has good matching performance.

3. Computational model

3.1. Governing equation

The Euler-Euler model is often used to simulate the flow of multiphase flow and the interaction between the various phases of the multiphase flow [8]. It has good adaptability to the internal flow of the drying equipment. In the analysis, the high-temperature air and sludge flows are treated as continuous flows, each of them is required to follow the mass, momentum and energy equations. The continuity equation is expressed as follow:

$$\frac{\partial \rho}{\partial \tau}+\frac{\partial(\rho u_x)}{\partial x}+\frac{\partial(\rho u_y)}{\partial y}+\frac{\partial(\rho u_z)}{\partial z}=0 \tag{1}$$

where ρ is the density of the high-temperature gas and sludge, and u is the velocity of each phase. The momentum equation of gas and fluid is:

$$\rho \frac{d\vec{u}}{d\tau} = \rho \vec{F_b} - \nabla \vec{p} + \mu \nabla^2 \vec{u} \tag{2}$$

where F_b is the volume force acting on the fluid system during each phase flow, p is the pressure, and μ is the viscosity coefficient of each phase. The drying of sludge involves heat transfer, and the energy equation of the heat transfer process is:

$$\frac{\partial}{\partial t}(\rho E) + \nabla \bullet \left(\vec{v}(\rho E + p)\right) = \nabla \bullet \left(k_{eff} \nabla T - \sum_j h_j \vec{J_j} + \left(\overline{\overline{\tau_{eff}}} \bullet \vec{v}\right)\right) + S_h \tag{3}$$

Where k_{eff} is effective thermal conductivity; $\vec{J_j}$ is the diffusion flux of component j. The first three terms on the right side of the equal sign indicate the energy transfer due to thermal conductivity, component diffusion, and viscous dissipation. S_h is the source item, including heat generated by chemical reaction heat and other user-defined volume heat sources.

3.2. Turbulence model

In the simulation analysis of CFD, it is necessary to take into account the viscosity of fluid. In this paper, the Reynolds number is calculated to be 5.13e5, so it can be known that the flow inside the drying equipment should be turbulent. The selected turbulence model is the realizable $k-\varepsilon$ model. And the turbulent kinetic energy k and the dissipation rate ε equations are:

$$\frac{\partial(\rho k)}{\partial t} + \frac{\partial(\rho u_j k)}{\partial x_j} = \frac{\partial}{\partial x_j}\left[\left(\eta + \frac{\eta_t}{\sigma_k}\right)\frac{\partial k}{\partial x_j}\right] + \eta_t \frac{\partial u_j}{\partial x_i}\left(\frac{\partial u_j}{\partial u_i} + \frac{\partial u_i}{\partial u_j}\right) - \rho\varepsilon \tag{4}$$

$$\frac{\partial(\rho u_i \varepsilon)}{\partial x_i} = \frac{\partial}{\partial x_i}\left[\left(\eta + \frac{\eta_t}{\sigma_\varepsilon}\right)\frac{\partial \varepsilon}{\partial x_i}\right] + c_1 \rho S \varepsilon - c_2 \rho \frac{\varepsilon^2}{k + \sqrt{\sigma_\varepsilon \varepsilon}} \tag{5}$$

The formula of η_t is:

$$\eta_t = \frac{1}{A_0 + A_S U^* \frac{k}{\varepsilon}} \rho k^2 \frac{1}{\varepsilon};$$

where $A_0 = 4.0$; $A_S = \sqrt{6}\cos$;$\Phi = \frac{1}{3}\arccos\left(\sqrt{6}W\right)$, $W = \frac{S_{i,j}S_{j,k}S_{k,j}}{\left(S_{i,j}S_{i,j}\right)^{\frac{3}{2}}}$,

$U^* = \sqrt{S_{i,j}S_{i,j} + \tilde{\Omega}_{i,j}\tilde{\Omega}_{i,j}}$, $\tilde{\Omega}_{i,j} = \overline{\Omega_{i,j}} - 3\varepsilon_{i,j,k}\omega_k$.

The constants c_1 , c_2 , σ_k , σ_ε in those equations have the following default values:1.44 ,1.92 ,1.0 ,1.3 .

3.3. Boundary conditions

Boundary conditions are the laws of the time-varying variables in the solution of the flow region. The reasonable boundary condition is very important for the simulation results. In this paper, the high temperature gas inlet and the feed inlet are set to the velocity inlet, the discharge port is set to the pressure outlet. The variables selected for turbulent flow are the hydraulic diameter and the turbulence intensity. The hydraulic diameter should be set according to each import and export, and the turbulence intensity is set to 5%.The boundary conditions are determined as shown in table 1.

Table 1. Boundary conditions

Attribute	Feed inlet	Air inlet	Discharge port
Boundary type	velocity-inlet	velocity-inlet	pressure-outlet
Velocity/(m/s)	18.51	18.24	
Temperature/K	371	576	530
Hydraulic diameter/m	0.8	0.2	0.8
Turbulence intensity/%	5	5	5

4. Simulation and analysis

By means of Fluent software, a series of temperature distribution and velocity distributions are obtained. The flow field inside the drying equipment can be displayed visually.

From Fig2.a it can be seen that when the air inlet is located under the drying equipment, a large range of low speed area will appear inside the drying equipment. The existence of this region will make the heat exchange between sludge and high-temperature gas inefficient, which is not conducive to the drying of sludge. Fig2.b tells that when the air inlet is set in the side below the drying equipment, the internal flow field has undergone great changes. The previous low-speed area is reduced a lot, and the overall speed of the material within the equipment improves a lot. Such a promotion for the drying of the sludge is very favorable, it will improve the drying efficiency of the sludge to a certain extent.

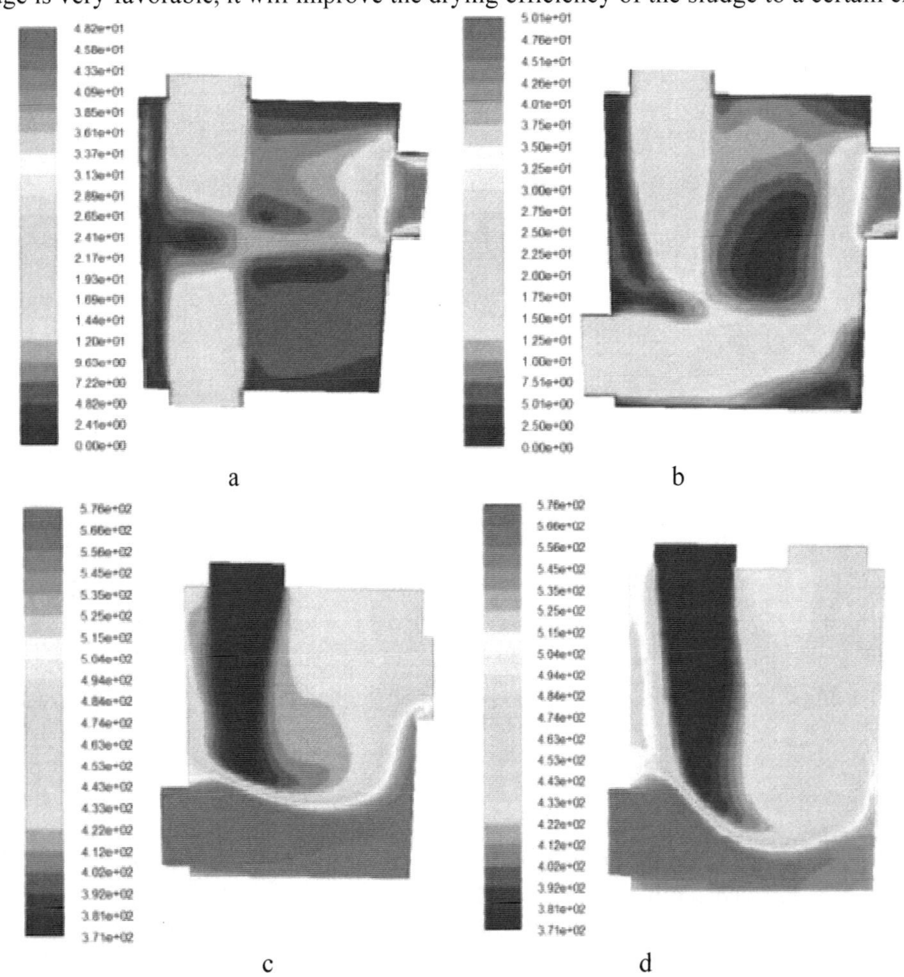

Figure 2. The impact of the air inlet and discharge port location

Fig2.c shows that when the material outlet is set on the side of the drying equipment, most of the area inside the drying equipment near the outlet is about 430K. Considering that the dried sludge will enter the cement kiln, this temperature may have some effect on the temperature field inside the cement kiln. It can be seen from Fig2.d that when the material outlet is located above the drying equipment, the internal temperature is raised a lot. Especially in the material outlet area, the temperature of most of the region is around 500K.This is not only conducive to the drying of sludge, but also can reduce the impact on the cement kiln.

5. Conclusion

According to simulation results, the final structure of the drying equipment is: the shape of the feed inlet is square and the size is 200mm; the high-temperature gas inlet is set side below the drying tower; the discharging port and the feed inlet are arranged on the top of the drying equipment tower. After the drying equipment is put into operation, the condition of equipment working is very good and the ability to deal with sludge increased significantly. The capacity of disposal sludge reached 2.5 tons / hour. At the same time, the operating cost of equipment is much lower than other equipments, which brings about a higher economic benefit.

References

[1] A.Zabaniotou,C.Theofilou. Green energy at cement kiln in Cyprus—Use of sewage sludge as a conventional fuel substitute [J]. Renewable and Sustainable Energy Reviews,2008,12 (2):531-541.
[2] Onaka,T.Sewage can make Portland cement: A new technology for ultimate reue of sewage sludge[J]. Water Science and Technology,2000,41(8):93-98.
[3] Wang Yating. Municipal Sewage Sludge Treatment and Disposal and Comprehensive Utilization [J]. Environmental Science And Management, 2011,36(1):90-94.
[4] LEI Xiao-ling,YE Fang-jian. Selection of Sludge Treatment and Disposal Scheme in Municipal Sewage Treatment Plant in Chongqing[J]. China Water and Wastewater, 2008,24(24)9-12:.
[5] Jamaleddine, Tarek J.Ray, Madhumita B.The drying of sludge in a cyclone dryer usingcomputational fluid dynamics[J].Drying Technology,2011,29(12):1365-1377.
[6] Jamaleddine, Tarek J.Ray, Madhumita B.Drying of sludge in a pneumatic dryer using computational fluid dynamics[J].Drying Technology, 2011,29(3):308-322.
[7] Ding Weiwei. Technical research on the Co-combustion of sludge in cement producing kilns[D]. Zhejiang University of Technology,2013.
[8] Han Zhanzhong.Fluent-Example and Analysis of Fluid Engineering Simulation[M].Beijing: Beijing Institute of Technology Press,2009.

IOP Conf. Series: Journal of Physics: Conf. Series **916** (2017) 012047 doi:10.1088/1742-6596/916/1/012047

Study on the Effect of water Injection Momentum on the Cooling Effect of Rocket Engine Exhaust Plume

Kan Yang [1,*], Yanhui Qiang [1], Chenghang Zhong [2], Shaozhen Yu [1]

[1]Naval Equipment Research Institute, Beijing, China
[2]Office of Military Delegates from Navy in 8359 Institute of CASIC, Beijing, China

*Corresponding author e-mail: yk258000381@126.com

Abstract. For the study of water injection momentum factors impact on flow field of the rocket engine tail flame, the numerical computation model of gas-liquid two phase flow in the coupling of high temperature and high speed gas flow and low temperature liquid water is established. The accuracy and reliability of the numerical model are verified by experiments. Based on the numerical model, the relationship between the flow rate and the cooling effect is analyzed by changing the water injection momentum of the water spray pipes. And the effective mathematical expression is obtained. What's more, by changing the number of the water spray and using small flow water injection, the cooling effect is analyzed to check the application range of the mathematical expressions. The results show that: the impact and erosion of the gas flow field could be reduced greatly by water injection, and there are two parts in the gas flow field, which are the slow cooling area and the fast cooling area. In the fast cooling area, the influence of the water flow momentum and nozzle quantity on the cooling effect can be expressed by mathematical functions without causing bifurcation flow for the mainstream gas. The conclusion provides a theoretical reference for the engineering application.

1. Introduction

In recent years, with the extensive application of large thrust carrier rocket, the use of liquid water vaporization heat absorption principle to reduce the rocket engine tail flame high temperature airflow impact and ablation damage has also been drawn more and more attention, and has achieved the engineering application [1].

To the rocket engine tail flame water injection, most of the current application is to suppress the large thrust carrier rocket in the launch pad ignition off the noise generated in order to avoid the rocket body itself or the load carried being damaged. In this field of research, domestic and foreign relevant scholars and engineering and technical personnel have organized a number of experiments, established the relevant numerical calculation model and in-depth study [2-4], and have acquired a large number of research results that have been applied in space launch. For the rocket ignition process of the impact and ablation, with the large thrust carrier rocket ignition tail after the increase in the radial and length profile, the damage to the ground equipment is more serious, so the basis of water injection noise the cooling effect of tail flame is becoming more and more obvious, and the research on the mechanism of water cooling is becoming more and more further.

The two-phase coupling mechanism between liquid flow and tail flame flow is complex and has many influencing factors. Such as water injection pipe hole shape, gas-liquid two-phase relative position, water velocity and droplet diameter, temperature and water injection angle. In addition, the experimental study on the water injection and cooling effect is carried out by M. J. Miller and the equation between the water injection volume and the cooling effect is obtained according to the collected experimental data. In reference [6], numerical simulation is carried out to simulate the water injection in the spaceflight field. In recent years, domestic research has made important progress in this field. Reference [7] proved that the water injection can effectively achieve the cooling of the gas flow field. In reference [8], the method of experiment and numerical calculation is used to study the two-phase flow field which is cooled by columnar water flow. It is concluded that the effect of water injection on reducing the impact and ablation of the tail rock of the rocket engine is significant, and the accuracy and reliability of the numerical calculation model are verified. In reference [9], the factors such as water injection velocity, liquid water temperature, and water injection pressure and droplet diameter are studied deeply, and the effective droplet diameter can be effectively cooled. The multi-stage water injection pipe used to carry out the gas flow field cooling effect is more effective.

In this paper, the cooling effect on the flow field of the columnar water flow is analyzed. The numerical results are verified by the scaling test. On this basis, the fitting data equation obtained by the experimental method in reference [5] is verified and analyzed influence of water injection location on flow field profile and cooling effect. And it can provide reference for optimal design of water injection cooling in space launch site.

2. Sub-scale model test and numerical calculation model

2.1. Water injection sub-scale model test

The water injection system designed with the launch space of the large launch vehicle should not only achieve the positive cooling effect but also be strictly limited to the noise intensity standard, so the reference gas flow noise reduction test conclusion is very practical significance for the practical application of the water injection cooling system. According to the reference [10] through the analysis of the effect of water spray noise analysis to the relative 60 ° angle of the implementation of water injection will be the best effect of jet noise suppression, drawing on the above conclusions, in the water injection test study, the same 60 ° angle water injection method. Figure 1 (a) shows the use of vertical test bed to eliminate the impact of gravity caused by the error, the height of the test bench to the engine after the exit plane as the origin, from the bottom of the impact platform height of 1.8m. Figure 1 (b) shows the use of double injection pipe water flow has a good symmetry, after the impact of water flow was water mist. It shows that the water injection of the double water pipe can cool the same position of the gas flow field, and will not tilt the tail flame flow field due to the impact of water flow.

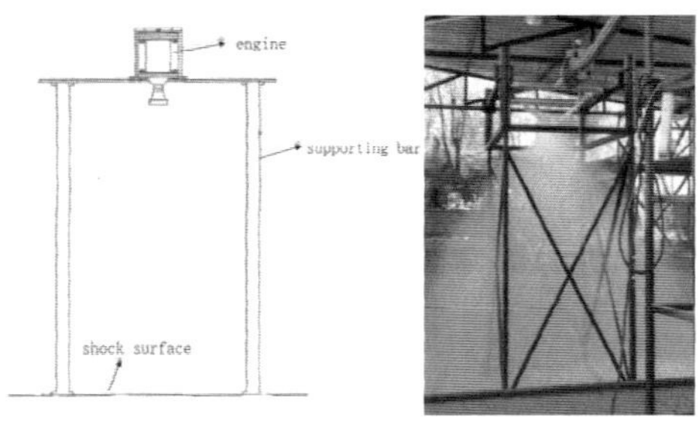

(a) (b)

Fig.1 Diagram of water injection with two pipes

The test engine and charge are shown in Figure 2 (a). Test using the engine compression ratio, according to the reference [11] scaled principles, design and dimensions of the combustion chamber of the engine Laval nozzle profile, the outlet velocity, temperature and the outlet pressure reaches the design requirements. The theoretical design value of the combustion chamber pressure curve of the model engine is 7 106 Pa and the Mach number of the outlet gas flow rate is 3.3. The measured engine thrust curve is shown in Fig. 2 (b). The engine outlet gas flow is 1.5kg / s.

(a) The engine model and propellant

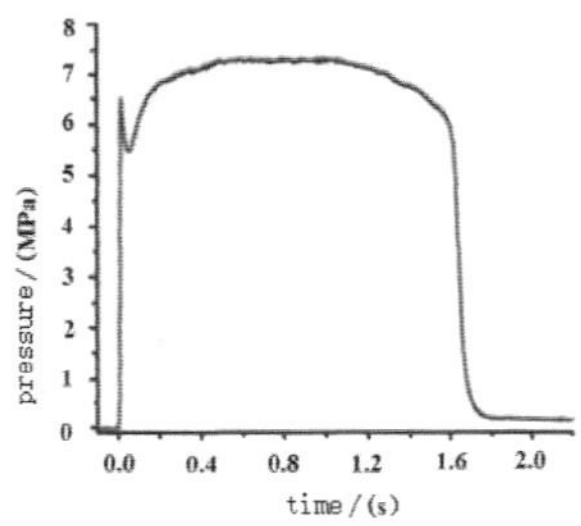

(b) Pressure curve in the combustor

Fig.2 The engine model and pressure curve

2.2. Numerical calculation model

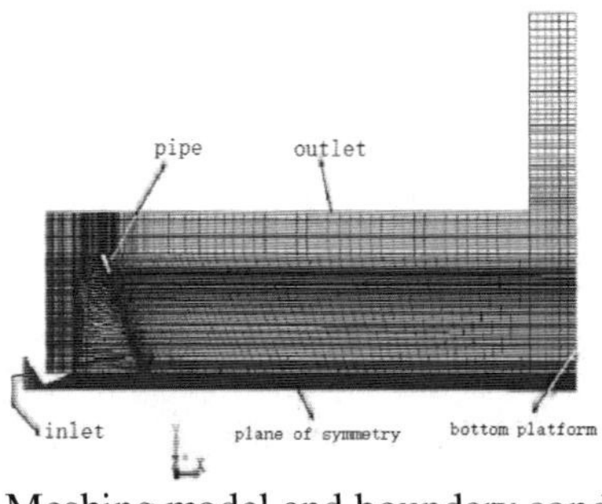

Fig.3 Meshing model and boundary conditions

The numerical calculation field is shown in Fig. 3, and the end of the combustion chamber (ie, the nozzle inlet) is set as the gas inlet, and the nozzle flow field and the external flow field are integrated. According to the symmetry using the 1/4 model. The boundary condition is set as: the engine nozzle is set as the pressure inlet, the inlet pressure curve input is the discrete curve of the measured curve, and the difference is calculated at each time pressure value. The bottom impact platform is set to solid wall. According to the test procedure, the water injection pipe is reserved in the model and set as the solid wall. The gas flow outside the flow field area is set to quiescent atmospheric conditions. The computational grid adopts structured hexahedral mesh, and the tail flame flow field is processed by the area and the water pipe wall, and the number of grid is 1.1 million.

According to the reference [12] numerical calculation method, the finite volume method and the turbulence calculation method are used to calculate the RNG $k-\varepsilon$ model. The near wall turbulence is calculated using the standard wall function. The Coupled algorithm based on pressure is used in the flow field calculation, and the coupling process of liquid water and gas flow is calculated by unsteady algorithm. The Mixture multiphase flow model in the Euler-Euler model is used to calculate the component diffusion and transport between the gas-phase gas component and the liquid water vaporization generated by the component transport model. The vaporization model is coupled to the flow field calculation in the process of vaporizing the phase transition process and energy transfer by adding the source term. The source is mainly involved in the vaporization of liquid water and the condensation of water vapor and the interphase energy transfer during vaporization.

3. Model validation

3.1. Free flow field calibration

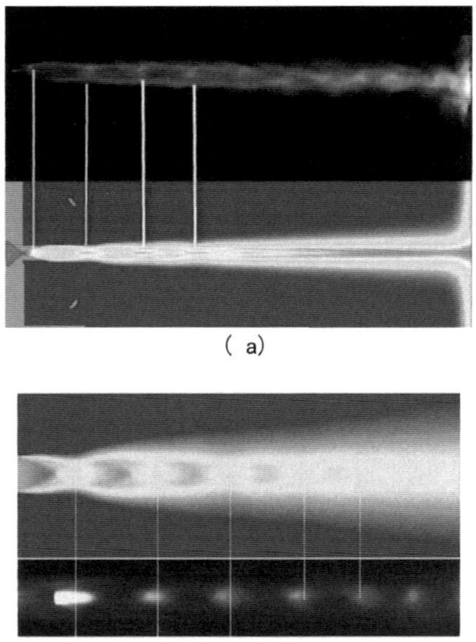

(a)

(b)

Fig.4 The free jet flow

Fig.4 shows the numerical results and the experimental results of the free-flow tail flame flow field. From Fig. 4 (a), the numerical results show that the wave structure and temperature distribution of the tail flame flow field are in good agreement with the experimental results. Figure 4 (b) is the first half of the wave structure of the distribution map, the figure can be clearly seen in the numerical results of the core area of the five high-temperature regions from the distribution position and shape consistent with the test results.

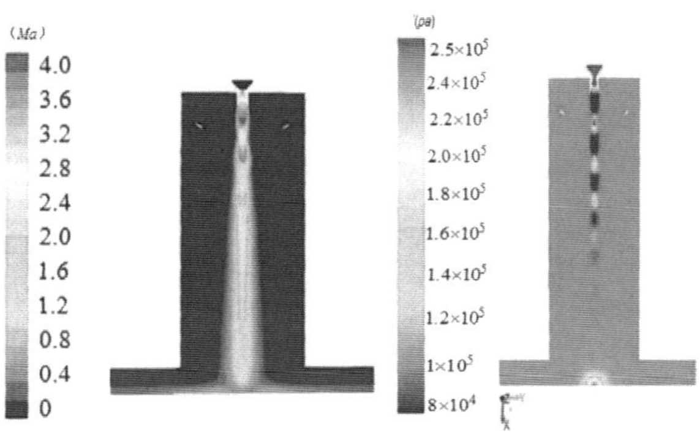

Fig.5 Contours of Ma and static pressure

Figure 5 shows the flow field Mach number distribution and pressure distribution cloud, from the figure can be seen free jet exit Mach number of 3.4 or so, in line with theoretical design values, and in the mainstream of the gas distribution and pressure distribution in line with high-speed gas jet feature. Therefore, in summary, the numerical calculation of high accuracy, through theoretical analysis of in-depth study.

3.2. Water flow correction of water injection

Fig.6 The jet flow with water injection

Figure 6 shows the high-speed photographic images of the tail-flow field after water injection. The water flow parameters are as follows: the flow rate of the outlet water is 16 m / s and the flow rate of the single water pipe is 1.2 kg / s. In the area after the impact of water flow, the shape and temperature distribution of the tail flame flow are consistent with the experimental results. In the second half of the mainstream area, due to a large number of water vapor generated cover the core area, so in the high-speed photography results in the second half of the flow field cannot be clearly displayed in the picture out, but the high-temperature area contrast image can also be Indicating that the temperature of the second half of the flow field is low, there is no high temperature area. It is consistent with the simulation results.

4. Result analysis

4.1. Analysis of Flow Field Parameters after Water Injection

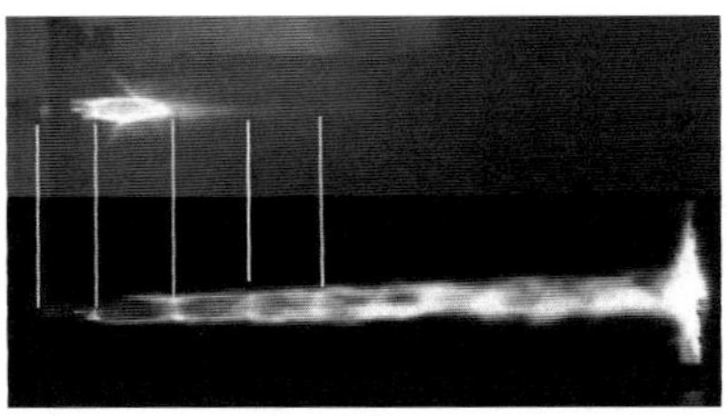

(a)

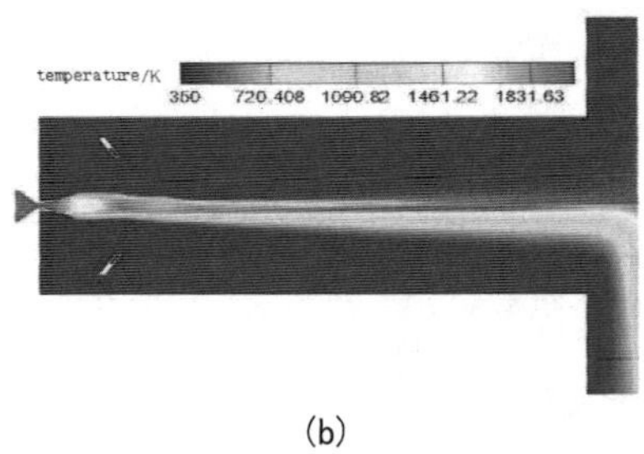

(b)

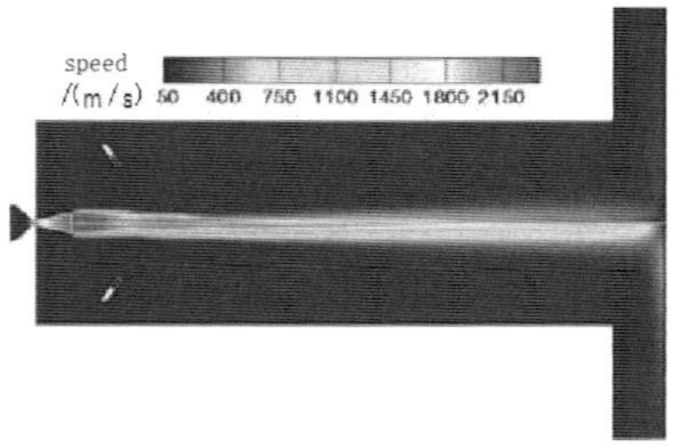

(c)

Fig.7 The jet flow with and without water injection

Figure 7 shows the distribution of the temperature and velocity of the flame at both the water injection and the non-water injection. Figure 7 (a) shows forming the spindle structure in the high temperature region of the flow field after injection. From the water injection position, the wave structure of the core region of the tail flame is destroyed, and as the development temperature of the flow gradually decreases, it can be seen that the length of the high temperature region is about 1/2 of the flow field of the non-water injection tail, and there is no obvious high temperature region. Cooling effect and influencing flow field profile. Figure 7 (b) and Figure 7 (c) for the two flow field temperature and velocity comparison, it can be drawn from the figure, after the high temperature flow field and the free jet flow field at the same location temperature and velocity distribution consistent. It is shown that the energy transfer between the gas and liquid phases is not transmitted in the radial direction in the cross section of the flow field, that is, the heat transfer rate in the gas flow field is much lower than the phase heat transfer rate.

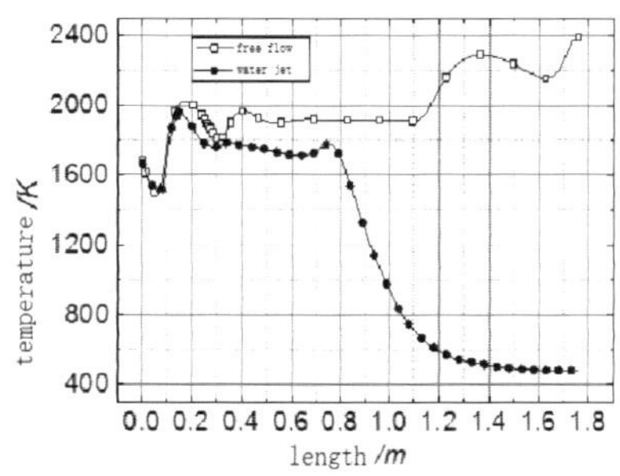

(a)

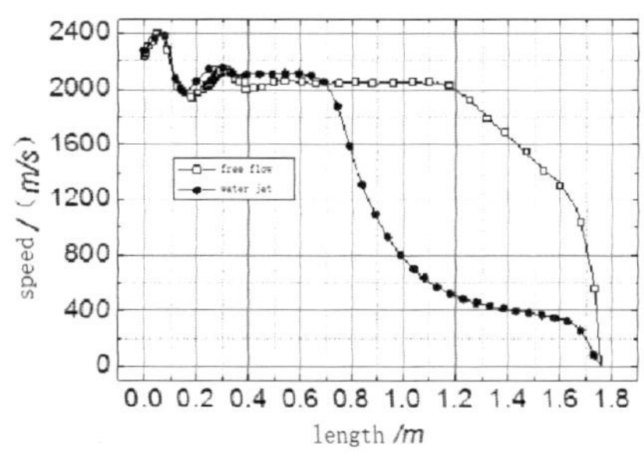

(b)

Fig.8 Comparison of flow parameters on the axis

Figure 8 shows the temperature and velocity curve on the flow field axis. From the figure, the temperature change effect of the flow field can be obtained intuitively. The temperature change on the axis of Fig. 8 (a) is slightly decreased from the intersection point, which can be called the slow cooling section. After 0.8m, the mainstream temperature drastically decreases. From 1.4m, the temperature of the gas stream is basically gentle, reaching a minimum value of about 550K, which can be regarded as a rapid decline in temperature. The same trend also appears in Figure 8 (b) for velocity changes. The rapid cooling and cooling area in the flow field is between 0.8m and 1.4m, which indicates that the water injection has obvious effect on the cooling and deceleration of the second half of the main stream.

4.2. Analysis of Factors Affecting Water Cooling

4.2.1. Analysis of Flow Field Calculation Results.

Table1. Operating conditions of water injection

Parameter	Case				
	1	2	3	4	5
Gas momentum/(kg·(m/s^2))	3966	3966	3966	3966	3966
Flow momentum/(kg·(m/s^2))	180	102.3	90	72	45
Momentum ratio *M*	22	39	44	55	88

Reference [5] through the test data collection analysis, obtained after the water flow cooling effect with the water flow associated with the amount. Therefore, the numerical calculation of the establishment of Table 1 in the five calculation conditions, and the temperature on the axis of the analysis. The gas momentum in Table 1 is based on the fuel flow parameters at the exit of the engine nozzle. The use of double injection pipe by changing the flow rate to change the water injection momentum. Momentum ratio M is the ratio of gas flow to water injection.

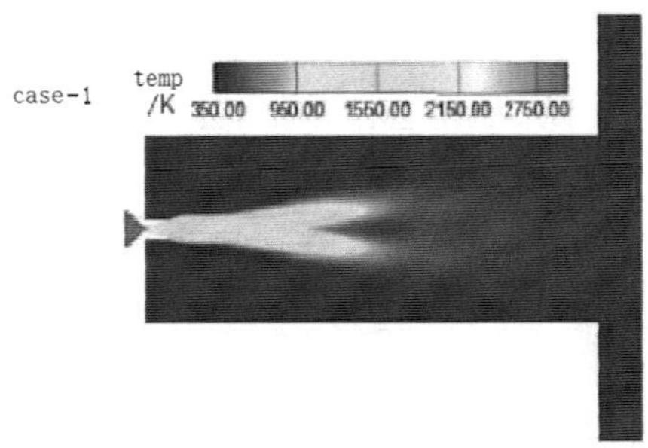

(a)

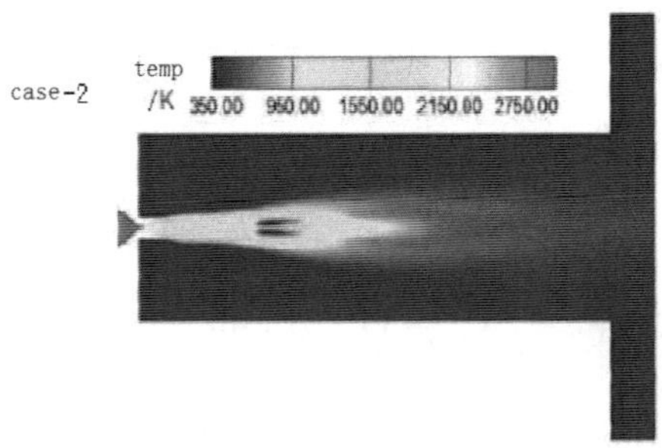

(b)

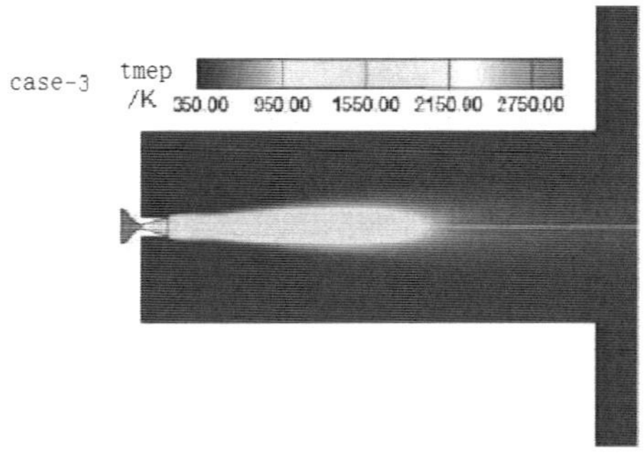

(c)

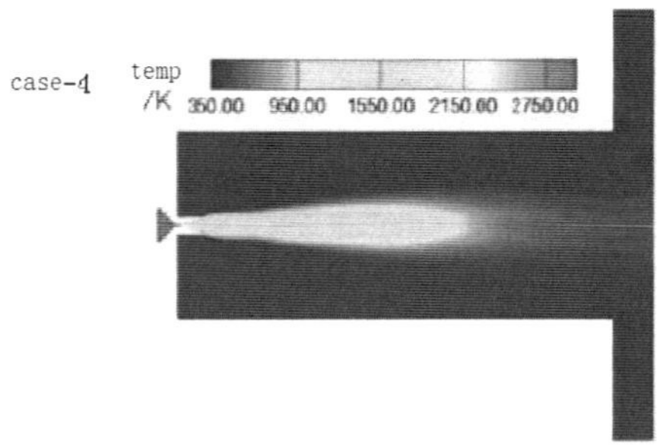

(d)

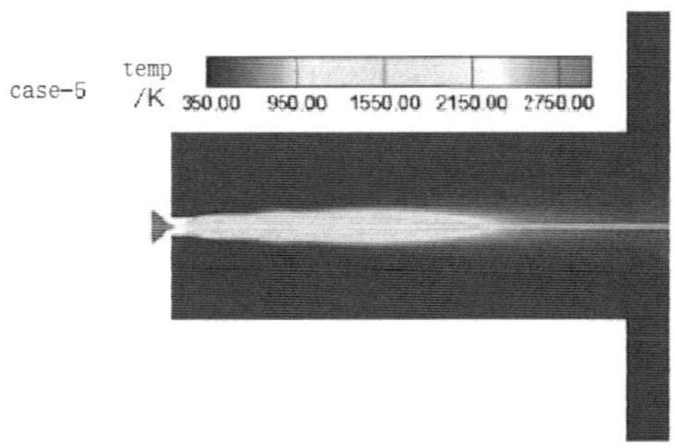

(e)

Fig.9 Contours of temperature for all operating conditions

Figure 9 shows the flow chart of the flow field in different calculation conditions. The graph shows that the plane is perpendicular to the plane of the water injection pipe. It can be concluded that the main gas flow is affected by the flow in the working conditions 1 and 2, and there has been the phenomenon of bifurcation flow. In case 2 the mainstream appears bifurcation flow and then polymerized into a share. In case 3, case 4 and case 5 the flow rate of the water flow is further reduced, the impact on the mainstream is weakened, the main flow is not in the bifurcation flow phenomenon, the post-flow field temperature distribution structure is basically the same, and the temperature distribution is also similar.

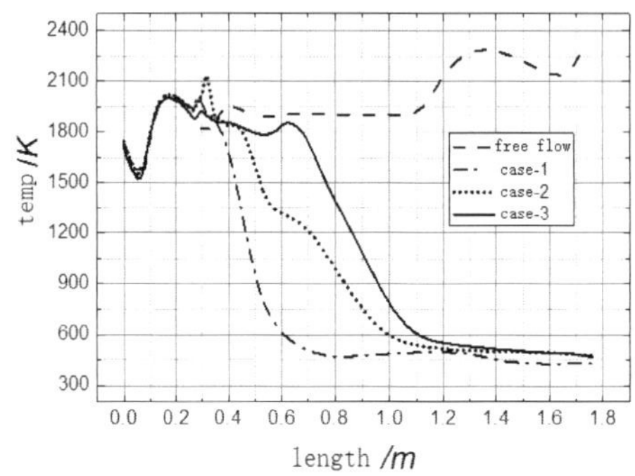

(a) Case 1-3

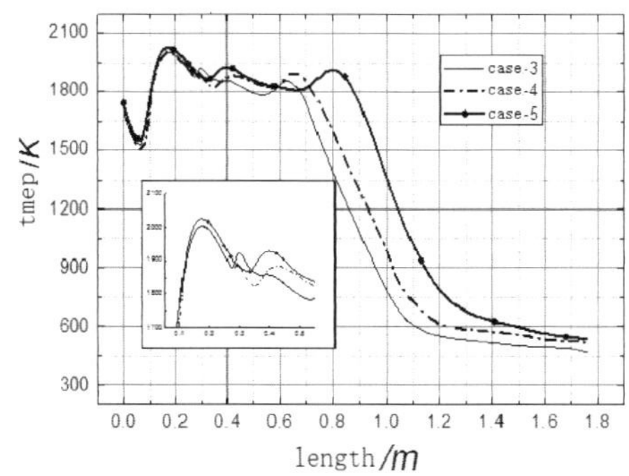

(b) Case 3-5

Fig.10 The temperature curve on the axis

Figure 10 (a) shows the temperature profile of the axis 1-3, and Figure 10 (b) shows the temperature profile of the case 3-5. From the curve of the temperature curve, it can be seen that the overall trend of temperature change is similar in several water injection conditions on the axis. After the water cooling effect on the axis of the two changes in the stage. That is, the rapid decline in temperature and temperature stability section. In the next stage of rapid temperature, from the point of view of the temperature change point of each calculation condition, the position of temperature mutation gradually shifted with the decrease of water injection momentum, and the temperature of the axis gradually increased. As the working conditions of the mainstream by the impact of separation into two shares, the temperature on the axis relative to the other four conditions, the difference is obvious. The condition 2 ~ 5 remained stable at the position of 1.1 m, the temperature changed diametrically opposite, the temperature dropped to around 450, and the change was slow, indicating that the heat transfer and mass transfer between the two phases did not work after this position. Obvious. The temperature of the flow field is increasing after 1.1m, and the temperature of the curve is in the range of 0.3m ~ 0.4m. In the case of 1 ~ 3, the temperature jumping point appears. Intersecting intersection, the impact between the two phases in this position to form a stagnation, the speed caused by lower temperature rise. The peak values in the working conditions 4 and 5 are not obvious, indicating that the effect of water flow on the mainstream of the gas is not obvious for the core area.

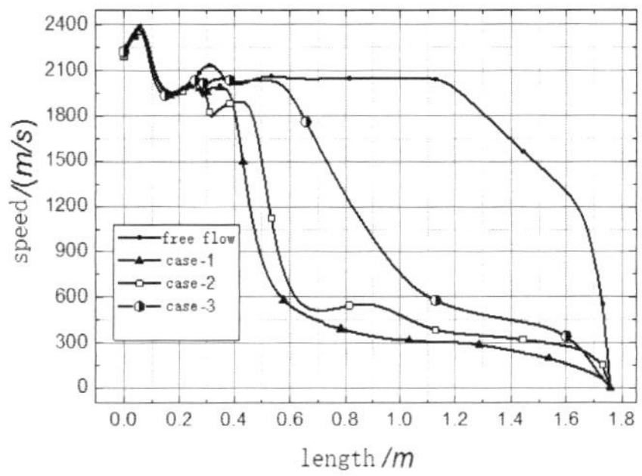

(a) Case 1-3

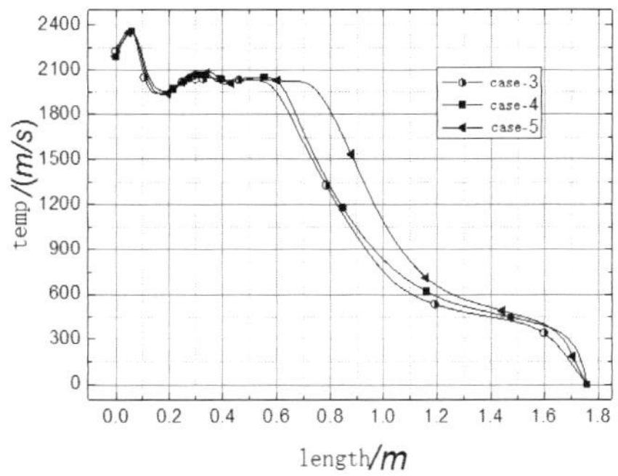

(b) Case 3-5

Fig.11 The velocity curves on the axis

Figure 11 shows that the trend of the axial velocity is valid for the temperature change of the above diagram. From the analysis of the velocity curve, the effect of the water spray on the gas flow field is quite obvious, and the purpose of reducing the temperature of the gas flow can be achieved. It can effectively reduce the impact of gas flow on the chute and ablation.

4.2.2. Influences of Water Injection on Cooling Effect. For the relationship between the flow rate and the cooling effect of the flow field, reference is made to Miller et al. [5], which concludes that the dimensionless relationship between the heat flux density and the two-phase momentum ratio M in the test conditions is:

$$Q_R = ae^{bM} \tag{1}$$

Where: $a = 0.850, b = -0.00116$ is the percentage decrease in the measured heat flux density.

According to the heat flow meter measurement principle and the Fourier law of heat conduction, the heat flux density expression is:

$$q'' = \frac{Q}{F} = -\lambda \frac{\Delta T}{\Delta X} \tag{2}$$

Where: Q is the unit time through the area F of heat; ΔT and ΔX were measured on both sides of the sensor temperature difference and the sensor length. In the test ΔX is considered as a constant, because the sensor itself using thermal insulation material to maintain the accuracy of heat flux measurement, so ΔT actually reflect the main point of the measured temperature changes, that is, heat flux density q'' is proportional to the measured temperature T.

Because the engine parameters and water spray parameters used in the experiment are different from those of the Miller test, this paper analyzes the calculation conditions by reverse verification method for the above dimensionless formula (1), if there is a relationship with the equation function, it is necessary that the temperature value of a certain point on the axis conforms to the above relation. If the point exists, it proves that the above equation has certain applicability, and also proves that the calculation result has certain reliability.

Analysis of the calculation of working conditions 1 to 5, because the value of the impact of M is the numerical value of the size of the index e^{bM}, so the case of 1 ~ 5 axis temperature according to the value of appropriate amplification or reduction of a certain proportion, there must be a data intersection. The specific conversion equation is as follows:

Substitute formula (1) into the axis temperature to obtain the corresponding conditions Q_{Rn}, and take the condition 5 as the reference parameter, then the following equation is used to calculate the ratio between the temperature values in different working conditions, that is:

$$\mathrm{R}_i = (1 - Q_{Ri}) / (1 - Q_{R5}) \quad (i = 1, 2, 3, 4) \tag{3}$$

The corresponding conditions on the axis of the temperature amplification of R_i times the curve shown in Figure 12.

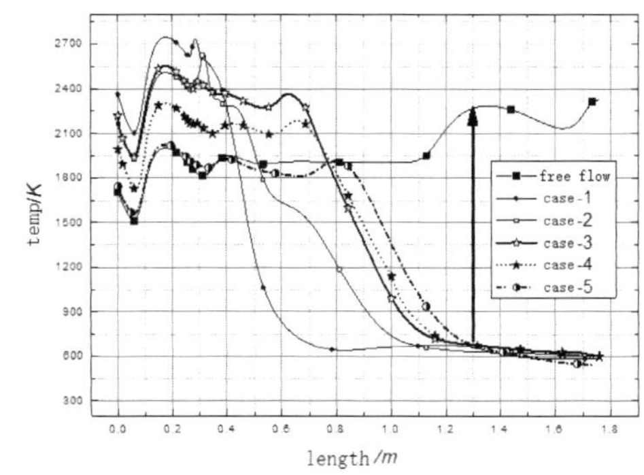

Fig.12 Adjusted temperature curves of the axis

It can be seen from Fig. 12 that there are common points in the vicinity of the 1.3m position except for the working condition 2, and the value of the working condition 2 is close to the common value, so it can be explained that in the analysis of the water flow rate (1), and then can be applied to (1) on the axis of the other location points to establish the corresponding parameter equation.

According to the temperature change of the axis in the working condition shown in Fig. 10, the temperature of the axis begins to decrease obviously after the 0.9m, and then the temperature point data of 0.9m are verified by the above equation. It is assumed that the temperature change after the position is in accordance with the equation (1), which is expressed as follows:

$$\ln Q_R = \ln a e^{bM} = \ln a + bM \tag{4}$$

For a certain point, the gas flow and water flow ratio M as an independent variable, the temperature change value for the dependent variable, the data connection between the different conditions should be a straight line, the slope is b .

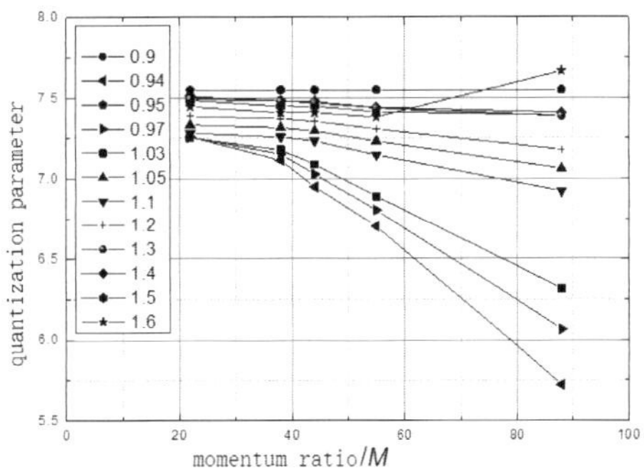

Fig.13 Relations of temperature on different points

Figure 13 is the logarithmic relationship between the equation 4 after the establishment of the curve, it should be noted that in the case of 1 gas flow was shocked into two shares, so the axis temperature changes compared to the latter four kinds of calculation conditions larger , So the curve in all the curves in the working conditions 1 to 2, the slope of the line with the main line slope of the difference between the larger, from the other four conditions, in addition to 1.6m position point of work-5 abnormalities, the other locations (1) The equation of function (1) can be used to determine the relationship between the temperature change and the relative momentum ratio on the axis. As can be seen from Fig. 12, the temperature change parameter a and b at different positions of the axis is different because of the difference in the slope, and the relevant numerical solution needs to be determined. In addition, the application condition of the formula (1) is limited to the value of the momentum ratio M greater than the condition 2 in the double water pipe water injection condition.

4.2.3. The effect of the number of nozzles on the cooling effect.

Table2 Operating conditions of water injection

Parameter	Case6	Case7	Case8
Gas momentum/(kg•(m/s^2))	3966	3966	3966
Single nozzle flow mass flow rate/(kg/s)	3.6	3.6	3.6
Mean flow velocity /(m/s)	16	16	16
Pipe quantity	2	3	4
Injection momentum /(kg•m/s^2)	115.2	172.8	230.4
Momentum ratio M	34	23	17

In order to verify the relationship between the cooling effect on the axis and the water injection momentum under the condition of the number of different water pipes and the same single nozzle water injection parameters, the calculation condition is established as shown in Table 2.

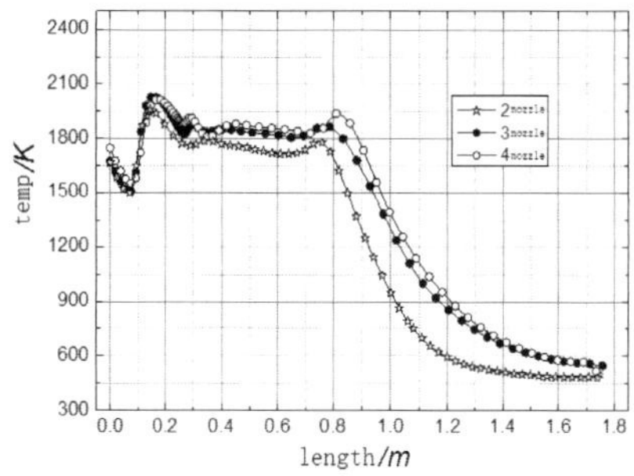

(a) Case 6-8

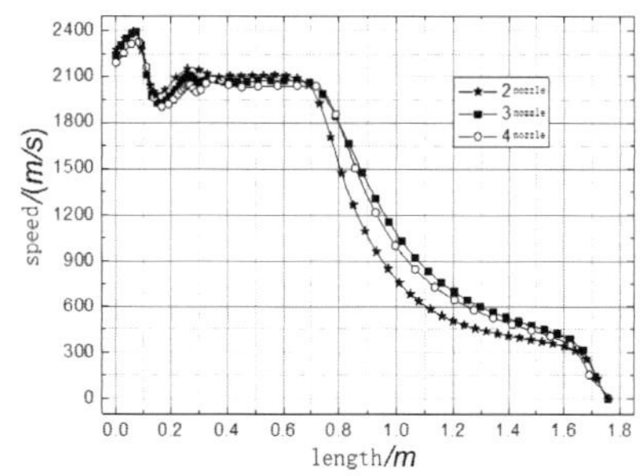

(b) Case 6-8

Fig.14 The curves of flow parameters on the axis

Figure 14 is the temperature and velocity curve of the flow field after the injection of water from 6 to 8, and the temperature of the axis increases with the increase of the number of water pipes in figure 14 (a), which is inversely proportional to the water injection momentum. With the increase in the number of water pipes, the gas flow is affected by the impact of water flow, by the side of the double water pressure side of the expansion side of the expansion of the deformation of a 3 nozzle or 4 nozzle under the state of uniform compression deformation, water flow into the mainstream gas The depth of the reduction, resulting in the above results, which also shows that the number of water pipes is also one of the factors affecting the cooling of the gas flow. Figure 14 (b) for the flow field axis velocity curve, from the 0.8m position began to rapid decline in the axis speed, the flow of the block effect is significant, the figure 4 due to the large amount of water flow, resulting in the axial flow rate slightly lower than the case 3.

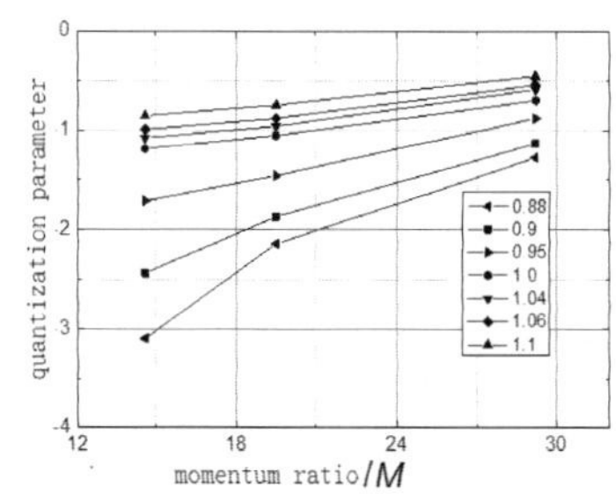

Fig.15 Relations of temperature on different points

Formula (4) transfers the temperature change of the axis to verify whether the calculated condition conforms to the function relation obtained by the formula (1), as shown in Fig. Through the curve slope change can be seen in the axial temperature changes 0.9m after the three conditions of temperature changes in accordance with the formula (1), from Figure 13 axis temperature curve shows 0.9m after the rapid decline in the temperature of the axis, the main heat loss main occurrence area. After 0.9 m for the relevant position, the value a and b can be obtained through the curve relationship.

4.2.4. Effect of small flow water injection on cooling effect.

Table3 Operating conditions of water injection

Flow parameter	Case 9	Case 10	Case 11
Gas momentum/(kg•(m/s2))	3966	3966	3966
Single nozzle flow mass flow rate /(kg/s)	0.9	0.6	0.45
Mean flow velocity /(m/s)	4	2.67	2
Pipe quantity	4	6	8
Injection momentum /(kg•m/s2)	14.4	8	7.2
Momentum ratio M	275	496	550

Water consumption is one of the main optimization design factors of water injection system. Reducing the amount of water will inevitably weaken the effect of water cooling. However, reducing the velocity of water injection can extend the action time after two-phase contact and increase the energy transfer. In order to test the effect of small flow water injection cooling, the total water injection rate is 3.6 kg/s, which is 2.4 times of the engine gas flow rate, increases the number of nozzles and changes the water flow and water velocity of each water spray pipe accordingly. The calculation conditions are shown in Table 3.

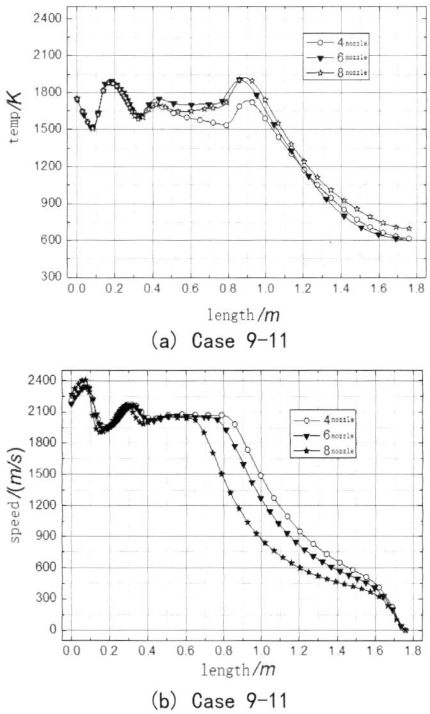

Fig.16 The curves of flow parameters on the axis

Figure 16 shows the temperature and velocity variations on the flow field axis. Figure 16 (a) The temperature of the axis decreases significantly after 0.3 m at the intersection, which is about 100 K lower than the temperature at the same position as in Figure 14 (a), and after the rapid descent of the temperature starts at 0.9 m, it is delayed by 0.1 m . In the temperature drop trend, the impact surface temperature above 600K. The results show that the flow of water flow and the decrease of flow rate lead to the relative growth of the two - phase flow at the same position, and the heat transfer between the two phase's increases and the cooling effect is enhanced in the first half of the flow field. The same analysis of the speed curve Figure 16 (b), with the increase in the number of water pipes, the main effect of the block for the more obvious gas. However, compared with Figure 14 (b) using a small flow of water, the flow field of the block at the same location greatly weakened. And the dominant position of the dominant velocity moves more forward.

The above shows that the use of small flow of water can achieve the cooling of the end of the flow field, but the water injection should be optimized to meet the thermal protection needs. In addition, as shown in Fig. 16 (a), the water injection of the Nozzle and the water injection in the 6 nozzles are crossed at the rapid descending section of the axial temperature and do not conform to the equation (1). It is shown that the formula (1) does not apply to the calculation of low flow water injection cooling.

5. Conclusion

The effective cooling and deceleration of the rocket engine wake can be realized by means of water injection, and the profile of the tail flame flow field can be reduced to realize the thermal protection of the ground diversion tank and the auxiliary launching device.

Liquid water in the rocket engine tail flame flow field, in the two-phase intersection after the existence of the main flame temperature and flow rate of slow decline and rapid decline in the area, the use of large water injection and water flow under the premise of the mainstream temperature eventually reached stable state, the use of small flow of water state, the main gas temperature drop, but cannot reach the final stable state.

The ratio of the mainstream temperature of the tail flame to the ratio of the gas flow and the flow rate of the injected water can be established. The calculated results showed that by injecting the water in the double water pipe and the two-phase flow field does not destroy the main flow, the expression is valid, and the number of nozzles can be increased. Similarly, the constant value in the expression should be determined based on the calculated result or the test result.

In the case of slow flow rate, the temperature of the axis decreases from the two-phase contact position, but in the rapid cooling section, the temperature is lower than the high-flow water injection cooling effect. It can be optimized design in the actual engineering applications.

References

[1] Liu Han. The LM-7 passed water injection test[J]. Aerospace China. 2015, 04: 13.

[2] Ignatius J K, Sathiyavageeswaran S, Chakravarthy S R. Hot-flow simulation of aeroacoustics and suppression by water Injection during rocket liftoff [J]. AIAA Journal, 2014, 53(1): 235-245.

[3] Vu B T, Bachchany N, Peroomianz O, et al. Multiphase modeling of water injection on flame deflector[C] 21st AIAA Computational Fluid Dynamics Conference. San Diego, USA, Jun 24-27, 2013.

[4] Chen Yu, Zhou Xu, Tong Li, et al. Estimation of rocket noise reduction by water injection[J]. Environmental Engineering, 2014, 30:172-175.

[5] M. J. Miller, J. H. Koo, F. M. Sickler ,et al. Effect of water to ablative performance under solid rocket exhaust environment [C]. 29th Joint Propulsion Conference and Exhibit. American Institute of Aeronautics and Astronautics. 1993.

[6] G. Pierre, F. Philippe, G. Laurent. Simulation of water injection into a rocket motor plume [M]. 35th Joint Propulsion Conference and Exhibit. American Institute of Aeronautics and Astronautics. 1999.

[7] Jiang Yi, Zhou Fan, Zhang Xue-wen. Experimental study on flow field of high temperature supersonic impinging jet injected by water[J]. Journal of Experiments in Fluid Mechanics,

IOP Conf. Series: Journal of Physics: Conf. Series **916** (2017) 012047 doi:10.1088/1742-6596/916/1/012047

2011, 25(4):32- 36.
[8] J Li, Y Jiang, et al. Cooling effect of water injection on a high-temperature supersonic jet[J]. 2015, 8 (11):13194-13210.
[9] M Rajagopal, D Rajamanohar. Modeling of an exhaust gas cooler in a high-altitude test facility of large-area ratio rocket engines[J]. Journal of Aerospace Engineering. 2015, 28(1):04014049.
[10] V Bakulev. Features of experiment research of supersonic jet noise reduction[C]. International conference on mechanics-seventh polyakhovs reading. RUS. Feb 2-6, 2015:1-3.
[11] Chen Jing-song, Ma Hong-ya, Lin Yu. Rocket-launched gas jet shrinkage test similarity parameter[J]. Acta Aerodynamica Sinica, 2005, (03):307 - 311.
[12] S Jai, A Vineet, H Ashvin, et al. Analysis of Flame Deflector Spray Nozzles in Rocket Engine Test Stands [M]. 46th AIAA/ASME/SAE/ASEE Joint Propulsion Conference & Exhibit. American Institute of Aeronautics and Astronautics. 2010.

IOP Conf. Series: Journal of Physics: Conf. Series **916** (2017) 012048 doi:10.1088/1742-6596/916/1/012048

Thermodynamic Model and Experimental Study of Oil-free Scroll Compressor

Bin Peng*, **Shengxian Zhao and Yaohong Li**

School of Mechanical and Electronical Engineering, Lanzhou University of Technology, Gansu, China

*Corresponding author e-mail: pengb2000@163.com

Abstract. In order to study the performance characteristics of oil-free scroll compressor, this paper is based on the basic equation of circle involute profile, and uses the differential geometry theory to calculate the variation law of pressure with volume. Based on the basic law of thermodynamics, the thermodynamic model of the oil-free scroll compressor is established by considering the heat transfer model and the gas leakage model, considering the mass, energy conservation equation and gas state equation. The change of the mass flow rate of the gas in each chamber is obtained by solving the established model by using the improved Euler method. The experiment results show that with the increase of frequency, the temperature, the displacement and the power show a clear upward trend. The thermodynamic model has some guidance and reference for the development and performance analysis of oil-free scroll compressors.

1. Introduction

With the increasing global climate problems and the growing shortage of non-renewable resources, energy conservation, emission reduction, new energy development and utilization has become one of the important subjects of research in the world today. Green development will become a new concept of development, while increasing the promotion of new energy vehicles plan. And as a new energy vehicle gas source of the oil-free scroll compressor will be rapid development. Oil-free scroll compressors are widely used in refrigeration equipment, air-conditioning systems and various gas compression applications due to the advantages of light weight, small size, high volume efficiency and simple structure [1].

Study on thermodynamics model for oil-free scroll compressor, there are many scholars of the relevant research work. Zhao is famed for a fuel cell vehicle spray oil-free scroll compressor for detailed performance analysis and research [2]. Fang Shiyi ignoring gas and scroll wrap heat exchange case, calculating thermodynamic process of a new oil-free scroll compressor [3]; Le Wang designed and developed for aerospace applications of refrigeration oil-free scroll compressor, established the thermodynamic model and study the performance of the prototype [4]; Yanagisawa studied the volumetric efficiency of oil-free scroll compressors in detail [5]; Brycer. Shaffer has designed a plastic and metal material of a new oil-free scroll compressor, and the comparative study of the performance, plastic material more sensitive to temperature, but demands for precision low while costs are also relatively low [6]; Sunder through method of test research heat transfer process about fixed and orbiting scroll, obtained the physical contact is important form of heat transfer in compression process [7]; Poi and Zhu through Two-dimension model to research two scroll plate heat transfer model [8]; Yi et al simulation the compression process of a car air-conditioning scroll compressor and verify the model by experiment [9, 10];

Based on the geometric model of oil-free scroll compressor, the first law of thermodynamic, conservation of mass and gas equation of State, considering the working chamber leakage model and heat transfer model, established the thermodynamic model of the prototype, and the thermodynamic model was validated by the test platform. The calculated results provide a theoretical basis and reference value for the performance analysis and structural optimization of oil-free scroll compressor.

2. Geometric model

The geometrical model of the oil-free scroll compressor is the basis for the analysis and study of the thermodynamic model, so it is necessary to study the geometric model of the scroll compressor.

Figure1shows the development of a new type of oil-free scroll compressor prototype structure.

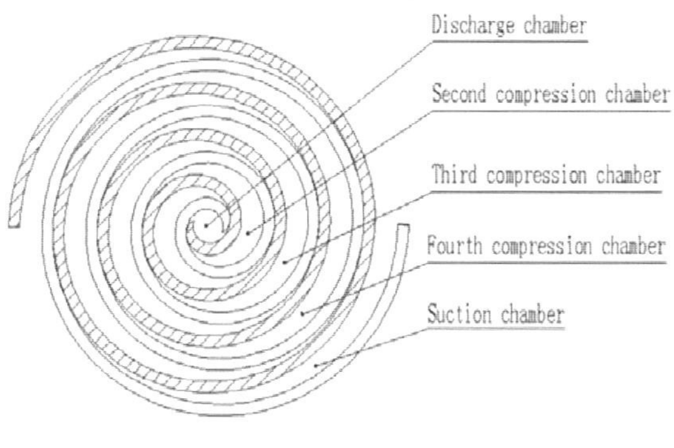

Figure 1. Oil-free scroll compressor.

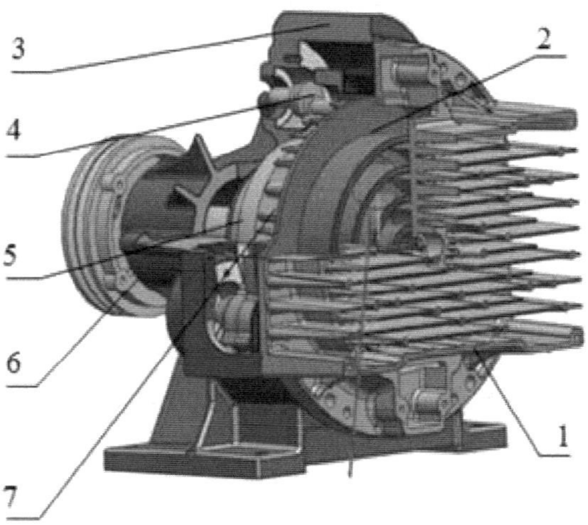

Figure 2. Oil-free scroll compressor profile.

1-Fixed scroll; 2-Orbiting scroll; 3-Frame; 4-Small crank; 5-Spindle; 6-Leather belt wheel and Small balance iron; 7-Large balance iron;

The circle involute equation is used to deduce the spiral equation of the scroll plate, the circle involute baseline equation:

$$\begin{cases} x = r_b\left(\cos\varphi + \varphi\sin\varphi\right) \\ y = r_b\left(\sin\varphi - \varphi\cos\varphi\right) \end{cases} \tag{1}$$

After the baseline is determined, the inner and outer profiles of the scroll are determined according to the characteristics of the normal equidistant curve.

The involute equation of inner circle:

$$\begin{cases} x = r_b\left[\cos(\varphi+\alpha) + \varphi\sin(\varphi+\alpha)\right] \\ y = r_b\left[\sin(\varphi+\alpha) - \varphi\cos(\varphi+\alpha)\right] \end{cases} \tag{2}$$

The involute equation of outer circle:

$$\begin{cases} x = r_b\left[\cos(\varphi-\alpha)+\varphi\sin(\varphi-\alpha)\right] \\ y = r_b\left[\sin(\varphi-\alpha)-\varphi\sin(\varphi-\alpha)\right] \end{cases} \tag{3}$$

According to the above-mentioned scroll profile equation, combined with the geometric parameters listed in Table 1, the use of the baseline method to generate scroll inner and outer profile shown in Figure 2.

Table 1. Scroll compressor geometric parameter.

Name	Symbol	Illustration
Type of spiral		Circle involute
Base circle radius	r_b	3.675mm
Offset distance	R_o	5.745mm
Height	h	36mm
Thickness	T	5.8mm
Ending angle	φ_e	8.49π
Discharge angle	θ_d	265°
Number of turns	N	4.25

3. Thermodynamics model

The thermodynamic model of oil-free scroll compressor is established to take full account of the effects of the suction heating and working chamber heat transfer as well as radial leakage and flank leakage on the thermodynamic model, the variety rule of the gas in actual working condition is obtained.

3.1. Basic equations

The change of the gas temperature in the working chamber with the rotation angle of the spindle [11]

$$\frac{\mathrm{d}T}{\mathrm{d}\theta}=\frac{1}{mc_V}\left\{-T\left(\frac{\partial p}{\partial T}\right)_v\left[\frac{\mathrm{d}V}{\mathrm{d}\theta}-\frac{v}{\omega}(\dot{m}_{\mathrm{in}}-\dot{m}_{\mathrm{out}})\right]-\sum\frac{\dot{m}_{\mathrm{in}}}{\omega}(h-h_{\mathrm{in}})+\frac{\dot{Q}}{\omega}\right\} \tag{4}$$

The mass of the gas in the working chamber varies with the spindle angle:

$$\frac{\mathrm{d}m}{\mathrm{d}\theta}=\sum\frac{\dot{m}_{\mathrm{in}}}{\omega}-\sum\frac{\dot{m}_{\mathrm{out}}}{\omega} \tag{5}$$

3.2. Heat transfer model.

There is a temperature difference in the environment will inevitably occur heat transfer phenomenon, the three ways of heat transfer are heat conduction, thermal convection, heat radiation. When the gas enters the suction chamber, the heat transfer process occurs in the chamber. In order to simplify the heat transfer model, neglecting the heat conduction and heat radiation, only consider the thermal convection phenomenon.

3.2.1. Suction heating. The suction process is calculated according to the convective heat transfer of the fluid in the tube, using the Dittus-Boelter equation, the surface heat transfer coefficient [12]

$$h = 0.023\frac{\kappa}{d_p} Re^{0.8} Pr^{0.4} \tag{6}$$

The temperature of the compressed gas entering the working chamber from the suction pocket can be calculated from equation (7):

$$T_s = T_p - (T_p - T_{s,o})\exp\left(-\frac{\pi\, d_p L_p h}{\dot{m} c_p}\right) \tag{7}$$

The value of the heat transfer between the gas and the gas pipe is:

$$\dot{Q} = \dot{m} c_p (T_p - T_{s,o})\left[1 - \exp\left(-\frac{\pi\, d_p L_p h}{\dot{m} c_p}\right)\right] \tag{8}$$

3.2.2. Working chamber heat transfer. The average convection of the spiral plate heat exchanger is used to simulate the heat transfer process between the gas and scroll tooth, the surface heat transfer coefficient is [13, 14]

$$h = 0.023\frac{\kappa}{D_{ef}} Re^{0.8} Pr^{0.4}\left(1 + 1.77\frac{D_{ef}}{R_{aver}}\right) \times \left\{1 + 8.48\left[1 - \exp(-5.35St)\right]\right\} \tag{9}$$

The equivalent diameter in this model is defined as the ratio of the working chamber volume to the scroll wrap:

$$D_{ef} = \frac{4V}{A} \tag{10}$$

The equivalent radius is defined as [15]:

$$R_{aver} = R_g\left[\frac{(\varphi_k - \pi/2) + (\varphi_{k-1} - \pi/2)}{2}\right] \tag{11}$$

3.2.3. Area of heat transfer. The expression of the area of heat transfer between the scroll wrap and plate is in the range of increasing angle φ_k to $\varphi_k - 2\pi$:

$$dA = h r_b (\varphi - \varphi_0) d\varphi \tag{12}$$

The expression of the heat transfer area at the bottom of the scroll is [16]:

$$dA = \frac{1}{2} r_b^2 [(\varphi - \varphi_{i0})^2 - (\varphi - \pi - \varphi_{o0})^2] d\varphi \tag{13}$$

Gas and scroll tooth, the bottom of the scroll heat exchange capacity in any working chamber can be calculated by the following formula [17]:

$$\dot{Q} = h\int_A \left[T_{scroll}(\varphi) - T(k, j)\right] \mathrm{d}A \tag{14}$$

3.3. Leakage model

There are two different leakage in the scroll compressor, namely, radial leakage through the axial gap and flank leakage through the radial gap. Leakage model shown in Figure 3.

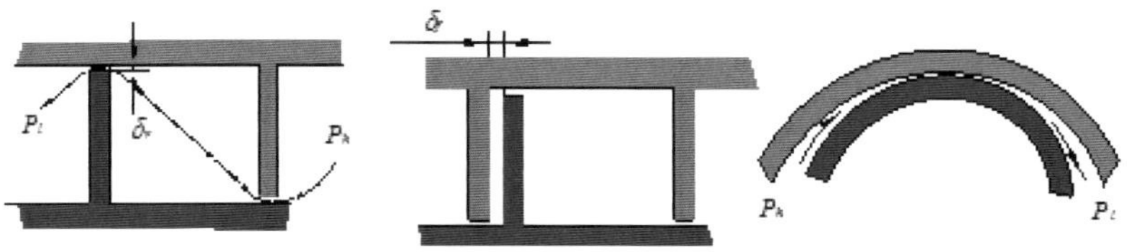

(a) Radial leakage (b) Flank leakage

Figure 3. Radial leakage and Flank leakage.

For two internal leakage of the compressor, the gap can be expressed as a high and low side pressure function, resulting in a radial leakage area [18]

$$A_r = \delta_r L_r \tag{15}$$

L_r For the radial leakage line length, the axial clearance δ_r can be expressed as a linear function with respect to the pressure ratio:

$$\delta_r = 1.1\times 10^{-6}\left(\frac{P_d}{P_s}\right) + 10^{-6} \tag{16}$$

Flank leakage area:

$$A_f = h\delta_f \tag{17}$$

Similarly, the radial gap δ_f can also be expressed as a linear function of the pressure ratio:

$$\delta_f = -9.615\times 10^{-5}\left(\frac{p_h}{p_l} - 1.67\right) + 20\times 10^{-6} \tag{18}$$

The leakage of the scroll compressor can be abstracted into a nozzle model in which the mass flow of the gas in the chamber is [19]

$$\dot{m} = \psi A_s\sqrt{2p_h\rho_h}\sqrt{\frac{\gamma}{\gamma-1}\left[\left(\frac{p_l}{p_h}\right)^{\frac{2}{\gamma}} - \left(\frac{p_l}{p_h}\right)^{\frac{\gamma+1}{\gamma}}\right]} \tag{19}$$

The flow rate is limited by the critical pressure ratio of the blocked flow conditions, the pressure ratio is:

$$\frac{p_l}{p_h} = \left(\frac{2}{\gamma+1}\right)^{\frac{\gamma}{\gamma-1}} \tag{20}$$

According to the character of the ideal gas compressible, the mass flow of the radial leakage can be determined by the following formula [20]:

$$\frac{p-p_a}{\rho g} = \lambda \frac{Lu_m^2}{4m2g} \tag{21}$$

Similarly, the mass flow of flank leakage can be represented by the following formula:

$$\frac{p-p_a}{\rho g} = \int_{-\phi}^{\phi} \lambda \frac{Rd\phi u_m^2}{4m2g} \tag{22}$$

4. Simulation analysis

In order to describe the thermodynamic model of oil-free scroll compressor in detail, this paper establishes the thermodynamic model of oil-free scroll compression based on the basic laws of thermodynamics, the mass and energy conservation equations and the gas state equation, considering the heat transfer model and the leakage model. The leakage of the oil-free scroll compressor is analyze-

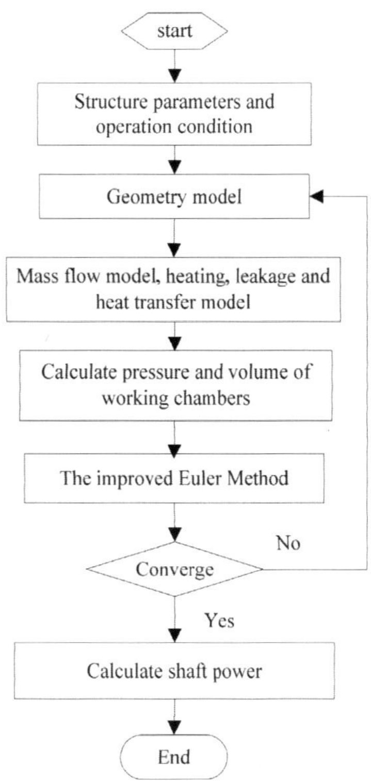

Figure 4. Flowchart of thermodynamic model.

End and studied deeply. The pressure-volume curve and the mass flow of leakage between the chambers are obtained by solving established thermodynamic model. As shown in the figure 4 is the flowchart of thermodynamic model of oil- free scroll compressor.

As shown in the figure 5, the curve of pressure with the volume. In the suction process, the volume continues increase to maximum of 213918 mm3, while the pressure is always maintained at a fixed value of 101.3 kPa. When the gas enters the compression chamber, the volume of the gas is further compressed so that the volume inside the chamber is gradually reduced and the pressure continues to increase. When the spindle angle reaches the exhaust angle, the gas enters the exhaust chamber. At this time, the volume of the chamber is 44885mm3, and the gas is gradually discharged from the exhaust port. As shown in the figure 6, the mass flow curve of the fourth compression chamber. The radial mass flow of the spindle in the process of one cycle is increasing, indicating that the leakage of the third compression chamber to the chamber is lesser than the chamber to the next working chamber. And the flank mass flow rate tends to rise slightly, the leakage of the upper chamber to the chamber is slightly lesser than the chamber to the next chamber.

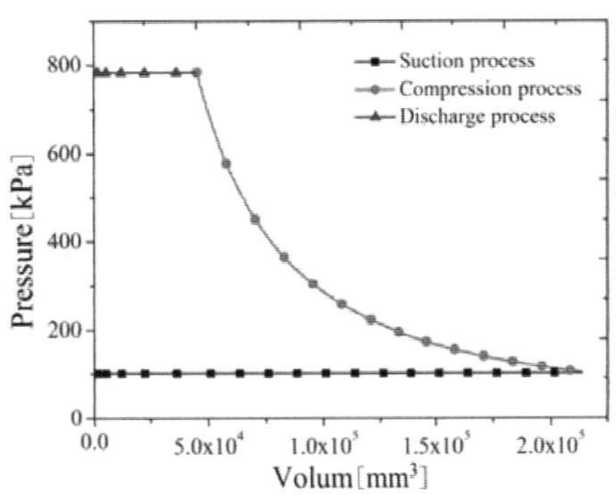

Figure 5. Pressure-volume curve.

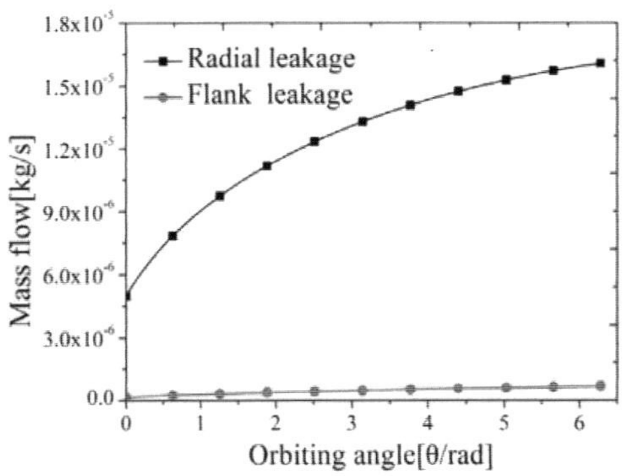

Figure 6. The mass flow in fourth chamber.

As shown in the figure 7, the third compression chamber mass flow curve. With the continuous rotation of the spindle, the mass flow of the radial leakage is declining, indicating that the leakage of the third compression chamber to the fourth compression chamber is lesser than the second compression chamber to the third compression chamber. While the mass flow of flank leakage remain essentially unchanged. As shown in the figure 8, the second compression chamber of the mass flow

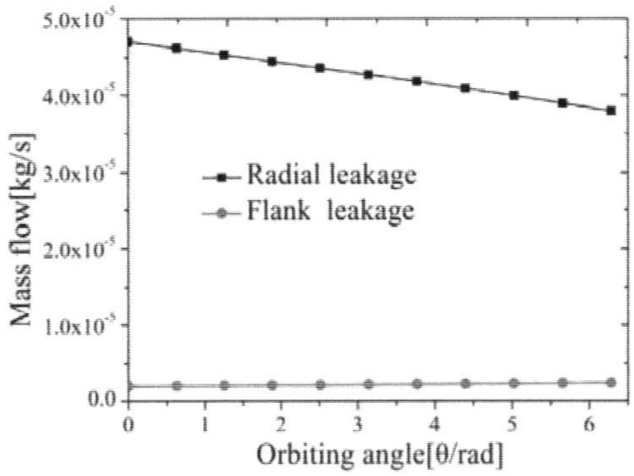

Figure 7. The mass flow in third chamber.

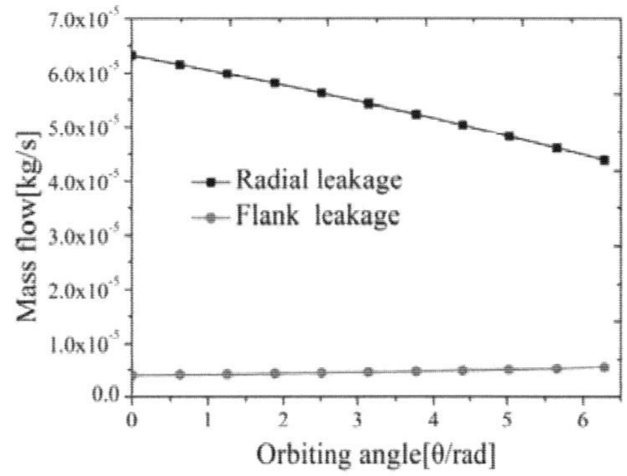

Figure 8. The mass flow in second chamber.

Curve. The mass flow of radial leakage is significantly higher than the fourth and third compression chambers. This is due to the continuous rotation with the spindle, the pressure of working chamber rise further, while the leakage to the third compression chamber is decreased. At this time, the mass flow of flank leakage is gradually increasing.

As shown in the figure 9, the central exhaust chamber mass flow curve. With the spindle angle increases, radial leakage to the second compression chamber leakage mass flow decreased. This is because the central exhaust chamber can only leakage downwards to the second compression chamber, resulting the mass is decreased. While the flank and radial leakage mass flow is declining until the spindle angle reaches the exhaust angle 1.472π, the central chamber began to exhaust, the mass of the chamber is quickly discharged the outside. As shown in the figure 10, the total mass flow curve. The mass flow of the above-mentioned chamber in radial and flank leakage are superimposed on each other, resulting in a curve change as shown in the figure 10. The mass flow of radial leakage is much larger than the flank leakage, and the variation of mass flow in radial direction is larger than flank leakage.

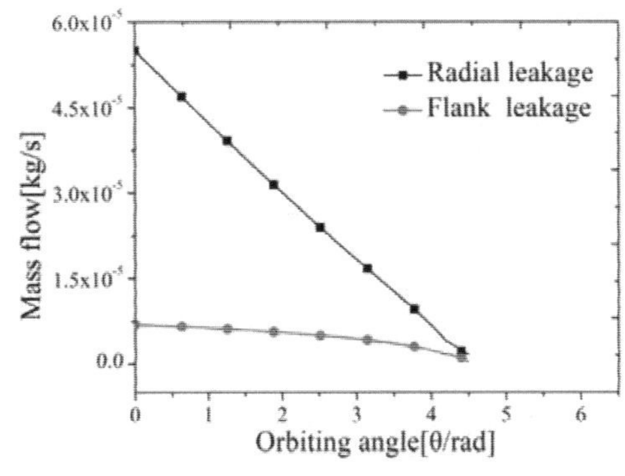

Figure 9. The mass flow in center exhaust chamber.

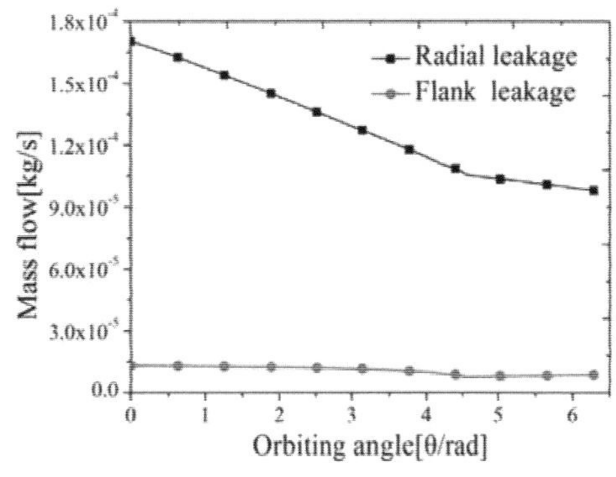

Figure 10. The mass flow of total leakage.

5. Experimental Verification

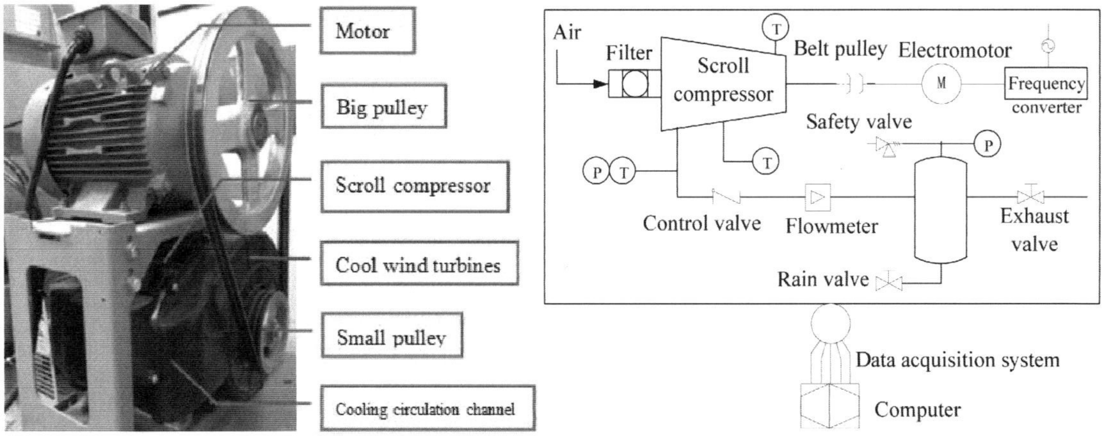

(a) Testing prototype (b)Testing system schematic

Figure 11. Testing platform.

In order to verify the correctness of the oil-free scroll compressor thermodynamic model and analyze the performance of the whole machine under different working conditions, a prototype of the new oil-free scroll compressor is developed. Figure 11(a) shows the test prototype and according to the test system schematic diagram of Figure 11(b), the test platform of oil-free scroll compressor is built and the simulation and test data are compared.

Figure 12 is in different pressure conditions, the temperature variety with the frequency. Under the same discharge pressure, with the frequency increases, the temperature gradually increased. Under the same frequency, with the discharge pressure increases, the discharge temperature also gradually increased. Figure 13 under the different pressure conditions, the displacement with the frequency. Under the same discharge pressure, with the frequency increases, the displacement gradually increased. Under the same frequency, the discharge volume is slightly reduced as the discharge pressure increases. The discharge pressure for the displacement of smaller impact, the change of displacement is little with different discharge pressure.

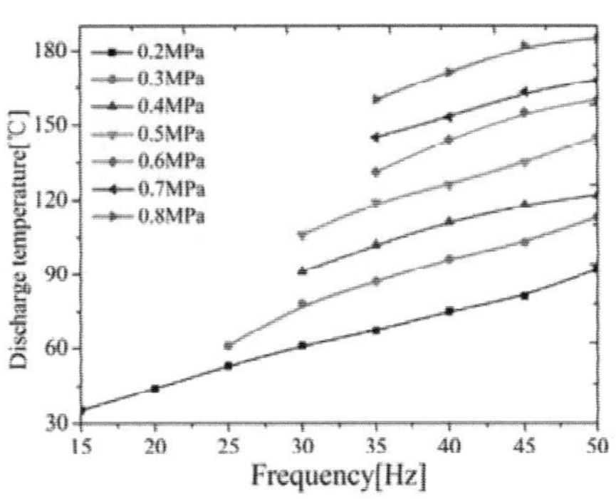

Figure 12. Discharge temperature vs. frequency.

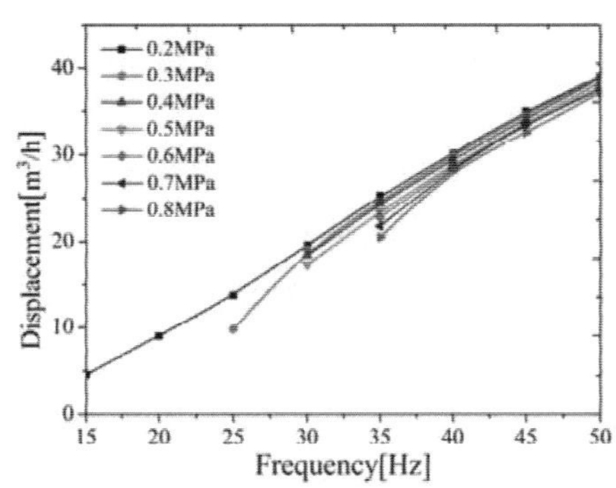

Figure13. Displacement vs. frequency.

Figure 14 compares the simulated and tested shaft power. With the increase of rotational speed, the values of the simulated shaft power and the measured values are increasing, and the errors are small, which verifies the correctness of the thermodynamic model. Figure 15 shows the variety in power with frequency. In the case of same discharge pressure, with the increases of frequency, the power continue to increase. Under the same frequency, the power gradually increases with the increase of discharge pressure.

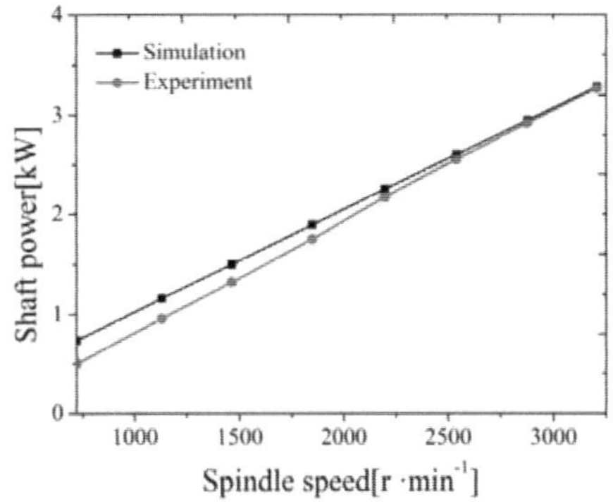

Figure 14. Simulation and test shaft power.

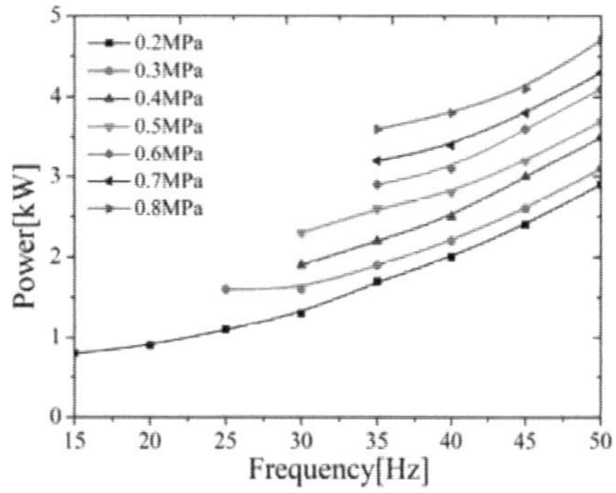

Figure15. Power with frequency.

The figure 16 shows the variety of voltage with spindle speed under different pressure conditions. Under the same discharge pressure, with the increase of the spindle speed, the voltage increases. Under the same spindle speed, discharge pressure increases, voltage is almost kept constant. The discharge pressure for voltage is smaller, the voltage variety is slight with different discharge pressure. Figure 17 shows the current with frequency under different pressure conditions. At the same discharge pressure, the current decreases with increase of frequency. In the same frequency conditions, the current increases with the discharge pressure increases. The variation of the current at low frequency is large, and the variation range at high frequency is small.

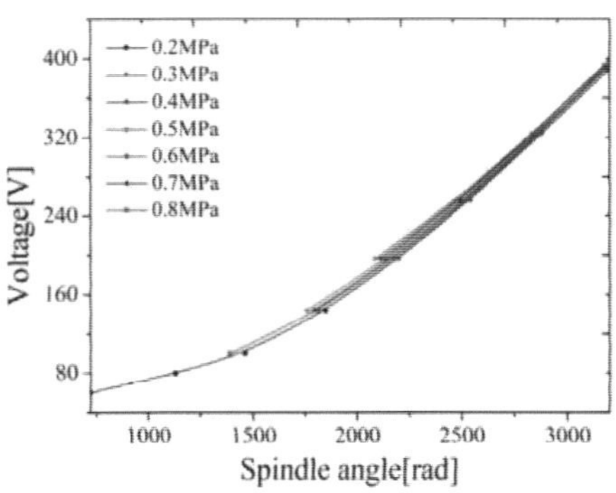

Figure 16. Voltage vs. spindle angle.

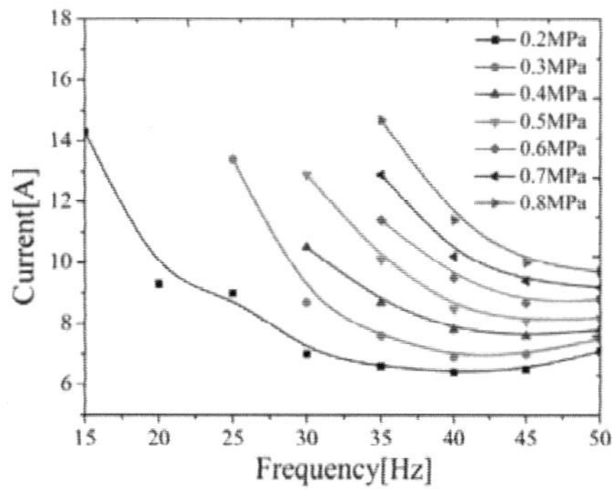

Figure 17. Current vs. frequency.

6. Conclusion

According to the oil-free scroll compressor, the relevant thermodynamics model is established. Based on the developed prototype, the performance of the prototype is tested by constructing the test platform. The following conclusions can be drawn:

1) Based on the geometric parameters and structural dimensions of oil-free scroll compressor, the geometrical model of prototype was established and the variation curve of pressure-volume was calculated.

2) Based on the conservation of energy, the law of conservation of mass and the gas state equation, the thermodynamic model of oil-free scroll compressor is established, considering the heat transfer model and leakage model.

3) The leakage mass flow of each chamber is obtained by solving the established thermodynamic model by using the improved Euler method. Based on the developed prototype model, the thermodynamic model is tested and verified.

4) The test results show that under different discharge pressure, the temperature and power change range is larger, and variation of displacement is small. The experimental data for new oil-free scroll compressor research and performance analysis has a certain guidance and reference.

Acknowledgments
This work supported by National Natural Science Foundation of China (Grant No. 51275226, 51675254), and Longyuan youth innovation team Funded projects.

References

[1] Peng Bin, Arnaud Legros, Vincent Lemort , Xie Xiaozheng, Gong Haifeng (2016). Recent Advances on the oil-free Scroll Compressor. Recent Patants on Mechanical Engineering, 9(1): 37-47.
[2] Yuanyang Zhao, Liansheng Li, Huagen Wu, etc. Theoretical and Experimental Studies of Water- injection Scroll Compressor in Automotive Fuel cell Systems[J]. Energy Conversion and Management, 2005, 46(9-10): 1379–1392.
[3] FANG Shiyi, LI Liansheng, SHU Pengcheng, et al. Study on working process of oil-free scroll-type air compressor [J]. China Mechanical Engineering, 2005, 16(2): 123-127.
[4] L Wang, Y Zhao, L Li, G Bu, etc. Research on oil-free hermetic refrigeration scroll compressor[J]. Proceedings of the Institution of Mechanical Engineers, Part A: Journal of Power and Energy, 2007, 221(7): 1049-1056.
[5] Yanagisawa T, Fukuta, Y Ogi. Performance of an oil-free scroll-type air compressor[C]. In: International Conference on Compressors and their Systems, City university, Londen, England, 1999: 279- 287.
[6] Bryce R. Shaffer, Eckhard A. Groll. Performance of the Use of Plastics in Oil-Free Scroll Compressors[C]. In: Proceedings of International Compressor Engineering Conference at Purdue, USA, 2012, 1267: 1-9.
[7] S. Sunder, Thermodynamic and Heat Transfer Modeling of a Scroll Pump, PHD Thesis, Massachusetts Institute of Technology, 1996.
[8] K. T. Oil, J. Zhu, Convective heat transfer in a scroll compressor chamber: a 2-D simulation, Int J Therm Sci 43(7): 2004, 677-688.
[9] F. Yi, E.A. Groll, J.E. Braun, Modeling and Testing of an Automobile AC Scroll Compressor. Part I. Model Development, Proceedings of 2004 International Compressor Engineering Conference at Purdue University, 2004, p. 1–8 (C082).
[10] F. Yi, E.A. Groll, J.E. Braun, Modeling and Testing of an Automobile AC Scroll Compressor. Part II. Model Validation Proceedings of 2004 International Compressor Engineering Conference at Purdue University, 2004, p. 1–8 (C083).
[11] Chen Yu (2000). Mathematical modeling of scroll compressors. Purdue University Doctor article.
[12] Incropera FP, Dewitt DP (1996). Fundamentals of heat and mass transfer. New York: John Wiley & Sons Publishers.
[13] Kakac S, Shah RK (1987). Handbook of single-phase convective heat transfer. New York: John Wiley & Sons Publishers.

[14] Rajavel R, Saravanan K (2008).Heat transfer studies on spiral plate heat exchanger. Thermal Science, 12(3): 85-90.
[15] K Jang, S Jeong (2006). Experimental investigation on convective heat transfer mechanism in a scroll compressor. International Journal of Refrigeration, 29(5): 744-753.
[16] K Jang, S Jeong (2006). Experimental investigation on convective heat transfer mechanism in a scroll compressor. International Journal of Refrigeration, 29(5): 744-753.
[17] MARCO C, EVANDRO L, CESAR J. A lump-parameter thermal model for scroll compressors including the solution for the temperature distribution along the scroll wraps[J]. International Journal of Refrigeration, 2015, 53:184-194.
[18] Halm, N.H (1997). Mathematical Modeling of Scroll Compressors. Purdue University Master article.
[19] FOX R, MCDONALD A. Introduction to fluid mechanics[M]. New York: John Wiley & Sons, Inc., 1992
[20] JANG K, JEONG S. Experimental investigation on convective heat transfer mechanism in a scroll compressor[J]. International Journal of Refrigeration, 2006, 29(5): 744-753.

One-dimensional numerical simulations of heat transfer on the piston effect

Songzi Wu

School of Urban Construction and Environmental Engineering, Chongqing University, Chongqing, China

e-mail:1043363296@qq.com

Abstract. Strong singularities in thermodynamic and transport properties appear in near-critical fluids, which give birth to the piston effect. In the present study, the piston effect is investigated numerically in a one-dimensional cavity filled with near-critical CO2, being heated at one side while the other side keeps at initial temperature. The thermodynamics model, which ignores velocity field and assumes uniform pressure field, is numerically solved by explicit finite difference method. The results are presented and discussed from two aspects. In the first part, the results show typical evolutions of temperature and density fields. The variations of temperature, density and pressure are results of competitions and eventual equilibrium between the heating piston effect and the cooling piston effect. In the second part, the results indicate influences of the critical proximity on the equilibrium. Results show that with the increase of critical proximity, piston effect gets weaker, equilibrium time gets longer, and equilibrium temperature becomes smaller.

1. Introduction

Strong singularities in thermodynamic and transport properties appear in near-critical fluids. Specifically, the isothermal compressibility, the isobaric expansion coefficient and specific heat capacity at constant pressure diverge while thermal diffusivity tends to zero [1]. As the result, theoretically, the process of heat transfer extremely slows down under zero gravity, namely critical slowing down, leading to a substantial increase in relaxation time of temperature field. However, in 1985, in an experiment measuring specific heat capacity at constant volume of supercritical SF6 under microgravity environment, researchers actually observed a critical speeding up. Because the phenomenon was exactly contrary to their expectation, many researches started to pay great attention to studying it [2].

Early studies mainly focused on explanations, theoretical analyses and experimental verifications of the abnormal phenomenon. In 1990, three groups put forward their explanations based on different approaches independently. Ounce et al. proposed a theoretical model based on thermodynamics theories [3]. Bourke et al. raised a numerical model based on thermodynamics theories [4]. Napoli et al. named the phenomenon the piston effect (PE), and presented the process of PE more directly through numerical simulation [5]. Afterwards, Napoli and Carlos, based on the abnormal physical properties, studied through matched asymptotic descriptions techniques [6]. Then they put forward the decoupling of temperature and density relaxations, and solved physical process of PE based on a one-dimensional partial heated model theoretically: after the wall of a sealed container filled with supercritical fluid being heated, an extremely thin thermal boundary layer forms near the heated boundary. Supercritical fluid has relatively higher thermal expansion coefficient and compressibility. As the result, thermal boundary layer expands and compresses the bulk fluid. By this means, energy is transferred from the boundary layer to the bulk fluid quickly through pressure wave, leading to a rapid temperature rise in the bulk fluid. As for experimental studies, researchers mainly measured temperature variations with thermistors or measured density variations with interferometers in early experiments, in order to verify the existence of PE and the decoupling of temperature and density relaxations. In 2006, Miura et al. made the first experimental observation of thermos-acoustic waves with ultra-sensitive interferometers [7]. Then Shen and Zhang [8] carried on a series of numerical simulations, studying the influences of thermal boundary conditions and initial conditions on thermos-acoustic waves. Besides, they also studied the propagation and reflection of thermos-acoustic waves.

In this paper, one-dimensional numerical simulations are performed by using thermodynamics model. The PE in a one-dimensional sealed cavity being heated at one side while the other side keeps at initial temperature is studied. Such a physical model results in the coexistence of heating piston effect and cooling piston effect. The competitions of the two kinds of PE is revealed by analyzing the evolutions of temperature and density fields. In addition, the influences of critical proximities on their interactions are discussed.

2. Physical model and governing equations

2.1. Physical model

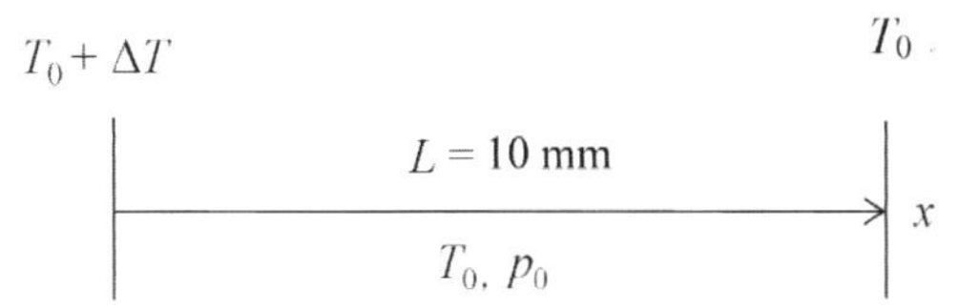

Figure 1.The physical model.

As shown in Fig.1, the physical model is a one-dimensional sealed cavity filled with CO_2 of which the length is L = 10 mm. Its initial temperature and pressure are T_0 and p_0 respectively. Left wall temperature suddenly rises by ΔT at initial time while the other side keeps at initial temperature T_0. In order to exclude the influences of natural convection, the gravity is ignored. In this paper, the initial conditions are chosen along the critical isochore ($\rho=\rho_c$), and the initial temperature is determined through the dimensionless reduced temperature ε, given by

$$\varepsilon = \frac{T - T_c}{T_c}$$

In this paper, the parameter ε measures the proximities to the critical point. The critical parameters of CO_2 are p_c = 7.38 MPa, T_c = 31 ℃, ρ_c = 468.6 kg m^{-3}, with p_c, T_c and ρ_c the critical pressure, temperature and density, respectively. The physical properties of CO_2 are obtained from NIST database [9].

2.2. Governing equations

Boukari *et al.* [4] put forward a thermodynamics model to describe the PE under zero gravity in a single-phase supercritical fluid in a sealed container. They ignored the velocity field in the fluid, and assumed that the pressure in the whole domain is uniform. This model has a relatively simple form and is convenient for programming and calculating. Therefore, it is widely accepted by researchers [10-12]. The energy equation for a one-phase fluid is given by

$$\rho c_p \frac{\mathrm{D}T}{\mathrm{D}t} = \nabla \cdot (\lambda \nabla T) + T\beta_p \frac{\mathrm{D}p}{\mathrm{D}t} + \Phi \tag{1}$$

with ρ the density, p the pressure, T the temperature, λ the thermal conductivity, β_p the isobaric expansion coefficient, Φ the viscosity dissipation term. When ignoring velocity and applying a uniform pressure field, a further reduced equation can be derived as following:

$$\rho c_p \frac{\partial T}{\partial t} = \nabla \cdot (\lambda \nabla T) + T\beta_p \frac{\partial p}{\partial t} \tag{2}$$

The equation of state is the linearized equation of state, expressed as

$$\rho = \rho_0 + \rho\left[\alpha_T (p - p_0) - \beta_p (T - T_0)\right] \tag{3}$$

Equations (2) and (3) are not close, so it's necessary to derive a pressure equation from mass conservation equation, which is given by [4]

$$p = p_0 + \frac{\int_V \rho\beta_p (T - T_0) dV}{\int_V \rho\alpha_T dV} \tag{4}$$

The equations stated above can be solved by discretization and iterative processes. Then we obtained the developments of temperature, pressure and density fields under the PE.

3. Numerical method

In this paper, the governing equations (2), (3) and (4) are solved by explicit finite difference method. Generally speaking, there are two kinds of finite difference methods, the explicit one and implicit one. Each has their own advantages [13]: explicit method is relatively easier to construct process and program, though its selection of time step is restricted by stability conditions, when grid size is pretty small, the time step should be small enough to have numerical stability; as for implicit method, there is no restriction for time step to guarantee numerical stability. However, programing and constructing are relatively more complex, and it will take more time for every single time step. For these reasons, explicit method is mainly used to solve unsteady situation, it is better for solving exact solution over time; implicit method is widely employed in solving stationary problems, its accuracy of process over time is not very important. This is an unsteady problem in essence, considering the high demand for accuracy, so it is proper to use explicit finite difference method in this study. As for discretization methods, taking computational accuracy and time consuming into account, the time derivative term in Eq. (4) is discretized with first-order forward difference scheme, and the spatial derivative terms are discretized with second-order central difference scheme. In addition, the grid size selected is $\Delta x = 0.01$ mm, and time step is $\Delta t = 0.001$ s or 0.005 s. This selection guarantees stability conditions of explicit method. Figure 2 gives the algorithm flowchart we used in simulations.

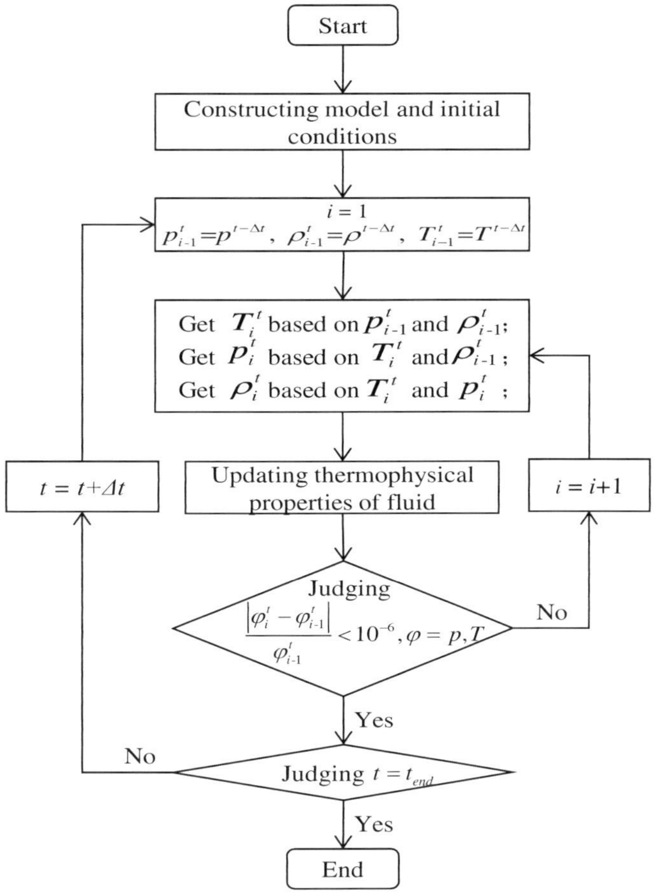

Figure 2. Algorithm flowchart in simulations.

4. Results and discussions

4.1. Typical evolutions of temperature and density fields

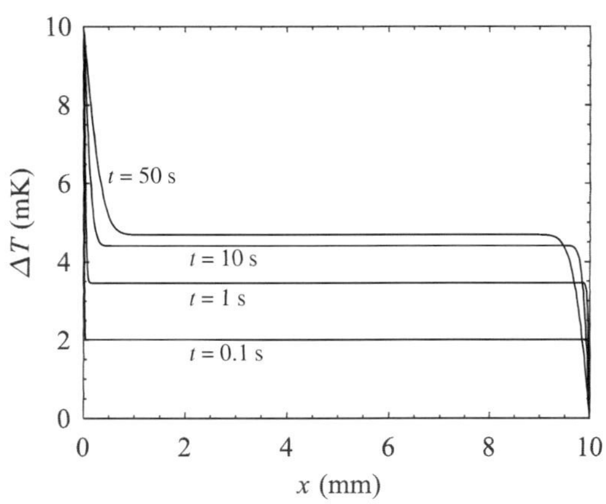

Figure 3. Spatial distributions of temperature at different time (ε = 0.001).

Figure 3 plots spatial distributions of temperature at different time for ε = 0.001. After the temperature at the left wall rises suddenly, heat diffuses into the fluid. A hot boundary layer (HBL) then appears near the left wall. Due to the diverging compressibility of near-critical CO_2, the HBL expands strongly and activates the *heating piston effect* (HPE), leading to a rapid, homogenous rise of bulk temperature. Besides, due to the right boundary keeping at initial temperature, the loss of heat results in the formation of temperature gradients, namely that a cold boundary layer (CBL) forms near the right boundary. Similarly, the contraction of the CBL causes the *cooling piston effect* (CPE), leading to a rapid, homogenous drop in bulk temperature. Therefore, the variations of bulk temperature in Fig.3 are results of competitions between the HPE and the CPE, where the horizontal part of the lines represents the bulk fluid. One should also bear in mind that since the CPE is a result of the HPE, the former is always weaker than the latter. Obviously, during 0.1 s to 50 s, the boundary layers develop gradually, bulk temperature becomes higher, but the rate of increase in bulk temperature drops gradually. The reason is that initially, the HPE is strong while the CPE is weak. However, as time goes by, bulk temperature gets higher. The HPE gradually becomes weak while the CPE gets strong. We can conclude that the deceleration in the change of bulk temperature results from the equilibrium between the HPE and the CPE. To study such an equilibrium, we could define the equilibrium time as the moment when the rise of bulk temperature per unit time (0.01 s) is smaller than 0.1% of the amount the bulk temperature rises during the first 0.01s. It should be noted that the term *equilibrium* used here doesn't refer to the thermal equilibrium of the whole system, but represents the balance between the HPE and the CPE.

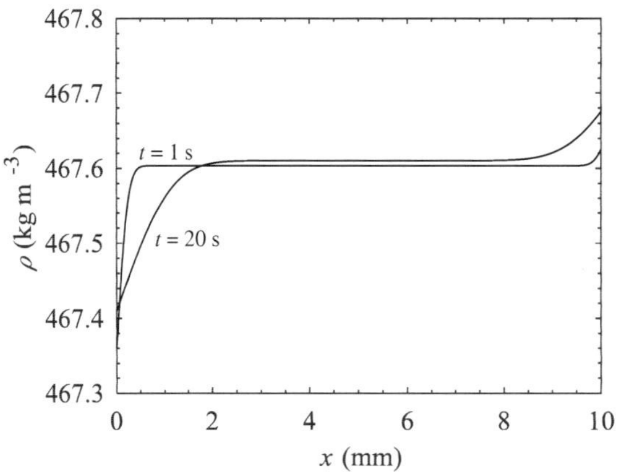

Figure 4. Spatial distributions of density at different time (ε = 0.05).

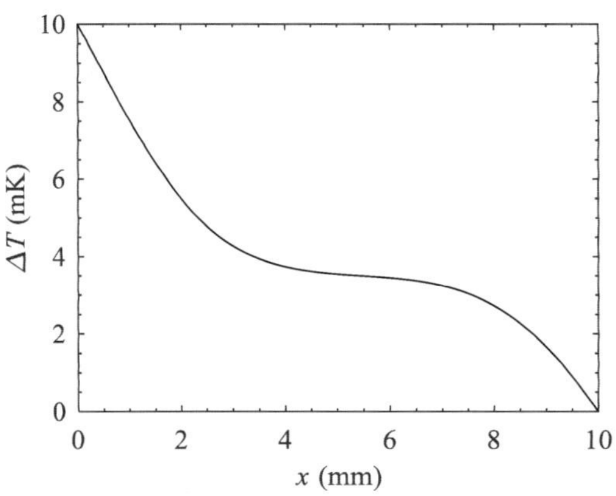

Figure 5. Temperature field at late time (ε = 0.05, t = 100 s).

In Fig.4, we show profiles of density field at different time for $\varepsilon = 0.05$. Heat diffusion and the two kinds of piston effects cause the variations in temperature, which also results in the change of density. The left HBL and right CBL along with the homogenous bulk fluid (horizontal part of the lines) are clearly observed in Fig.4. Indeed, after the sudden rise of the temperature at the left wall, the HBL expands and its density drops, activating the HPE. As the result, the bulk fluid is compressed and its density increases. Similarly, the contraction of CBL cause the CPE, leading to the decrease in bulk density. Since HPE is always stronger than CBL, the overall effect is compression and the bulk density increases gradually. Obviously, during 1 s to 20 s, the bulk density gets higher.

Figure 5 plots the temperature field at late time for $\varepsilon = 0.05$. It is observed that, as time goes by, the influences of the piston effect disappear, and both of the boundary layers develop through diffusion and merge to each other. Eventually, temperature field should be linear distribution.

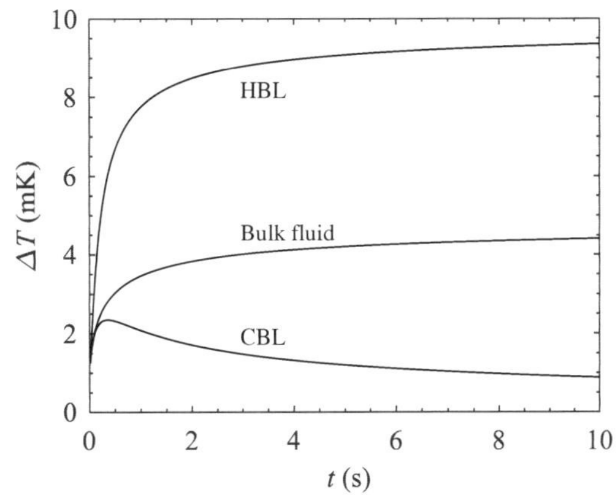

Figure 6. Time evolutions of temperature at different spatial points ($\varepsilon = 0.001$), where the words indicate the locations of the points.

We plot time variations of temperature at three typical spatial points in Fig.6. It shows that, after the sudden rise of the temperature at the left wall, the temperature of the point in the HBL rise rapidly during short time. Then, the HBL expands, activating the HPE. As the result, the temperature of the point in the bulk fluid rises fast and evenly. Meanwhile, the fluid near the right wall first goes through a brief temperature rise due to the HPE, which drops then through heat diffusion. As time goes by, the interaction between HPE and CPE gradually approaches to equilibrium, and the increase rate of bulk temperature decreases.

4.2. Influences of the critical proximity on the equilibrium

Table 1. The measured equilibrium time and temperature for different critical proximities.

Critical Proximity (ε)	Equilibrium Time (s)	Equilibrium Temperature (mK)
0.001	3.24	4.04
0.005	11.9	3.78
0.01	19.5	3.63
0.05	68.5	3.29

Now we discuss the behaviors of equilibrium for different critical proximities. Table 1 gives the measured equilibrium time and temperature for different critical proximities. It is observed that with the increase of critical proximity, piston effect gets weaker, equilibrium time gets longer, and equilibrium temperature becomes lower. The reasons are presented below. The fluid at large critical proximity has a small compressibility. And since piston effect is closely related to the compressibility of the fluid. So under the same temperature disturbances, the intensity of HPE/CPE decreases with increased critical proximity. Larger the critical proximity, weaker the HPE, and smaller the rise amount of bulk temperature. As the result, the CPE also becomes weak due to the coupling effect of the bulk temperature and critical proximity. Both the HPE and the CPE get weak at large critical proximity, since the temperature disturbances are identical, we can conclude that it takes more time for HPE and CPE approaching to equilibrium. So the phenomenon stated above occurs.

5. Conclusion

In this paper, we conduct a series of explorations into the piston effect in a sealed cavity filled with near-critical CO_2 under zero gravity, with one heated boundary and the other boundary keeping at initial temperature. The results are presented and discussed from two aspects. In the first part, the results show typical evolutions of temperature and density fields, which are indeed the competitions between the heating piston effect and the cooling piston effect. In fact, the sudden rise of temperature activates the HPE, then the HPE triggers the CPE. Initially, the HPE is strong while the CPE is weak. As time goes by, the HPE gradually becomes weak while the CPE gets strong, eventually they approach to an equilibrium. In the second part, the results indicate influences of the critical proximity on the equilibrium. Results show that with the increase of critical proximity, piston effect gets weaker, equilibrium time gets longer, and equilibrium temperature becomes smaller.

References

[1] Carles P. A brief review of the thermophysical properties of supercritical fluids[J]. Journal of Supercritical Fluids, 2010, 53(1-3): 2-11.
[2] Nitsche K, Straub J. The critical hump of Cv under microgravity, results from D-Spacelab experiment "Wärmekapazität"[C]. Proceedings of the 6th European Symposium on Material Sciences under Microgravity Conditions, 1987: 109-116.
[3] Onuki A, Hao H, Ferrell R A. Fast adiabatic equilibration in a single-component fluid near the liquid-vapor critical point[J]. Phys Rev A, 1990, 41(4): 2256-2259.
[4] Boukari H, Shaumeyer J N, Briggs M E, et al. Critical speeding up in pure fluids[J]. Phys Rev A, 1990, 41(4): 2260-2263.
[5] Zappoli B, Bailly D, Garrabos Y, et al. Anomalous heat transport by the piston effect in supercritical fluids under zero gravity[J]. Phys Rev A, 1990, 41(4): 2264-2267.
[6] Zappoli B, Beysens D, Garrabos Y. Heat Transfers and Related Effects in Supercritical Fluids[M]. Springer, 2015.
[7] Miura Y, Yoshihara S, Ohnishi M, et al. High-speed observation of the piston effect near the gas-liquid critical point[J]. Physical Review E, 2006, 74(1): 010101.
[8] Zhang P, Shen B. Thermoacoustic wave propagation and reflection near the liquid-gas critical point[J]. Phys Rev E Stat Nonlin Soft Matter Phys, 2009, 79(6 Pt 1): 060103.
[9] Lemmon E W, Huber M L, Mclinden M O. NIST reference fluid thermodynamic and transport properties–REFPROP[J]. NIST standard reference database, 2002, 23: v7.
[10] Yufei Mao, Liejin Guo. Numerical simulations of the piston effect in supercritical fluids[J]. Progresses in natural science, 2006, 16(4).
[11] Masuda Y, Aizawa T, Kanakubo M, et al. One dimensional heat transfer on the thermal diffusion and piston effect of supercritical water[J]. International Journal of Heat and Mass Transfer, 2002, 45(17): 3673-3677.
[12] Wei L, Yuan D, Wu B, et al. A Simulation for the Piston Effect in Supercritical Carbon Dioxide with the Non-flow Model[J]. Journal of Thermal Science, 2013, 22(5): 472-477.
[13] Shiming Yang, Wenquan Tao. Heat transfer, the fourth edition [M]. Higher educational press, 2006.

IOP Conf. Series: Journal of Physics: Conf. Series **916** (2017) 012050 doi:10.1088/1742-6596/916/1/012050

Study on Gas-liquid Falling Film Flow in Internal Heat Integrated Distillation Column

Chong Liu

Langfang Teachers University; Langfang Hebei 065000 China

Abstract: Gas-liquid internally heat integrated distillation column falling film flow with nonlinear characteristics, study on gas liquid falling film flow regulation control law, can reduce emissions of the distillation column, and it can improve the quality of products. According to the distribution of gas-liquid mass balance internally heat integrated distillation column independent region, distribution model of heat transfer coefficient of building internal heat integrated distillation tower is obtained liquid distillation falling film flow in the saturated vapour pressure of liquid water balance, using heat transfer equation and energy equation to balance the relationship between the circulating iterative gas-liquid falling film flow area, flow parameter information, at a given temperature, pressure conditions, gas-liquid flow falling film theory makes the optimal parameters to achieve the best fitting value with the measured values. The results show that the geometric gas-liquid internally heat integrated distillation column falling film flow heat exchange area and import column thermostat, the average temperature has significant. The positive correlation between the heat exchanger tube entrance due to temperature difference between inside and outside, the heat flux is larger, with the increase of internal heat integrated distillation column temperature, the slope decreases its temperature rise, which accurately describes the internal gas-liquid heat integrated distillation tower falling film flow regularity, take appropriate measures to promote the enhancement of heat transfer. It can enhance the overall efficiency of the heat exchanger.

Keywords: internal heat integrated distillation column; gas-liquid falling film flow; liquid water balance; heat transfer efficiency.

1. Introductions

Distillation is a distillation tower vapor-liquid contact devices, there are two main types of plate column and packed column, according to the mode of operation can be divided into continuous distillation and batch distillation. The internally heat integrated distillation column consists of evaporator, condenser and several parts thermostat, the internally heat integrated distillation column steam enters from the bottom of the column, the gas and the evaporated liquid drop of countercurrent contact phase contact, the drop in the volatileness(low boiling) components to gas phase transfer, difficult volatile gas phase (high boiling point) component continued to decline in liquid medium transfer, gas phase is close to the top of the tower, the higher the concentration of volatile components,

IOP Conf. Series: Journal of Physics: Conf. Series **916** (2017) 012050 doi:10.1088/1742-6596/916/1/012050

while the decline in the closer tower bottom liquid, the more volatile rich, so as to achieve the purpose of separation. The gas rising from the top of the tower into the condenser, the liquid part of the condensate as reflux liquid return to the top of the tower into the distillation tower, the remainder is removed as the distillate.

The internal gas-liquid heat integrated distillation tower from the tower bottom liquid flowing part of liquid into the reboiler, heating evaporation into a gas phase back to tower, the other part of the liquid as residue removal, evaporator heat transfer area increases, resulting in gas-liquid film flow phenomenon, the basic principle of distillation is the liquid mixture multiple partial gasification and condensation, the components of different volatility characteristics, to achieve the purpose of separation distillation unit operation. According to its operation method can be divided into: simple distillation, flash distillation and special rectification. In recent years the emergence of super gravity distillation technology, using hundreds or thousands of times the gravity generated by high-speed rotation the super gravity field instead of the conventional gravity field, greatly enhance the gas-liquid mass transfer process, the height of mass transfer unit is reduced by 1 orders of magnitude. The huge tower equipment into a height less than 2 meters high gravity Rectification machine, increase efficiency, and reduce the volume. The internally heat integrated distillation column has changed the traditional model of distillation tower equipment, as long as the indoor factory can realize the continuous distillation process. The development of the society can save steel resources, extend the service life of the earth's resources. For the development of enterprises, can save space and space resources, reduce pollution emissions, improve product quality, improve management mode, reduce the labor intensity of production, increase production safety. However, the design of internal heat integrated distillation tower manufacturing technology is facing many problems: one is the manufacturing material must withstand low temperature, high pressure, corrosion resistance, abrasion resistance; two is the structure of heat exchanger is not only to ensure safe and efficient operation, and convenient maintenance. The three is to select the correct combination of theory and practice have a great influence on the heat transfer efficiency of the internally heat integrated distillation column The medium; four is the study of internally heat integrated distillation column liquid falling film flow and heat transfer theory, gas-liquid film flow drop theory is the fundamental basis of the distillation tower design, is the basis for optimization design of distillation column. In this regard, this paper studies the internal gas-liquid heat integrated distillation tower falling film flow phenomena, obtained by gas-liquid distillation column reduction film flow in saturated vapor pressure of liquidwater balance, using heat transfer equation and energy equation to balance the relationship between the circulating iterative gas-liquid falling film flow area, flow parameter information, at a given temperature, pressure conditions, the theory of gas liquid falling film flow optimal parameter values to achieve the best fit with the measured value, and the validity of the conclusion.

2. Theoretical analysis of liquid vapor balance in distillation column gas-liquid falling film flow

2.1. Physical model of gas liquid falling film flow in internal heat integrated distillation column

Figure 1 introduced the structure principle of internal heat integrated distillation tower, the structure of internal heat integrated distillation tower main evaporator, and condensers thermostat three heat exchanger, evaporator and condenser coexist in a large shell. Based on the blast furnace as heat source, medium gas and liquid propane as gasification medium for gas, each of which two kinds of refrigerant in the evaporator, the condensation heat transfer flow and temperature regulator, internally heat integrated distillation column in the heating process, the gas quantity control ignition temperature change, the change of temperature inside the heat integrated distillation tower flue, the internally heat integrated distillation column gas flow reverse regulation in the regulation process, often takes several hours to achieve changes in temperature feedback, the influence of time delay of control, resulting in internal heat integrated distillation column liquid falling film flow, in order to reduce the control delay of distillation tower, distillation tower to improve the control response speed, first in the heating mode, the pushing time, switching time and other aspects of discrete variable structure system control,

control regulation, analysis of liquid water balance mechanism, reservoir dynamic reserves of each period direct calculation of water drive gas using automatic fitting method dynamic, gas-liquid falling film regulation.

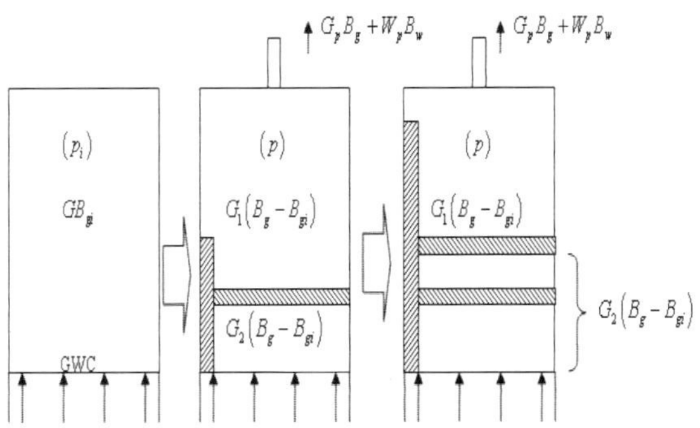

Fig. 1 Schematic diagram of internal heat integrated distillation column

Figure 1 shows that the high temperature distillation liquid first into the thermostat, the temperature is TWA1, the flow inside the heat pipe in heat exchanger temperature regulator, outlet temperature of TWA2, then enters into the evaporator, further heat flows in the heating pipe of the evaporator, finally discharged from IFV, the temperature is TWA3; low temperature distillation liquid into the first the condenser, the temperature is T1, in the condensation heat exchanger tube flow and endothermic gasification, the outlet temperature is increased to T2, and then enter the thermostat flush liquid distillation tube bundle heat, the temperature increased further to design temperature T3; propane in the evaporator endothermic gasification, liquefaction in the condenser heat, so repeated cycles of heat transfer the saturation temperature constant is TPR.

2.2. .Liquid equilibrium model for gas-liquid falling film flow

Assume that a period of water area by water is divided into n independent regions $G_{21}, G_{22} \cdots G_{2i} \cdots G_{2n}$, each region of the gas containing water were $W_{e21}, W_{e22} \cdots W_{e2i} \cdots W_{e2n}$ respectively, as shown in Figure 1. When production is not water pressure reducing, water gas expansion is not into the seal area, the expansion of the volume of material balance were $V_{21}, V_{22} \cdots V_{2i} \cdots V_{2n}$. Test independent areas secure seal area:

$$\sum_{i=1}^{n} \Delta V_i = \sum_{i=1}^{n} G_{2i}\left(B_g - B_{gi}\right) + \sum_{i=1}^{n} W_{e2i} \tag{1}$$

And

$$\Delta V = G_2\left(B_g - B_{gi}\right) + W_{e2} \tag{2}$$

The equation of mass balance without water seal zone is:

$$G_p B_g + W_p B_W = G_1\left(B_g - B_{gi}\right) + W_{e1} + \Delta V \tag{3}$$

As you can see, in different periods of water influx in W_{e1}, W_{e2}, G_1, G_2 are in the air, change, but in the normal production of gas well without strong drainage period, the change tendency of the gas G_1 decreases, increasing water volume G_2, W_{e2} increased continuously, until the water flooded wells.

The mass balance equations of the physical model of the gas-liquid falling film flow in the internal heat integrated distillation column are obtained by combining formula (5) and (6):

$$\begin{aligned} G_p B_g + W_p B_W &= G_1 \left(B_g - B_{gi}\right) \\ &+ W_e + G_2 \left(B_g - B_{gi}\right) \end{aligned} \tag{4}$$

Since the water invasion amount W_e is consistent with the function of water invasion volume factor (ω) and the degree of recovery (R), the conventional method is used to calculate:

$$\ln \omega = B \ln R \tag{5}$$

In the formula: ω is the volume coefficient of water invasion, $\omega = \dfrac{W_e - W_p B_w}{GB_{gi}}$; R is the degree of recovery. By formula (7), it can be obtained:

$$1 - \omega + \frac{G_2}{G_1} = \frac{B_g}{B_{gi}} \left(1 - \frac{G_p}{G_1} + \frac{G_2}{G_1}\right) \tag{6}$$

The state parameters of the refrigerant in and out of each heat exchanger influence each other:

$$\varphi = \frac{1 - \dfrac{G_p}{G_1} + \dfrac{G_2}{G_1}}{1 - \left(\dfrac{G_p}{G_1}\right)^B + \dfrac{G_2}{G_1}} \tag{7}$$

Set, $A = \dfrac{10^8}{G_1}$, $C = G_2$, Then formula (10) becomes:

$$\varphi = \frac{1 - G_p A + CA}{1 - \left(G_p A\right)^B + CA} \tag{8}$$

In the formula, φ is regarded as relative pressure; A is the reciprocal of G_1 without water reserve; B is the water penetration intensity of W_{e1}; and C is the closed reservoir G_2.

3. Automatic fitting method of mass balance

The internally heat integrated distillation column of the three heat exchangers are separated independently establish a one-dimensional model, analyses the essence of automatic fitting algorithm is the parameter identification problem, namely, to seek the best fit with the measured values of the optimal parameters of the theoretical value, the deviation is minimum, it can be expressed as:

IOP Conf. Series: Journal of Physics: Conf. Series **916** (2017) 012050 doi:10.1088/1742-6596/916/1/012050

$$E=\sum_{i-1}^{n}\left[\varphi_{\text{theory}}(A,B,C)-\varphi_{\text{Measured}}\right]^{2} \tag{9}$$

In formula: φ_{real} is used to measure the apparent pressure in the process of production; φ_{theo} is calculated as the theoretical value of (12) formula; E is the objective function

Formula (12) is an onlinear least squares problems. Using the automatic fitting method for fitting, a reasonable set of parameters of the objective function to minimize DFF, this paper uses the Complex method for automatic fitting. Because of two kinds of gas hydrate in the lattice filling performance caused by different initial conditions of hydrate formation change, change the V-H balance, therefore, in the evaporator heat integrated distillation tower, heat exchanger tube immersed in liquid propane, establish the balance system, the flow in the heat exchanger tube, tube outside heat absorption. The gasification of propane in the evaporator heat exchange tube is divided into 10 sections, each section for the following equation and relationship:

$$\phi_{\text{WA-EVAP}(i)}=qm_{\text{WA}}(h_{\text{WA-EVAP}(i)}-h_{\text{WA-EVAP}(i-1)});i=2\text{-}11 \tag{10}$$

$$\phi^{*}_{\text{EVAP}(i)}=k_{\text{EVAP}(i)}A_{\text{EVAP}(i)}\Delta T_{\text{EVAP}(i)};i=2\text{-}11 \tag{11}$$

$$\phi_{\text{WA-EVAP}(i)}=\phi^{*}_{\text{EVAP}(i)};i=2\text{-}11 \tag{12}$$

In the condenser, vapor flow equilibrium hydrate heat absorption tube gasification, gasification steam heat. In Alkyl condensation in the heat exchange tube, the condenser in the heat exchanger tube is divided into 50 sections, each section for the following equation and relationship:

$$\phi_{\text{LNG-COND}(i)}=qm_{\text{LNG}}(h_{\text{LNG-COND}(i)}-h_{\text{LNG-COND}(i-1)});i=2\text{-}51 \tag{13}$$

$$\phi^{*}_{\text{COND}(i)}=k_{\text{COND}(i)}A_{\text{COND}(i)}\Delta T_{\text{COND}(i)};i=2\text{-}51 \tag{14}$$

$$\phi_{\text{LNG-COND}(i)}=\phi^{*}_{\text{COND}(i)};i=2\text{-}51 \tag{15}$$

In the thermostat, water convection heat transfer in the tube, no liquid water, the decomposition temperature strongly depends on the water content of hydrate in gas phase, the thermostat in the heat exchanger tube is divided into 10 sections, each section has the following equation and relationship:

$$\phi_{\text{LNG-THERM}(i)}=qm_{\text{LNG}}(h_{\text{LNG-THERM}(i)}-h_{\text{LNG-THERM}(i-1)});i=2\text{-}11 \tag{16}$$

$$\phi^{*}_{\text{THERM}(i)}=k_{\text{THERM}(i)}A_{\text{THERM}(i)}\Delta T_{\text{THERM}(i)};i=2\text{-}11 \tag{17}$$

$$\phi_{\text{LNG-THERM}(i)}=\phi^{*}_{\text{THERM}(i)};i=2\text{-}11 \tag{18}$$

When calculating each heat exchanger independently, the decomposition temperature of hydrate decreases with the decrease of water content under the given pressure, and the independent calculation results need to satisfy the following formula of total energy conservation:

IOP Conf. Series: Journal of Physics: Conf. Series **916** (2017) 012050 doi:10.1088/1742-6596/916/1/012050

$$\phi_{\mathrm{WA}} = qm_{\mathrm{WA}}(h_{\mathrm{WA1}} - h_{\mathrm{WA3}}) \tag{19}$$

$$\phi_{\mathrm{LNG}} = qm_{\mathrm{LNG}}(h_{\mathrm{LNG3}} - h_{\mathrm{LNG1}}) \tag{20}$$

$$\phi_{\mathrm{WA}} = \phi_{\mathrm{LNG}} \tag{21}$$

$$\begin{aligned}\phi_{\mathrm{WA\text{-}THERM}} &= qm_{\mathrm{WA}}(h_{\mathrm{WA1}} - h_{\mathrm{WA2}}) = \phi_{\mathrm{LNG\text{-}THERM}} \\ &= qm_{\mathrm{LNG}}(h_{\mathrm{LNG3}} - h_{\mathrm{LNG2}})\end{aligned} \tag{22}$$

In the above, ϕ represents the heat exchange obtained by using the energy equation, and ϕ^* represents the heat transfer obtained by the heat transfer equation. qm represents the mass flow of the refrigerant, and h represents the enthalpy of the refrigerant.

The entrance temperature of TWA1 gas-liquid internally heat integrated distillation column is determined by the local climate conditions, the discharge temperature of TWA3 is too low, to meet the environmental requirements, it is also a constant effect on the internally heat integrated distillation column gas-liquid pressure pWA changes on the physical properties of seawater is very small, it can be it is a constant; propane TPR temperature is not too high, otherwise it will contain the shell pressure put forward higher requirements, a one-time increase in the manufacturing cost of IFV. This equation has not closed, need to set a known quantity, the gas-liquid heat integrated distillation column discharge temperature thermostat Twa2 is a known quantity, the reason is the water flow is larger, the computational convergence is very sensitive to the change of Twa2, if Twa2 is unknown, Twa2 will give the initial value in the calculation of the beginning, the assumption of value must be very small difference between the true value and to make the calculation process converges smoothly, otherwise it will easily lead to the divergence of calculation results.

4. Parameter information calculation of gas liquid falling film flow in distillation column

The distribution model of heat transfer coefficient of building internal heat integrated distillation tower, the gas-liquid distillation falling film flow in the saturated vapor pressure of liquid water balance, at a given temperature and pressure conditions, by describing the heat transfer equation and energy equation:

$$h = 90q^{0.67}M^{-0.5}P_r^n\left(-\lg P_r\right)^{-0.55} \tag{23}$$

$$n = 0.12 - 0.2\log_{10} R_p \tag{24}$$

$$Nu = 0.021Re^{0.82}Pr^{0.5}\left(\frac{\rho_w}{\rho_b}\right)^{0.3}\left(\frac{C_p}{C_{pb}}\right)^n \tag{25}$$

By using the heat transfer equation and the balance relation of energy equation, the parameters such as the area, flow and other parameters of the gas-liquid falling film flow are solved by cyclic iteration. The parameters and their setting values are obtained, as shown in Table 1.

Table 1. Parameters and setting of gas liquid falling film flow in distillation column

Parameters	T_{WA1} /K	T_{WA2} /K	T_{WA3} /K	T_{LNG1} /K	T_{LNG3} /K	qm_{LNG} /kg·s^{-1}	T_{PR} /K	p_{LNG} /MPa	p_{WA} /MPa
Setting	243.15	243.15	248.15	134.15	235.15	92	223.15	145.0	0.42

According to the above parameters setting, according to the temperature difference between the heat integrated distillation column, analyzed the falling film flow characteristics of liquid vapor, liquid falling film flow theory makes the optimal parameters to achieve the best fit with the measured values of the heat integrated distillation tower falling film process is described as follows:

1) Temperature TWA1, gas-liquid distillation falling film flow known TWA2, TWA3, LNG, TLNG3 and TLNG1 temperature, flow rate of qmLNG, according to the energy conservation of the entire IFV system can be obtained by gas-liquid distillation falling film flow rate of qmWA, according to the relationship of the thermostat in seawater and LNG energy conservation by temperature TLNG2;

2) As an initial value for the ATHERM area of heat thermostat, gas-liquid distillation falling film flow in assuming the thermostat in the uniform temperature for /2 (TWA2+TWA3), according to temperature thermostat in the process is divided into 10 sections, the heat exchange section, the value of temperature all known;

3) The heat transfer area of the first 9 sections is solved by the equilibrium relation between the heat transfer equation and the energy equation, and the heat exchange area is obtained by subtracting the heat exchange area of the first 9 sections with the initial total area;

4) Using the heat transfer equation and the heat balance equation to judge the heat balance relation, if not satisfied, then the heat exchange area of the thermostat is updated and returned 3) to be recalculated;

5) For the evaporator heat transfer area of initial value of AEVAP, propane saturation temperature is TPR, known, according to the temperature of the sea water, the evaporator heat pipe is divided into 10 segments, for the whole heat transfer section, the temperature of each point is known;

6) The calculation of the heat exchange area of a coherent heater, the heat transfer equation and the equilibrium relation of the energy equation are used to solve the heat transfer area of the evaporator by cyclic iteration;

7) For the condenser heat transfer area of initial value, propane saturation temperature is TPR, known, according to the LNG temperature, the condenser heat pipe is divided into 50 segments, for the whole heat transfer section, the temperature of each point is known;

8) The calculation of the heat exchange area of the coherent heater, the heat transfer equation and the balance equation of energy equation are used to solve the heat transfer area of the evaporator by cyclic iteration.

5. Experimental test analysis

In order to test this method on the internally heat integrated distillation column gas-liquid drop application performance analysis of membrane flow characteristics in the experimental analysis, Figure 2 shows the evaporator temperature, liquid propane and distillation temperature (TWA, TPR and T) along the curve in the horizontal coordinate diagram. FG is a dimensionless position, the corresponding Figure 2 from left to right, the evaporator heat exchange tube along the position; for rectifying the liquid from the condenser entrance to the exit of the heat exchanger tube along the position of propane in the housing. The uniform temperature is constant, so it is a straight line.

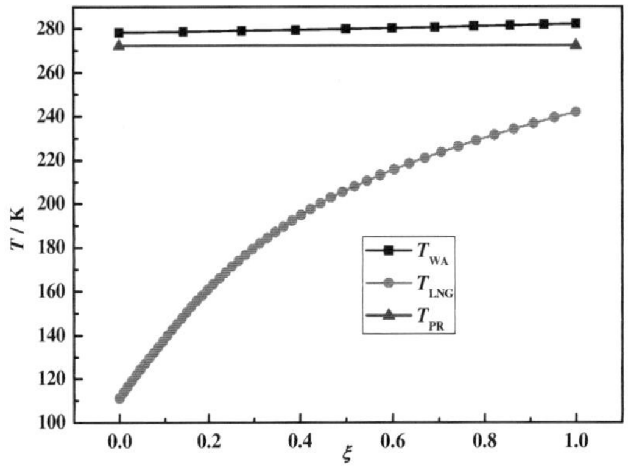

Fig. 2 Temperature variation curve of internal heat integrated distillation column

It can be seen that the distillation liquid in the evaporator temperature gradually decreased from right to left, due to the design requirements, the overall temperature of the distillation liquid drop is small, resulting in distillation liquid flow, liquid falling film distillation the same temperature in the evaporator is also very small decline; propane saturation temperature is close to the set value, the basic rectification the liquid in the evaporator temperature, higher set propane saturation temperature, IFV is conducive to the safe operation and can reduce the manufacturing cost of disposable liquid; distillation temperature gradually increased from left to right, in the entrance section of the temperature rise of a steep slope, which is due to the heat transfer in the entrance section of large temperature difference, heat is also high. With the heat transfer temperature decreases, the slope decreases gradually. The temperature rises with liquid distillation.

Figure 3 shows the temperature distribution of heat transfer temperature difference between inside and outside diameter, the abscissa β is a dimensionless position for the diameter of the heat transfer temperature, the corresponding Figure 3 from left to right thermostat tube along the position. It shows the distillation temperature of the liquid from the entrance to the exit channel along the position.

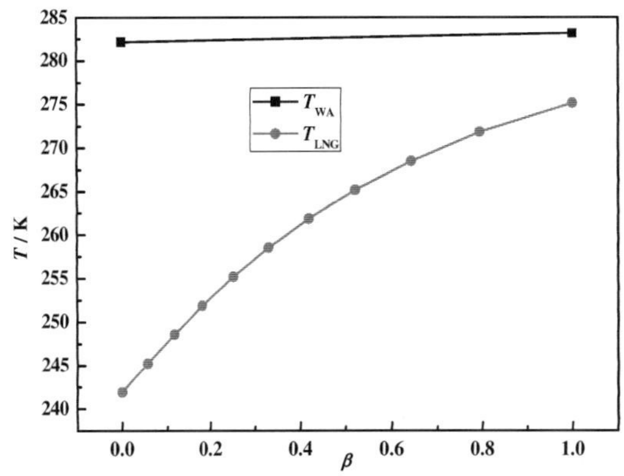

Fig. 3 Temperature distribution of refrigerant in thermostat

Analysis shows that the geometry of gas-liquid internally heat integrated distillation column falling film flow heat exchange area and import column thermostat, the average temperature has a significant positive correlation between the entrance section of heat pipe temperature difference between inside and outside, the heat flux is larger, with the increase of internal heat integrated distillation column temperature, the slope decreases gradually rise the temperature map internally heat integrated distillation column corresponding to the temperature; the subscript WA said water, PR propane, EVAP said COND said evaporator, condenser, THERM thermostat, heat exchange efficiency of the evaporator, condenser temperature regulator and a low of two appropriate heat transfer enhancement measures. The heat transfer efficiency of IFV. To enhance the overall heat transfer coefficient of the evaporator increases from left to right, mainly because from right to left, the water temperature gradually decreased, resulting in inside and outside the tube heat transfer coefficient decreased, and the heat transfer in the evaporator The flow of heat transfer inside and outside of the pipe with temperature variation is consistent, asymptotically linear curve of heat transfer of the evaporator, an increase of about 1000 $W{\cdot}m^2{\cdot}K^1$; the heat transfer coefficient is increased gradually from left to right, the reason is that the LNG temperature increased gradually, leading to the surface of the heat exchanger tube temperature increases, the flow of heat transfer by condensation heat exchanger the heat pipe and the pipe heat transfer coefficient increased, while the heat transfer coefficient is also increased; the heat transfer coefficient of the thermostat from right to left in the entrance section gradually reduced, reducing slope, after the flat, the main reason is increasing with the internal heat integration of distillation temperature, exceed the critical temperature natural gas, heat capacity at constant pressure decreases with the increase of temperature, the thermostat heat transfer coefficient is mainly affected by the internal heat integrated distillation column pressure influence the change of heat capacity with temperature, changes appear as shown in Figure 4.

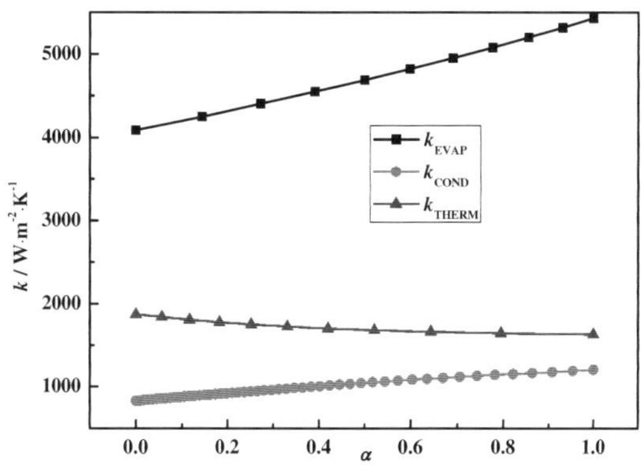

Fig. 4 Gas liquid falling film flow in internal heat integrated distillation column

It is similar to Figure 2, from right to left - internally heat integrated distillation falling film flow temperature drop is very small, so the geometric calculation process of gas-liquid heat exchange area in thermostat falling film import and the average temperature as the qualitative temperature influence on the result of calculation is very small, low temperature liquid falling film the temperature gradually increased from left to right, because the entrance section of heat exchange tube temperature difference between inside and outside, the corresponding heat flux is larger, with the increase of temperature liquid falling film, the slope decreases its temperature rise. The research results show that the geometry of gas-liquid internally heat integrated distillation column falling film flow heat exchange area and import column thermostat the average temperature has a significant positive correlation between the entrance section of heat pipe temperature difference between inside and outside, the heat flux is larger, with the increase of internal heat integrated distillation column temperature, the slope decreases its temperature rise, which accurately describes the internal heat integrated distillation. The characteristics of the gas liquid falling film flow in the tower are improved, and the heat transfer enhancement measures should be taken to promote the overall heat transfer efficiency.

6. Conclusion

This paper studies the internal gas-liquid heat integrated distillation tower falling film flow phenomenon, according to the distribution of gas-liquid mass balance internally heat integrated distillation column independent region, distribution model of heat transfer coefficient of building internal heat integrated distillation tower, the gas-liquid distillation falling film flow in saturated vapor pressure of liquid level scale, using heat transfer equation and energy equation the balance between the cycle of iterative solution of gas-liquid falling film flow area, flow parameter information, at a given temperature, pressure conditions, gas-liquid flow falling film theory makes the optimal parameters to achieve the best fitting values with the measured values. The results show that the geometric gas-liquid internally heat integrated distillation column falling film flow heat exchange area and import distillation column thermostat, the average temperature has significant positive correlation, can accurately describe the gas-liquid heat integrated distillation falling film flow regularity, take appropriate Strengthening heat transfer measures to promote the overall heat transfer efficiency, thereby reducing the pollution emissions from distillation tower, and improve product quality.

Acknowledgments
Fund Project: Langfang Teachers University science research project: "Investigation of gas-liquid falling film flow phenomena in internally heat-integrated distillation Colum"
(NO. LSLB201402)

References
[1] Abdukader Ablimit, Zhang Yonghong, Zhang Zengpeng, Liu Chenjiang. Recent Advances of the Access to Indolinones via C—H Bond Functionalization/Cyclization Cascade Strategy. Chin. J. Org. Chem., 2016, 36(5): 875-888.
[2] WENG Xiaomin, HU Haitao, LAI Zhancheng, etal. Analysis of influence of pore density on heat transfer and pressure drop characteristics of wet air in metal foams. CIESC Journal, 2016, 67(6): 2218-2223.
[3] Wang Liang, Li Zhana, Wan Kang, et al Progress in Diazo Compounds Mediated Transition-Metal-Catalyzed C—H Alkylation. Chin. J. Org. Chem., 2016, 36(5): 889-912.
[4] ZHANG Jili, CHEN Jingdong, MA Zhixian, WANG Yonghui. Effect of evaporation temperature on boiling heat transfer in horizontal ribbed and embossing finned tube pool. CIESC Journal, 2016, 67(6): 2230-2238.
[5] Tedesco Triccas L,Burridge JH,Hughes AM,et al.Multiple\ssessions of transcranial direct current stimulation and upper\sextremity rehabilitation in stroke:A review and meta-analysis[J].Clin Neurophysiol, doi:10.1016/j.clinph.2015.04.067.
[6] Han CE,Kim S,Chen S,et al.Quantifying arm nonuse in individuals poststroke [J].Neurorehab Neural Re,2013,27(5):439-447. Taub E,Uswatte G,Bowman MH,et al.Constraint-induced\smovement therapy combined with conventional neurorehabilitation techniques in chronic stroke patients with plegic hands:a case series [J].Arch Phys Med Rehab,2013,94 (1):86.

Thermal distortion compensation of a high precision umbrella antenna

Pengpeng Wang *, **Feng Wang, Tian Shi and Bo Wang**

Xi'an Institute of Space Ratio Technology, Xi'an, China

*Corresponding author e-mail: huohuoxjtu@163.com

Abstract. Ribs of umbrella antennas act as cantilever beams, which make it difficult for the holding of reflector shape, and especially in thermal environment, thermo-elastic distortion of the ribs would largely degenerate the system performance. Combined with development of a high precision umbrella antenna, thermal distortion compensation under in-orbit temperature field is conducted. Distortion compensation of a single rib model is carried out firstly by temperature field optimization of the fixed end, and distortion compensation of the whole antenna is accomplished by genetic algorithm. Simulation results show very good compensation effect on reflector thermal distortion, while root mean square of the reflector thermal distortion reduces from 0.37mm to 0.27mm.

1. Introduction

Umbrella Antennas also named radial rib antennas are widely used in tracking and data rely satellites and deep space exploration spacecraft's, such as the TDRS, the Galileo Jupiter Probe of the USA and the Louche satellite of Russia [1, 2]. Main reflectors of umbrella antennas are composed of deployable composite ribs and metal mesh. While design frequency is low as in S and UHF band, the antenna could perform well without much mesh adjustment work, however, as the design frequency comes to Ku band or above, distortion of the reflector is no longer tolerable, and tensile cable networks are introduced in the mesh reflector. By tension of the front and back cables, connect points of the mesh reflector are pulled to the ideal reflector surface. Compared with other types of deployable antennas, umbrella antennas are simple in structure configuration, reliable in deployment, and are very suitable for center feed layout. However, as ribs of umbrella antennas act as cantilever beams, which make it difficult for the holding of reflector shape, and especially in thermal environment, thermo-elastic distortion of the ribs would largely degenerate the system performance.

To control thermal distortion of the reflectors, there are two commonly used methods. The first one is to limit temperature range of the antenna by thermal controls such as SLI (single layer insulation), MLI (multilayer insulation) and heaters. The second one is by shape control strategies in orbit, which modifies shapes of the reflector by piezoelectric ceramics, shape memory alloys or other smart actuators. Active shape controls have more direct control effects, but measurement of reflector distortion in outer space, actuators suitable for aerospace environment and high efficient control algorithm design are blocks waiting to be removed. So far, active shape controls are mainly conducted at simulation or laboratory level, no applications in aerospace have reported yet [3-6].

Combined with development of a high precision umbrella antenna, thermal distortion compensation under in-orbit temperature field is conducted. Distortion compensation of a single rib model is carried out firstly by temperature field optimization of the fixed end, and distortion compensation of the whole antenna is accomplished by genetic algorithm. Simulation results show very good compensation effect on reflector thermal distortion, while root mean square of the reflector thermal distortion reduces from 0.37mm to 0.27mm.

2. Thermal distortion of high precision umbrella antenna

As the antenna be launched into orbit with satellite and deployed, the most important influence factor of outer environment on antenna thermal distribution is the solar irradiation. As a result of earth self-rotation, temperature of the antenna will change from high to low daily. Besides, as the solar incident angle changes periodically, there are three typical solar days, the first one is summer solstice, the date the sun radiates vertically at the Tropic of Cancer. The second one is winter solstice, the date the sun radiates vertically at the Tropic of Capricorn. And the last one is Autumnal Equinox or Spring Equinox, the date the sun radiates vertically at the equator. Also considering degeneration of thermal control materials, there are six typical days for antenna thermal analysis, these are BOLSS, BOLWS, BOLEQ, EOLSS, EOLWS and EOLEQ, in which BOL and EOL refer to begin of life and end of life respectively.

For thermal analysis of the umbrella antenna, 24 simulation steps are set each day with an interval of an hour. Figure 1 shows temperature distribution of the reflector main structure at EOLSS twenty four o'clock satellite time, at which the satellite is in shadow of the earth, and temperature of the reflector reaches its minimum, while temperature of the tensile cables come to -157℃. Figure 2 is the corresponding reflector distortion map, as the tensile cables shrink at low temperature, the reflector deforms as a trend of close up, and far end deformation of the ribs are nearly 1mm.

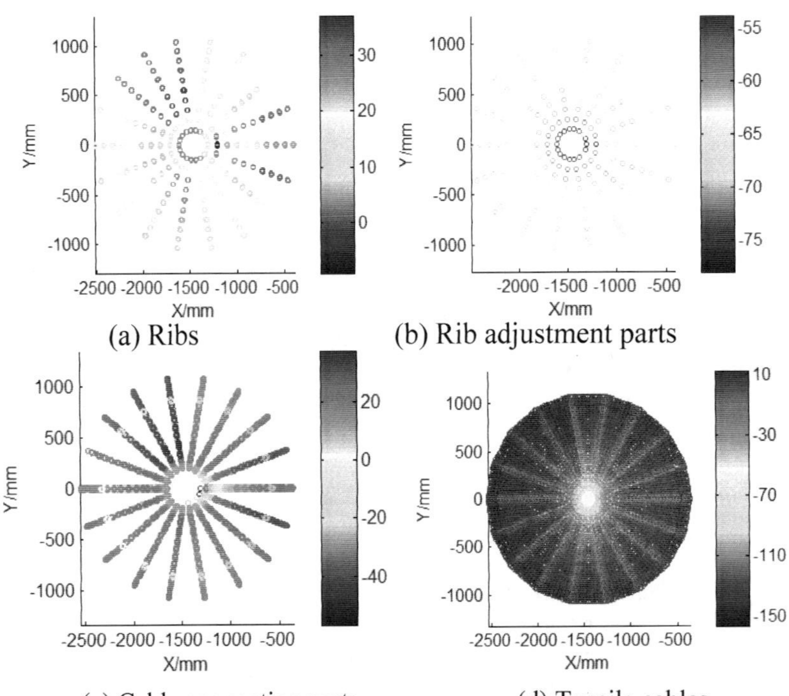

Figure 1. Thermal map of the reflector

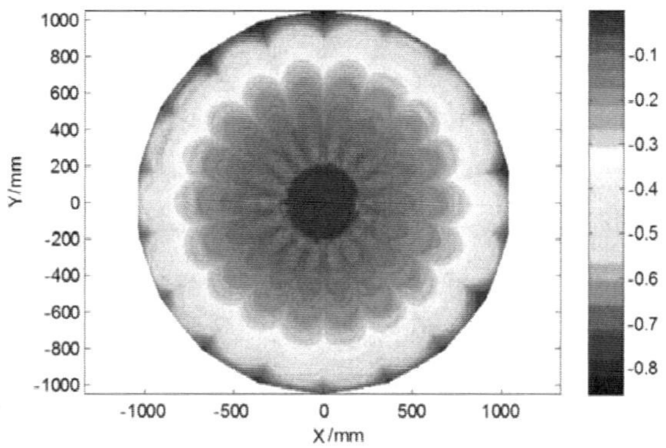

Figure 2. Thermal distortion map of the reflector

Thermal distortion of the reflector expressed in RMS is summarized in Figure 3, As a result of thermal control degeneration, thermal distortion RMS in EOL cases are generally greater than that of BOL cases, and the maximum thermal distortion RMS of the reflector is 0.37mm. Thermal distortion of the reflector has already exceeded the thermal distortion budget, and further thermal distortion optimization has to be done to meet the thermal distortion requirement.

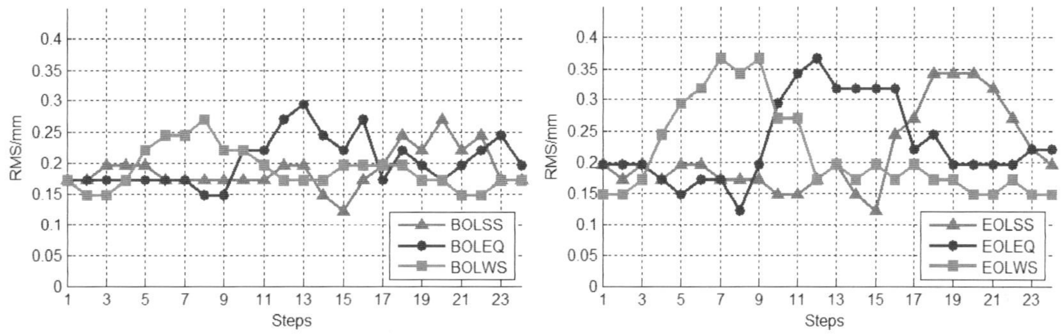

(a) Beginning of Life (b) End of Life

Figure 3. Thermal distortion RMS curve of the reflector

3. Thermal distortion compensation of a single rib model

As shown in Figure 4a, temperature distribution is the same at the front and back side of the rib under uniform temperature field, and the inner force induced by thermal load can be given as:

$$F_1=EA\alpha\Delta T \quad F_2=EA\alpha\Delta T \quad M=0 \tag{1}$$

Where F1 and F2 are axial tension or compression forces at the front and back side of the rib, EA is the tensile stiffness coefficient of the rib, α and ΔT refer to thermal expansion coefficient and temperature rise of the rib respectively, M is the bending moment induced, but as axial tensile forces are the same at the front and back side of the rib, M should be zero.

If by external temperature rising, temperature at the front and back side of the rib is modified to be different, the inner force induced by thermal load can be given again as:

$$F1=EA\alpha\Delta T1 \quad F2=EA\alpha\Delta T2 \quad M=Ea.\alpha\,(\Delta T1-\Delta T2)\,h \tag{2}$$

Where h is distance between the front and back side of the rib to bending neutral surface. As axial tensile forces at the front and back side of the rib are no longer the same, the bending moment M will result in rotation of the rib, as shown in Figure 4b. So by changing temperature at the front and back side of the rib, thermal distortion of the rib could be compensated.

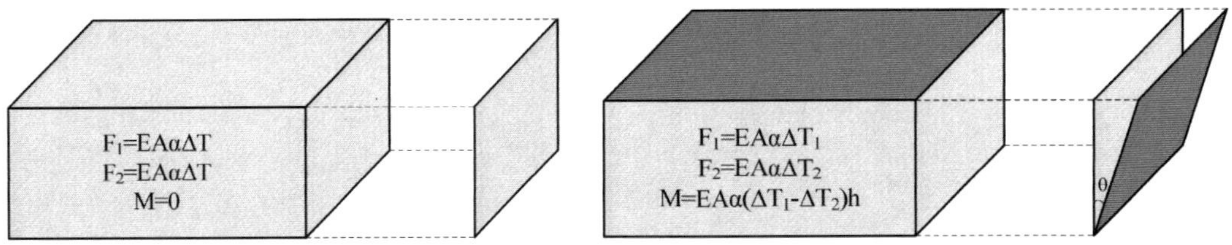

(a) Uniform temperature (b) Non-uniform temperature

Figure 4. Sketch of rib deformation under thermal load

Thermal distortion compensation of a single rib model under the uniform temperature filed of +60℃ is conducted, and the simulation results are shown in Figure 5. In Figure 5a, without thermal distortion compensation, the maximum deformation at far end of the rib is 0.12mm. However, as shown in Figure 5b, after thermal distortion compensation, the maximum deformation at far end of the rib reduces to 0.01mm while temperature at the front side of the rib rises +20℃.

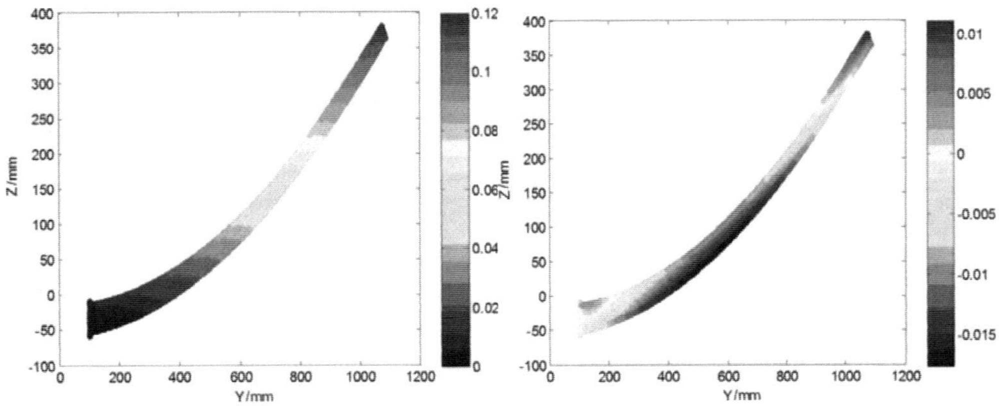

(a) Without compensation (b) With compensation

Figure 5. Thermal distortion compensation of a single rib model

Figure 6 shows relationship curve between far end thermal deformation and the temperature rise, the deformation and temperature rise show very well linear relationship. Far end of the rib could deform 0.0054mm while the temperature raises 1℃. And for further application, Calibration of the deformation and temperature rise should be done firstly by test to get the exact mechanical-thermal coupling coefficient.

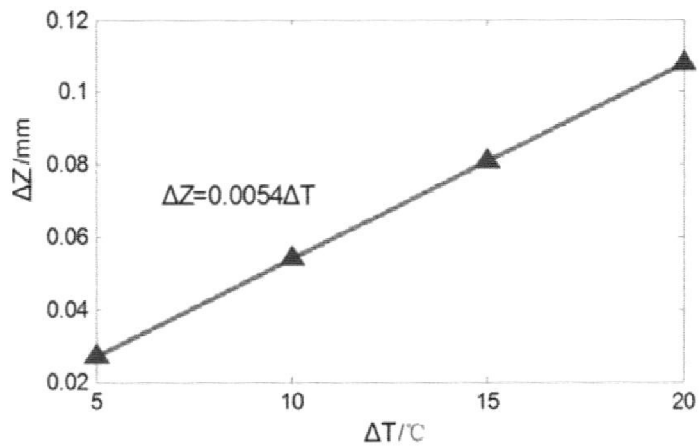

Figure 6. Relationship curve between far end thermal deformation and temperature rise

4. Thermal distortion compensation of the antenna

Sketch of a rib mounting on the antenna is shown in Figure 7, the high precision umbrella antenna consists of 18 composite material ribs, after be launched into orbit, drive motor will operate and drive the deploy arm to rotate with the rib until the antenna is fully deployed, then the deploy arm will be locked by the lock unit. The deploy arm, the rib and the lock unit could be simplified to be a one-dimension rotation system, length changes of the lock unit will induce rigid rotation of the deploy arm about its rotate axis, and result in further close up or close down of the reflector. As the rib is relatively long, small rotation angle at the near end would result in large deformation at the far end. By former simulation and test, the amplification factor is 35.7, which means that if length of the lock unit changes 1μm, far end displacement of the rib would be 35.7μm.

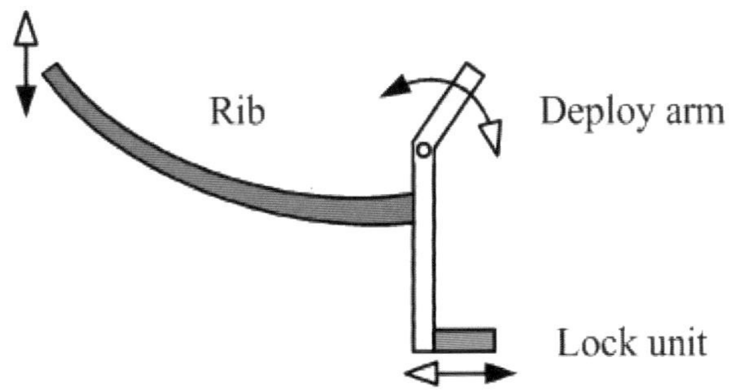

Figure 7. Sketch of a rib mounting on the antenna

Displacement amplification of the rib increases difficulties to design the lock unit, which requires it to be dimension stable. However, it also supports a new approach for thermal distortion compensation of the reflector, and by exact length control of the lock unit, a higher level displacement control could be achieve at the antenna level.

The focal point of the reflector thermal distortion compensation in orbit is how to find the corresponding temperature rises of the 18 lock units. As a commonly used intelligent algorithm, the genetic algorithm could find a relatively best solution in a short time, and is very suitable for multiple solution problems just as the thermal distortion compensation. By binary encoding, each chromosome includes 18 genes, and each gene corresponds to a lock unit with the digits of 7, which mean that the temperature rise could change from 0℃ to 127℃.

By predefinition of the design space, the reflector thermal distortion compensation problem could be simplified as a non-constraint optimization as below [7,8]:

$$\text{find Min}\{\text{RMS}\} \tag{3}$$

The equilibrium equation of thermal distortion analysis is:

$$\mathbf{KX} = \mathbf{F_T} \tag{4}$$

While K refers to the global stiffness matrix, FT is the node equivalent thermal load matrix, and X is the node displacement matrix.

Thermal distortion of the reflector can be finished by the following steps:

Finite element modeling of the antenna to obtain the global stiffness matrix K;

Apply a uniform temperature field T0 on the antenna finite element model and get the node displacement matrix XT0 by linear static analysis;

The node equivalent thermal load matrix FT0 can be determined by $\mathbf{F_{T0}} = \mathbf{KX_{T0}}$, and the corresponding node equivalent thermal load vector FT0k (k=1~18) of the kth lock unit can be separated from FT0 according to the node id, so mechanical-thermal coupling coefficient Dk of the kth lock unit can be defined as:

$$\mathbf{D_k} = \frac{\mathbf{F_{T0k}}}{\mathrm{T_0} - \mathrm{T_{ref}}} \tag{5}$$

Suppose the thermal case needs to be distortion compensated is Ts, and by the same process of step 2 and 3, the corresponding node displacement matrix XTs and node equivalent thermal load matrix FTs can be obtained, and thermal distortion RMS of the reflector without compensation is:

$$\mathbf{RMS_{Ts}} = \sqrt{\frac{\sum_{m}^{n} \mathbf{X}_{\mathbf{Tsk}}^{2}}{n - m + 1}} \tag{6}$$

Where number of the reflector nodes is n-m+1.

By genetic algorithm, set temperature rises of the 18 lock units to be design variables Tn={T1,T2,...,T18}, then the added node equivalent thermal load matrix of the kth lock unit result from temperature rise Tn is:

$$\mathbf{P_{Tnk}} = \mathbf{D_k}(\mathrm{T_n} - \mathrm{T_{ref}}) = \frac{\mathrm{T_n} - \mathrm{T_{ref}}}{\mathrm{T_0} - \mathrm{T_{ref}}} \mathbf{F_{T0k}} \tag{7}$$

The overall added node equivalent thermal load matrix is:

$$\mathbf{P_{Tn}} = \{\mathbf{P_{Tn1}}, \mathbf{P_{Tn2}}, \cdots, \mathbf{P_{Tn18}}\} \tag{8}$$

Then the thermal-mechanical equilibrium equation with distortion compensation is:

$$\mathbf{KX_{Ts}} = \mathbf{F_{Ts}} + \mathbf{P_{Tn}} \tag{9}$$

The node displacement matrix XTs and the corresponding thermal distortion RMSTs can be updated by equation 6.

Repeat steps 5 N times for genetic iteration to get a relatively minimum of RMS.

During genetic algorithm iteration, population at each generation is set to be 40, the number of generation is 30, probability of crossover is 0.8, and the probability of migration is 0.005. Relationship curve between thermal distortion RMS of the reflector and generations is shown in Figure 8, thermal distortion RMS of the reflector reduces as the generation increases, and thermal distortion RMS reaches its minimum at 0.27mm after 30 generations, the thermal distortion RMS has reduced about 27% compared with the former result. And the final temperature rises of the 18 lock units are summarized in Table 1.

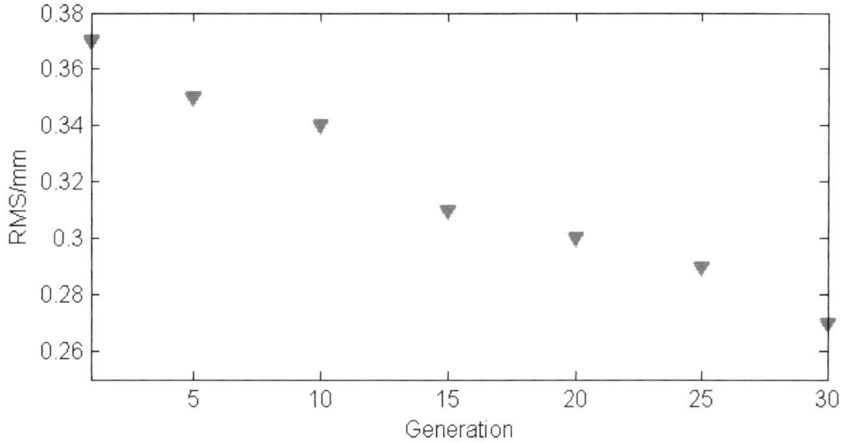

Figure 8. Relationship curve between thermal distortion RMS and generation

Table 1. Final temperature rises of 18 lock units

No	Temperature rise /℃	No	Temperature rise /℃	No	Temperature rise /℃
1	88	7	117	13	31
2	82	8	101	14	87
3	48	9	113	15	24
4	73	10	43	16	18
5	65	11	45	17	103
6	117	12	61	18	109

5. Conclusion

Combined with development of a high precision umbrella antenna, thermal distortion compensation under in-orbit temperature field is conducted. Based on thermal compensation simulation of a single rib model, thermal distortion compensation of the whole antenna is carried out. The proposed thermal distortion compensation method has the following merits: Firstly, thermal control method is mature in aerospace engineering, thin-film heaters and sheathed heaters are widely used before and could supply good source of thermal energy. Secondly, the control algorithm is robustness in nature, and could find a relatively best solution in a short time. And finally, local control of the rib and lock unit could realize the exact and rough control of reflector respectively.

However, there are still several aspects need to be further researched before the thermal compensation method to be used in aerospace environment, such as: The thermal compensation depends on thermal expansion of the lock units, and temperature rises mean that the compensation can only makes the reflector to be close up. Thermal isolation of the thermal expansion units from surrounding units and reflector thermal distortion measurement in space environment still need further investigation and validation.

References

[1] R Blyth, D Haldeman, TDRSS multiple access telecommunications service, The 8th Communications Satellite Systems Conference, Orlando, Florida, USA, 527 (1980) 317-327.
[2] G Tsuyuki, R Reeve, A summary of thermal analysis support to the Galileo high gain antenna

IOP Conf. Series: Journal of Physics: Conf. Series **916** (2017) 012051 doi:10.1088/1742-6596/916/1/012051

deployment anomaly recovery effort, The 6th AIAA/ASME Joint Thermophysics and Heat Transfer Conference, Springs, Colorado, USA, 2003 (1994) 1-11.
[3] S Matunaga, J Onoda, Actuator placement with failure consideration for static shape control of truss structures, AIAA Journal. 33 (1995) 1161-1163.
[4] E Kiely, G Washington, J Bernhard, Design and development of smart microstrip patch antennas, Smart Materials and Structures. 7 (1998) 792-800.
[5] HS Yoon, G Washington, An optimal method of shape control for deformable structures with an application to a mechanically reconfigurable reflector antenna, Smart Materials and Structures. 19 (2010) 1-9.
[6] H Tanaka, MC Natori, Shape control of space antennas consisting of cable networks, Acta Astronautica. 55 (2004) 519-527.
[7] X Wang, W Zheng, Y Hu, Active flatness control of membrane structures using fuzzy logic integrated genetic algorithm, The 48th AIAA/ASME/ASCE/AHS/ASC Structures, Structural Dynamics, and Materials Conference, Honolulu, Hawaii, USA, 1844 (2007) 1-11.
[8] DK Anthony, AJ Keane, Robust-optimal design of a lightweight space structure using a genetic algorithm, AIAA Journal, 41 (2003) 1601-1604.

IOP Conf. Series: Journal of Physics: Conf. Series **916** (2017) 012052 doi:10.1088/1742-6596/916/1/012052

Measurement of Tank Cooling Airflow Based on Array Sensors

Hui Zhou[1, a,*], Yan Han[1, b], Jianguo Wang[2, c] and Pizhuang Zhang[1, d]

[1]School of Information and Communication Engineering, North University of China, Taiyuan, 030051, China
[2]North Automation Control Technology Institute, Taiyuan, 030006, China

[a]*Corresponding author: ncitnacy@nuc.edu.cn
[b]hanyan@nuc.edu.cn
[c]15935155270@163.com
[d]zhpzh@163.com

Abstract. Researching on the cooling airflow characters of tank will be helpful for optimizing the design of cooling system, and will be of great importance to improve the performance of armoured vehicles. According to the test requirements of tank under the actual working conditions, we studied and designed the proposed cooling airflow measurement system. The most important thing was that we finished the assembly of array sensors without any damage to tank. In addition, according to national standard we set the quantity and locations of the sensors, which was on the premise of avoiding the influence to airway. In this paper, there was description of cooling airflow signal processing; and there was analysis of air pressure distribution which was presented simultaneously by three-dimensional surface graph and plane graph at the exhaust port of tank engine compartment according to fitting algorithm.

Key words: Tank Cooling Airflow; Array Sensors; Assembly; Signal Processing; Fitting Algorithm; Air Pressure Distribution.

1. Forewords

In modern warfare armored combat systems are required with greater mobility, for which performance of tank engine compartment is the dominant factor [1]. Researching on cooling airflow characters will be helpful for optimizing the layout of engine compartment and the design of cooling system, and for reducing thermal load of engine compartment. The researches will be of great importance to improve the performance of armored vehicles [2]. Therefore, there is an urgent need for analysis methods of tank engine cooling airflow performance used in engineering applications.

Measurement system of cooling airflow characters in armored vehicle power compartment was designed in [2]. The system was used to measure the air velocity and flow at the cooling air inlet with different speeds of the cooling fan.

In addition, numerical simulation has been developed rapidly as an effective method for airflow and heat transfer researching in recent years [3~5], however, its application in research on tank airflow and heat transfer is not mature enough, it is necessary to solve a series of theoretical and technical problems. The study of airflow and heat transfer in engine compartment must be subject to its own characteristics of tank and be special treated [1].

Different from numerical simulation, we have carried out the measurement of pressure parameters under actual vehicle conditions. In [2], there was measurement of parameters at the air inlet, different from which, we measured the parameters of cooling air at the exhaust side of the engine compartment for further calculation of heat taken away from the engine compartment by cooling air and heat dissipation efficiency of engine compartment.

2. The composition of array sensors measurement system

According to the test requirements of tank under the actual working conditions, which were poor conditions like wide temperature range, electromagnetic interference, and inconvenient assembly for tank measurement in dynamic process, we studied a feasible and reliable test method for tank parameters and has carried out studies such as the design of the cooling airflow acquisition system, the arrangement and assembly of array sensors. Besides we made certain emergency measures and spares which improved the feasibility and reliability of the long time and effective outfield tests in power supply, circuit connections, signal acquisition and storage.

2.1. The Overall Structure of System

Cooling airflow measurement system based on array sensors was mainly composed of a host computer, array sensors, air pressure conditioning circuits, data acquisition and storage system, power supply system, conversion connection system. The overall structure of the system was shown in Fig.1.

2.2. Components of System

We used special software in the host computer to configure sampling frequency, sampling time, sampling volume, sampling channels and other relevant parameters of data acquisition and storage system before the measurement; to preserve and segment data after the measurement for later data processing; or to format.

The power supply system, data acquisition and storage system, conversion connection system, these three parts were fixed in a sealed box made of iron which was used for anti-electromagnetic interference. We used shielded multi-core cable for configuration of host computer to the data acquisition and storage system, for power supply to sensors and air pressure conditioning circuits, and for signal transmission of air pressure conditioning circuits and data acquisition and storage system.

A 12V / 60AH battery was used for power supply system, and there was a spare one.

64-channel storage technology was used for multi-channel data acquisition and storage system which was specially developed for the measurement system, and was very small in volume to save more space for assembly.

The specially developed circuits in conversion connection system were used to change electrical parameters provided by the battery to electrical parameters which sensors, chips and circuits needed.

Air pressure conditioning circuits were fixed inside a specially made aluminum box which was sealed for anti-electromagnetic interference and accurate measurement. The signals of micro-pressure sensors were transmitted to air pressure conditioning circuits, then the signals of circuits were transmitted to the data acquisition and storage system, so that noise from environment and circuits could be reduced.

It was difficult for assembly of the measurement system as any damage to the tank body should not be allowed; of course drilling a hole was not allowed. In this case, we designed mounting fixture specifically to conform to the idle screws of tank; the sealed box made of iron was designed to be the same size as the original spare box of tank, and then we made the replacement; at the same time, some reasonable anti-vibration measures were added for protection.

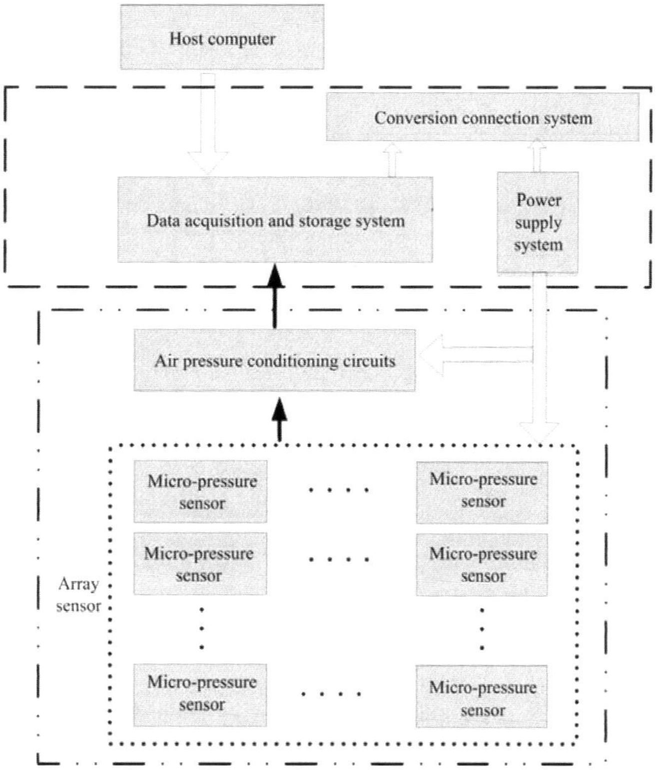

Figure 1. Overall structure of the system

3. Design of array sensors

Under the dynamic conditions of tank, we eliminated measurement of air hood and adopted the distributed measurement. We selected multiple measuring points of the rectangular cross section at the cooling air exhaust side of tank engine compartment, and then air pressures in different regions were tested at the same time. After the air parameters acquisition by multi-sensor, data were studied with algorithms for tank cooling airflow.

Since the airflow conditions of the inlet and outlet would not be obstructed, it was very important to set the quantity and locations of sensors, and the assembly and fixation were also important.

According to [7], the number of lines (parallel to the small side) and the number of measuring points on each line are both not less than 5 in the case of rectangular section; if the aspect ratio of rectangular section differs greatly from 1, there should be more than 5 lines.

It was important to note that more sensors would inevitably influence the cooling air duct, fewer sensors leaded to limited air information of the exhaust side for fewer samples.

Taking all the situations above into account, we finally set six lines, four measuring points each line. The sensor positions were arranged according to" Log-Tchebycheff". The sketch map of array sensors assembly was shown in Fig.2.

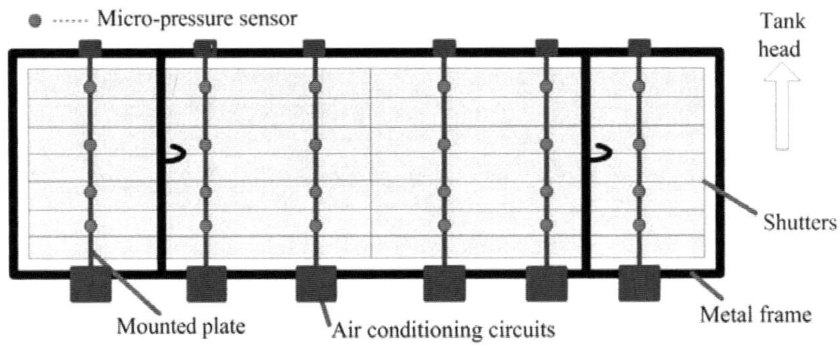

Figure 2. The sketch map of sensor assembly at the cooling air exhaust side of engine compartment

Sensors could not be installed directly at the cooling air exhaust of engine compartment. We have studied a variety of schemes, finally, designed a frame according to the vehicle size as the bracket for array sensors. The frame, made of metal and with the shape of glyph in Chinese letter" Purpose", was embedded at the outer edge of exhaust hood; two small bridges in the middle of the frame were firmly hooked to the shutters respectively by two hook-type metal firmware, in this way the metal frame was fixed at the exhaust side of engine compartment. There were six special mounted plates arranged on the metal frame. Four micro-pressure sensors were fixed on each mounted plate, and their corresponding air conditioning circuits were fixed into an aluminium sealed box at the end of each mounted plate for anti-electromagnetic interference and less measurement error. Then all the 24 micro-pressure sensors with the range of ± 6kPa were fixed to the position of the tank exhaust port shutter and every sensor inlet was just vertical to the shutter flat.

4. Data processing and analysis

4.1. Signal Processing of Tank Cooling Airflow

During the process of cooling airflow signals which we collected in the tank tests, basic low-pass filter was used as method one, however, energy of the filtered signal was much smaller than that of the original one, which means there was energy loss. We thought out a new signal processing method as method two, which was named as "residual energy method" [6].

The original signal, the processed results with two methods were placed together in one figure of the same scale for comparison, as shown in Fig.3. Some details of the signal were restored by method two, and there was a much better correspondence of the air pressure alteration and the cooling fan alteration, and of course there were better restored signals. The rationality and validity of the very signal processing method were verified [6].

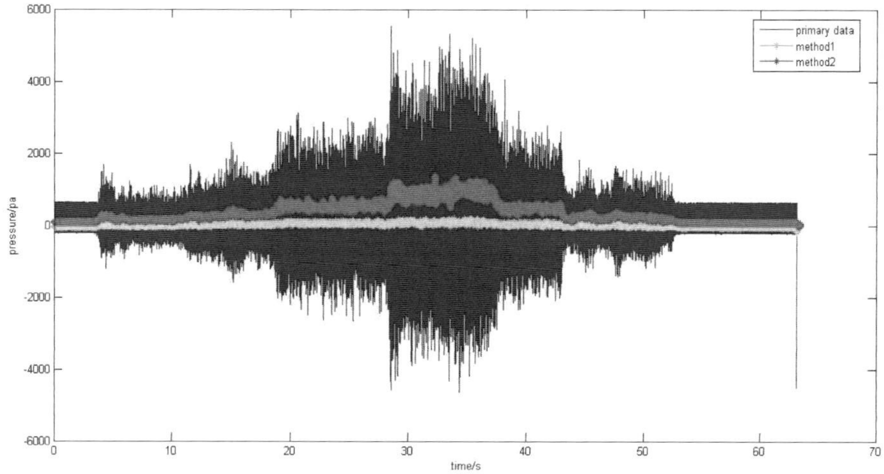

Figure 3. Comparison of the original signal and the results of the two methods

4.2. The Fitting Analysis of the Measurement System

We have made many groups of tests for accurate, effective characters of tank cooling airflow. In this paper, several randomly selected sets of data were analyzed which were the data of tank with three different revolving speeds (speed 1 <speed 2 <speed 3) in time1 and time 2. Here time 1 was 1s in data acquisition time and time 2 was 2s. The 24 sets of signals of array sensors were processed by "residual energy method", and then the processed data were linearly fitted according to the sensor positions and rectangular cross section of exhaust hood of engine compartment. The origin of coordinate system (0,0) for fitting was the left rear corner of tank, referring to Fig.2, X value increased to the right side of the vehicle and Y value increased to the front direction. The fitting results were presented simultaneously by three-dimensional surface graph and plane graph according to fitting algorithm, shown from Fig.4 to Fig.9.

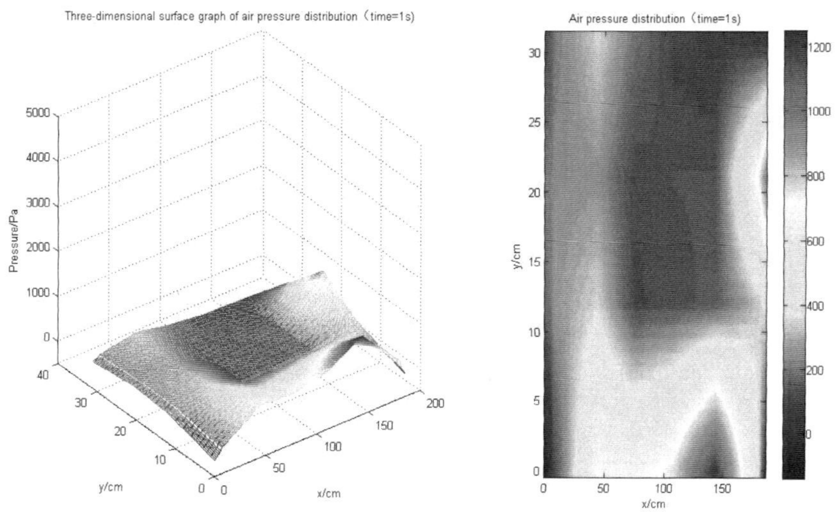

Figure 4. Air pressure distribution of speed 1 in time 1(1s)

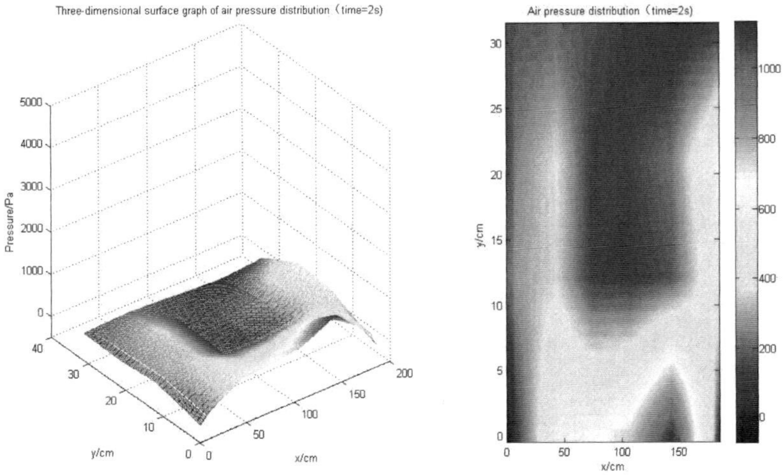

Figure 5. Air pressure distribution of speed 1 in time 2(2s)

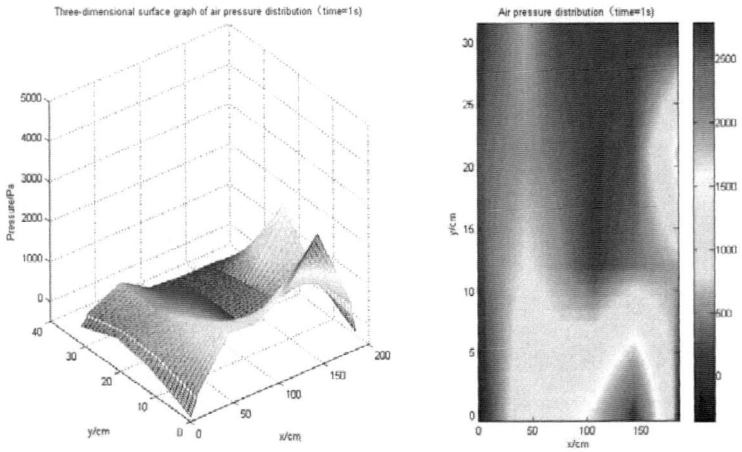

Figure 6. Air pressure distribution of speed 2 in time 1(1s)

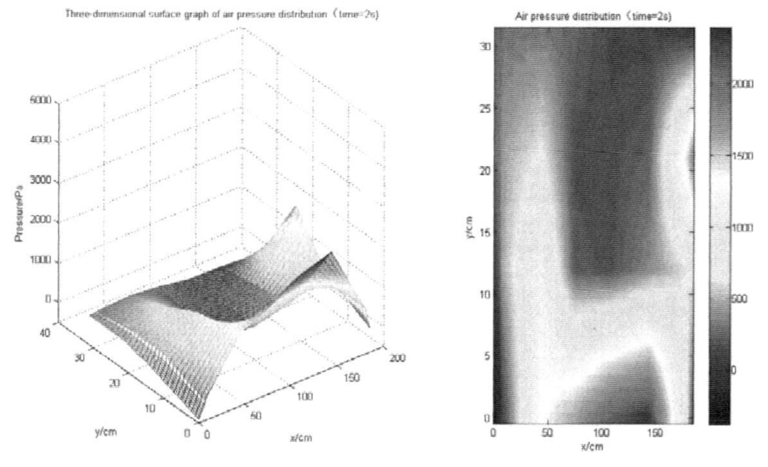

Figure 7. Air pressure distribution of speed 2 in time 2(2s)

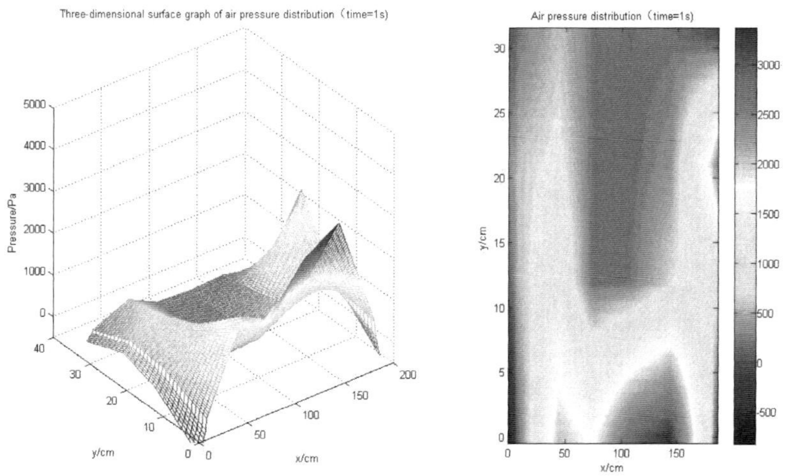

Figure 8. Air pressure distribution of speed 3 in time 1(1s)

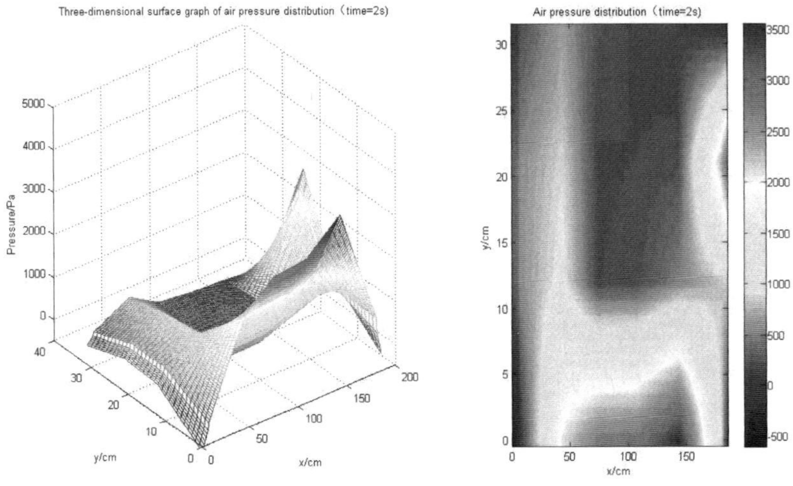

Figure 9. Air pressure distribution of speed 3 in time 2(2s)

We could see from the results above that pressure values were roughly the same in one speed even at different acquisition time, so were the pressure distributions; at the same time point, the greater the speed, the higher the pressure value, and the pressures have different increases in different positions; but in general, despite of speeds and time, there was one thing in common from the graphs above, namely, the pressure values in regions of smaller Y were larger than that of other regions, just because the cooling fan was at the end inside the tank, when the fan operated, close to which there would be greater pressure.

5. Conclusion

According to the test requirements of tank under the actual working conditions, we studied and designed a feasible and reliable cooling airflow measurement system based on array sensors to cope with poor conditions like wide temperature range, electromagnetic interference, and inconvenient assembly. In this paper, there were description of cooling airflow signal processing and analysis of air pressure distribution in different speeds and different time by linear fitting algorithm.

Nevertheless, we should keep trying to accomplish further theoretical deduction and algorithm implementation of heat and dissipation efficiency of engine compartment. These are the contents that need to be improved and supplemented later.

In a word, all the results show that we finished tank tests under the poor working conditions with cooling airflow measurement system based on array sensors, which is of great help to the measurement of tank parameters.

Acknowledgments

This work was supported by Shanxi Natural Science Foundation (2015021099). The authors would like to thank Dr. Youcheng Wu for excellent technical support.

References

[1] X. X. Liu, Y. K. Cao, X. P. Bi, "Review of research on air flow and heat transfer in engine compartment of tank." Binggong Xuebao, v28, n8, p1011-1016, August 2007

[2] X. F. Shen, Z. X. Ma, X. X .Liu, "Measurement System Design of Cooling Airflow Character in Armored Vehicle Power Compartment and Measurement Method Study of Flux," RAILWAY LOCOMOTIVE&CAR, Vol.23 Suppl.1, p156-159, Nov 2003

[3] J. R. Bao, X. P. Bi, "Numerical Study of the Influence Factors of Airflow and Heat Transfer in

Engine Compartment of Tank." Journal of Academy of Armored Force Engineering March 2005 Vol.19 No.1
[4] Y. Matsushima, T. Takeuchi,I. Kohri, "Prediction method of engine compartment air flow using CFD analysis," JSAE review, v21, n2, p197-203, April 2000
[5] W. Eichlseder,G. Raab, J. Hager, M. Raup, "Use of simulation tools with integrated coolant flow analysis for the cooling system design," SAE Technical Papers, 1997
[6] H. Zhou, Y. Han, J. G. Wang, P. Z. Zhang, "Research on Processing of Cooling Airflow Fluctuated Signals in Tank," in press
[7] GB/T 10178—2006/ISO 5802:2001 Industrial fans—Performance testing in situ

IOP Conf. Series: Journal of Physics: Conf. Series **916** (2017) 012053 doi:10.1088/1742-6596/916/1/012053

Experimental Investigation on Solid Particle Flow Characteristics in Particle Curtain Heat Exchanger

Cong Wen*, **DongLin Chen and Wei Hu**

Changsha University of Science and Technology Changsha, Hunan Province 410114, China

*Corresponding author: 2472877706@qq.com

Abstract. The particle curtain heat exchange, which is based on the rapid heat equilibrium theory of gas-solid two-phases, has attracted more and more attentions of researchers due to its flexible system arrangement, excellent heat transfer performance and real-time adjustable heating surface etc. In order to provide data for optimal design and operational control of the curtain exchanger, experiments were conducted to investigate flow characteristics of sand particles and geometry of the curtain under different conditions such as air velocity at the inlet, initial thickness of particle curtain, diameter of sand particle and mass flow rate of particles etc.

Key words. Particle curtain heat exchanger; solid particle; flow characteristic; experimental investigation

1. Introduction

Based on the principle of gas-solid two-phase rapid thermal equilibrium, the particle curtain heat exchanger is attracting more and more attentions due to its flexible system arrangement, outstanding heat transfer capacity, and real-time adjustable heat-transfer capacity. With the micron size of silica sand as its heat carrier, the heat exchanger passes flue gas surplus heat to the heat carrier. Then by passing heat from the heat carrier to combustion air to realize the maximized recycle of the application of flue-gas heat in preheating combustion air [1-5].

The research of the gas-solid two-phase flow theory is relatively mature at home and abroad, but little researches on the flow characteristics of falling particles in horizontal airflow, mostly concentrated on the experiment. But most of the experiments are limited to a single inlet velocity, single particle diameter or single mass flow rate in the horizontal airflow [6-13]. The content and depth of the research is rather simple, the gas-particle two phase flow and heat transfer characteristics and fouling performances in a particle curtain heat exchanger still need to be further deepened. At the same time, the gas-particle two-phase flow and the heat transfer process in the particle curtain heat exchanger are very complicated.

IOP Conf. Series: Journal of Physics: Conf. Series **916** (2017) 012053 doi:10.1088/1742-6596/916/1/012053

2. The experimental equipment's and method

2.1. The Experimental Equipment's

The experimental equipment's of the flow property of solid particle curtain in a particle curtain heat exchanger is shown in Figure 1. The device consists of the heat-exchanger itself, particle feeding system, particle collecting system, gas reducing uniform distribution, the heat exchanger itself is a channel of 2000mm long, 800mm wide and 1000mm high.

2.2. The Geometric Feature Parameters of the Particle Curtain

As shown in Figure 2, in order to quantitatively analyze the structural characteristics of the falling particle curtain, we defines the following geometric feature parameters related to it as follow: the left entrance of measuring pocket is taken as the origin, the horizontal direction of heat transfer exit as x-axis, the vertical direction of heat transfer as y-axis:

(1) Front/back boundary of particle curtain: the last /first boundary that airflow contacts with the particle curtain;

(2) The longitudinal thickness of the particle curtain xb : the distance between the front and back boundary of particle curtain at the location of the same height in heat transfer channel;

(3) The vertical distance from the top of heat exchange chamber h;

(4) Particle curtain's horizontal offset of the back edge x1: the distance between the particle curtain's back boundary after falling and before falling at the location of the same height in heat transfer channel;

(5) The horizontal offset of falling point x2: the distance between the particle curtain's center lines of droppoint after falling and before falling at the location of the same height in heat transfer channel;

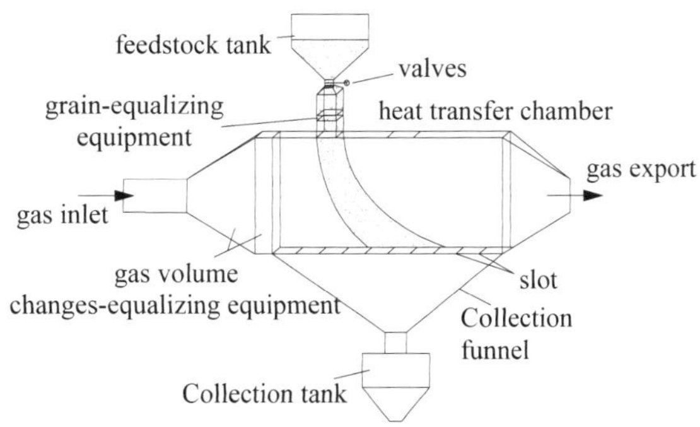

Fig.1 Schematic diagram of the particle curtain heat exchanger based on the single-stage unit.

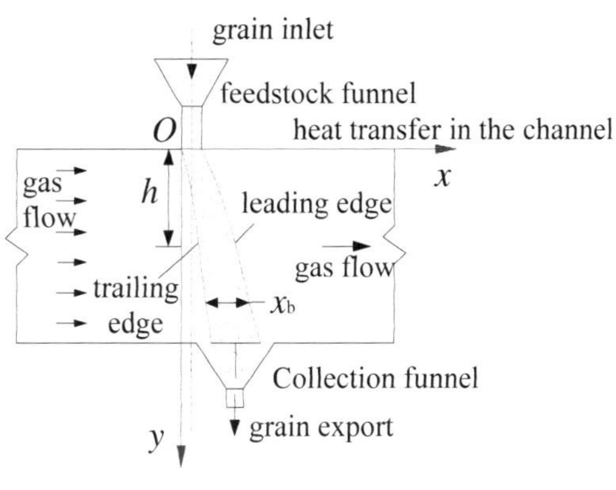

Fig.2 Schematic diagram of geometric feature parameters of the particle curtain

2.3. The experimental factors and design

2.3.1. The experimental factors. By adopting orthogonal design, this experiment is mainly researched the influence rules of four factors , i.e. , inlet velocity v0, the initial thickness of particle curtain b0 and particle diameter dp to particle curtain's horizontal offset of the back edge x1, the horizontal offset of falling point x2 and airflow velocity of particle curtain's back Vg edge, etc.

2.3.2. The experimental design. In order to prevent the out entrainment of small particles from the heat exchange chamber (this phenomenon is called elutriation), the maximum velocity vmax is generally less than 2m/s, so the ranges of inlet velocity takes 1.0~2.0m/s; the initial thickness of particle curtain takes 60~180mm; when the solid particle size<200μm, particles enter the flue gas or air can reach thermal equilibrium instantly(~10-1s), heat transfer efficiency is high, therefore, the ranges of particle takes 120~300μm.

2.4. The experimental method

Five measurement holes with a diameter of 25mm were installed at equal intervals on the straight line that 250mm from the entrance of the heat exchange chamber and 850mm from the exit, the distance between every measurement hole is 166mm.To get the particle curtain's front and back flow velocity distribution, the hot wire anemometer with precision of ±0.05m/s of VT110 model is used to measure and record air velocity values of measuring point 1-5. The intelligent anemometer of XY-5000PA model and standard pitot tube are used to measure and record air velocity of measuring point 6-10. In the heat exchange chamber, transparent lattice thin films were glued on the front wall of it, and using color pens to record particle curtain's front and back boundary point in every measuring point, the distance between the front and back boundary point of particle curtain is particle curtain's thickness in its location; the distance between particle curtain's back boundary after falling and before falling is particle curtain's horizontal offset of the back edge; after the falling of all particles, measuring the distance between the particle curtain's center line of droppoint after falling and before falling, we can get the horizontal offset of falling point. At the same time, in order to guarantee the accuracy and credibility of experimental data, the high-speed digital camera of 1076 frames/s was used to capture the image of falling particles. Under this experimental condition, the thickness value of particle curtain, particle curtain's horizontal offset of the back edge and particle curtain's horizontal offset of falling point were gained through a weighted average of these data extracted from the photo and the measured values.

3. Experimental results and analysis

3.1. The impact of inlet velocity on the flow property of solid particle

3.1.1. The longitudinal thickness of the particle curtain and the horizontal offset of the back edge. Air velocity in this experiment is adjusted by controlling the forming complete sets of vetilator's rotation rate through frequency convertor. Particle curtain's longitudinal thickness and horizontal offset of the back edge under different inlet velocity are shown in Figure 3 and 4 respectively. As you can see in the two figures, when the inlet velocity increased from 1.45m/s to 1.88m/s, the longitudinal thickness of the particle curtain at the location of the same height decreases, while the horizontal offset of the back edge is of the opposite, the largest increase reached 25.61% and appeared on the point that with a vertical distance of 832mm from the top of the heat exchange chamber. This is because the gas-particle inter-phase momentum interaction degree increases with the increase of initial airflow speed, with it the particle's horizontal movement speed will also become bigger, therefore, the horizontal offset of the back edge increases with the increase of inlet velocity.

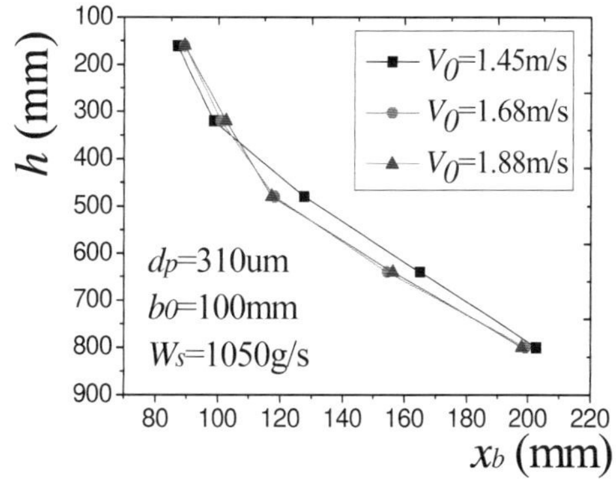

Fig.3 The longitudinal thickness of particle curtain under different air velocity

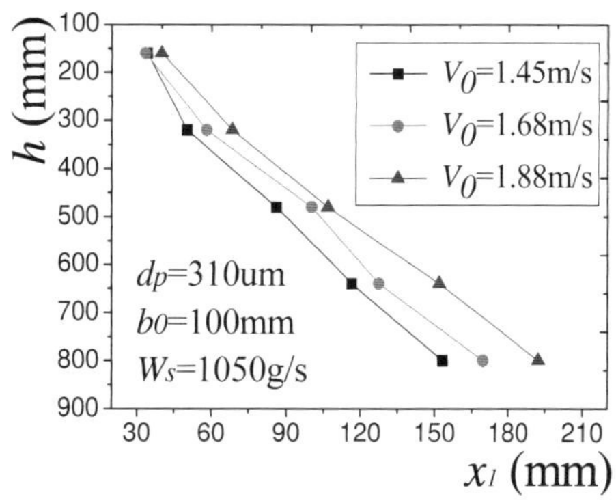

Fig.4 The trailing edge horizontal offset of particle curtain under different air velocity

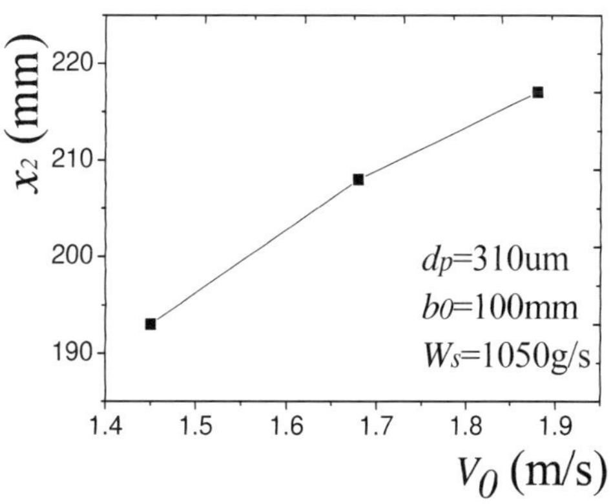

Fig.5 The placement horizontal particle curtain under different air velocity

3.1.2. Particle curtain's horizontal offset of falling point. Particle curtain's horizontal offset of falling point under different inlet velocity is shown in Figure 5. As you can see in the figure, the horizontal offset of falling point increases with the increase of inlet velocity. The reason is that when the particle curtain falling inside the heat exchange chamber, the contact of gas and solid particles is a process of exchange of momentum. When the horizontal movement speed becomes bigger, the greater the gas-particle inter-phase momentum interaction degree is, the bigger the particles horizontal velocity component will be, and the bigger the particle's horizontal displacement at the location of the same height will be.

3.2. Effect of the initial thickness of particle curtain on the flow property of solid particle

3.2.1. The longitudinal thickness of the particle curtain and the horizontal offset of the back edge. The initial thickness of particle curtain is defined as the thickness when the heat exchange particles just entered the heat exchange.

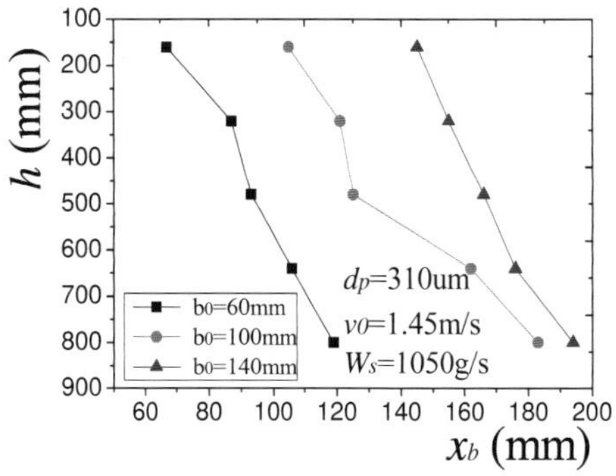

Fig6 The longitudinal thickness of particle curtain under different initial thickness of particle curtain

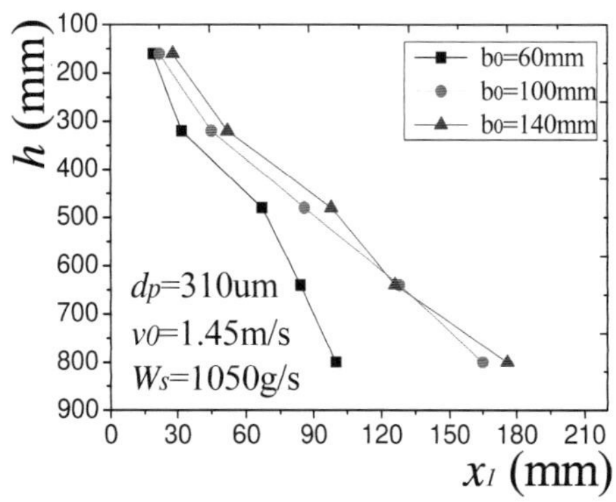

Fig.7 The trailing edge horizontal offset of particle curtain under different initial thickness of particle curtain

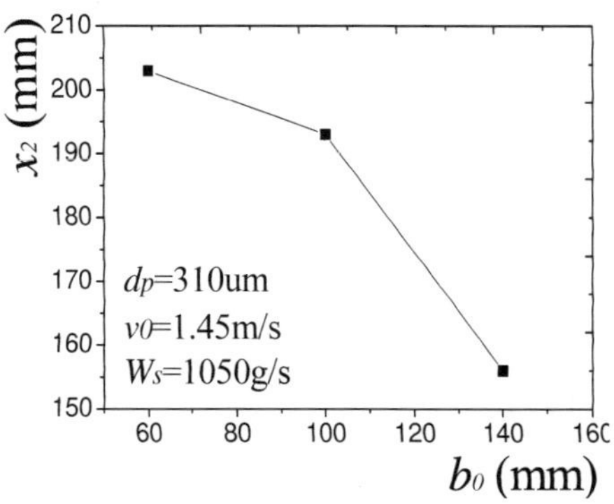

Fig.8 The placement horizontal particle curtain under different initial thickness of particle curtain

Channels. In this experiment, the initial thickness of particle curtain is changed by changing falling width of effective particles on the bottom floor screen cloth in the unfirming device of particles. The longitudinal thickness of the particle curtain and the horizontal offset of the back edge in different initial thickness of particle curtain are shown in Figure 6 and 7 respectively. As you can see in the two figures, when the initial thickness of particle curtain increased from 60mm to 140mm, the longitudinal thickness and horizontal offset of the back edge increases with it, the largest increase are 63.03% and 76% respectively, and these all appeared on the point that with a vertical distance of 832mm from the top of the heat exchange chamber. This is because when the falling particles' mass flow rate remain the same, the smaller the initial thickness of particle curtain is, the more compact the distribution of particles will be. The greater the degree of interaction among falling particles when they move horizontally is, the smaller the particle curtain's moving forward distance will be, therefore, the horizontal offset of particle curtain decreases with the decrease of the thickness of particle curtain.

3.2.2. Particle curtain's horizontal offset of falling point. Particle curtain's horizontal offset of falling point in different initial thickness of particle curtain is shown in Figure 8. As you can see in the figure, with the increase of the initial thickness of particle curtain, the particle curtain's horizontal offset of falling point gradually decreased. The reason is that when the initial thickness of particle curtain increases, under a certain amount of particle mass flow rate, the sparse degree among particles will become bigger, the air drag force on particles will become smaller, consequently, the particles' horizontal movement velocity components decreases, causing the decrease of particles' horizontal displacement.

3.3. Effect of particle diameter on the flow property of solid particle

3.3.1. The longitudinal thickness of the particle curtain and the horizontal offset of the back edge. In this experiment, silica sand particles of different average particle diameter dp were obtained by filtering with different aperture of screen cloth. The longitudinal thickness of the particle curtain and the horizontal offset of the back edge with different particle diameters are shown in Figure 9 and 10 respectively. As you can see in the two figures, when particle diameter dp increases from 215μm to 310μm, the longitudinal thickness and horizontal offset of the back edge have a decrease tendency, the largest decrease are 20.78% and 24.48% respectively, and these all appeared on the point that with a vertical distance of 832mm from the top of the heat exchange chamber. The reason is that the smaller the particle diameter is, the larger the particles' specific surface will be, it will make interaction area of gas-particles two-phase become bigger. That is to say, the greater the gas-particle inter-phase momentum interaction degree is, the bigger the particle's horizontal velocity component will be, then the larger the particle curtain's horizontal offset of the back edge will be. The horizontal velocity also increases the complex speed of the particles, leading to a gradual increase of the particle concentration in the cross section of the particles. When the airflow passes through the particle curtain, the particles will spread out, therefore, the smaller the particle diameter is, the larger the longitudinal thickness of the particle curtain will be.

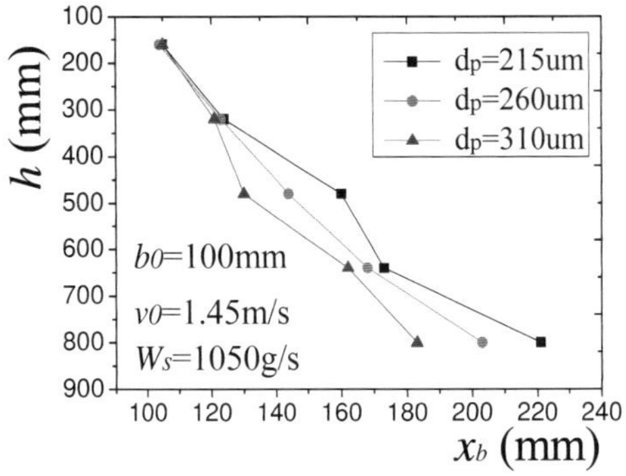

Fig.9 The longitudinal thickness of particle curtain under different particle size

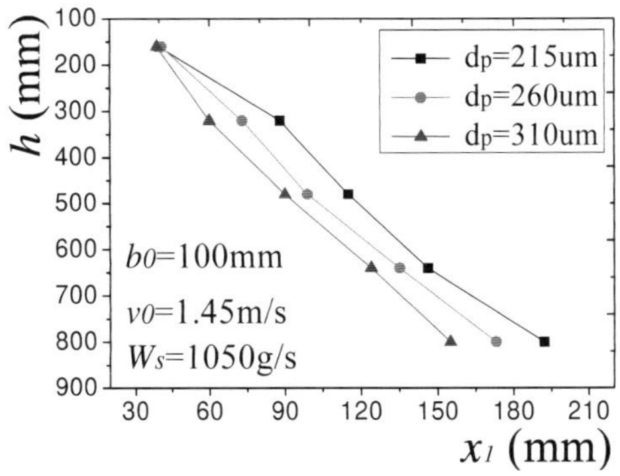

Fig.10 The trailing edge horizontal offset of particle curtain under different particle

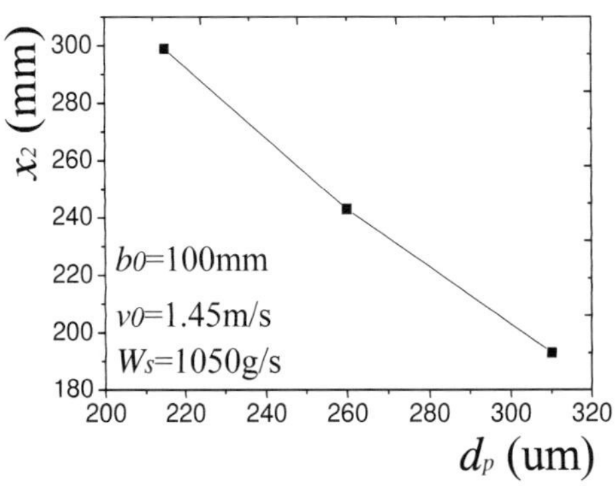

Fig.11 The placement horizontal particle curtain under different particle size

3.3.2. Particle curtain's horizontal offset of falling point. Particle curtain's horizontal offset of falling point at different particle diameters dp is shown in Figure 11. It can be seen from the figure that with the increase of particle diameter, particle curtains horizontal offset of falling point decreases gradually. The reason is that when the particle diameter increases, the specific surface area of the particle decreases, which makes the area of momentum exchange between the airflow and the particle decreases. The concrete manifestations are: the force of airflow on the particles in the horizontal direction decreases, resulting in a decrease of particles' velocity component in the horizontal direction and the reduction of speed in the horizontal direction eventually reduces the Horizontal offset.

4. Conclusion

Based on the above experimental results, the following conclusions can be drawn from the experimental study on the flow property of solid particle in the particle curtain heat exchanger:

(1) The longitudinal thickness of the particle curtain increases with the initial thickness of particle curtain, and decreases with the increase of inlet velocity, particle diameter and particle mass flow rate.

(2) Particle curtain's horizontal offset of the back edge increases with the increase of inlet velocity and the initial thickness of particle curtain, and decreases with the increase of particle diameter and particle mass flow rate.

(3) Under different inlet velocity, different initial thickness of particle curtain, different particle diameter and different particle mass flow rate, the longitudinal thickness of the particle curtain and particle curtain's horizontal offset of the back edge all increase with the increase of particle curtain's falling height.

(4) Particle curtain's horizontal offset of falling point increases with the increase of inlet velocity, and decreases with the increase of initial thickness of particle curtain, the particle diameter and particle mass flow rate.

Acknowledgments

I would like to express my gratitude to all those who helped me during the writing of this thesis. I gratefully acknowledge the help of my supervisor, Mr.Chen, who has offered me valuable suggestions in the academic studies. In the preparation of the thesis, he has spent much time reading through each draft and provided me with inspiring advice. Without his patient instruction, insightful criticism and expert guidance, the completion of this thesis would not have been possible.

References

[1] CHEN Donglin. Tubeless heat exchanger. Chinese Patent: 201120261405.2,2011-07-22.
[2] CHEN Donglin. Tubeless heat exchanger and waste heat recovering system [P]. Chinese Patent: 201110206635.3,2011-07-22.
[3] CHEN Donglin, ZOU Chan, DU Yang. Steady heat transfer characteristics of a particle curtain based heat exchanger [J]. Thermal Power Generation, 2014,43(3):37-42.
[4] CHEN Donglin, WANG ganquan, WANG an. Design and experimental investigation of the uniforming device of particles from particle curtain heat exchanger[J]. Journal of Electric Power Science and Technology, 2014, 29(1): 75-80.
[5] CHEN Donglin, YANG Jianbo. Design and experimental investigation in the airflow uniform device for particle curtain heat exchangers[J]. Journal of Electric Power Science and Technology,2015,30(1):93-98.
[6] S Ouyang, Q M Mao, M Rhodes, et al. Short contact time gas-solid systems[J]. Reviews in Chemical Engineering, 2003, 19 (2) :133-228.
[7] LI Na, WANG Jiajun, FENG Lianfang. Research progress of granular flow and fluidization behavior of coarse cohesive particles. Chemical Industry and Engineering Progress, 2014,33(2):298-303.
[8] K Ogata, K Funatsu, Y Tomita. Experimental investigation of a freely falling powder jet and the air entrainment[J]. Powder Technology,2001,115(1):90-95.
[9] Zhang Xiaoqiang, Zhang Dongfeng, Wang An. Transportation characteristics of gas-solid two-phase in a long-distance pipeline[J]. 2015,21(8):196-202.
[10] J J Kelly, P O'Donnell. Dynamics of granular material in rotary drier and coolers[J]. Institution of Chemical Engineers Symposium Series, 1968,29(2) : 34-44.
[11] Gong Yan, Yu Guangsuo, Guo Qinghua. Experimental study of the particle deposition characteristics in an entrained flow gasifier[J]. Chemical Engineering Science,2015,138(11): 291-302.
[12] Yan F, Rinoshika A. Characteristics of particle velocity and concentration in a horizontal self-excited gas–solid two-phase pipe flow of using soft fins[J]. International Journal of

IOP Conf. Series: Journal of Physics: Conf. Series **916** (2017) 012053 doi:10.1088/1742-6596/916/1/012053

Multiphase Flow, 2012, 41:68-76.
[13] ZHOU Yuegui, PENG Jun, ZHU Xian. Experimental research on gas-solid two-phase flow characteristic for flue gas desulfurization in circulation fluidized bed[J]. Journal of Power Engineering, 2009,29(6):559-564.

PCB-level Electro thermal Coupling Simulation Analysis

Runjing Zhou [a], **Xuchen Shao** [b]

College of Electronic Information of Inner Mongolia University Hohhot, China

[a]zhourj@163.com
[b]1576421980@163.com

Abstract. Power transmission network needs to transmit more current with the increase of the power density. The problem of temperature rise and the reliability is becoming more and more serious. In order to accurately design the power supply system, we must consider the influence of the power supply system including Joule heat, air convection and other factors. Therefore, this paper analyzes the relationship between the electric circuit and the thermal circuit on the basis of the theory of electric circuit and thermal circuit.

Keywords. Electro thermal coupling; Joule heat; Voltage distribution; Thermal distribution.

1. Introduction

When PCB is working, Joule heat can change the Power supply design and the thermal design of system with high density device. And the high temperature can even cause smoke or fire. So it is necessary to combine electricity and heat to analyze electro thermal coupling.

2. Contents of the Study

Part 2.1 introduces the method of electro thermal coupling; Part 2.2 analyzes the interactions of thermal and circuit.

2.1. The Method of Electro thermal Coupling

This chapter briefly discusses the theory of electro thermal coupling.

Voltage distribution equation in steady state is represented by the formula (1):

$$\nabla\left(\frac{1}{\rho(x,y,z,T)}\nabla\varphi(x,y,z)\right)=0 \tag{1}$$

Among them, $\rho(x,y,z,T)$ is Resistivity, $\varphi(x,y,z)$ is Voltage distribution.

Solid and fluid thermal distribution equation in steady state is represented by the formula (2):

$$\nabla\left[K(x,y,z)\cdot\nabla T(x,y,z)\right]=-P(x,y,z) \tag{2}$$

IOP Conf. Series: Journal of Physics: Conf. Series **916** (2017) 012054 doi:10.1088/1742-6596/916/1/012054

$$\sigma c_p \bar{v}(x,y,z) \cdot \nabla T(x,y,z) = \nabla (K_f \nabla T(x,y,z)) \tag{3}$$

$P(x,y,z)$ Include device heat and Joule heat. The device heat is available from the device technical manual. Joule heat can be calculated by formula (4):

$$P_{Joule}(x,y,z) = \bar{J} \cdot \bar{E}(x,y,z) \tag{4}$$

Among them, $\bar{J}$ is current density, $\bar{E}(x,y,z)$ is electric field distribution in PDN.

Effect of temperature on electrical properties formula (5) to calculate the new resistivity:

$$\rho = \rho_0[1+a(T-T_0)] \tag{5}$$

Among them, ρ_0 is the resistivity at temperature T_0 ; α is the factor that the resistivity is affected by temperature. As shown in Fig.1, the resistivity of the conductor increases as the temperature increases, and finally increases the DC voltage drop in PDN. Fig.2 shows the relationship between electricity and heat. This paper summarizes an analysis method of electro thermal coupling, as shown in Fig.3.

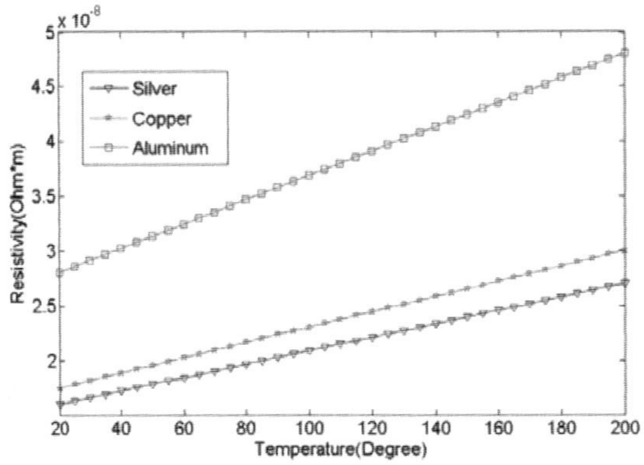

Fig.1 The resistivity of silver, copper and aluminum affected by temperature

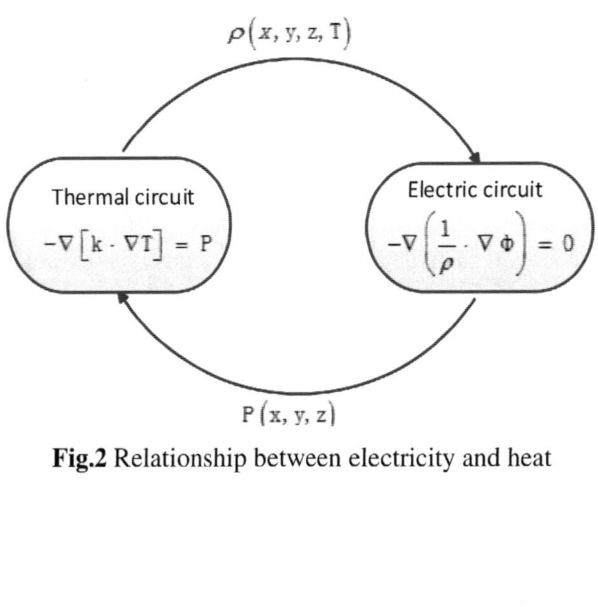

Fig.2 Relationship between electricity and heat

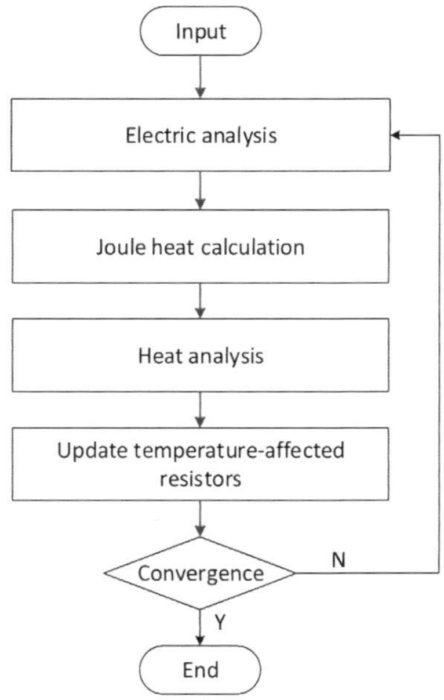

Fig.3 A method of electro thermal analysis

2.2. Electro thermal Coupling Analysis

In this paper, Strati GX is used as the research object as Fig.4 shows. The system can be powered from a DC power supply from 14V to 20V. There are four experiments: Electrical analysis; Thermal analysis; Electro thermal coupling analysis without fan; Electro thermal coupling analysis with fan.

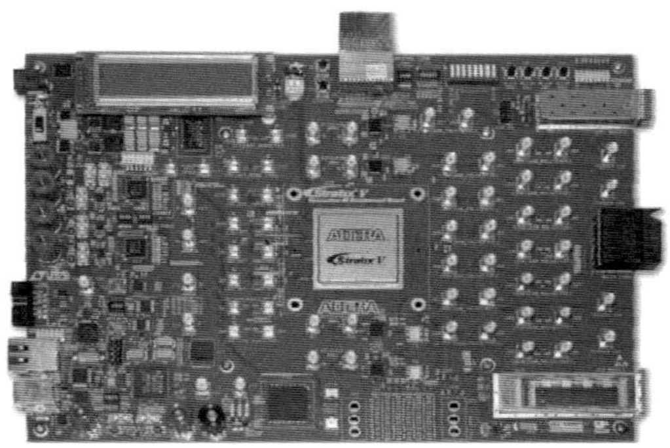

Fig.4 The appearance of Strati GX

2.2.1. The effect of Thermal circuit on Electric circuit. (1) The effect of Temperature on Power supply

Table 1. Lists the Actual Voltage in the experiment.

IOP Conf. Series: Journal of Physics: Conf. Series **916** (2017) 012054 doi:10.1088/1742-6596/916/1/012054

Table 1. Actual Voltage of E, ET_without Fan and ET_with Fan (unit: V)

Name	E	ET_without Fan	ET_with Fan
SINK_U29_2P5V_GND	2.48073	2.47867	2.47923
SINK_U19_2P5V_GND	2.48117	2.47918	2.47967
SINK_U29_S5GX_VCC_GND	0.810279	0.805775	0.808855

Voltage has reached the power supply requirements if we only consider the Actual Voltage. But the difference is not fully demonstrated by the actual voltage. So we also need to discuss the difference by Voltage Margin.

(2) The effect of Temperature on Voltage Margin

Table 2. Lists the Voltage Margin in the experiment.

Table 2. Voltage Margin of E, ET_without Fan, ET_with Fan (unit: V)

Name	E	ET_withoutFan	ET_withFan
SINK_U29_2P5V_GND	0.105734	0.103667	0.104233
SINK_U19_2P5V_GND	0.10617	0.104184	0.104674
SINK_U29_S5GX_VCC_GND	0.0027792	-0.0017251	0.00135501

The above table shows: the results that we only consider the circuit is better than the actual Voltage Margin. Although the effect of temperature on DC power supply is less than 5mV in the PCB-level Electro thermal Coupling Analysis, this value is sufficient to influence the judgment of whether the power supply system design meets the threshold requirements. The power supply performance of the system will be overestimated without the electro thermal coupling analysis.

(3) The effect of Temperature on Current Density

The current density on PCB mainly includes Current density of plane and Current density of via.

1) Current density of plane

According to the experimental analysis, we can find that the maximum current density is located on the S5GX_VCC (0.85mV) power supply of U2. Table 3. Shows the data results of plane current density. And Fig.5~Fig.8 shows the distribution of Current Density.

Table 3. Current density of plane (unit: A/mm2)

Net Name	Start Layer	End Layer	E	ET(without Fan)	ET(with Fan)
S5GX_VCC	Plane$L18_PWR4	Signal$ BOTTOM	233.6299	248.2353	237.9395

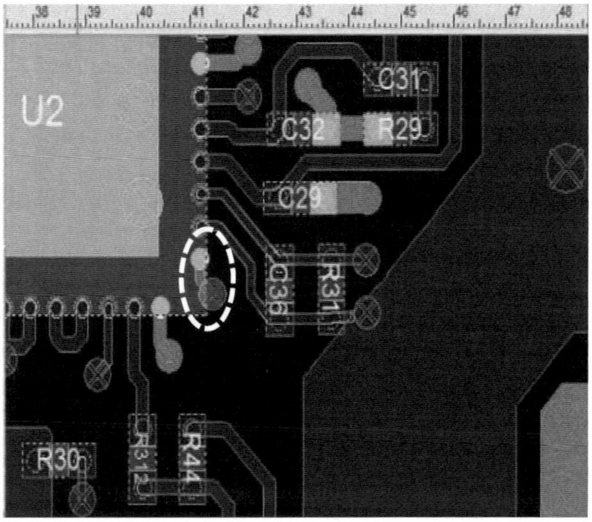

Fig.5 S5GX_VCC (0.85mV) of U2

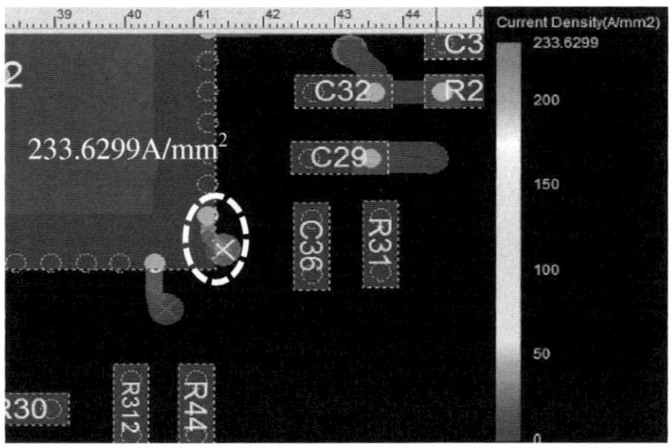

Fig.6 E

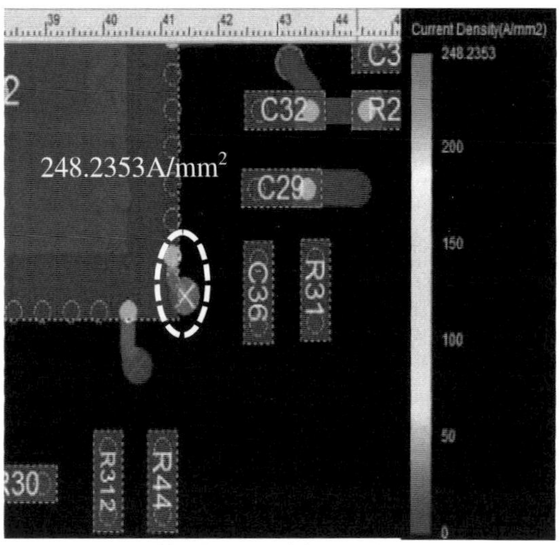

Fig.7 ET (without Fan)

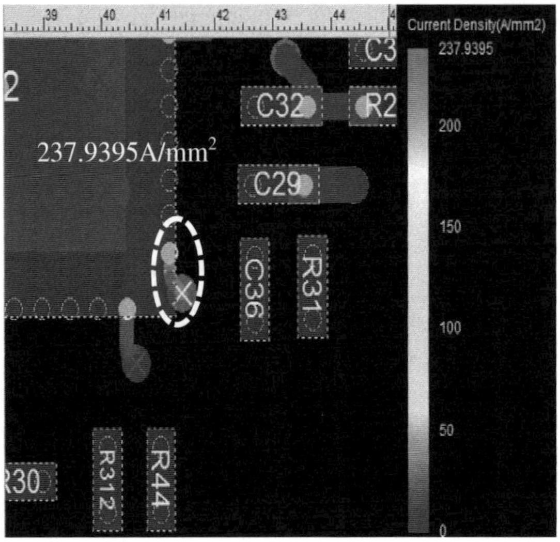

Fig.8 ET (with Fan)

We can see from the above analysis: the result of electrical analysis without electro thermal coupling analysis is "The maximum current density is 233.6299A/mm2". Because it does not exceed the threshold experience value, it is likely to come to a wrong conclusion "current density design is qualified". The results of ET (without Fan) is "The maximum current density is 248.2353A/mm2", which is more than 240A/mm2. So the design is not qualified. The result of ET (with Fan) is 237.9395A/mm2, which is qualified.

2) Current density of via

Table 4. Current density of via (unit: A/ mm2)

Net Name	Start Layer	End Layer	E	ET(without Fan)	ET(with Fan)
S5GX_VCC	Plane$L18_PWR4	Signal$ BOTTOM	282.2795	299.9253	287.9061

In general, the current density of via is 20% greater than the current density of plane in the same case. The results of ET (without Fan) is 299.9253A/mm2, which is more than 288A/mm2. So the design is not qualified. The result of ET (with Fan) is 287.9061A/mm2, which is qualified.

2.2.2. The effect of Electric circuit on Thermal circuit. (1) The effect of Route on Temperature Distribution

Material of PCB is composed of FR-4 and copper, the route will affect the temperature distribution of system.

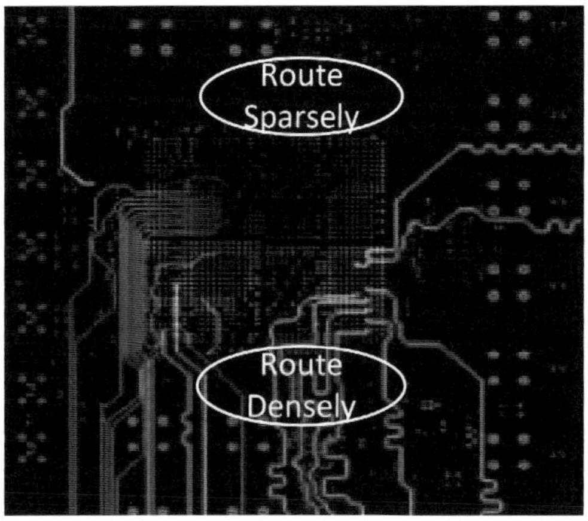

Fig.9 Route distribution of PCB

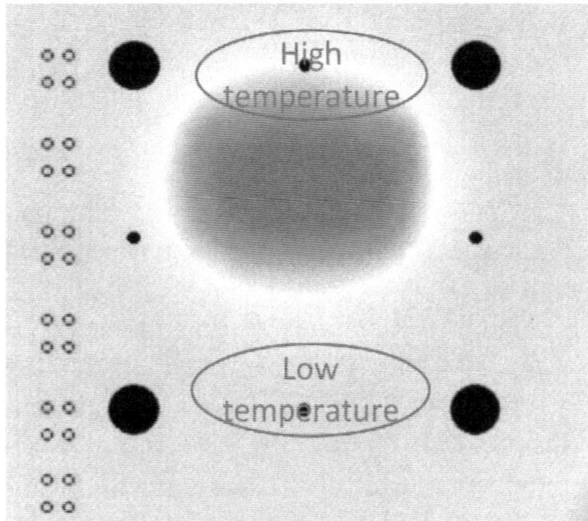

Fig.10 Temperature distribution

Fig.9 shows that the route on the top of the chip is sparse. So the heat transfer is poor with less wiring and low copper coverage. Fig.10 shows that the temperature on the top is higher.

(2)The effect of Joule heat on Temperature Distribution

The Joule heat generated by the current can affect the thermal conductivity of the material, which in turn affects the thermal distribution of the system and the design of the power supply.

Table 5 shows the data results and Fig.11 shows the distribution of temperature. We can see that the effect of Joule heat is 1.41℃.

Table 5. Temperature change (unit :℃)

	Thermal analysis	Electro thermal analysis	Effect of Joule heat
The maximum temperature	95.42	96.81	+1.41

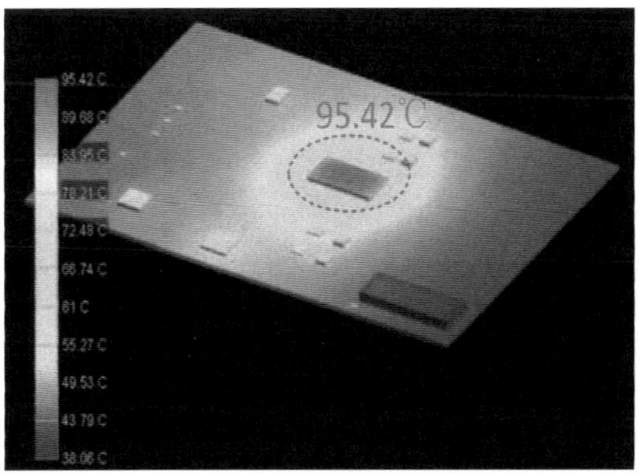

Fig.11 Devince temperature of Thermal v couplinganalysis

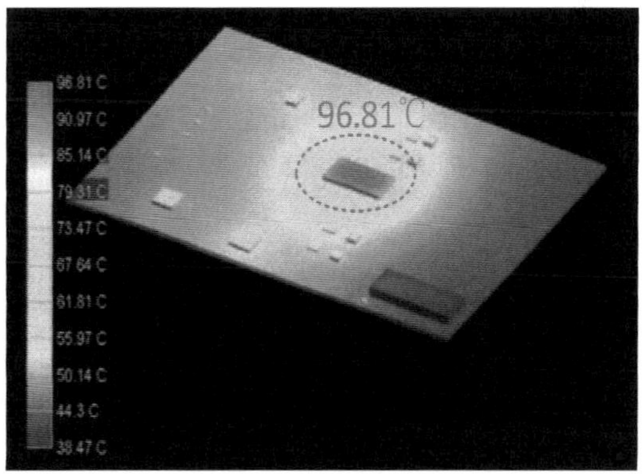

Fig.12 Devince temperature of Electro thermal analysis

Summary

This paper analyzes the effect of temperature on supply voltage, voltage margin, and current density. By comparing the results of electrical analysis and electro thermal coupling analysis, we can see that the result with electro thermal coupling analysis is worse than the electrical analysis. The result of

IOP Conf. Series: Journal of Physics: Conf. Series **916** (2017) 012054 doi:10.1088/1742-6596/916/1/012054

effect of Electric circuit on Thermal circuit is similar. Therefore, if we do not consider the electro thermal coupling, we will make a wrong judgment on the system's power supply performance. And the result also fully illustrates the necessity of electro thermal coupling analysis.

References

[1] Bar-Cohen A, Srivastava A, Shi B. Thermo-electrical co-design of three-dimensional integrated circuits: Challenges and opportunities [J]. Computational Thermal Sciences, 2013, 5(6): 441-458.

[2] Shang Q, Li X, Mao J. Transient thermal analysis of global interconnects based on transmission lines [A]. In 2012 Asia-Pacific Symposium on Electromagnetic Compatibility (APEMC) [C]. 2012: 245-248.

[3] Shao Y, Li X C, Mao J F. An electrothermal model of interconnects based on a transmissionline network [A]. In Microwave Conference Proceedings[C]. 2012: 1247-1276.

[4] Lu T, Jin J M. Electrical-Thermal Co-Simulation for DC IR-Drop Analysis of Large-Scale Power Delivery [J]. IEEE Transactions on Components Packaging & Manufacturing Technology, 2014, 4(2): 323-331.

[5] R ocas E, Collado C, Orloff N D, etc. Passive Intermodulation Due to Self-Heating in Printed Transmission Lines [J]. IEEE Transactions on Microwave Theory & Techniques, 2011, 59(2): 311-322.